SMART GRID

SMART GRID

Applications, Communications, and Security

Edited by
Lars Torsten Berger
Krzysztof Iniewski

A JOHN WILEY & SONS, INC., PUBLICATION

Published by John Wiley & Sons, Inc., Hoboken, New Jersey
Published simultaneously in Canada

Library of Congress Cataloging-in-Publication Data:

Berger, Lars Torsten.
Smart grid : applications, communications, and security / Lars Torsten Berger & Krzysztof (Kris) Iniewski.
p. cm.
Includes bibliographical references and index.
ISBN 978-1-118-00439-5 (cloth)
1. Electric power distribution–Automation. 2. Electric power distribution–Data processing. 3. Electric power distribution–Security measures. 4. Telecommunication systems. I. Iniewski, Krzysztof. II. Title.
TK3226.B355 2012
621.319'130285–dc23

2011024331

Printed in the United States of America

10 9 8 7 6 5 4 3 2 1

CONTENTS

PART II COMMUNICATIONS

PART IV CASE STUDIES AND FIELD TRIALS

PREFACE

Today's power grid is a system that supports electricity generation, transmission, and distribution operations. It is composed of a few central generation stations and electromechanical power delivery systems operated from control centers. Power flows mainly from the central stations toward the medium and low voltage customers. One prerequisite for grid stability is the balance between energy consumption and generation. Presently, energy generation follows the time variant consumption. With the increasing integration of renewable energy generation facilities, for example, utilizing solar and wind power, the energy flow is in some cases reversed. Energy may flow from the customer into the grid, leading to a more complex grid structure. Besides, many renewable resources reveal an intermittent and unpredictable nature of energy supply. This makes their integration a challenging task and requires an upgrade to the aging electricity infrastructure.

Fostered by growing governmental support, it is expected that the world's energy infrastructure will undergo a major transformation in the coming decade. This transformation is commonly referred to as *Smart Grid*. The evolving Smart Grid combines the electrical power infrastructure with modern distributed computing facilities and communication networks. It is a collection of complex, interdependent systems whose key functions include *reliable and efficient power delivery* facilitated through wide-area situational awareness, *peak energy curtailment* through demand response schemes, *widespread integration of intermittent renewable energy resources* through real-time control and energy storage, and the shift from a largely fossil fuel driven transport system to *electric transportation*.

One important concept within the Smart Grid is *demand response* (DR). It allows managing electricity consumption in response to supply conditions and electricity demand. To enable DR a remote *advanced metering infrastructure* (AMI) is currently being implemented in various locations worldwide. At the heart of AMI are smart meters, which are among others capable of measuring and recording usage data in real time. The smart meters have bidirectional communication links to the utility's central server. This allows transmission of the recorded metering data, and the reception of configuration as well as energy pricing information from the utility. Communication can be established using an IP access network, over digital subscriber line, power line, or wireless network infrastructure. A common flavor of DR schemes counts on household appliances, such as refrigerators or air conditioning units that are informed in real time about the energy price. Dependent on the user's choice, these household appliances could respond to a price increase through a reduction of power usage. Recent studies on DR and the tightly related concept of time-of-use pricing reveal that energy cost awareness alone can lead to changing habits and therewith to a reduction of energy usage on the order of 14%. Dynamic pricing may also be used to prevent transmission and distribution network bottlenecks. This postpones upgrading costs and decreases the risk of grid instabilities.

Furthermore, it is largely believed that *plug-in electric vehicles* (PEVs) will form an important part of the Smart Grid. On the one hand, charging periods can, to a certain degree, be scheduled to times of abundant electricity availability. On the other hand, on-board batteries can be used to administer electricity to the grid in periods of high demand or low production.

This book gives a profound introduction to the various aspects of Smart Grids that have already started to influence many areas of our lives. It contains four major parts: *Applications*, *Communications*, *Security*, and *Cases Studies and Field Trials*.

In Part I, it starts out with a detailed introduction to Smart Grid applications, spanning the transmission, distribution, and customer side of the electricity grid. Issues like fault detection, isolation and restoration, wide area monitoring protection and control, as well as demand response/demand side energy management, and the integration of plug-in electric vehicles are discussed, to name a few.

In general, grids become smart as electrical devices are empowered to collect and exchange information. Advances in communication technology are key. Wireless, wireline, and optical communication solutions are discussed in Part II from the physical layers up to sensing, automation, and control protocols running on the application layers.

Due to the immense importance of the electricity supply in our everyday lives, it is crucial that Smart Grids are at least as reliable as the energy grids we know today. Smart Grid security has therefore to penetrate every aspect of Smart Grid deployments preventing nonintentional faults, like instabilities and natural disasters, as well as intentional faults (e.g, due to cyber attacks). Part III deals with cyber security, raising awareness of security threats, reviewing ongoing cyber security standardization, and presenting methods for authentication and encryption key management.

The book is rounded-off with Part IV, presenting self-contained chapters on Smart Grid case studies and field trials. These chapters allow the reader to benefit from lessons learned in situations where the Smart Grid of tomorrow has already become a reality.

Not only the Case Study and Field Trial chapters, but all chapters are written as far as possible in a self-contained manor. Additionally, chapters are cross-referenced, allowing each reader to encounter a personal reading path. We hope you enjoy the diverse and rich contents contributed by experts from industry and academia, making this book one of the first of its kind in the world of the Smart Grid.

Lars Torsten Berger
Krzysztof (Kris) Iniewski

CONTRIBUTORS

Claus Amtrup Andersen, EURISCO, Denmark

Peter Bach Andersen, Centre for Electric Technology (CET), Technical University of Denmark (DTU), Copenhagen, Denmark

Ana García Armada, Department of Signal Theory and Communications, University Carlos III, Madrid, Spain

Lars Torsten Berger, BreezeSolve, Valencia, Spain

Paul Choudhury, BC Hydro, Vancouver, British Columbia, Canada

Panayotis G. Cottis, School of Electrical and Computer Engineering, National Technical University of Athens, Athens, Greece

Jacob Dall, EURISCO, Denmark

Athanasios E. Drougas, School of Electrical and Computer Engineering, National Technical University of Athens, Athens, Greece

Xiaoming Feng, ABB Corporate Research, Raleigh, North Carolina, USA

Pedro Marín Fernandes, Cisco Systems, Lisbon, Portugal

María Julia Fernández-Getino García, Department of Signal Theory and Communications, University Carlos III, Madrid, Spain

Nigel Fitzpatrick, Azure Dynamics, Vancouver, British Columbia, Canada

Steffen Fries, Siemens AG, Corporate Technology, Germany

Dieter Gantenbein, IBM Research, Zurich, Switzerland

Juan José García Fernández, Department of Signal Theory and Communications, University Carlos III, Madrid, Spain

Víctor P. Gil Jiménez, Department of Signal Theory and Communications, University Carlos III, Madrid, Spain

Einar Bragi Hauksson, Centre for Electric Technology (CET), Technical University of Denrnark (DTU), Copenhagen, Denmark

Hans-Joachim Hof, Munich University of Applied Sciences, Munich, Germany

Kris Iniewski, CMOS Emerging Technologies, Vancouver, British Columbia, Canada

Bernhard Jansen, IBM Research, Zurich, Switzerland

Mats Larsson, ABB Corporate Research, ABB Switzerland Ltd., Baden-Dättwil, Switzerland

Alberto Leon-Garcia, Department of Electrical and Computer Engineering, University of Toronto, Toronto, Ontario, Canada

Wenyuan Li, BC Hydro, Vancouver, British Columbia, Canada

Wolfgang Mahnke, ABB Corporate Research, Industrial Software Systems, Ladenburg, Germany

Anthony Metke, Motorola, USA

Salman Mohagheghi, Engineering Department, Colorado School of Mines, Golden, Colorado, USA

Hamed Mohsenian-Rad, Department of Electrical and Computer Engineering, Texas Tech University, Lubbock, Texas, USA

Petros I. Papaioannou, School of Electrical and Computer Engineering, National Technical University of Athens, Athens, Greece

Anders Bro Pedersen, Centre for Electric Technology (CET), Technical University of Denmark (DTU), Copenhagen, Denmark

Angeliki M. Sarafi, School of Electrical and Computer Engineering, National Technical University of Athens, Athens, Greece

Wenbo Shi, Department of Electrical and Computer Engineering, The University of British Columbia, Vancouver, British Columbia, Canada

Troels B. Sørensen, Department of Electronic Systems, Aalborg University, Denmark

James Stoupis, ABB Corporate Research, Raleigh, North Carolina, USA

Jun Sun, BC Hydro, Vancouver, British Columbia, Canada

Alec Tsang, BC Hydro, Canada

Vincent W. S. Wong, Department of Electrical and Computer Engineering, The University of British Columbia, Vancouver, British Columbia, Canada

Wilsun Xu, Department of Electrical & Computer Engineering, University of Alberta, Edmonton, Canada

PART I

APPLICATIONS

1

INTRODUCTION TO SMART GRID APPLICATIONS

Xiaoming Feng, James Stoupis,
Salman Mohagheghi, and Mats Larsson

1.1 INTRODUCTION

Smart Grid encompasses the entire electric energy conversion, transmission, distribution, and utilization cycle. It consists of advanced actuators, sensors, communication infrastructure, IT systems, advanced monitoring, control, and decision making applications. There are multiple objectives for developing and deploying Smart Grid technologies. The key objectives are as follows:

- To improve efficiency and economy in energy conversion, transmission, distribution, storage, and utilization
- To enhance security and safety in system operation by increasing the observability and controllability of the power grid
- To improve the reliability and availability of the power supply to customers
- To enable and promote the integration and utilization of renewable and sustainable energies

Smart Grid: Applications, Communications, and Security, First Edition.
Edited by Lars Torsten Berger and Krzysztof Iniewski.

- To enable and facilitate demand side participation to increase asset utilization and return on investment
- To maintain and improve the quality of power delivery to increasing share of digital loads

Smart Grids are enabled by technologies in the following key categories:

- Actuators (generators, capacitors, reactors, distributed generators, energy storages)
- Sensors (voltage, current sensors, power flow measurements, pressure, acoustic, temperatures, and more advanced sensors such as synchrophasors)
- Controllers (devices to effect changes in the actuators, such as FACTS [flexible AC transmission systems] controllers, capacitor controllers, energy storage systems, etc.)
- Communication systems
- Applications (decision making logic)

Smart Grids rely on many advanced applications to deliver benefits to the customers, grid operators, and other societal stakeholders. These applications are the equivalent of word processors, email programs, spreadsheet programs, and internet browsers for a personal computer, which would not be of much use to most people if it were not for these "killer applications." By applications we refer to any programs, algorithms, calculations, data analysis, and other software that execute either continuously or periodically on some computing platform on a single device, a subsystem, or at the system level, that process historical, real-time, and/or forecast data to provide output that is used to alter the operating state of the power system with the objective to improve security, efficiency, reliability, and/or economy.

In this chapter we describe some of the most talked about applications or application areas that are considered by most power engineering professionals as core applications for Smart Grid.

- Volt and var control (VVC)
- Fault detection, isolation, and restoration (FDIR)
- Demand response (DR) management
- Distributed energy resource (DER) integration and management
- Wide area monitoring, protection, and control (WAMPC)

The first three applications are for the electric distribution systems, which are medium voltage networks that distribute electric power at typically 35 kV or below from the distribution substation to the service transformers at customer sites. Traditionally, the level of instrumentation and automation on the distribution systems is low, in contrast to the high voltage bulk power transmission systems. The worldwide initiatives toward Smart Grid are placing a major focus on the distribution system automation due to the fact that distribution systems represent the last frontier for power system modernization.

The distributed energy resource integration and management application is applicable to both the distribution systems and transmission systems. At the distribution system level, the distributed energy resources are small scale wind generators, community energy storages, commercial or residential photovoltaic cells, microturbines, fuel cells, and plug-in electric vehicles. At the transmission system level, distributed energy resources are the mega wind farms, solar farms, geothermal power plants, and so on.

Collectively these applications address the key objectives of Smart Grid, with respect to energy efficiency, service reliability, consumer participation, and sustainability. VVC aims to reduce the energy loss and manage peak demand on the distribution system. FDIR manages the detection, isolation, and restoration of power to affected customers after the occurrence of a fault (short circuit) on the system so as to minimize the interruption of service to customers. DR and DER management manages the dispatch of demand response and distributed energy resources to reduce the power demand during peak hours and/or shift some of the demand to hours of the day when the demand is low, thus improving the load factor (utilization efficiency) of transmission and distribution assets. The wide area monitoring, protection, and control (WAMPC) is an application for the bulk power transmission grid that improves the situational awareness of the power system operators and provides systemwide protection, in contrast to device or component protection, through fast acting controls, to maintain the integrity of the system and reduce the risk of large scale blackout.

1.2 VOLTAGE AND VAR CONTROL AND OPTIMIZATION

1.2.1 Introduction

Distribution systems are the medium voltage networks (typically 10 kV to ~35 kV) that deliver electric power from the substations to the consumer locations where the voltage is stepped down to service voltage suitable for end user consumption. Voltage and var control (VVC) on the distribution system narrowly refers to the control of voltage regulation devices and reactive power compensation devices for the purpose of reducing consumer demand and energy loss, and maintaining voltages at various points in the distribution network within acceptable limits. Voltage regulation devices typically include the OLTC (on load tap changers) of the substation transformers and the tap changers on the voltage regulators on the distribution feeders. Reactive power devices consist of the switchable capacitor banks on the feeder in the substation. With the integration of distributed energy resources and energy storage devices, the control scope of VVC will be expanded to include the reactive power and even the active power of dispatchable distributed generations and the charging and discharging schedule of energy storages, including

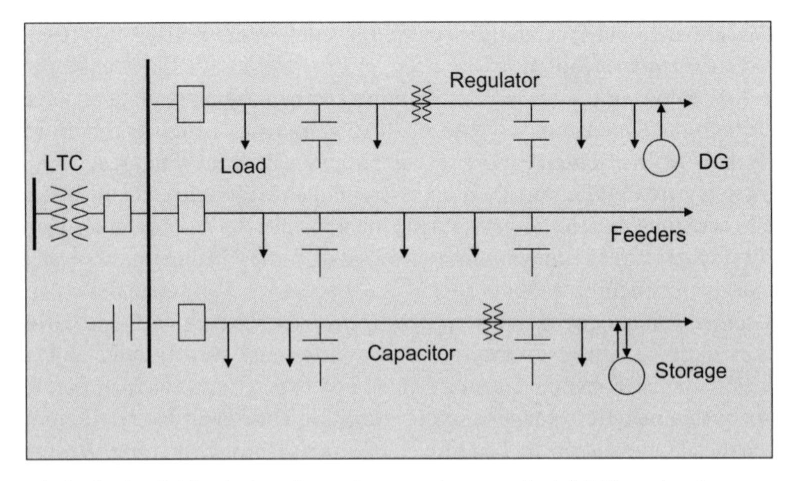

Figure 1.1. Controllable devices for voltage and var control (VVC) and voltage, var, and watt control (VVWC).

those associated with plug-in electric vehicles. Some people in the industry refer to the expanded control as volt, var, and watt control (VVWC) to highlight the inclusion of real power. Figure 1.1 shows the controllable devices and their typical locations in a highly simplified distribution system.

1.2.2 Devices for Voltage and Var Control

Voltage and var are fundamental concepts in AC (alternating current) transmission and distribution systems. The two primary devices for volt and var control are on load tap changers for transformers and switched capacitor banks.

A basic transformer has two windings wound around a laminated magnetic sheet core that forms a closed magnetic circuit. One of the windings is called the primary winding and the other the secondary winding. The two windings are insulated from each other and AC energy transfer from one winding to another is accomplished through magnetic coupling via the magnetic circuit. The ratio of voltages across the two windings is determined by the turn ratio. When one of the windings contains multiple taps, each tap corresponds to a different turn ratio. An on load tap changer can change the tap while supplying load. A typical tap changer has 33 taps with +/– 10% voltage regulation.

A voltage regulator is a special type of transformer with tap changing capability. Its basic function is not voltage step up or step down, but voltage regulation. It is an auto transformer with multiple taps.

Var stands for reactive volt-ampere. The sinusoidal current through a reactive circuit element like an inductor lags the sinusoidal voltage by 90 degrees across the reactive element. For a capacitor, the current leads the voltage by 90 degrees. Since the voltage and current are 90 degrees out of phase, a reactive circuit element does not consume any energy; it simply stores and releases electric energy periodically at twice the frequency of the sinusoidal voltage and current. But the reciprocating energy flow between the reactive element and energy source causes voltage drops and energy losses due to the resistance on the conductors. Power engineers call this reciprocating power reactive power to differentiate it from the real power that is consumed by circuit elements like a resistor. By convention, an inductive element is said to consume reactive power and a capacitive element is said to generate reactive power. Most circuit elements in transmission and distribution systems and loads are inductive in nature; they consume var (reactive power). Shunt capacitors are installed at selected locations to supply the required var. Capacitors can be switched on or off to control the amount of var supplied to the system, depending on the reactive power needs that change with time.

Distributed generators can also vary their reactive power output and be used for var control. Many distributed generators are connected to the distribution grid through power electronics interfaces. The inverter converts DC voltage to AC voltage with the right voltage and frequency. The inverter has the capability to supply or consume reactive power as needed. The advantages of var control by distributed generators are their fast response and continuous control.

1.2.3 Voltage Drop and Energy Loss in Distribution System

The voltages on a distribution system must be kept within a narrow deviation from the nominal service voltage to provide satisfactory service and avoid damage to customers' equipment. The allowable deviations are specified in the United States by the ANSI C84.1 (1) to be 5% of the nominal voltage. For 120 volt service, this means the acceptable voltage range is 114 to 126 volts.

Voltage drop occurs when current flows through impedances. The voltage drop over a conductor with impedance $Z = R + jX$ by current I is approximately

$$|\Delta V| = RI\cos\theta + XI\sin\theta = RI_d + XI_q$$

where the impedance of the conductor Z has two components—the resistance R and reactance X; $\cos\theta$ is the power factor of the power at the receiving end of the conductor; and I_d and I_q are the real and imaginary parts of the current corresponding to the real and reactive powers transmitted. X is generally several times greater than R for distribution lines. Voltage drop can be reduced by reducing the reactive component I_q without affecting the real power transmitted.

The energy loss ΔE over the conductor is proportional to the product of the conductor resistance and the square of the current magnitude:

$$\Delta E = RI^2 = R(I_d^2 + I_q^2)$$

On a typical distribution system, due to the absence of distributed generation, the active power must be supplied by the utility; there is very little to no control over

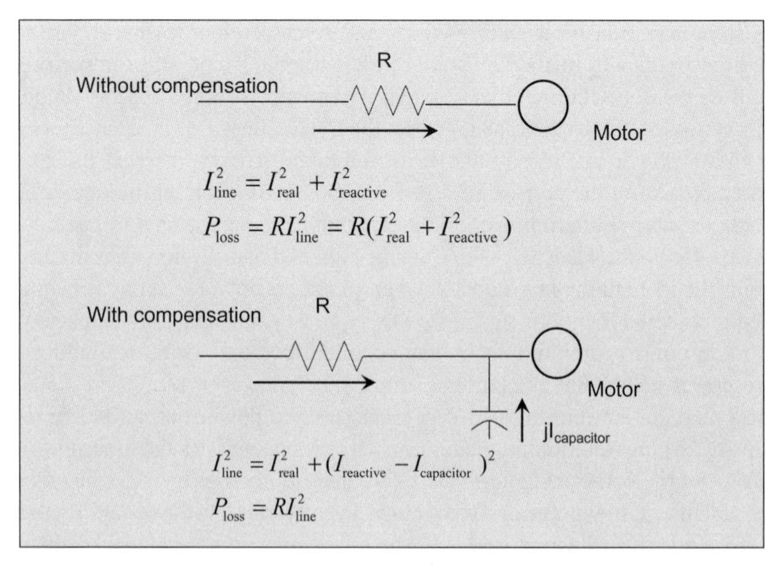

Figure 1.2. Capacitor's effect on reactive power (current) flow and loss.

the reactive power distributed. The only available option is to use reactive power compensation to reduce the reactive power (current) flows on the system. As illustrated by Figure 1.2, the installation of a capacitor at the site of the load reduces the reactive power (current) flow through the distribution conductor, which results in lower energy losses.

In real distribution systems, due to the installation and operation expenses, only a limited number of capacitors are installed. Also due to the discrete nature of capacitor switching, the compensation by capacitor banks is not exact and cannot match the load variations throughout the day; the turning on and off needs to be scheduled to achieve the best compensation results. Since the loading on the feeders varies with the hour of the day, day of the week, or month of the year, setting the right amount of reactive compensation is not a simple task and requires consideration of the actual real-time loading condition on the feeder and the locations and sizes of the installed capacitor banks.

1.2.4 Load Response to Voltage Variations

The actual power consumption of loads connected to the distribution system could be different from the nominal or design value, depending on the actual voltage condition. These loads can be characterized by the following model with different values for the parameter α:

$$P_{actual} = P_{design} \left(\frac{V_{actual}}{V_{design}} \right)^{\alpha}$$

The most common types of loads are the constant impedance ($\alpha = 2$) and the constant power ($\alpha = 0$) loads. The power consumption of a constant impedance load varies quadratically with the voltage. The resistive heater commonly found at home is an example of a constant impedance load. If a heater is designed (rated) to consume 1.0 kWh at the rated voltage, it will consume 1.02 kWh if the operating voltage is 1.01 times the design voltage. The power consumption of a constant load remains invariant, within limits, of voltage deviations from the design voltage. When the voltage drops, the load will draw more current so that the power stays the same. In the world, any load may be a combination of the constant impedance and constant power loads.

A common utility practice for peak management is the so-called conservation voltage reduction (CVR). The objective of CVR is to reduce the total power consumption on the distribution system, either for the purpose of relieving reserve shortage or reducing coincident peak. CVR is motivated by the observation that on most distribution systems, the power consumption, in aggregate, tends to drop as voltage profiles in the system are lowered. The effectiveness of CVR is measured by the conservation voltage reduction factor (CVRf), which is the ratio of incremental percentage power reduction to one percentage of voltage reduction. CVR is most effective when all the loads are constant impedance. Ignoring energy loss, the theoretical maximum value of CVRf is 2.0. On a real distribution system, a CVRf of 0.6 or 0.7 is more common.

1.2.5 Benefit Potentials of Voltage and Var Control

Currently, a significant amount (about 10%) of electric energy produced by power plants is lost during transmission and distribution to consumers. About 40% of this total loss occurs on the distribution network. In 2006 alone, the total energy losses and distribution losses were about 1638 billion and 655 billion kilowatt-hours, respectively. A modest 10% reduction in distribution losses would therefore save about 65 billion kilowatt-hours of electricity.

As the demand for electricity grows, new power plants will have to be built to meet the highest peak demand with additional capacity to cover unforeseen events. The peak demand in a system usually lasts less than 5% of the time (i.e., just a few hundred hours a year). This means that some power plants are only needed during the peak load hours and their potential is utilized relatively infrequently. By active demand management on the distribution system, through demand response and voltage and var optimization, the peak demand on the whole electric grid can be reduced. This eliminates the need for expensive capital expenditure on the distribution, transmission, and generation systems. Even very modest reductions in peak demand would yield huge economic savings. For the United States in 2008, for example, the noncoincidental peak demand (i.e., the separate peak demands made on the electrical system recorded at different times of the day) was about 790 GW. With every 1% reduction in the peak demand there would be a reduced need to build a 7900 MW power plant.

1.2.6 Voltage and Var Control Approaches

A VVC solution typically includes the following components:

- Capacitor banks
- Capacitor switches
- Capacitor controllers
- On load tap changers for substation transformers
- Tap changers for voltage regulators
- Controllers for tap changers
- Current, voltage, and power flow sensors
- Communications infrastructures
- Control decision computing platform
- Application for control decisions
- Supporting IT infrastructure, such as supervisory and data acquisition (SCADA) system or distribution management system (DMS)
- Supporting applications (network modeling, load forecasting, state estimation, power flows, etc.)

There are several different types of VVC solutions. Depending on the types of solutions, some of the supporting components, such as SCADA, DMS, and the associated supporting application, may not be required.

Historically, the voltage and var control devices are regulated in accordance with locally available information. A capacitor bank may be switched on or off based on time or temperature, presumably due to their correlation with the daily load variation. A capacitor bank may also be switched based on the voltage or current measurements at the location where the capacitor bank is installed. This solution approach is easy to implement and does not require communication infrastructure. On a feeder with multiple voltage regulation and var compensation devices, each device is controlled independently, without regard for the resulting consequences of actions taken by other control devices. This practice produces sensible control actions from the perspective of each local controller, but conflicting and unstable results for the entire feeder or substation. When communication is not available, the VVC system is basically operating in the dark; the states of the majority of the controlled devices are not visible to the system operator and the performance of the solution is difficult to verify. Some solutions attempted to overcome some of the shortcomings by coordinating the control settings and arming delays at deployment time. The main limitations of this approach are limited effectiveness and inflexibility to system change. This feedback control based on local measurements can be referred to as a Tier 1 solution.

To overcome the limitations of Tier 1 VVC control, we can take advantage of the increasing availability of two-way communications. Rather than making the control decision for each controlled device independently, the control decisions for all the interacting control devices are made at a central computer by control algorithms that consider all measurements and the potential interactions of control actions being contemplated. The central computer may be a substation computer or a dedicated regional VVC computer server. The basic form of this solution approach requires two-way communications in order to collect measurement information and transmit control commands to the field devices. We refer to this solution approach as the Tier 2 solution. A Tier 2 solution does not require the support of a DMS and requires limited network model information. The control decisions for a Tier 2 solution are generally made based on system state feedback and heuristics.

Heuristic methods often use experience-based rules and available measurements to control capacitors and voltage regulators locally. For instance, planning engineers review the reactive load curve to set the time of the day when a capacitor should be turned on or off, or set the voltage level and dead band for voltage regulator tap changing. These control methods are simple and the principle is generally correct from a local perspective. However, as there is no model-based analysis as the foundation, the heuristic-based control strategy only focuses on the effectiveness of local control. The control decisions from different devices are not coordinated at the network level and often the local controls have conflicting results. In addition, heuristic-based control is mostly triggered by events; that is, such methods can only be executed in the situation where operating limits are likely to be violated and it does not aim to proactively reduce energy loss and system demand in normal operation.

Ideally, information should be shared among all voltage and var control devices. Control strategies should be evaluated comprehensively so that the consequences of possible actions are consistent with optimized control objectives. This could be done centrally using a substation automation system or a distribution management system. This approach is commonly referred to as integrated voltage and var optimization (VVO), which we refer to as Tier 3 VVC. A Tier 3 solution is best implemented in a control center computer due to its dependence on the support of distribution SCADA and DMS and the related applications such as load forecasting, state estimation, and distribution power flow. A Tier 3 solution is generally based on model predictive control (MPC) design. To support model predictive control, the measurements of the distribution system from a SCADA system are combined with recent meter data from AMI (advanced metering infrastructure) and historical time series data available from meter data management systems (MDMSs) to perform load forecast and filtering for the feeder loads. These filtered loads are then used to produce an updated system model using state estimation techniques. With the updated system network model, the load forecast, and the load models, a MPC control problem can be formulated, where the control objective is the minimization of either energy loss or demand and the control variables are the switching status of capacitor banks and the tap position of tap changers.

A general problem formulation of the Tier 3 VVC solution is:

Minimize weighted sum of (power loss + power demand + voltage violation + current violation) subject to

Power flow equation constraints
Voltage constraints
Tap change constraints (operation ranges)
Shunt capacitor change constraints

The power loss and power demand in the objective are measured in kilowalts (kW) or megawatts (MW), depending on the size of the system being controlled. The terms for voltage violation and current violation in the objective are usually included with a sufficiently large penalty coefficient to ensure compliance with service quality requirement. The power flow equation constraints are included to ensure feasibility of the solution. For North American distribution systems, a multiphase power flow model is generally used to account for the imbalance in distribution system design as well as consumer loading. The MPC formulation is a mixed integer nonlinear optimization subject to nonconvex constraints. The high dimensionality of the constraints results from the thousands of multiphase nodes in even moderate distribution circuits. The MPC VVO problem is a challenging mathematical programming problem. However, major progress has been made in the industry in recent years to solve this problem for large realistic distribution systems with speed suitable for online application [1, 2]. More research is still needed to go from VVC to VVWC (voltage, var, and watt control), which will include the control of active power and reactive power of distributed energy resources (DERs) and various energy storage systems (ESSs). It is an even more ambitious goal to include demand response with VVWC in one integrated control optimization framework.

1.2.7 Communication Requirements

A Tier 1 VVC solution basically has no communication requirement for decision making, since the information used is local. However, to increase visibility to the system operator about the status of the controlled devices, the status information should be reported to the distribution SCADA system. This requires very low bandwidth, since only status information is communicated and it is generally reported by exception; that is, communication is needed only when the status is changed. The requirement on communication latency is not high. A latency of a few seconds is desirable but a longer latency of tens of seconds is also acceptable.

The Tier 2 VVC solution requires communicating the status of controllable devices and the voltage, current, and power flow measurements at selected points on the distribution circuits to the substation or regional computer. The number of points is quite limited. The measurements are to be refreshed at the minutes interval. The latency for transmitting the control command should be low, in the range of tens of seconds. It is helpful to keep in mind that VVC is not designed for and is not capable of very fast control actions. This is due to the voltage flicker problem that could be caused by frequent switching of capacitors or changing the tap changer positions. The cycle of data scan could be made as short as a few minutes, but the cycle of control is longer, usually more than 5 minutes. The data values from multiple scans are usually smoothed before being acted upon by the VVC logic so that VVC will not react to momentary and transient system conditions.

A Tier 3 VVC solution has the highest requirement for communication infrastructure. The quality of a MPC VVC solution is dependent on the quality of the underlying distribution system model and the load forecast. The system model can be improved by having power, voltage, and current sensors at more locations on the distribution system. At the present time, the number of sensors on each feeder is quite limited. The information provided by more sensors can improve the quality of distribution state estimation (DSE), which provides the power flow model used by a Tier 3 MPC VVC solution. The load forecast can be produced from historical meter data readings. Power demand data measured by smart meters installed at customer sites can be read at very short intervals. The present industry practice is to read the power demand at 15 minute intervals. These interval demand readings are typically communicated via a 900 MHz meshed radio local area network to meter data aggregators in the field, which then communicate the data to the head end server in the utility back office via public or private wide area network. The meter readings are managed by a meter data management system (MDMS) that can be integrated to the distribution management system in the utility's control center. The number of meters in a typical utility company can range from a few hundred thousand to a couple of million. It is easy to see that collecting so many data points even at 15 minute intervals can be a challenge to the communication infrastructure.

1.2.8 Inclusion of New Controllable Resources

The classic VVC problem includes only capacitor and tap changer controls. New controllable resources are emerging in distribution systems that will introduce additional complexity and opportunity for more sophisticated optimization. The new controllable resources include dispatchable distributed generation and energy storage.

Unlike capacitor banks, distributed generators (DGs) inject real power as well as reactive power into the distribution system. DGs can influence both the real power flow and the reactive power flow, thus providing more capability for loss minimization and demand management. Figure 1.3 illustrates the effect of using the DG to change the real and reactive power flows on distribution circuits.

Energy storage devices, such as plug-in hybrid electric vehicles (PHEVs) and community energy storage systems (CESs) can absorb power from the grid (in charging mode) and feed power back to the grid (in discharge mode). This capability opens up additional opportunity for loss minimization and demand management. Given that some energy is lost in a charge and discharge cycle (cycle efficiency < 1.0), only well scheduled operation of the energy storage will yield operational benefits. The amount of charge/discharge and speed of charge/discharge depends on

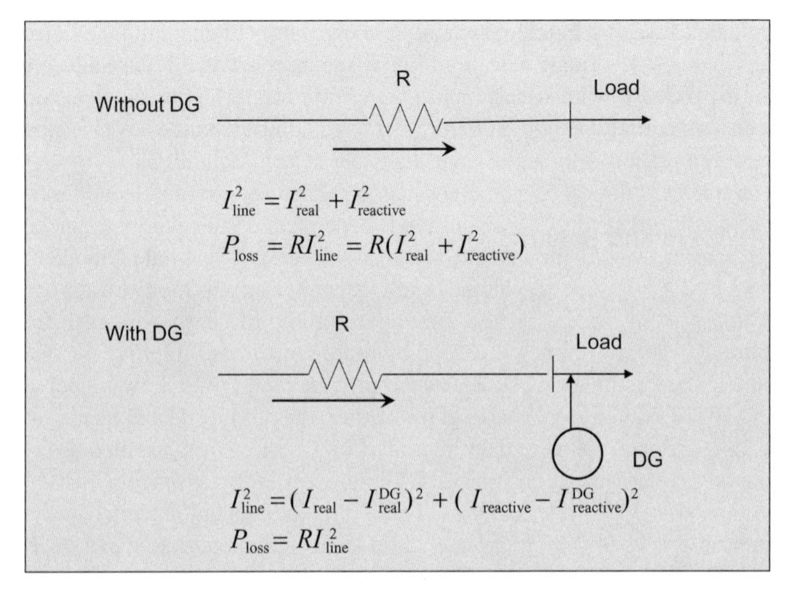

Figure 1.3. Distributed generator's effect on real and reactive power flow and loss.

the state of charge (SOC) of the energy storage device. The inclusion of energy storage devices in VVC or VVWC is another challenge that needs to be overcome in order to realize the benefits of emerging energy storage technology.

1.2.9 Interaction with Other Applications

We should keep in mind that VVC is just one of several important applications concurrently running in distribution systems. As such, any VVC solution should consider how it will interact with other applications such as fault detection, location, isolation, and restoration (FDIR), and demand response management (DRM) to avoid unintended consequences. Upon the detection of a fault by FDIR application, Tier 2 and Tier 3 VVC execution should be suspended until the restoration is completed and the system topology is updated. For demand reduction, VVC, in contrast to deamdn response or direct load control, is considered less intrusive to customers and is usually taken as the first option. If further demand reduction is still needed, different types of DR programs (see Section 1.4 on DR) will be invoked in increasing order of intrusiveness to customers.

1.3 FAULT DETECTION, ISOLATION, AND RESTORATION (FDIR)

Fault detection, isolation, and restoration (FDIR) is a phrase commonly used when discussing Smart Grid and communications. The automation of field switching devices, which entails the deployment of microprocessor-based intelligent electronic devices (IEDs) with long-range high-speed communication devices, paired with automated FDIR software schemes directly contributes to significant reliability and customer service enhancements. In this section, FDIR benefits and schemes are described in detail.

1.3.1 Drivers and Benefits of FDIR

The major driver and benefit for utilities deploying FDIR is enhanced reliability, which eventually also leads to improvements in overall system operation and efficiency. The automation of field switching devices leads to faster fault isolation and restoration of unaffected customers. Thus, customer satisfaction is a side benefit of FDIR. Many utilities have established a target maximum restoration time to be achieved with this level of automation, such as 1 or 5 minutes. Metrics are typically used to measure reliability for utilities, and regulatory commissions typically use these metrics to pressure utilities into improving reliability on underperforming distribution circuits.

1.3.2 FDIR Background

Electric utilities have traditionally used a customer telephone-based trouble call system to detect when a power outage has occurred on the distribution system. The trouble call system attempts to determine the cause of the outage, typically a failure or operation of one or more components of the distribution circuit (e.g., cable/conductor insulation failure, transformer failure, fuse operation, recloser or circuit breaker operation). Subsequently, a maintenance crew is dispatched by operations personnel to find the fault location, isolate the fault from unaffected customers, and repair the damaged component, which can take several hours.

The automation of field switching devices, such as reclosers, switches, and even circuit breakers, including the installation of long-range communication devices at the remote locations, leads to a significant reduction in the time to isolate the fault and restore unaffected customers. This time reduction leads to significant savings for the utilities, both in terms of reducing the amount of lost revenue from customers and the cost of fines from regulatory bodies.

The majority of faults that occur on distribution networks today are the result of electrical insulation failures of the underground cables or overhead conductors, and only involve a single phase. Also, the majority of distribution circuits today are radial, meaning that power flows from the substation transformers (and circuit breakers) out to the loads in the field. Most distribution feeders have a main three-phase line as the "spine" of the circuit, typically overhead with three-phase reclosers and switches spaced along the main line for protection and isolation of faulted sections during maintenance. Fused laterals feed loads off the main line, reflecting a tree-like architecture with laterals acting as "tree branches." Normally, open tie switches are used to connect main lines of adjacent feeders, allowing for manual feeder reconfiguration by the maintenance crews when a fault has occurred. The proliferation of automated switching devices, tie points, and other field devices will lead to more interconnected distribution systems with multiple alternate sources available for post-fault reconfiguration.

Several FDIR solutions exist on the market today. These are typically categorized as either field-based or control center-based FDIR schemes. The field-based schemes use distributed intelligence methods, such as embedding the intelligence in the IEDs themselves or in substation devices, in order to achieve isolation and restoration. The control center-based scheme uses the network model of the distribution management system (DMS) to determine the most effective isolation and restoration switching actions and transmits control commands accordingly. For both scheme types, the communication media deployed are typically unlicensed radio, licensed radio, cellular, fiber optic, distribution line carrier, and even satellite communications as outlined in Chapters 5 to 8.

1.3.3 Field-Based FDIR Schemes

The field-based FDIR solutions can be subdivided into schemes that involve communication between IEDs or between IEDs and substation devices, and schemes that do not involve any communications at all [3, 4].

1.3.3.1 Noncommunication-Based Field Schemes. When communications is not available or not deployed on distribution feeders, automatic capabilities of substation circuit breakers, reclosers, or switches must be relied upon to clear and isolate the fault. Reclosers are fault-interrupting devices that typically appear outside the substation on the main feeder lines and laterals. Switches are load current-interrupting devices that also appear on the main feeder lines and laterals.

The type and combination of switching devices deployed determines how the fault is cleared. If a substation circuit breaker is deployed with nonsectionalizing switches along the main feeder line, then the circuit breaker clears all feeder faults. If a substation circuit breaker is deployed with sectionalizing switches along the main feeder line, then the switches isolate any faults automatically during the dead time of the circuit breaker reclosing sequence. If a substation circuit breaker is deployed with one or more reclosers along the main feeder line, then a breaker or recloser isolates the fault. The breaker isolates any fault between itself and the most upstream recloser, while any downstream recloser isolates a fault between itself and the next downstream recloser. For all the above cases, the devices mentioned are capable of performing upstream isolation only. If any restoration is to occur, then the maintenance crew must also manually isolate a fault downstream by opening the next downstream switching device and then close a tie point switching device from an alternate feeder for temporary restoration.

For the above schemes, only fault detection and upstream isolation can be performed automatically, meaning that downstream isolation and restoration must be performed manually by the utility maintenance crews. A loop scheme with reclosers and without communications is sometimes deployed by utilities, where two feeders are tied together through a normally open point. The tie point IED typically receives voltage signals from both sides, providing a measurement of the presence of voltage from both connected feeders. This IED uses loss of voltage and timers to determine when it should close. If fault current is present when it recloses, then the tie recloser is immediately reopened. In some cases after reclosing the tie recloser, there could be a midpoint recloser between the tie recloser and the fault location, in which case the midpoint recloser would trip first and the unaffected customers between the midpoint and tie reclosers would remain with power.

When it comes to reliability, the above FDIR schemes are much less effective than when communication is used. These schemes are inefficient in that the fault isolation and restoration typically takes significantly more time than for communication-based schemes. Also, the sum of outage costs eventually far exceeds the cost of deploying communication-based schemes.

1.3.3.2 Communication-Based Field Schemes. Communication-based field FDIR schemes can be broken down into two different types: (1) peer-to-peer (IED-to-IED) schemes and (2) substation computer-to-IED schemes. Both types of schemes significantly enhance the reliability of the deployed feeders by isolating faults and restoring unaffected customers much more quickly than schemes that do not use communications. Also, because communication is used, more complex distribution networks with multisource multibackfeed capabilities can be supported, as opposed to the aforementioned, more basic, two-source loop schemes.

The peer-to-peer communication schemes are typically used for true distributed FDIR, where the feeder IEDs located at the field switching devices perform all the intelligent control actions, communicating with IEDs at other switching devices to exchange data and/or control commands (see Figure 1.4), via the DNP3 protocol or the IEC 61850 standard (see Chapter 10 for details). After a fault occurs, the nearest upstream switching device performs upstream isolation. Subsequently, this device's IED communicates data and/or control commands to the next downstream switching devices, as well as other IEDs in the system, in order to achieve downstream isolation and restoration. Before any restoration would occur, a capacity check would be performed to ensure that the alternate feeder would be able to pick up the load of the unaffected customers. The communication between IEDs may be only with "nearest neighbors" or could be as comprehensive as one IED broadcasting pertinent data to all other IEDs in the system. The advantage of deploying this type of system is that the reaction time to faults is typically much faster than centralized schemes where the feeder IED data must be transmitted back to a substation or utility control center for processing.

The substation computer-based schemes employ a master–slave communications architecture, where the substation computer is the master controller sending commands to the feeder IEDs (see Figure 1.5). For this scheme, the logic is

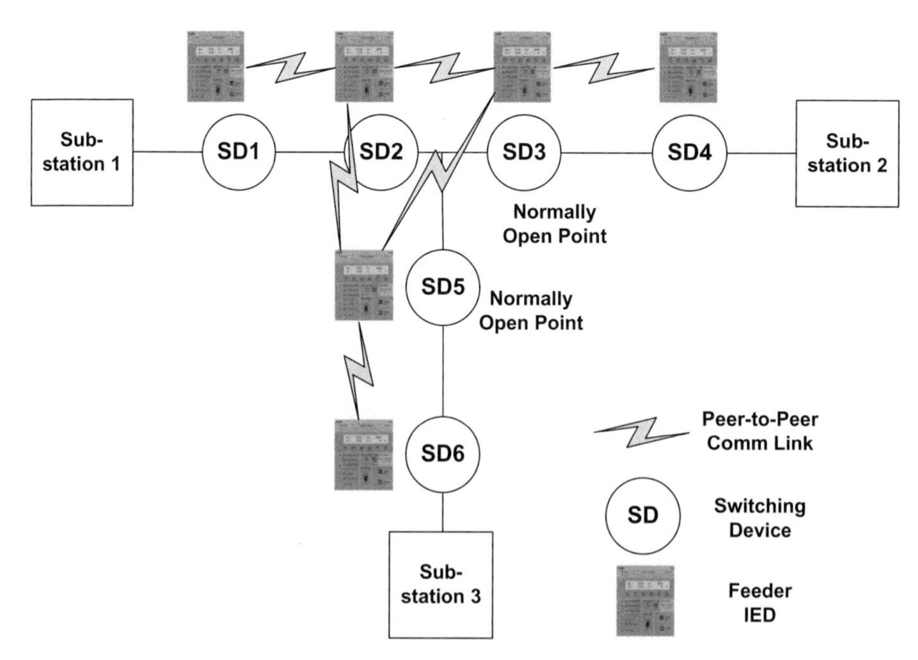

Figure 1.4 Peer-to-peer communication-based FDIR scheme.

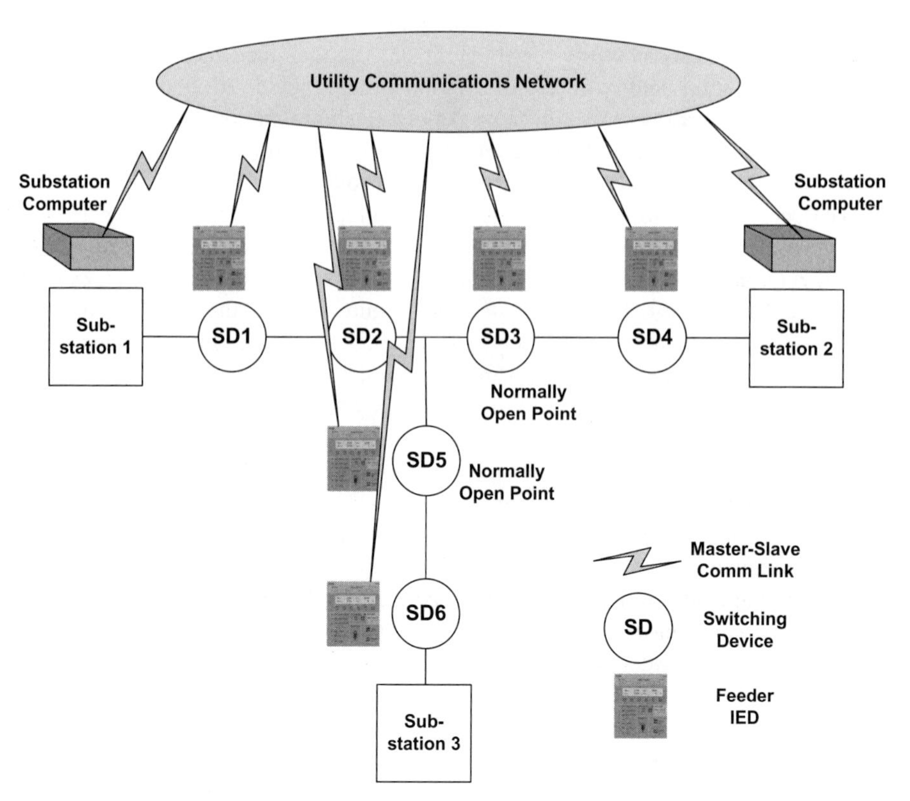

Figure 1.5. Substation computer-based FDIR scheme.

centralized and resides in the substation computer, while the feeder IEDs basically act as end node devices that transmit feeder data back to and receive control commands from the substation computer. Protocols typically used for these schemes are Modbus, DNP3, or the IEC 61850 standard. With most of these schemes, the substation computer contains a simple connectivity model of the feeder topology, including switching devices, lines, and loads. The substation computer continually or periodically receives data from the feeder IEDs, depending on the communication system deployed. After a fault occurs, the substation computer detects the faulted feeder segment from the feeder IED data, allows the upstream IED and switching device to isolate the fault from one side, determines the optimal switching actions to take for downstream isolation and restoration, and eventually transmits the control commands to the appropriate IEDs to perform these actions. As with the peer-to-peer scheme, a capacity check is first performed to ensure that unaffected customers on the faulted feeder can adequately be restored.

1.3.4 Control Center-Based FDIR Schemes

Control center-based FDIR schemes differ from the field schemes in that a network model maintained at the utility control center is used to determine the optimal isolation and restoration field switching actions to restore power to unaffected customers. The network model typically contains detailed distribution circuit component information, such as geographic location, component type, component rating, impedance data, involved phases, and other pertinent information. It is contained in the distribution management system (DMS) software, which typically receives field data from the utility supervisory control and data acquisition (SCADA) system. The automated field switching devices communicate directly or indirectly (e.g., via substation computers or gateways) with the SCADA system (see Figure 1.6). In the past, communication to the field devices was not available, and the switching actions had to be carried out manually by communicating with maintenance crews via voice communications.

A network load flow analysis is performed as part of the FDIR scheme after a fault occurs, to determine if any voltage or current violations will occur as a result of the switching actions. When an automatic recloser, sectionalizer switch, or circuit breaker operates to lock out in the field due to a fault, and the updated device status is transmitted back to the SCADA/DMS, an analysis is performed in the DMS (including the load flow) using the network model to determine the optimal switching actions to isolate the fault downstream and for restoration of the unaffected customers. Data from the utility trouble call system is also used in this process. The analysis provides a switching plan to restore from one or more alternate feeder sources. The switching actions are then automatically transmitted to the field devices through the SCADA communication system. However, some utilities may prefer to have operator intervention to supervise the switching actions that are transmitted to the field devices.

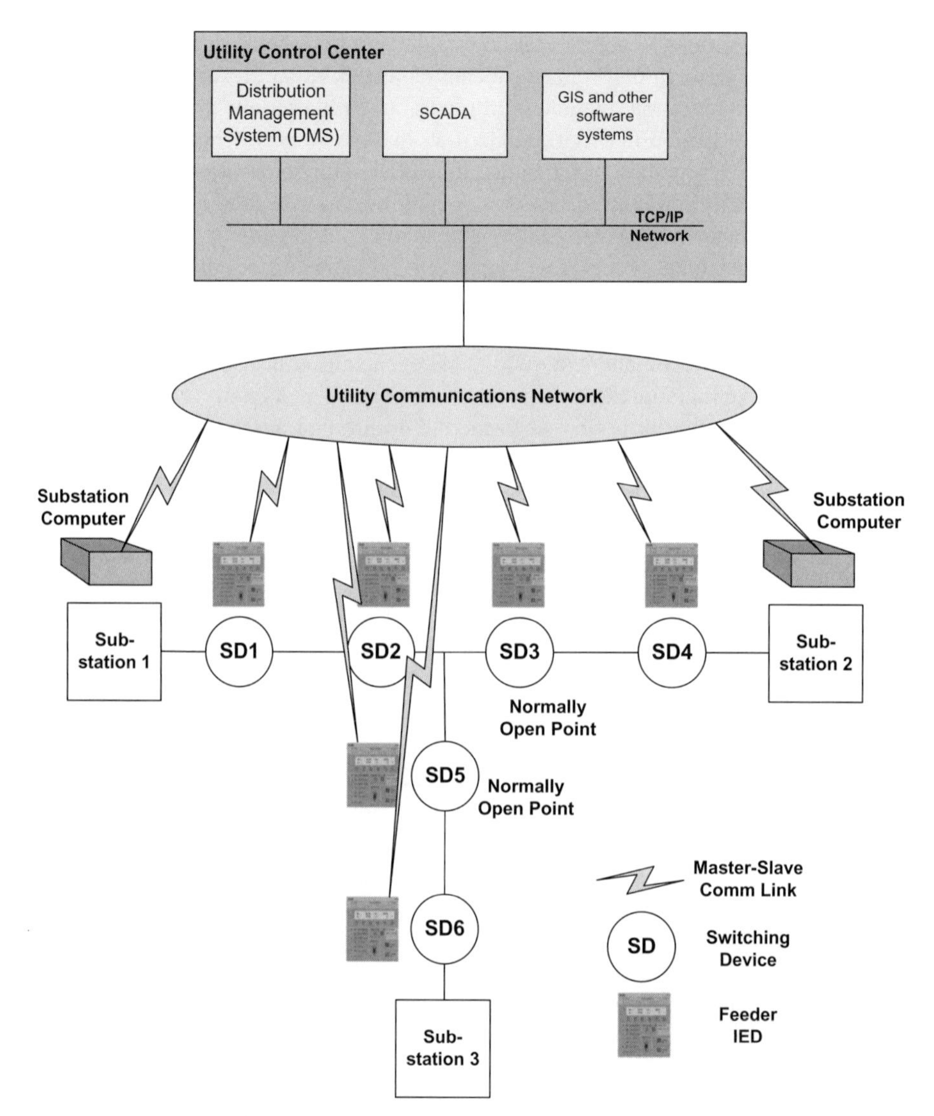

Figure 1.6. Control center-based FDIR scheme.

1.3.5 Reliability: Present and Future

Today, the reliability of power distribution systems is being greatly enhanced with the deployment of automated FDIR schemes across some utility networks. This trend will continue in the future. However, new technologies are being introduced today that will significantly enhance reliability even further in the future.

The proliferation of demand response (DR) programs and distributed energy resources (DER) will provide complementary resources to FDIR schemes. DER devices will provide extra generation capacity, and thus, an alternate source from which customers can be restored after a fault occurs. Demand response programs will allow utilities to reduce system loading during peak periods of operation, resulting in the capability of alternate feeders to pick up more loads during restoration. The gathering of automated meter data at the utility control center will also enhance the outage management process, allowing utilities to determine outage causes and locations faster, ultimately resulting in faster restoration.

Coordinating the FDIR schemes with voltage and var optimization methods and load reconfiguration schemes will also significantly impact reliability. Coordination with voltage and var optimization methods will greatly enhance system efficiency and minimize losses after a fault has occurred and the network has been reconfigured. Load reconfiguration entails switching loads from one feeder to another to avoid feeder overload. Coordination with this type of scheme will provide the most optimal switching scenario should a fault occur after feeders have been reconfigured due to overload.

1.4 DEMAND RESPONSE (DR)

Demand response can be viewed as an additional variable for control and management of the Smart Grid. Demand responsive loads can be adjusted or interrupted in order to provide load reduction during the hours of peak demand in the distribution grid. Introduction of low-cost smart sensors that can be deployed across the distribution grid and the availability of two-way secure communication networks connecting the utility control center to the end users are some of the main technical drivers for incorporating DR in the Smart Grid paradigm. This integration would improve the reliability of the power network, enhance the transparency and efficiency of the energy market, lead to mutual financial benefits for the utility and the consumers, and last but not the least, reduce emissions of the generating plants and alleviate the environmental impacts by allowing a more effective utilization of the current grid capacity.

Generally, demand response refers to a series of actions initiated by the utility, the end-use customers, or both in order to reduce the energy consumption level of individual customers, thereby lowering the total system demand. These actions often take place during the hours of peak energy consumption in the grid, where the utility gets close to its generation capacity, and tries to find the midway solution between having to acquire the possible generation shortage through spot market and shedding the excess load—the former leading to inefficient financial management and the latter negatively affecting the reliability of the grid and the availability of service. For this purpose, the utility monitors and forecasts system demand for the present and future time intervals, in order to detect the possible shortage of supply well in advance so it can be addressed by issuing a demand response event. A typical DR event may consist of a desired schedule (start time and end time), the desired demand reduction level, or a list of targeted customers and other attributes as necessary depending on the type of the DR programs selected.

1.4.1 Types of DR Programs

1.4.1.1 Utility-Initiated Programs. In this category, the utility initiates the event and notifies the customers of the event's attributes, which can be demand reduction signals (duration, amount, etc.) or updated electricity rates, to which the customer reacts by adjusting consumption level. The former would lead to involuntary reduction of demand, while the latter would encourage customers to voluntarily reduce their demand to maximize their savings.

The involuntary demand reduction, also referred to as incentive-based DR, is classified into different categories depending on the type of DR signal and the way it is enforced. Every utility may have its own definitions and design parameters for various programs; however, in the broadest sense, the most common programs are:

- *Direct Load Control (DLC).* The utility would send a DR command to the controllable appliances, such as heaters, thermostats, or washers/dryers, to turn them off. This type of program is more suitable for residential customers with remotely controllable electronic devices and is one of the main drivers for home automation (HA) networks. The devices would stay turned off for the duration of the event, after which they can be turned back on, either remotely or manually by the user. In a further enhanced version of this program, the utility would shift the operation of certain appliances, such as washer/dryers or chargers, from peak hours to nonpeak hours. This is often referred to as demand shifting to highlight the inherent difference with simple demand reduction.
- *Interruptible/Curtailable Load (I/C).* Here, the signal sent to the customers is a request—and not a command—to reduce their demand by a certain level. Compliance is not compulsory but—depending on the terms of the contract—may incur penalties if the request is rejected. This category is more suitable for larger scale customers (i.e., industrial plants and commercial buildings).
- *Emergency Demand Reduction.* This is similar to the previous class of programs, with the exception that it is mainly used during emergency conditions (i.e., very short advance notice), as one step before issuing load shedding actions. Under this program, larger scale customers provide ancillary services to the utility by behaving as a virtual spinning reserve that can reduce its demand upon request.

Similarly, the voluntary demand reduction, also known as price responsive DR, is initiated by the utility—by issuing updated electricity prices; however, it is executed by the customer as a part of its local demand-side management (DSM), and it is more in the form of demand shifting from peak hours to nonpeak hours. The customer may subscribe to different time-based rate structures, namely, time of use (TOU), critical peak pricing (CPP) or real-time pricing (RTP). By adjusting customer load patterns or increasing price responsiveness, large-scale implementation of time-based rates can reduce the severity or frequency of price spikes and reserve shortages, thereby reducing the potential need for incentive-based programs [5].

1.4.1.2 Customer-Initiated Programs. This program is oriented toward proactive customers who make bids to reduce their demand in exchange for financial incentives. The bids would be sent to the utility—along with attributes such as demand reduction available, duration, or start time,—where they are analyzed and shortlisted according to the utility's financial and/or technical criteria.

Figure 1.7 illustrates the demand response engine that may be implemented at the control center in the distribution management system (DMS) environment. The engine would receive the load forecast, load flow data and meter data from external modules, and generate the DR signals/commands to be transmitted to different DR resources. The information on the terms of DR contracts is provided to the engine by

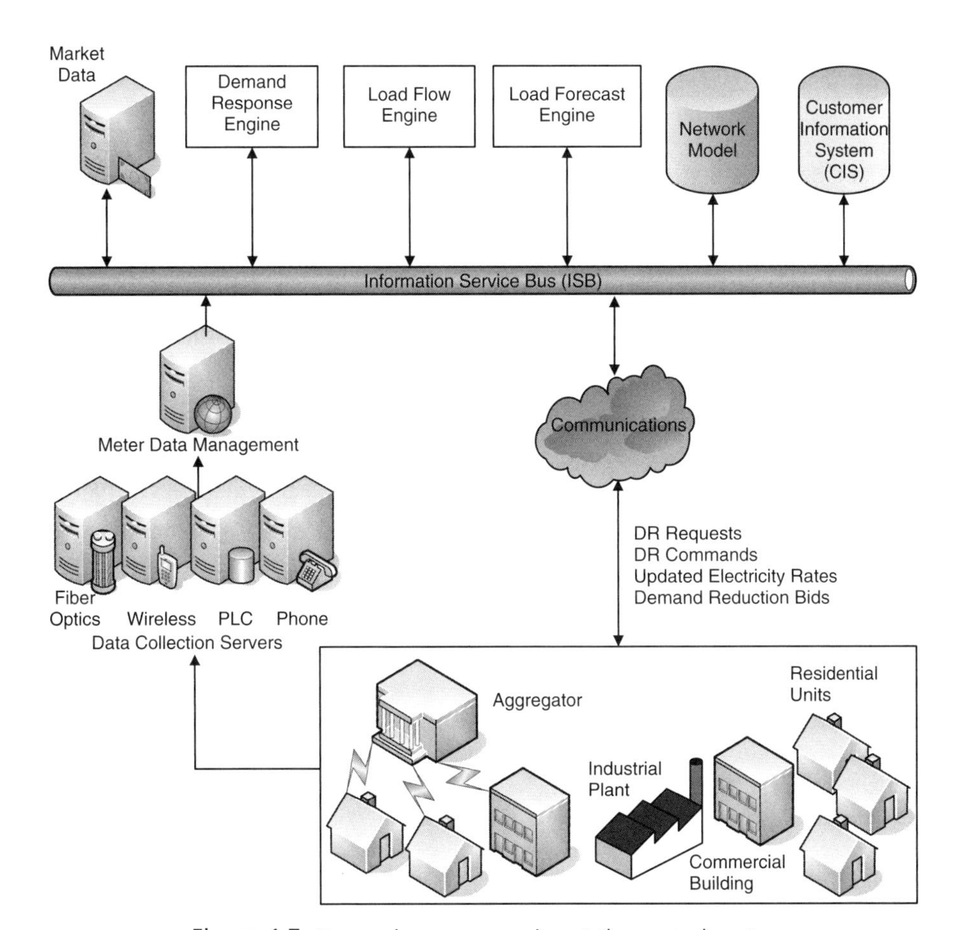

Figure 1.7. Demand response engine at the control center.

the customer information system (CIS). Data transmission to and from the resources can be via two separate networks or through the same communication channels.

1.4.2 Communication Requirements

DR signals are transmitted to the customers through communication media, which can be fiber optics, power lines, telephone lines, or wireless networks (see Chapters 5 to 8). These signals can be the updated electricity rates, demand reduction requests, or turn on/off control commands. While traditional DR managers at the customer sites were based on manual or semiautomated solutions with a human operator in the loop, modern systems are shifting toward full-automation capabilities, where a local energy manager within the building automation or home automation network can regulate the energy usage of various appliances.

In voluntary demand reduction programs, the customer's usage is reported back to the utility via the advanced metering infrastructure (AMI)/automatic meter reading (AMR) system and the customer is billed accordingly. Involuntary reduction programs require a faster two-way communication since the utility needs to validate the response of the individual customers to the DR event issued in order to determine whether the shortage of supply is met or more drastic measures such as emergency DR or load shedding are needed. This could be done by measuring the pre-event and post-event consumption levels, or by receiving indicator signals from the intelligent energy managers on the consumption status of various consumers.

Regardless of the communication media used, the network has to provide sufficient throughput and be dependable. Most DR signals are not very sensitive to network delays of a few seconds, as long as the data is received on time for calculations and processing (usually considered on the order of tens of seconds to a few minutes). This leaves most transmission technologies as viable options.

Cyber security is another requirement of the communication media used. The two-way communications between customers and the utility have to be secure with respect to cyber intrusions and unauthorized access. Proper selection of communication media and/or the communication protocol adopted can solve many of these potential threats.

1.4.3 Statistical Reliability of Demand Response

Higher success rates of DR events would benefit the utility by not having to turn to alternative options such as emergency demand reduction and/or load shedding. In order to achieve this, the utility might perform statistical analysis on the historical data available from the customers' load profiles and consumption patterns so it can associate an attribute with each customer, or group of customers, indicating the probability that they would comply with a DR event. Theoretically, by using these compliance attributes, the utility on any given date/time can select the customers that are more likely to adhere to a specific DR event. It should be noted, however, that although the bulk meter measurements at any customer location are automatically provided to the utility, the detailed consumption patterns broken down in terms of the devices and appliances in use may potentially violate the privacy of the individual customers, and therefore the issue must be handled delicately by the utility.

1.5 DISTRIBUTED ENERGY RESOURCES (DERs)

Distributed energy resources, sometimes also known as distributed generation (DG), embedded generation, dispersed generation, or decentralized generation, refer to generation units within the distribution grid or on the customer side of the network [6]. These sources of energy often generate localized power for loads in their proximity. This would reduce the congestion on the system as well as power losses since less power needs to be transmitted to remote load points. Moreover, the sizes and the number of power lines that need to be constructed will be reduced. This is in contrast with the traditional approach of generating power, where most of the electricity was produced in large-scale centralized facilities such as fossil-fuel-based, nuclear, or hydro power plants. While the latter approach had better economies of scale, it could lead to higher environmental impacts. Furthermore, the decentralized approach would fit better in the new liberalized electricity markets with the ever-increasing need for higher reliability and availability of service. This is further driven by the recent interest and endeavor in the utilization of renewable sources of energy, on top of the technological advances in designing more efficient small-scale generation units and subsequent cost reduction of micro generation at the residential level.

No unique definition exists for the size of the DER unit. In fact, units ranging from a few kilowatts in size to 100 MW or more have been considered in the literature as a DER [6]. These units can be owned and operated either by the utility [e.g., dispersed large-scale combined heat and power (CHP) units, or large-scale wind or solar farms] or by the customers [e.g., rooftop photovoltaic (PV) units, micro gas turbines (MGTS), diesel engines, small-sized wind turbines, and suchlike].

Based on the interface and the capabilities, the DER units can be broadly classified into two groups:

- *Conventional Rotary Units.* These units contain a rotating mass with inertia, such as a synchronous machine or an induction machine. Examples are the micro gas turbine, the reciprocating engine, the pumped hydro storage (PHS), and wind turbines directly connected to the grid via a gear-boxed induction generator. These units can sometimes provide bidirectional flow of power and may be able to absorb power (e.g., when in the motoring mode).
- *Electronically Coupled Units.* These units contain power electronics circuitry to regulate a DC voltage generated by the source of energy (wind, solar, storage, etc.) and consequently convert it to AC voltage through inverters. They have faster response times compared to the previous type and can limit the current injected into the grid during startup or disturbances.

1.5.1 Operation and Control

1.5.1.1 Dispatching DER Units. DER units that run on fuels or use the previously stored energy can be dispatched for a given period of time. These units can be dispatched either remotely by the utility through an engine running at the DMS level, or locally by the individual owners. They can therefore assist in the generation dispatch of the distribution grid and at times of peak load—of course, subject to their operational constraints. However, other units that utilize renewable energy resources such as solar and wind are normally non dispatchable, meaning they can only provide power as long as the source of energy is available (i.e., sun shining or wind blowing). These units are often set to operate in the maximum power point tracking (MPPT) mode, indicating that they are tuned to extract as much energy as available. Nevertheless, when operating in hybrid designs in parallel with energy storage systems such as batteries, they can be turned into dispatchable units.

It should be noted that the DER units connected to the grid via DC/AC inverter circuits are capable of injecting reactive power into the grid regardless of whether or not the main source of energy is present, which makes them efficient sources for reactive power control.

1.5.1.2 Mode of Operation. DER units can operate in the grid connected mode or islanding mode. In the former, the unit must adhere to the frequency dictated by the grid and must operate within the predefined operational limits [7]. In this case, the unit is often controlled in the load-following mode, where it tries to support local loads by providing the setpoint active and/or reactive power values. On the other hand, islanding mode occurs when the DER unit is responsible for not only supporting the local load but also maintaining the frequency of the network to which it is connected. In case more than one DER unit operates in the islanding mode (e.g., a microgrid), one (normally) larger unit would maintain the frequency while the other units would operate in the load-following mode.

During a fault or disturbance in the network, the DER unit connected to the grid might have to stay connected and temporarily supply reactive power to the system for a predefined duration of time before it goes to the islanding mode. This is referred to as low voltage ride through (LVRT) or fault ride through (FRT), where the DER is intended to provide post-fault voltage support, and the corresponding requirements, (e.g., duration of connection) are driven by government policies and national grid codes.

Lastly, a DER unit with surplus generation can inject its excess power into the grid. This energy can be used by the grid operator to supply other loads during peak demand, or if available during light load conditions it can be stored in energy storage systems such as PHS for future use. Such energy buyback plans can create additional revenue for owners of DER units, and can be an incentive for higher penetration of distributed generation, for instance, rooftop PV units for residential households. Figure 1.8 illustrates how different resources including DER units in various modes of operation and DR resources can function under a grid management scheme.

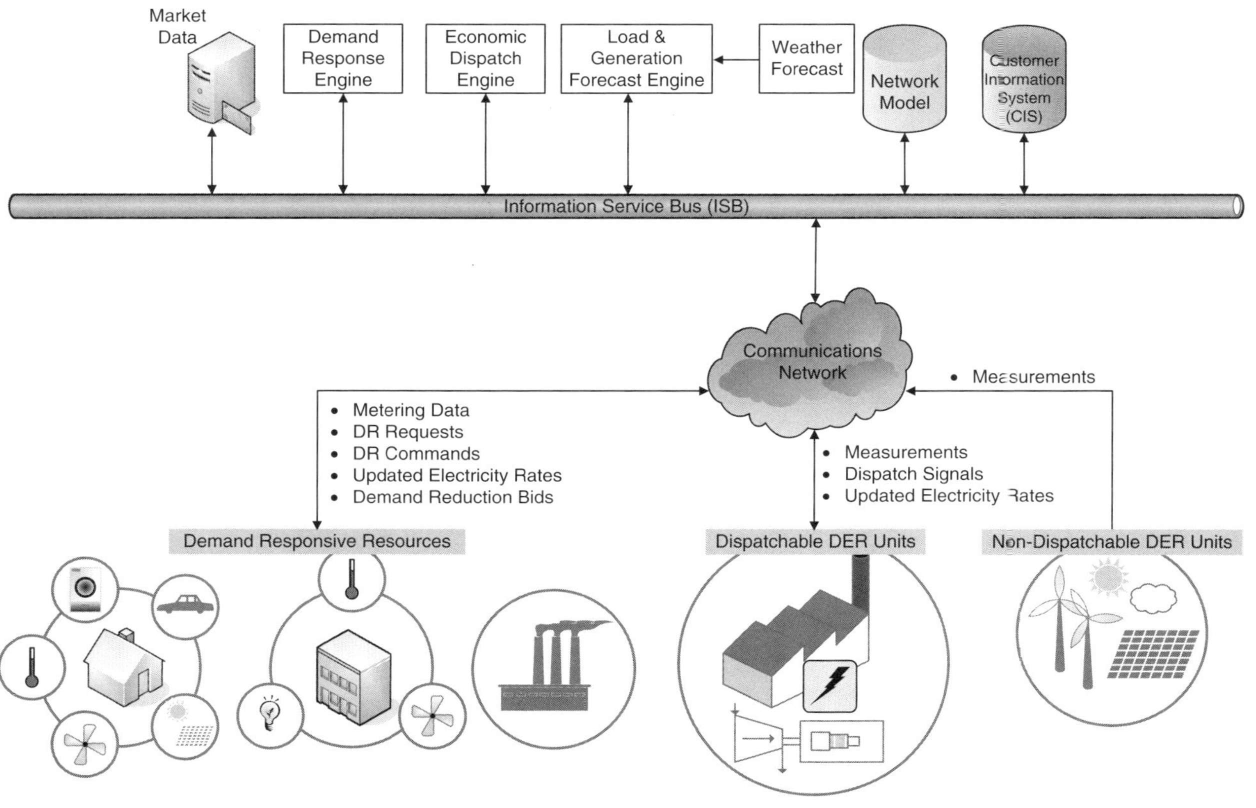

Figure 1.8. Schematic diagram of a grid management solution.

1.5.2 Communication Requirements

Communications between the DER units and the utility can be done through the SCADA system or other public communication links for smaller units. Typical signals exchanged are dispatch signals (in terms of active and/or reactive power set-points), turn on/turn off signals, disconnection commands, measurements, and other related data. The dependability and security requirements are high because of the potential impact that, for example, a lost communication channel or an integrity violation by an unauthorized entity can have on the overall performance of the system.

Also, depending on the mode of operation, the sensitivity to transport lags may differ. Direct dispatch of DER units requires near-real-time data exchange between the utility and the generation units, while other standalone modes of operation might only necessitate a status change report to the utility based on a report by exception scheme.

1.5.3 Sustainable Power Grid

Deployment of demand responsive loads and distributed energy resources in the distribution system is the first step toward making the grid more sustainable. Localized generation by the DER and demand reduction and shifting using DR allow the utility to utilize the current capacity of the grid as much as possible and therefore postpone capacity expansion projects (CAPX) that are normally associated with considerable environmental concerns.

The advanced metering and control system technologies adopted by the utilities enable them to seamlessly integrate the available DER and DR resources in their overall grid management solutions. Standardization of the communications and signaling systems have paved the way for higher penetration of autonomous energy resources in a naturally multivendor environment. Efforts have been made to include technologies such as demand response in future building codes and standards. By being equipped with DR capabilities and small-scale power generation resources, buildings of the future may be used not just for reliability but as a true alternative supply of energy to the grid. As technical advances are made, microgeneration technologies are likely to move toward lower energy payback times, thereby creating more incentives for higher penetration levels, particularly at the residential level.

Energy storage systems distributed across the grid will enable an efficient cycle of energy consumption and generation with continuous charge/discharge for reusing the available energy. This is estimated to be further enhanced by the widespread usage of electric vehicles (see Chapter 2), and the foreseeable introduction of bidirectional power exchange capabilities between the vehicle and the grid in the future.

1.6 WIDE-AREA MONITORING, CONTROL, AND PROTECTION (WAMCP)

Power system stability can be classified as the ability of the power system to approach and operate at a stable equilibrium with acceptable grid voltages and system frequency. To ensure power system stability, a wide range of dynamic phenomena need to be considered, ranging from generator angular dynamics on a subsecond time scale, via frequency dynamics on a time scale of seconds, to voltage and load dynamics on a time scale of tens of seconds to minutes [8]. Most blackouts over the last decades have been caused by the failure to preserve one or more forms of power system stability [9].

Classical supervisory control and data acquisition systems/energy management systems (SCADA/EMS) typically have a measurement update rate of several seconds or even minutes and thus need to assume that the power grid is operating in steady state, and are therefore not suitable to monitor or control such phenomena. Wide-area monitoring, control, and protection (WAMCP) extends the time resolution of classical SCADA/EMS down to the subsecond time scale, making it capable of monitoring and reacting to dynamic instabilities in the grid. The information provided by a wide-area monitoring (WAM) system simply aims at providing information, in most cases in real time, about the stability issues in the system, whereas a wide-area control and protection system aims at real-time control of such instabilities and to protect the power system against possible consequent blackouts.

1.6.1 Structure of a Wide-Area Monitoring, Control, and Protection System

The core idea of the wide-area monitoring, control, and protection (WAMCP) system is the centralized processing of the data collected from various locations of a power system, aiming at the evaluation of the actual power system operation conditions with respect to its stability limits [10]. Figure 1.9 shows a rough outline of the automation system architecture for power networks. Directly connected to the physical network assets such as transformers, generators, and power lines, there are lower level control devices called intelligent electronic devices (IEDs). Typical functionalities of an IED are to control and operate circuit breakers, or to execute protection algorithm detect faults, or simply to acquire measurements from sensors placed on the primary equipment. The switchgear and grid connections in a substation are organized in so-called bays, that contain all the equipment connecting the substation, for example, to a single power line or generator. The IEDs of different bays are connected by a local area network (LAN) to the substation automation and monitoring systems (SAMSs). The SAMSs typically also included gateway functionality to allow remote control actions to be accepted from, and to transmit status and measurement data to the SCADA/EMS over a wide-area network (WAN). On the network level, operator workstations are present for wide-area monitoring control and protection as well as SCADA/EMS. Finally, the enterprise level includes functionality for the operational, maintenance, and expansion planning and interfaces to other enterprise computing systems, such as financial accounting systems and energy trading systems.

The positioning of the WAMCP system in relation to other control and monitoring systems is shown in Figure 1.9.

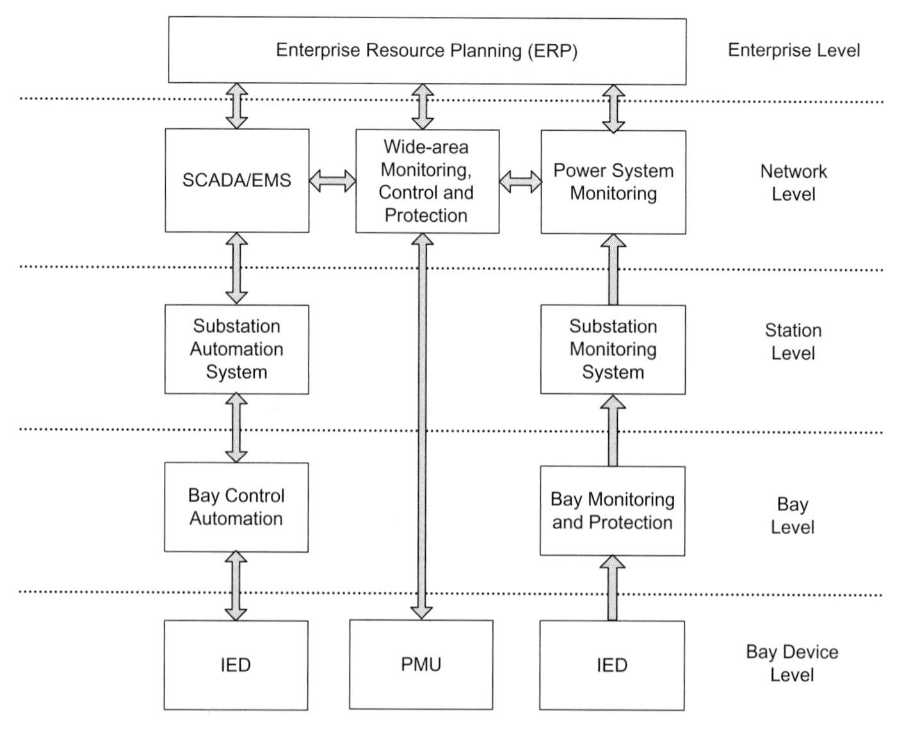

Figure 1.9. WAMCP and its connections to other control and monitoring systems in the power network.

Although the particular application range of WAMCP is quite wide, depending on the addressed phenomena, the fundamental principles remain the same. The structure of a typical WAMCP system is shown in Figure 1.10. See reference [11] for information about a commercially available WAMCP system. The employed hardware and software can be explained based on the stages of the data handling applicable to WAM:

- Data acquisition, carried out by the phasor measurement units (PMUs)
- Data delivery through a wide-area communication system
- Data processing, through the system protection center (SPC)

A WAMCP additionally features automatic actions for blackout prevention and stabilization. This includes the following additional steps:

- Command data delivery through a wide-area communication system
- Command execution through bay level systems

1.6.1.1 Data Acquisition. To properly observe the system dynamics, the needed measurements should have the following characteristics: they must be taken from

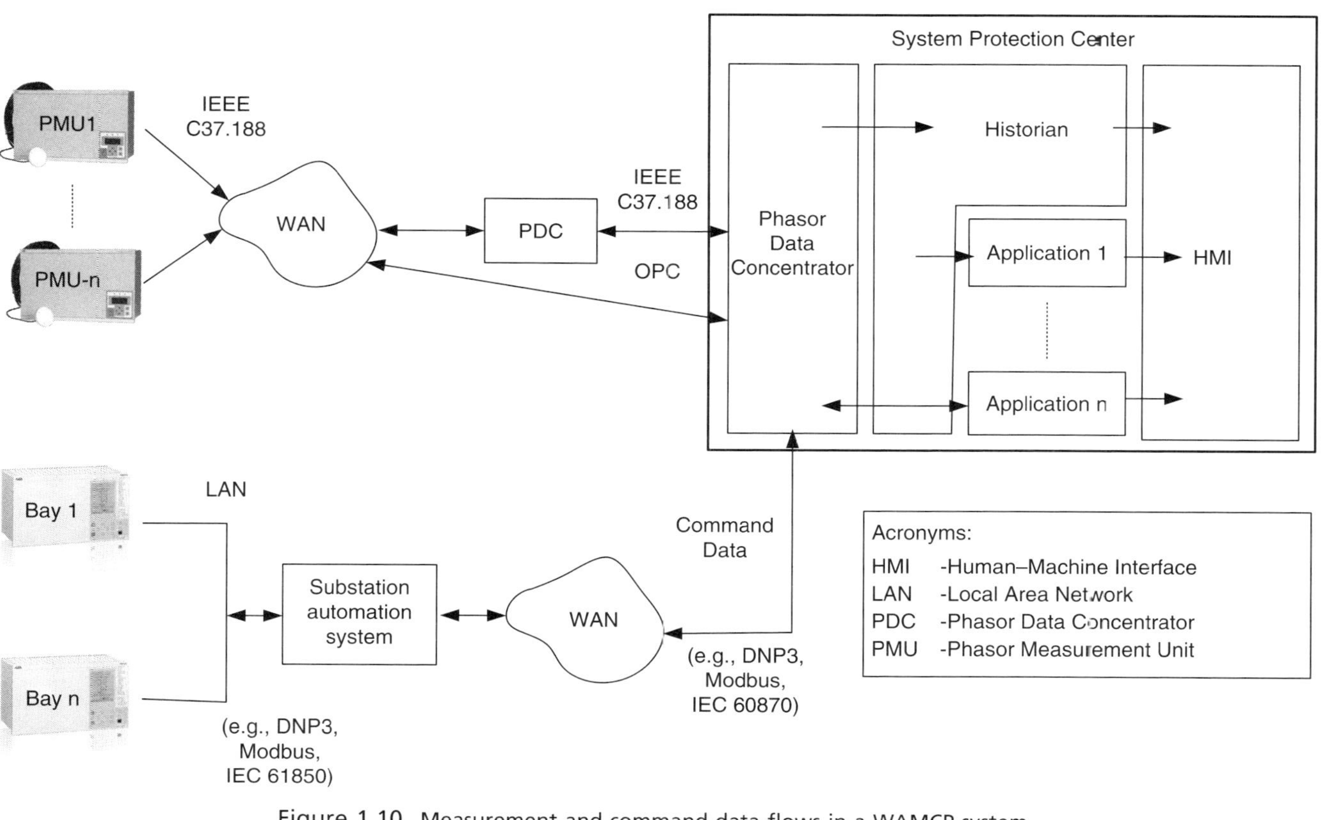

Figure 1.10. Measurement and command data flows in a WAMCP system.

different network locations, with high enough sampling rate and at the same time instant. The selection of quantities to be measured by PMUs and the locations for the installation of the PMUs are dependent on many factors. The most significant criteria are the instability phenomena, which are to be mitigated, followed by the practical economic aspects such as utilization of the existing communication infrastructures. From these initial considerations the requirements on sample rate and response time of the measurements and associated telecommunications can be derived. Basic guidelines are listed in Table 1.1.

PMUs require accurate synchronization of the measurement instant. For this purpose they use the global positioning system (GPS) time synchronization signal, which can provide synchronization accuracy to within 1 *u*s, corresponding to an accuracy of the phase angle measurement of around 0.02 degree. This is adequate for the applications listed in the next section. Typical data sample rates are 10–60 Hz and each data packet is time stamped with minimal latency and transmitted to the phasor data concentrator (PDC). Most PMUs are capable of measuring several voltage and current phasors as well as frequency and frequency derivative; see, for example, reference [12] for a detailed description of a commercial PMU, and references [13] and [14] for a description of the standardization of their functionality and communication interface.

1.6.1.2 Data Delivery. Transmission of the measured data is arranged by means of a wide-area communication network. The data delivery stage is divided into two phases: firstly there is the communication between the PMU and a phasor data concentrator (PDC), which is carried out using the Synchrophasor protocol IEEE C37.118 [13] or the legacy protocol IEEE 1344 [14]. The data packets contain measurement value, time stamp, and quality information as well as status information about the PMU itself. Secondly, the PDC collects the data and assembles collections (snapshots) of synchronously sampled phasor measurements that it makes available for the data processing layer. Currently, there is no widely accepted standard for the communication between the PDC and the System Protection Center (SPC). For this purpose, some systems have used the Synchrophasor protocol and other systems OPC.

1.6.1.3 Data Processing. The data are then collected in the central computer equipped with the appropriate software, which is called the System Protection Center (SPC). The software packages can be divided into the following:

- Auxiliary software common to all installations of WAMCP, such as software for automatic data archiving, alarm and event handling, and interfaces for data exchange with SCADA/EMS systems. Additionally, there is application server and operating system software to coordinate the execution of the various algorithms and auxiliary functions.
- Phasor data concentrator (PDC) that acquires data from the PMUs and makes it transparently available to the applications in the form of phasor snapshots.

TABLE 1.1. List of Considered Real-Time WAMCP Applications with Communication Requirements

Function	Uni-/ Bidirectional	Maximum Delay	Sample Rate	References
SITUATIONAL AWARENESS				
PMU-assisted state estimation	Unidirectional	30 s to 2 min	30 s to 2 min	[25, 26]
PMU-based state estimation	Unidirectional	0.1 s	0.1 s	[27]
Dynamic state estimation	Unidirectional	0.02–0.1 s	0.02–0.1 s	[28, 29]
Phasor data visualization	Unidirectional	1 s to 1 min	0.1 s	
FREQUENCY STABILITY				
Predictive underfrequency load shedding	Bidirectional	0.1 s	0.1 s	[30, 31]
Adaptive islanding	Bidirectional	0.1 s	0.1 s	[32]
OSCILLATORY STABILITY				
Local power oscillation monitoring	Unidirectional	10–30 s	0.1 s	[33–37]
Wide-area power oscillation monitoring	Unidirectional	0.1 s	0.1 s	[38–40]
MIMO small-signal model identification	Unidirectional	Hours	0.1 s	[41, 42]
Wide-area power oscillation damping	Bidirectional	0.1 s	0.1 s	[43, 44]
TRANSIENT STABILITY				
Precalculation transient stability control	Bidirectional	30 s to 2 min	30 s to 2 min	[45, 46]
Closed-loop transient stability control	Bidirectional	0.02–0.1 s	0.02–0.1 s	[47, 48]
Line thermal monitoring	Unidirectional	5 min	1–30 s	[49]
VOLTAGE STABILITY				
Local voltage stability monitoring	Unidirectional	10–30 s	0.5–5 s	[50, 51]
Wide-area voltage stability monitoring	Unidirectional	0.5–5 s	0.5–5 s	[52]
Wide-area voltage stability control	Bidirectional	0.5–5 s	0.5–5 s	[53]
Wide-area FACTS and HVDC control	Bidirectional	30 s to 2 min	30 s to 2 min	[54, 55]
Cascading failure control	Bidirectional	0.5–5 s	0.5–5 s	[23, 24]

- Human–machine interface (HMI) visualizing the phasor measurements and the results of WAMCP applications to the user (typically a planner or operator).
- History data access (HDA) provides a set of standard interfaces that allow WAMCP applications to access historical archives of measurements and for users to retrieve and store the data in a uniform manner.
- Wide-area monitoring, control, and protection algorithms as outlined in the next section.

1.6.1.4 Command Data Delivery. The SPC sends command data to the substation monitoring systems at the station level over a wide-area communication network, for example, a TCP/IP network. Preferably it shall be dedicated to this purpose to ensure command execution within specified time intervals. If this is not possible, care must be taken to make the control and protection algorithms robust against time delay and jitter in the transmission. Commonly used protocols are DNP3, IEC 60870-5-101, and Modbus [15], which allows addressing of individual pieces of equipment on the bay level and encoding of command orders.

1.6.1.5 Command Execution. Command execution is performed by the bay level control equipment, which communicates with the substation automation system over a local area network on the station level. Common legacy protocols are DNP3, Modbus, and SPA with the newer IEC 61850 gaining rapid acceptance [16] (for details see Chapter 10).

1.6.2 Overview of WAMCP Applications

Most WAMCP applications aim at monitoring and controlling one or more stability-related phenomena in the power grid. Instabilities shall be detected early and operators informed, or automatic control and protection shall be activated to preserve the stability in a timely manner to avoid blackouts. This section surveys the various applications for which WAMCP has been considered.

1.6.2.1 Enhanced State Estimation. Enhanced state estimation was one of the first proposed applications of the PMU data. PMU assisted SCADA/EMS state estimators are extended with the capability to benefit from PMU data. This improves accuracy and convergence properties of the conventional state estimators, which, however, still operate on a time scale that is too slow to monitor most dynamic stability phenomena in a grid.

However, if enough PMUs have been installed to provide complete observability of the network state only through the use of the PMU measurements, the state estimation problem becomes linear. Due to the high sample rate of the input data and the low computational complexity of linear state estimation, it can be executed on a subsecond time scale, enabling the use of advanced wide-area control and protection algorithms. However, currently no grid operator in the world has reached this degree of PMU deployment. Furthermore, dynamic state estimation has been proposed, which extends the conventional static state estimation problem with the estimation of, for example, generator speeds and rotor angles.

1.6.2.2 Frequency Stability Control and Protection. Frequency instability denotes the ability of a power system to operate with the (average) system frequency within normal operating limits. Failure to do so may cause damage to generation and/or load side equipment. Thermal generating plants are particularly sensitive to operation with under- or overfrequency, and special precautions are taken to ensure that such plants are not operated with large frequency deviations. Normal operating practice stipulates that enough spinning reserve should be online at all times to ensure that the system frequency remains within some normal operating band following the outage of the largest synchronous unit in the system. When more severe disturbances occur, for example, loss of a station (all generating units), loss of a major load center, or loss of AC or DC interconnection, emergency control measures may be required to maintain frequency stability. Emergency control measures to be used by a wide-area control system may include [17]:

- Disconnection of load (i.e., load shedding)
- Emergency HVDC power transfer control (i.e., fast rescheduling of active power section)
- Disconnection of generation (i.e., generation rejection)
- Fast generation reduction through fast valving or water diversion
- Controlled opening of interconnection to neighboring systems to prevent spreading of frequency problems
- Controlled islanding of local system into separate areas with matching generation and load

1.6.2.3 Voltage Stability Control and Protection. Voltage stability concerns the capability of the power network to sustain active and reactive power flows with a stable voltage profile [18]. The typical root cause of voltage instability is too high loads, sudden weakening of the transmission grid, or the activation of current limiters in generation units. Voltage instability will present itself as abnormally low or abnormally high voltages. To enhance and recover voltage stability using a wide-area control system emergency control, the common actions are [17]:

- Load shedding
- Shunt capacitor or reactor bank switching
- Disabling or modification of tap changer control logic
- Adapting generator voltage and active power generation setpoints
- HVDC power transfer control

1.6.2.4 Oscillatory Stability Control. Oscillatory stability, also known as small-signal angle stability, denotes the ability of the power system to maintain synchronism when subjected to small disturbances, such as the background noise provided by short-term load variations [19]. It is closely related to the stability of the resonance modes of rotating masses swinging against each other. Both local modes, where only a single machine participates, and interarea modes; where machines or machine groups in different parts of the system swing against each other, have been reported. Local modes are faster (0.7–2 Hz) than the interarea modes (0.1–0.7 Hz). Two known causes of poor damping of interarea modes are weak or stressed tie lines or voltage regulation with excessively high gain. Power system stabilizers and various types of FACTS devices, controlled based on a local frequency or power measurement, are the most widely used control measures to damp such oscillations. To enhance oscillatory stability a wide-area control scheme may employ modulation of generator voltage using wide-area power system stabilizers or modulate FACTS or HVDC control settings. The use of wide-area information improves performance and robustness of wide-area control.

Countermeasures that can be used by a wide-area control and protection scheme are, for example:

- Modulation of the generator terminals using power system stabilizers
- Modulation through supplementary HVDC controls
- Modulation through supplementary static-var compensator controls
- Modulation of active power loads

1.6.2.5 Transient Stability Protection. Transient stability denotes the ability of all synchronous units to remain in synchronism when subjected to large disturbances [20]. The disturbances that endanger transient stability typically start with a fault at one point in the network, creating a very low voltage in the surrounding area. Therefore, since nearby generators can no longer transmit their generated energy through the faulty section, this energy is instead stored as rotational energy in the masses of the generators and consequently results in acceleration. Unless this acceleration is arrested, synchronism of the units will be lost and the units will be disconnected from the network by local protection. Such loss of synchronism can occur to individual units, a group of units in the same power plant, or regions of the network. Transient stability is most likely to be an issue when the network transfers are high or the network is loosely meshed. Transient stability control aims at reducing the rotational speed of the accelerating units before synchronism is lost. Countermeasures that can be used by a wide-area control and protection scheme are, for example [17]:

- Reduction of generator mechanical power, for example, by generation rejection (tripping) and fast valving
- Increase in the electrical power, for example, by brake resistors, shunt compensation, DC links, and FACTS devices

The most important measurement used in transient stability control is the generator angles. This can be estimated from grid-side phasor measurements of voltage phase angle, generator terminal current, and the generator's transient reactance [21].

1.6.2.6 Cascading Overloads Control. In addition to the power system stability related issues discussed in the previous sections, another common cause of blackouts is cascading thermal overloads. In such scenarios the (thermal) loadability of single pieces of equipment, most commonly power lines or generators, is exceeded and the load on that component is transferred to another component, which in turn is overloaded, leading to cascading overloads. Although traditional operational practice, which is normally based on the so-called N-1 criterion [22] that stipulates that the outage of any single component should not jeopardize the security of the system, should prevent this in practice, it has been observed that this scenario is a quite common cause of large blackouts. Wide-area monitoring and control has been suggested as an effective tool to mitigate cascading overloads [23, 24].

1.6.3 Stabilizing and Emergency Control Actions

Wide-area control and protection relies on various primary power system components to execute control action. This section summarizes the most commonly used means of control and the special considerations needed for each of them.

1.6.3.1 Generation Rejection. Generation rejection involves opening of a generator breaker, which will disconnect the generator from the network. Typically, this can be achieved in less than 50 ms. However, a fast change in electrical output of a generator may result in overspeed, thermal stresses, and a reduction in the shaft life due to shock-initiated fatigue. For this reason, generation rejection is best used in hydro plants, which are mechanically more rugged than their thermal counterparts. The effect on the power system is an instantaneous reduction of the electrical power injected into the system. The system frequency (derivative) decreases and can be used in case of overfrequency. It is, however, important to note that following rejection, a generator can no longer contribute to voltage control, and therefore this action may worsen voltage problems. Also, the control is discrete in the sense that the effect is determined by the load on the generator before the rejection.

1.6.3.2 Turbine Fast Valving. Turbine fast valving is executed through rapid opening or closing of steam valves in a thermal power plant. It can typically be carried out in a time frame quicker than 1 s. A special consideration that needs to be taken into account is that a rapid change in electrical output of a generator may result in overspeed, thermal stresses, and a reduction in the shaft life due to shock-initiated fatigue. In this case the power plant remains connected to the power system and can be used for voltage control and to supply a reduced amount of active power to the system. Therefore, fast valving is a less intrusive control action than generation rejection. Some thermal plants, such as nuclear plants, have limited capability of rapidly changing their output and are therefore not often used for this purpose. The effect on the power system is a fast reduction of the electrical power injected into the system, leading to a decrease in the system frequency derivative.

1.6.3.3 Fast Generation Startup. Many system operators have access to emergency generation, most often gas turbines, that can quickly be started in a time of typically less than 100 s. Apart from the obvious effect of the injection of active power, there is also the added benefit of an increase of frequency as well as voltage control capability.

1.6.3.4 Active Power Rescheduling. Active power rescheduling, which adjusts the production of individual generation units, can be achieved by directly adjusting the reference values for generator governor controllers or indirectly through actions on the automatic generation control (AGC) [8]. However, many thermal plants and nuclear units in particular have limited capability of rapidly changing their output.

1.6.3.5 Load Shedding. Load shedding usually involves opening of feeder circuit breakers to large individual loads or an entire distribution feeder. This can be done rapidly, typically in less than 50 ms. The selection of the feeders to disconnect may involve considerations such as priority of load and special market contracts. Customers who allow the system operator to disconnect them may receive reduced tariffs or be remunerated in other ways. The effect on the power system is that the system load admittance decreases, which in general leads to voltage increase and frequency rise. The effect of the control is determined by the load on the disconnected feeder, which may need to be monitored in real time.

1.6.3.6 Generator Voltage Reference Adjustment. Generator voltage set-points can be quite rapidly changed with a response time of a few seconds or even tens or hundreds of milliseconds for modern units with static excitation systems. Generator capability curves must be considered. The effect on the power system is a redistribution of reactive power production, which can be used to relieve a generator of reactive load and thereby avoid armature and field current limitations.

1.6.3.7 HVDC Fast Power Change/Reversal. HVDC fast power change/reversal involves changing the reference value for active power transfer of HVDC links. The response time is quick, typically less than 1 s. To be most effective, the link shall connect two different synchronized areas and the effect on the neighboring power system and the characteristics of the DC cable must be considered. The almost instantaneous change in the power balance can be used to increase or decrease frequency as necessary.

1.6.3.8 Shunt Switching. Shunt components such as capacitor banks and reactors can be rapidly connected or disconnected by opening or closing their circuit breakers. It can be done rapidly, in less than 50 ms. However, capacitor banks may need 10–20 minutes to discharge following disconnection before they can be reconnected again. The effect on the power system is an increase of voltage when a shunt capacitor is connected or a shunt reactor is disconnected, and correspondingly a decrease of voltage when a shunt reactor is connected or a capacitor is disconnected. The effectiveness of the control mainly depends on the size of the connected equipment but also on the voltage at the time of operation.

1.6.3.9 Controlled Opening of Interconnection and Islanding. Controlled opening of interconnection involves opening of breakers of tie lines to neighboring power systems. It is also a rapid action that can be executed in less than 50 ms. The effect on the neighboring power system area should be considered. When the breakers are opened, the power system will be split into different islands. This has the effect that the import/export of active and reactive power between two areas is stopped. This is a commonly used method to prevent a blackout from spreading in an interconnected power system. Quite often there is a dramatic change in the power balance of the resulting islands and further emergency controls may need to be applied to ensure the stability of each of them. The effectiveness of the control depends on the flow on the interconnection before opening, which needs to be monitored in real time.

1.6.3.10 Tap Changer Blocking or Reversal. Tap changer blocking or reversal involves disabling or modifying the control logic of distribution tap changes. This can be used to mitigate the detrimental effect of tap changers on system stability. The effectiveness of the control depends on the characteristics of the load in the distribution system.

1.6.4 Implementation Aspects of WAMCP Systems

Many different kinds of WAMCP have been described in the scientific literature and a few also implemented in practice [17]. This section evaluates such previously described WAMCP in terms of three aspects: the implementation architecture, the decision making paradigm, and triggering characteristics. The advantages and disadvantages of each type of scheme are discussed in the following sections. The most important desired properties of a WAMCP are:

- *Dependability*, which refers to the certainty that the WAMCP operates when required, that is, in all cases where emergency controls are required to avoid a collapse.
- *Security*, which refers to the certainty that the WAMCP will not operate when not required, that is, does not apply emergency controls unless they are necessary to avoid a collapse.
- *Selectivity*, which refers to the ability to select the correct and minimum amount of action to perform the intended function, that is, to avoid using disruptive controls such as load shedding if they are not necessary to avoid a collapse.
- *Robustness*, which refers to the ability of the WAMCP to provide dependability, security, and selectivity over the full range of dynamic and steady-state operating conditions that it will encounter.

Table 1.1 shows a selection of approaches for frequency and voltage stability control in power systems. The methods are described in terms of their implementation architecture, the employed triggering conditions, and their decision making paradigm in the following sections.

1.6.4.1 Architecture

PURELY CENTRALIZED. A purely centralized system has all decision making functions concentrated in one place. This means that communication to and from all substations is necessary before corrective action can be executed in remote substations. The advantages mentioned below mainly result because a wide-area view of the system is possible and the drawbacks are due to the limitations introduced by the need for remote measurement and control. The advantages are:

- High security
- High dependability
- High selectivity

The disadvantages are:

- Slow response due to communication delay
- Necessity of high speed, synchronous communication
- Vulnerability to communication failures
- Average robustness

The dependability of a purely centralized system can be high, since it has a wide-area view and can identify stability problems throughout the system. A centralized system can monitor voltages (and frequencies) even at buses without measurement equipment, using state estimation techniques. The dependability can also be high since it can identify situations that appear serious when looking only at a single measurement value in isolation: for example, if a low voltage is experienced at a bus close to a generator which still has reactive capabilities enough to support the voltage. Also, the selectivity can be high, since control actions can be optimally scheduled using the wide-area measurements.

The centralized system can also provide good robustness since the complete state information can be used to find the correct control actions regardless of the current state. On the other hand, it is sensitive to communication failures. Unless redundant measurement equipment is placed so that the network remains observable even with the loss of measurement signals, the system may be unable to function at all. A secondary drawback is the additional delay introduced by communication. For some applications, fast action is necessary. One example would be the case of frequency stability control, for which the communication delay can be translated directly into performance.

CENTRALIZED WITH LOCAL BACKUP. To be able to react to disturbances that make the system collapse rapidly, in comparison to the communication delays or when the centralized system is nonfunctional due to communication failures, it may be necessary to complement a fully centralized system with local systems as backup. Local underfrequency and undervoltage load shedding are simple and reliable candidates for these backup systems. The supervisory, centralized, control layer should be tuned in such a way that corrective control is taken before the local layer acts. Since the supervisory layer has access to centralized information, it can coordinate and provide better selectivity than a purely local system and the local layer can still provide protection for fast disturbances or in the presence of communication failures.

HIERARCHICAL. Hierarchical schemes use a combination of centralized and local intelligence, aiming to preserve the fast response of the local schemes but augment it with some supervisory or coordinative layer based on a wide-area view. Examples of such schemes are adaptive protection schemes where the settings of local undervoltage and underfrequency relays are determined by a supervisory control layer and control schemes based on cooperating software agents. The advantages are:

- Some degree of coordination can be achieved.
- Fast response is possible for some tasks, for which the decision making can be based only on local criteria.
- Medium speed asynchronous communication is necessary.
- Modest computational requirements exist.

This structure combines the fast response of the local schemes with the ability of coordination that can be achieved by the centralized schemes.

PURELY LOCAL. A purely local scheme uses only local measurements for its decision making. Therefore, it cannot base its decisions on a wide-area view. Rather, it must use measurements of some key quantities and base its decisions on a comparison with threshold values based on offline tuning. The tuning of these threshold values requires good knowledge of the system and often involves offline simulations. Since the tuning must be made offline it must be made with a worst-case scenario in mind. This has two major implications. First, the scheme must be able to handle the worst-case scenario, meaning that it may be overly conservative in other cases. Second, it is not trivial to identify the worst-case scenario since it depends on future operating conditions being correctly foreseen. For the same reason, the tuning of such schemes should ideally be reviewed periodically after commissioning. Underfrequency and undervoltage load shedding are the most well-known examples of this type. The advantages are:

- Fast response
- Mature technology
- Low computational requirements
- No communication necessary

The main drawback is the low selectivity with the local schemes. The main advantages of local approaches are that they are inexpensive and simple to implement. However, the only means of coordination is the tuning parameters of the local controls and the selectivity properties are therefore poor. For some control actions such as load shedding and capacitor switching the local controllers can be quite robust; however, for other control actions like generator voltage setpoints global information is necessary for efficient control and the local approach is therefore not useful.

1.6.4.2 Triggering Conditions. The triggering conditions define when the WAMCP operates and is comprised of two parts:(1) a set of process variables or diagnostics that are monitored and (2) the decision process that decides if the WAMCP is to be activated. There are three commonly used approaches: event-based, response-based, or timer-based triggering as described in the following sections.

EVENT BASED. Event-based schemes act on a feedforward principle by monitoring topological information from the power system such as the status of lines and generators and can take predetermined actions based on this information. These schemes are faster than response-based schemes but less robust, since they can handle only such disturbances that can be foreseen during the design stage. During operation, the system is monitored for these disturbances. The actions can be preprogrammed using only offline information or precalculated using online information. The advantages of event-based schemes are:

- High security
- Fast response

The main disadvantage of an event-based scheme is low dependability.

Typically, the security of event-based schemes is high since they act only for predefined actions, which have been identified as serious at the design stage; however, they cannot act for any other disturbance than those foreseen. This results in low dependability.

RESPONSE BASED. Response-based schemes act on a feedback principle by monitoring electrical measurement signals and deduce the severity of a disturbance from these. This way, the present security level in the system and which stabilizing actions to take are decided. Underfrequency and undervoltage load shedding are the most well-known examples of this type. These systems are inherently slower than event-based schemes since they must wait for the effect of a disturbance to be visible in the electrical measurement signals before they become known to the protection scheme. On the other hand, they are more robust than event-based schemes since they can react also to disturbances that were not foreseen during the design stage. The main advantage is the high dependability: the disadvantages low security and delayed response.

TIMER-BASED. Timer-based schemes are operating continuously and are triggered by timers at regular intervals, typically at the same rate as the sample rate of the phasor measurements. For example, this is generally applied in wide-area control schemes to enhance oscillatory stability.

1.6.4.3 Decision Making Paradigm. Once the WAMCP has been activated, some decision logic must be applied to decide on emergency actions to take. The two dominant approaches are the rule-based (heuristic) or algorithmic approaches.

RULE BASED. In rule-based schemes the decisions are made on the basis of a predefined rule base, typically of an if-then-else type. Rule-based schemes may at first glance appear more robust than an optimization-based scheme since they normally do not directly rely on a process model. However, a certain model is implicitly assumed when selecting the rule base and tuning the decision thresholds. There are some standard rule-based protection schemes that appear versatile and in principle applicable to any system. For example, local relays for undervoltage load shedding, tap changer blocking, and underfrequency load shedding are robust controls that normally require only two tuning parameters to be set for each relay. However, since many relays have to be installed, the number of tuning parameter will be very large. Additionally, these schemes are based purely on local measurements.

The advantages of rule-based schemes are:

- Fast response due to low computational requirements is available.
- Intuitive and therefore operator and engineering knowledge can easily be included in the design.

The main drawbacks are.

- Low robustness
- Low dependability
- Low selectivity

The main advantages of the rule-based schemes are that they are relatively simple to implement and that their action will easily be predictable since the rule base is intuitive. Also, there exist formal methods to verify that a rule base is complete and noncontradictory in terms of the design criteria specified, also with the failure of a few measurement signals. However, no guarantee that the design criteria specified in the engineering phase are adequate can be obtained in the same way. As the system in which it has been installed evolves, either by changing loading patterns or by new installations, considerable reengineering may have to be done and the rule base should be reviewed at regular intervals. The rule base needs to be tailor-made for each installation, which requires much engineering and good knowledge of the system in which it is to be installed.

ALGORITHMIC. Algorithmic schemes are usually based on a static or dynamic system model and phasor measurements to perform state estimation to maintain a model of the system updated in real time. This model can be used to analyze the effect of control and predict the future system state. The selectivity of algorithmic schemes is good, since they can compare the effect of one control to that of another using the system model and can use that information for coordination. When there is also access to centralized information, this method can also be used to coordinate controls in different locations. Since the schemes can account for changes in the system configuration or load/generation pattern in the network, they usually have excellent selectivity and robustness. The advantages of algorithmic schemes are:

- High selectivity
- High security
- High robustness

The main disadvantage is most often the high computation complexity, which may introduce delay.

1.6.4.4 Communication and Response Time Requirements. The communication and response time requirements of WAMCP applications can be broken down into different types of requirements as follows [17]:

- Requirements of the time resolution of the data, which can be expressed as maximum allowed between two consecutive samples of the data
- Requirements in terms of maximum time delay with which the data is available in the SPC
- Whether uni- or bidirectional communications are required

Table 1.1 lists these requirements for various WAMCP applications that have been proposed in the literature.

REFERENCES

[1] X. Feng, W. Peterson, F. Yang, G. M. Wickramasekara, and J. Finney, "Smart grids are more efficient," *ABB Rev.*, 3, 2009.

[2] X. Feng, W. Peterson, F. Yang, G. M. Wickramasekara, and J. Finney, "Implementation of control center based voltage and var optimization in distribution management system," IEEE PES PSCE 2010, New Orleans, LA, USA.

[3] J. Stoupis, et al., "Restoring confidence: Control-center and field-based feeder restoration," *ABB Rev.*, pp. 17–22, Q3/2009.

[4] J. Northcote-Green and R. Wilson, *Control and Automation of Electrical Power Distribution Systems*, CRC Taylor & Francis Group, 2007, pp. 149–163, 251–264.

[5] "Assessment of demand response and advanced metering," Staff Report, Federal Energy Regulatory Commission (FERC), August 2006.

[6] T. Ackermann, G. Andersson, and L. Söder, "Distributed generation: A definition," *Electric Power Syst. Res.*, 57, pp. 195–204, 2001.

[7] IEEE Standard P1547.2, *IEEE Standard for Interconnecting Distributed Resources with Electric Power Systems*. IEEE, 2008.

[8] P. Kundur, *Power System Stability and Control*. McGraw Hill, 1994.

[9] A. Atputharajah and T. K. Saha, "Power system blackouts—literature review," *Proceedings of Industrial and Information Systems (ICIIS)*, 2009, pp. 460–465.

[10] M. Zima, M. Larsson, P. Korba, C. Rehtanz, and G. Andersson, "Design aspects for wide-area monitoring and control systems," *Proc. IEEE*, 93(5), pp. 980–996, 2005.

[11] ABB Power Technology Systems, "PSG 828—Data Sheet," 2005.

[12] ABB Automation Products, *RES 521—Technical Reference Manual*, 2008.

[13] *IEEE Standard for Synchrophasors for Power Systems*, IEEE Power Systems Relaying Committee Std., 2006, IEEE Std. C37.118.

[14] *IEEE Standard for Synchrophasors for Power Systems*, IEEE Power Systems Relaying Committee Std., 1995, IEEE Std. 1344–1995.

[15] J. Makhija, "Comparison of protocols used in remote monitoring: DNP 3.0, IEC 870-5-101 & Modbus," Electronics Systems Group, IIT Bombay, India, Tech. Rep., 2003.

[16] K. Brand, V. Lohmann, and W. Wimmer, *Substation Automation Handbook*. Utility Automation Consulting Lohmann, 2003.

[17] CIGRE, "System protection schemes in power networks," CIGRE Task Force 38.02.19, Tech. Rep., 2000.

[18] T. van Cutsem and C. Vournas, *Voltage Stability of Electric Power Systems*, Power Electronics and Power Systems Series. Kluwer Academic Publishers, 1998.

[19] G. Rogers, *Power System Oscillations*. Springer-Verlag, 1999.

[20] M. Pavella, D. Ernst, and D. Ruiz-Vega, *Transient Stability of Power Systems: A Unified Approach to Assessment and Control*. Springer, 2000.

[21] A. Del Angel, P. Geurts, D. Ernst, M. Glavic, and L. Wehenkel, "Estimation of rotor angles of synchronous machines using artificial neural networks and local PMU-based quantities," *Neurocomputing*, 70, pp. 2668–2678, October 2007. Available at: http://portal.acm.org/citation.cfm?id=1316076.1316105.

[22] ENTSO-E, "Operation handbook—operational Security," ENTSO-E, Tech. Rep., 2004.

[23] M. Zima and G. Andersson, "Wide area monitoring and control as a tool for mitigation of cascading failures," *2004 International Conferences on Probabilistic Methods Applied to Power Systems*, 2004, pp. 663–669.

[24] S. Talukdar, D. Jia, P. Hines, and B. Krogh, "Distributed model predictive control for the mitigation of cascading failures," *Decision and Control, 2005 and 2005 European Control Conference. CDC-ECC '05. 44th IEEE Conference on*, 2005, pp. 4440–4445.

[25] I. W. Slutsker, S. Mokhtari, L. A. Jaques, J. M. G. Provost, M. B. Perez, J. B. Sierra, F. G. Gonzalez, and J. M. M. Figueroa, "Implementation of phasor measurements in state estimator at Sevillana de Electricidad," *Proceedings IEEE Power Industry Computer Application Conference*, 1995, pp. 392–398.

[26] R. F. Nuqui and A. G. Phadke, "Hybrid linear state estimation utilizing synchronized phasor measurements," *Proceedings IEEE Lausanne PowerTech*, 2007, pp. 1665–1669.

[27] A. G. Phadke, J. S. Thorp, and K. J. Karimi, "State estimation with phasor measurements," *IEEE Trans. Power Syst.*, 1(1), pp. 233–238, 1986.

[28] W. Miller and J. Lewis, "Dynamic state estimation in power systems," *IEEE Trans. Automatic Control*, 16(6), pp. 841–846, 1971.

[29] W. Gao and S. Wang, "On-line dynamic state estimation of power systems," *Proceedings of the North American Power Symposivm (NAPS)*, 2010, pp. 1–6.

[30] M. Larsson and C. Rehtanz, "Predictive frequency stability control based on wide-area phasor measurements," *Proceeding IEEE Power Engineering Society Summer Meeting*, 1, pp. 233–238, 2002.

[31] R. M. El Azab, E. H. S. Eldin, and M. M. Sallam, "Adaptive under frequency load shedding using PMU," *Proceedings of 7th IEEE Internatrnal Conference on Industrial Informatics INDIN 2009*, 2009, pp. 119–124.

[32] T. Ohno, T. Yasuda, O. Takahashi, M. Kaminaga, and S. Imai, "Islanding protection system based on synchronized phasor measurements and its operational experiences," *Proceedings of IEEE Power and Energy Society General Meeting*, 2008, pp. 1–5.

[33] R. Doraiswami and W. Liu, "Real-time estimation of the parameters of power system small signal oscillations," *IEEE Trans. Power Syst.*, 8(1), pp. 74–83, 1993.

[34] P. Korba, M. Larsson, and C. Rehtanz, "Detection of oscillations in power systems using Kalman filtering techniques," *Proceedings of IEEE Conference on Control Applications CCA 2003*, 1, pp. 183–188, 2003.

[35] M. G. Anderson, N. Zhou, J.W. Pierre, and R.W. Wies, "Bootstrap-based confidence interval estimates for electromechanical modes from multiple output analysis of measured ambient data," *IEEE Trans. Power Syst.*, 20(2), pp. 943–950, 2005.

[36] M. Glickman, P. O'Shea, and G. Ledwich, "Estimation of modal damping in power networks," *IEEE Trans. Power Syst.*, 22(3), pp. 1340–1350, 2007.

[37] R. A. Wiltshire, G. Ledwich, and P. O'Shea, "A Kalman filtering approach to rapidly detecting modal changes in power systems," *IEEE Trans. Power Syst.*, 22(4), pp. 1698–1706, 2007.

[38] H. Ghasemi and C. Canizares, "On-line damping torque estimation and oscillatory stability margin prediction," *IEEE Trans. Power Syst.*, 22(2), pp. 667–674, May 2007.

[39] D. J. Trudnowski and J.W. Pierre, "Overview of algorithms for estimating swing modes from measured responses," *Proceedings of IEEE Power & Energy Society General Meeting PES '09*, 2009, pp. 1–8.

[40] M. Larsson and D. S. Laila, "Monitoring of inter-area oscillations under ambient conditions using subspace identification," *Proceeding of IEEE Power & Energy Society General Meeting PES '09*, 2009, pp. 1–6.

[41] I. Kamwa, "Using mimo system identification for modal analysis and global stabilization of large power systems," *Proceeding of IEEE Power Engineering Society Summer Meeting*, 2, pp. 817–822, 2000.

[42] P. Korba and K. Uhlen, "Wide-area monitoring of electromechanical oscillations in the Nordic power system: Practical experience," *IET Generation, Transmission & Distribution*, 4(10), pp. 1116–1126, 2010.

[43] I. Kamwa, R. Grondin, and Y. Hebert, "Wide-area measurement based stabilizing control of large power systems-a decentralized/hierarchical approach," *IEEE Trans. Power Syst.*, 16(1), pp. 136–153, 2001.

[44] K. Mekki, A. F. Snyder, N. Hadj Said, R. Feuillet, D. Georges, and T. Margotin, "Damping controller input-signal loss effects on the wide-area stability of an interconnected power system," *Proceedings of IEEE Power Engineering Society Summer Meeting*, 2, pp. 1015–1019, 2000.

[45] Y. Fang and Y. Xue, "An on-line pre-decision based transient stability control system for the ertan power system," *Proceedings of PowerCon 2000*, 1, pp. 287–292, 2000.

[46] M. Koaizawa, M. Nakane, K. Omata, and Y. Kokai, "Actual operating experience of on-line transient stability control systems (TSC systems)," *Proceedings of IEEE Power Engineering Society Winter Meeting*, vol. 1, pp. 84–89, 2000.

[47] P. Kundur, G. K. Morison, and L. Wang, "Techniques for on-line transient stability assessment and control," *Proceeding of IEEE Power Engineering Society Winter Meeting*, 1, pp. 46–51, 2000.

[48] D. Ernst and M. Pavella, "Closed-loop transient stability emergency control," *Proceeding of IEEE Power Engineering Society Winter Meeting*, 1, pp. 58–62, 2000.

[49] M. Weibel, W. Sattinger, P. Rothermann, U. Steinegger, M. Zima, and G. Biedenbach, "Overhead line temperature monitoring pilot project," *CIGRE Publication B2-311*, 2006.

[50] K. Vu, M. M. Begovic, D. Novosel, and M. M. Saha, "Use of local measurements to estimate voltage-stability margin," *IEEE Trans. Power Syst.*, 14(3), pp. 1029–1035, 1999.

[51] R. Balanathan, N. C. Pahalawaththa, U. D. Annakkage, and P. W. Sharp, "Undervoltage load shedding to avoid voltage instability," *IEE Proc. Generation, Transmission and Distribution*, 145(2), pp. 175–181, 1998.

[52] M. Larsson, C. Rehtanz, and J. Bertsch, "Monitoring and operation of transmission corridors," *Proceedings of IEEE Bologna Power Tech*, 3, 2003.

[53] M. Larsson and D. Karlsson, "Coordinated system protection scheme against voltage collapse using heuristic search and predictive control," *IEEE Trans. Power Syst.*, 18(3), pp. 1001–1006, 2003.

[54] M. Larsson, C. Rehtanz, and D. Westermann, "Improvement of crossborder trading capabilities through wide-area control of facts," *IREP Symposium 2004, Bulk Power System Dynamics and Control VI, Cortina D'Ampezzo, Italien*, 2004.

[55] G. Hug-Glanzmann, "Coordinated power flow control to enhance steady-state security in power systems," Ph.D. dissertation, Power Systems Laboratory, ETH Zurich, Switzerland, 2008.

2

ELECTRIC VEHICLES AS A DRIVER FOR SMART GRIDS

Nigel Fitzpatrick and Alec Tsang

2.1 INTRODUCTION

A number of automotive vehicle manufacturers are offering plug-in electric vehicles (PEVs) or plug-in hybrid electric vehicles (PHEVs) for personal and commercial use with batteries ranging in capacity from 16 to 28 kWh. It is expected, though there will be "opportunity charging" at stops: vehicles will primarily be charged at homes or in commercial fleet yards. The energy storage capacity of these vehicles is a significant fraction of the energy used by a house in a day, and there has been much discussion on the interaction of vehicles with the grid. An analogous situation will prevail in commercial operations where the storage capacity of a fleet is of the same order of magnitude as the daily use of the operations which it supports. The vehicles would typically require their power in off-peak periods and are seen as potentially improving the economics of a grid system that is often operating below peak capacity.

Smart Grid: Applications, Communications, and Security, First Edition.
Edited by Lars Torsten Berger and Krzysztof Iniewski.

The environmental specifications in niches that need electrical power, including vehicles, have changed with time, as have related technologies. Fuel cells, as an example, have not stood still but neither have inverters, electric motors that enable them, engines that compete, and batteries that both enable and compete. And there are changing strategic supply and environmental issues, including climate change, which impact the way we look at fossil fuel, nuclear, and renewable supplies. Our technologies evolve and interact in a changing environment, and the selection we make for each market niche will change from decade to decade. As with life forms, survival and growth of technology in one niche will mean it is available for another niche later.

In a sense we have benefited from the late arrival of the lithium-ion battery, as engineers had to cope with heavier batteries and focused engineering effort on light efficient motors, light vehicle structures, and efficient inverters. It was a low energy density battery that led to the hybrid option, which only needs a small low impedance, high power density battery. The success of the hybrid led to recognition that there was a major market for batteries in plug-in road vehicles.

This chapter compares the interaction of electric vehicles with the grids of a utility with rapid response hydro storage and an energy constrained utility that has an ancillary services market. We will suggest that to lower the costs and maximize the economic and environmental benefits of plug-in vehicles there may be an opportunity for a second life of the batteries in grid-side stationary use. It may be this reuse of batteries that allows locally generated renewable energy to be stored and thus reduce both grid and electric vehicle demand.

2.2 PLUG-IN ELECTRIC VEHICLES AND HYBRIDS

The early history of electric vehicles has been described by, among others, Wakefield [1, 2] Poulton [3], and Cairns and Hietbrink [4]. The latter dates the first practical electric vehicle to Robert Davidson of Aberdeen, Scotland in 1837.

Wakefield tells us that, in 1912, when there were around 30,000 electric vehicles on the roads in the United States, a third of these, 10,000, were actually commercial vehicles. Cairns and Hietbrink report a larger fraction in 1916 when 10,000 electric vehicles were manufactured and 4000 of these were for commercial use.

Personal vehicles disappeared but commercial electric road vehicles were produced in the United States until at least 1942 when Walker Electric in Chicago closed its doors. Walker had even supplied Harrods in London as well as customers as far away as Christchurch, New Zealand. In North America, electric vehicles continued on as "off-road" vehicles, such as fork lift trucks, golf carts, and mining vehicles. In Europe, commercial road electric vehicles survived. Cairns and Hietbrink report 45,000 electric vehicles in use in the 1960s for milk, post office, and city delivery. Today, a few thousand lead-acid electric commercial vehicles continue in the United Kingdom as modern lithium-ion vehicles are launched. The primary reason for the downfall of personal electric vehicles was the limited range resulting from the difference between the energy density of gasoline and batteries. Another factor was the relative rate of refueling.

Gasoline has a lower heating value of 42.5 MJ/kg or 11,800 Wh/kg and is used with a best efficiency of 34–38% in the lean burn Atkinson cycle engine in the Toyota Prius, say, a net 4250 Wh/kg. Diesel fuel at 11,944 Wh/kg can attain 42% efficiency or a net 5016 Wh/kg. Gasoline station pumps can deliver 40 liters/minute or 29.2 kg/minute. Using 4250 Wh/kg, this translates to energy transfer at a power of 7.4 MWe against the 7 kWe for Level 2 home electric vehicle charging and a maximum of 200 kWe for Level 3 public charging. Public charging is up to 35 times slower for a PEV than is fueling for a gasoline vehicle.

Commercial electric vehicles have held their place using lead-acid batteries with energy densities of 30–35 Wh/kg. The reason was their relatively low life cycle cost. For a century the dominant battery technology for vehicle traction was the flooded lead-acid cell using either tubular or plate anodes. Road vehicles launched from the 1960s to the 1990s were similar in vehicle range to those offered in 1890. Oil scarcity and air quality issues developed interest by the government, and in 1976 the U.S. Congress passed Public Law 94-413 calling for the development of electric and hybrid vehicles [5].

Air quality concerns in California impacted internationally. In November 1988 in Toronto, at EVS9, the 9th Electric Vehicle Symposium, a councilman from Los Angeles, Marvin Braude, was keen to reduce ground level emissions. Braude challenged attendees to deliver electric vehicles in Los Angeles. Two Canadian companies and a Swedish company (Clean Air Transport [6]) were selected to deliver prototype vehicles to Los Angeles in 1989. Magna developed and delivered the electric G-van, and a plug-in hybrid was delivered by Unique Mobility (UQM) in which Alcan Aluminium Limited of Montreal had bought an interest [7].

Both UQM and Magna were using the same flooded cell technology with a single point watering system. The batteries were made up of 18, six-volt, 205 Ah, Chloride 3ET 205 wet batteries, also common to the Lucas–Chloride Griffon built on a Bedford platform [2].

The 6-volt module was comprised of three 2-volt lead acid cells, which generated hydrogen if overcharged, and had a practical energy density of around 30 Wh/kg, a power density of around 100 W/kg, and a life of 2 years, though careful users such as China Light and Power in Kowloon were able to obtain a decade of life, and one Griffon, operated on a regular run, had covered an exceptional 60,000 miles on a single set of batteries.

Range limitations hampered the Los Angeles thrust for clean air and the Alcan/UQM plug-in hybrid shown to Marvin Braude was too early.

2.3 HYBRIDS

Early vehicle engineers were faced with large electric motors or the low energy density of batteries. Ferdinand Porsche's patent on the "Mixte" transmission, filed in 1897, led in due course to a successful series hybrid while Pieper [8] described a parallel hybrid with the electric motor coaxial with the engine. Pieper's vehicle could be charged externally and was a plug-in hybrid electric vehicle. Victor Wouk [9] reviewed the status of the hybrid electric vehicle (HEV) and included a British Lucas–Chloride parallel plug-in hybrid described in more detail by Harding et al. [10]. It had a "pure electric" vehicle range and an onboard charger is shown in a schematic in both papers.

The plug-in hybrid electric vehicle (PHEV) was a step to the commercially viable "unplugged" hybrids of today and may now be a step again to a more efficient future. In North America it was only in the heavy bus field that there was a move toward the production of hybrids [11] before Toyota launched the Prius in Tokyo in December 1997. This is despite Toyota's direction on hybrids being well known in the early 1990s [12].

The Honda Insight hybrid was delivered both in the United States and Canada just ahead of the Prius, with the first being delivered in the United States in December 1999, just as Canada's federal government was concluding a two-year, country-wide process to determine low-cost options to tackle climate change. The "Transportation Table" had input from the energy and automotive industries and other stakeholders as well as the provinces of Canada. A specific study for the Table [13] actually showed that *both* hybrid light and heavy duty vehicles were options to tackle greenhouse gas emissions. However, the final climate change mitigation options that came out of the Table for action did not include hybrids [14]. At the time this blind eye was turned there were not only already tens of thousands of light hybrid vehicles on the road in Japan, but the Canadian led Orion IV hybrid transit bus was *already* being offered commercially and five were in service with the New York City Transit Authority.

2.4 THE GENERAL ELECTRIC DELTA CAR

Ragone [15] pointed out that "the major hurdle to be overcome—is that of the energy storage system" and provided a very clear way of linking the acceleration and range needs of a vehicle to battery properties. Ragone describes how in the future, power limited cells might be coupled with energy limited cells to meet vehicle demands with a battery/battery hybrid.

In 1971 Brown [16] described a battery/battery hybrid electric vehicle that used high power density nickel cadmium, Ni-Cd, to provide power and lead-acid to provide range. Brown was able to demonstrate that the hybrid battery system had a dramatic increase in range over that obtained from lead-acid alone. A Ni-Cd battery is better suited than lead-acid to accept the power from a regenerative braking system. If suitably coupled, this energy gain adds to the energy efficiency gain obtained by running the lead-acid battery at a lower current density.

Even today we can read Brown's abstract with interest:

> Electric vehicles that mix with normal traffic must provide bursts of high torque as they accelerate from standstill to operating speed. Battery systems that are sized for

> sustained driving experience excessive drain during each acceleration from rest. A system providing a separate small battery for acceleration and the means to isolate the main battery from brief peak current demands is presented in this paper along with the benefits in weight, cost, and range that result. Operation of the General Electric Delta research vehicle with and without the dual battery system demonstrated the distinct advantages of the hybrid battery system.

As Ragone said, the best lead-acid battery of the time was simply not suitable for vehicle use and vehicle performance was compromised. Ni-Cd can undergo many more cycles than lead-acid and has low impedance due to the high conductivity of its potassium hydroxide electrolyte. Brown proved that both acceleration and range needed a different set of battery properties.

The batteries suitable for hybrids and hybrid systems appeared first in buses and later in light vehicles. We are now mainly familiar with the hybrid engine battery systems that were launched commercially with the low impedance nickel metal hydride battery where a hydrogen anode replaced Brown's highly toxic cadmium anode.

2.5 BATTERIES, ULTRACAPACITORS, AND SEMI AND FULL-FUEL CELLS

The low energy density of commercially available batteries has resulted in a natural selection process acting on other parts of the vehicle, and components have necessarily become more efficient, lighter, and less expensive.

Batteries that can only be discharged are called primary batteries while those that can be charged and discharged are called secondary batteries. Before there was a grid they could only be charged from a primary battery. It is the secondary batteries that are important to our grid discussion. Vehicle range requirements have forced consideration of primary range extenders as alternatives to the internal combustion engine. We will show that most range extenders benefit from being hybridized with a secondary battery.

Ultra- or supercapacitors are electrochemical devices where electrons are stored at surfaces in chemical layers and are capable of great power but have a very low energy density, much less then 10% of that of a lead-acid battery. But they have been used in hybrid vehicles and have been considered in conjunction with batteries as they would offer the benefit that we saw GE tried to obtain in the Delta car. But their presence does not add to the grid-side benefits of electric vehicles.

The refuelable metal–air family includes lithium, aluminum, magnesium, zinc, and iron as well as the hydrogen fuel cell, since hydrogen is classed chemically as a metal. The metal "dissolves" or ionizes at the anode and reacts with hydroxyl ions. These systems draw oxygen from the air and weigh less than those with metal oxides. Both aluminum–air and zinc–air have been demonstrated in hybrid road vehicles. Vehicle manufactures have developed hybrid and plug-in hybrid preproduction hydrogen fuel cell vehicles. Larger secondary batteries are being deployed to make these technologies viable as plug-in hybrid vehicles.

Nickel metal hydride batteries (Ni-MH) have a high cycle life and evolved from a battery with a nickel oxide cathode known as a nickel hydrogen cell which carries its hydrogen stored in cylinders and has such remarkable cycle efficiency that it is used today in low earth orbit where solar panels charge these batteries 16 times a day. This means 5000 cycles of charge and discharge are expected each year of operation.

Ni-MH serves the needs of the hybrid electric vehicle well enough. The significant difference from lead-acid gave much hope in the 1990s and Solectria and General Motors both demonstrated [17] Ni-MH in their electric vehicles. Two decades later, Ni-MH has the same challenges of cost and self-discharge as Ni-Cd. But it enabled the launch of the first Prius and Insight, which intensified interest in and development of other battery systems. Dell and Rand [18], in reviewing the status of batteries, note the move to nonaqueous lithium ion. It is now the system of choice and we will cover here some of the further evolution expected from new materials research. The ideal battery goal is a cell chemistry that meets all of the vehicle's needs; Ragone's approach described above was followed by, among others, Kalhammer et al., [19], when they projected how the development of lithium-ion (Li-ion) batteries might enable the arrival of electrically driven vehicles including the full performance battery electric vehicle (FPBEV). Other properties, of course, are important, such as cycle life, calendar life, and temperature stability, but consideration of power and energy density begins the process of selection.

Much development effort in North America was focused on the range limitation of electric vehicles and energy density. Power density was only addressed solidly by the U.S. Department of Energy after the launch of the Toyota Prius in 1997, when the key needs of neglected hybrid batteries were then much discussed.

The Ni-MH battery has double the energy density of the lead-acid battery and was put into electric vehicles on a trial basis in the United States by Solectria in both their Force and Sunrise. GM also put it into a few of their EV1s or Impacts. But Ni-MH has both a high cost and high self-discharge. Toyota engineers saw that it was very suitable for a hybrid where the small size deals with cost and self-discharge is not an issue—and can even help as a little spare battery capacity will always be available for regenerative braking.

In 2000, Tamor pointed out [20] that for a hybrid the battery must be efficient; that is, the losses during charge and discharge must be minimal. This means that the voltage polarization and impedance losses must be small over the required charge rates. Tamor showed that battery efficiency reduction can result in the vehicle range halving. Checkel [21] reported detailed results on 1992 plug-in hybrids. Checkel's team had measured the variation of the round-trip efficiency as a function of depth of discharge of the battery. The efficiency can be very poor, and, as a consequence, "conventional hybrid" vehicle producers choose to oscillate their hybrid battery around an optimum state of charge.

Table 2.1 presents targets that the U.S. Department of Energy [22] posted in 2003 for electric and hybrid vehicle batteries. In this table the high "power density" specification of the hybrid is equivalent to Tamor's efficiency imperative in that it implies low impedance. Goals for plug-in hybrid batteries were published in 2007 by the U.S. Advanced Battery Consortium. Interestingly, the calendar life target had

TABLE 2.1. Department of Energy Advanced Battery Goals in 2003[a]

Property	Electric Vehicle Battery Targets	Hybrid Electric Vehicle Battery Targets
Calendar life	10 years	10 years
Cycle life	1000 cycles @ 80% depth of discharge	200,000 cycles for 25 Wh pulses 50,000 cycles for 100 Wh pulses
Cost	<$150/kWh (75 $/kWh desired)	<$150/kWh (75 $/kWh desired)
Specific energy	150 Wh/kg	75 Wh/kg
Energy density	230 Wh/L	100 Wh/L
Specific power	300 W/kg	750 W/kg
Power density	460 W/L	1000 W/L

[a]Extract from public data.

risen to 15 years while a minimum round-trip efficiency of 90% is specified. Table 2.1 illustrates the enormous gap between pure electric and hybrid vehicle battery needs.

The targets for both the power density and partial state of charge cycles, developed after the arrival of the Toyota Prius, are modest compared with the capability of the ultracapacitor, which was already well known as something that might be combined with a battery in a managed energy storage pack [23].

The Transport Development Centre (TDC) in Montreal had earlier supported the development of both a pure electric and a hybrid delivery van, the Marathon C360 [24], and was impressed by the improvement that hybridization made to the secondary battery life. Alcan internally funded the demonstration of an alkaline telecom unit vehicle range extender in a plug-in hybrid version of the UQM Electrek. Results were reported in Vancouver at a 1989 SAE meeting by Parish et al. [25]. The Ontario Ministry of Energy then supported the demonstration of the aluminum–air hybrid in a UQM converted Chrysler Minivan. The results of this novel plug-in hybrid work were reported at EVS 10 in Hong Kong by Lapp et al. [26] and at EVS 11 in Florence by Lapp and Dawson [27].

The hydrogen fuel cell system has made a contribution to the progress of hybrids. Polymer electrolyte membrane (PEM) hydrogen fuel cells emerged after the Apollo program. Ballard Power Systems took up the challenge and developed higher power PEM fuel cells that could operate a vehicle with an initial hope that no added battery was needed.

The Canadian government evaluated both hydrogen and aluminum fuel cells [28] for submarines and showed that both could be used for charging the secondary batteries on the submarines. It was concluded [29] that the hydrogen PEM fuel cell is the better solution for large underwater vehicles where neutral buoyancy allows pure hydrogen to be easily stored on board in heavy metal hydrides and safely displace the normal ballast.

By 2007 five Ford Focus hybrid fuel cell vehicles had demonstrated a 260–320 km (162–200 mile) driving range. But there was another step on display: the

Ford Edge plug-in fuel cell hybrid. It had a smaller fuel cell and a 130 kg Li-ion battery packing 15.6 kWh. The fuel cell reliability and life have further increased. In a release that year, Ford said: "This provides another 200 miles of range for a total of 225 miles with zero emissions. Individual experiences will vary widely and can stretch out the time between fill-ups to more than 400 miles."

In early 2009 Ricardo Engineering confirmed [30] that a way forward for fuel cells in light road vehicles was as the range extender for a plug-in electric vehicle such as Ford had demonstrated. In August 2009 Mercedes Benz announced the B-Class F-CELL, which has only a 1.4 kWh lithium-ion battery and said it had a "range of about 400 kilometers and short refueling times of around three minutes." With the plug-in, Ford Edge was reaching a range of 400 miles in 2007; a >500 mile range fuel cell plug-in vehicle is now in sight if technologies are combined.

Will a world of light electric road vehicles needing intercity road range provide a significant niche for fuel cells? In the mass market light vehicle niche, the hydrogen or metal fuel cell would compete with combustion technologies and may enable, or displace them, depending on the dollars to reduce carbon now easily foretold with transportation greenhouse gas models like GHGenius [31].

All the lower carbon fuels are improved by hybridization [32]. Range has been an issue with natural gas and propane vehicles. If hydrogen became a vehicle fuel it would be hybridized [33]. Indeed, it is currently being demonstrated in a fleet of buses in Whistler with 47 kWh Li-ion batteries providing power for acceleration. PHEVs and HEVs will thus widen the range of fuels that might be considered and one day every vehicle may have a secondary battery ready to interact with the grid.

2.6 LITHIUM ION

Lithium melts at 180.5 °C and when it does so can reduce oxygen containing organic compounds such as the propylene and ethylene carbonate used in cells. Early lithium cells had metal anodes. Li-ion batteries may have oxide cathodes working via an electrolyte against carbon anodes containing lithium atoms which have entered the carbon structure and are safely isolated one from the other. But rapid overcharge of a cell can result in a local plating out melting and reaction. Also, oxygen can be released as a gas. These safety issues are easier to manage if the cells are small, as their thermal management is easier because of the shorter distances involved for heat transfer. A Li-ion vehicle battery is presently made up of small capacity cells (also called "small format" cells) in the range 6–10 Ah. To get a meaningful capacity there have to be a number of battery "strings" in parallel and a good deal of string and cell balancing control. Today, lithium-ion batteries offer up to four times the energy density and ten times the power–power density of lead-acid. They have yet to achieve the very high 205 Ah capacity of the simple low-cost Chloride 30 Wh/kg battery discussed earlier.

Emerging new electrode and electrolyte materials are showing great promise for both safety and performance. Davidson et al. [34] reviewed ongoing developments in new materials for Li-ion and described new electrode and electrolyte materials that show great promise for combining enhanced safety with outstanding performance. As an example of electrolyte evolution, oxygen-free dinitriles are being tried as solvents with better thermal properties; and organic ionic liquids are being blended with earlier organic liquids. A growing range of oxides are under examination; for example, $LiNi_{0.5}Mn_{1.5}O_4$ offers 4.7 volts, with relatively low cost and cycling stability at high temperature. Options to replace the early carbon anode are also expanding, and silicon, which is thermally less reactive than carbon, offers very high capacity.

In summary, the materials of the Li-ion anode, cathode, and electrolyte are changing and larger format cells will become safer. With this change there will be fewer parts and fewer control devices in each battery. Costs will fall and recycling will become easier.

2.7 CELL VOLTAGE, RELIABILITY OF STACKS, AND IMPACT OF INVERTERS

Cells are stacked in series and the required operating stack voltage impacts stack reliability directly. As an example, if two cells in series are each reliable 90% of the time, then the pair are reliable only 81% of the time. The fewer the number of cells in series, the more likely a stack is to survive. If in Table 2.2 the stack is specified to have a voltage of 312 volts under load, then the table shows the significant difference in reliability between cells of different chemistries when one in a thousand or one in ten thousand cells fail.

With the voltage of lithium cells being very much higher than for lead-acid batteries, the reliability of battery stacks can be expected to be significantly higher. This is a contrast with other electrochemical technologies, such as hydrogen, that have been proposed for electric drive vehicles. The multiple strings of lithium cells in

TABLE 2.2. Stack Reliability

	Stack Voltage		312 Volts	
	IF Reliability of Cells		99.90%	99.99%
	Cell Voltage	*N* Cells	THEN Stack Reliability	
Lithium	3.6	87	91.69%	99.14%
Lead-acid	2	156	85.55%	98.45%
Al/air	1.5	208	81.21%	97.94%
Ni-MH	1.25	250	77.90%	97.53%
Hydrogen	0.5	624	53.56%	93.95%

parallel that are used while large format cells are developed contribute further to reliability. Parish et al. [25] coupled a semi-fuel cell to a battery via an efficient DC/DC inverter. The stack voltage was significantly reduced in a road electric vehicle with a then high 160 km range. This is an alternative approach to reliability, and today medium duty and heavy hybrid vehicles typically have bus voltages higher than the battery voltage and so can make use of the batteries produced in quantity for the light vehicle market.

2.8 BATTERY MASS FRACTION, ENERGY, POWER, BENEFITS AND A PENALTY

In 1982 the Office of Technology Assessment of the U.S. Congress received a report [35] which describes hybrid personal and commercial vehicles equipped with a range of battery technologies where the "battery fraction" (the battery weight fraction of the vehicle's loaded weight) is 13–19% and there is discussion of a range of battery fractions of 10–32%. The report mentions hybrid vehicle fuel costs and assumes that part of the fuel came from the grid; these vehicles were PHEVs.

Cairns and Hietbrink [4] showed that all the electric vehicles of that time operated between 0.2 and 0.35 Wh/kg•km on the standard SAE urban J227a "C" cycle. They developed a relationship between urban ranges and "battery mass fraction," where the vehicles reported on vary from 20% to 50% battery fractions. At the time, no vehicle had had a battery with more than 40 Wh/kg but a chart in the paper shows 200 km would be attainable with a 30% battery fraction and a battery energy density of 120 Wh/kg.

When GM launched the EV1 and Ford the Ranger in the 1990s they did so with lead-acid batteries and their performance was not dissimilar to that noted by Cairns and Hietbrink. Contrast this with Table 2.3 comparing the Mitsubishi IMiev, the Nissan Leaf, and the Ford Transit Connect Electric.

These production electric vehicles are being *launched* with ranges of 130–160 km with only the battery mass fraction that the early lead-acid *hybrids* had. With the energy density improvements and cost reductions described above to come, if today's electric vehicles find niches there will be growth.

TABLE 2.3. Comparison of Three Production PEVs

	Battery						
Product	(kWh)	(Wh/kg)	Battery Weight	Curb Weight	Test Add 150 kg	Battery(%) Curb Weight	Battery(%) Test Weight
Imiev	16	106	151	1080	1230	14.0%	12.30%
TCE	28	94	298	1800	1950	16.6%	15.30%
Leaf	24	120	200	1589	1739	12.6%	11.50%

The penalty of large battery mass fractions is that the weight must always be carried even for short trips. Matching battery size to mission is easier for a telecom or utility fleet than for the mass personal vehicle market. Falling costs and improving energy density would reduce this penalty.

2.9 VEHICLE CLASSES, NICHES, AND CONSTRAINTS

Table 2.4 presents the greenhouse gas contribution by vehicles in Canada from data published by the Natural Resources Canada (NRCan). The table defines in more detail some vehicles with high impact and some with negligible impact. Vehicle distributions across North America will be similar.

In Canada, medium and heavy trucks produce 34.8% of road vehicle greenhouse gas (GHG) emissions and light vehicles 62.4%. Note that we can say that, in Canada, motorcycles, school buses, urban transit, and intercity buses make a negligible contribution, as they contribute only 2.9% between them.

Vehicles used both in and between cities require technologies that can operate on highways and in stop–go traffic. This group contributed 54% of the road GHG emissions in 2005 and comprises:

- Cars
- Passenger light trucks (including suburban utility vehicles and minivans)

Vehicles used primarily in cities operating in stop-and-go traffic contributed 16.3% of the road GHG emissions in 2005 and comprise:

- Freight light trucks
- Medium trucks including shuttle buses on the same chassis

TABLE 2.4. Greenhouse Gas Contribution by On-Road Vehicles in Canada

GHG by Mode (Mt of CO_2E)	1990	2005	Total Growth 1990–2005	Percentage in 2005
Small cars	25.1	22.4	−10.9%	16.2%
Large cars	27.1	23.0	−15.0%	16.7%
Passenger light trucks	14.9	29.0	94.6%	21.1%
Freight light trucks	6.7	11.5	71.2%	8.3%
Medium trucks	9.4	11.0	17.1%	8.0%
Heavy trucks	14.9	37.0	147.5%	26.8%
Motorcycles	0.2	0.3	54.9%	0.2%
School buses	1.0	1.1	8.6%	0.8%
Urban transit	2.1	1.9	−9.9%	1.4%
Intercity buses	0.7	0.6	−8.9%	0.5%
	102.1	**137.8**	**34.9%**	**100.0%**

Source: Government of Canada data [22].

Vehicles that operate primarily between cities contributed 26.8% of the road GHG emissions in 2005 and comprise:

- Heavy trucks

Electric vehicles are returning in urban delivery fleets with daily short ranges. For example, 30 two-ton electric vehicles deliver mail in New York City. These vans are powered by a lead-acid battery. The trucks are recharged via dedicated 220 V electrical outlets at postal service facilities [36]. The niche is expected to grow as costs fall. Today, Azure is offering the Ford Transit Connect delivery van in Europe and North America, while Smith Electric Vehicles, a supplier to fleets in the United Kingdom, is expanding into the U.S. market with heavier delivery trucks. All are using Li-ion batteries.

2.10 MESSAGES FROM FULL-CYCLE MODELING, ENERGY SECURITY, AND AIR QUALITY

American studies show that plug-in hybrids and hybrids give equivalent greenhouse gas saving. According to Samaras and Meisterling [37], plug-in hybrids need cleaner power stations to give more benefit than hybrids. Williams [38], agrees and suggests a carbon pricing signal will be needed before the plug-in hybrid improves beyond the conventional hybrid.

In Canada the transportation climate change model GHGenius [39] now covers both plug-in hybrids and pure electric vehicles [40]. With the assumptions made, PHEVs are better for GHG reductions than HEVs in Quebec, Manitoba, and British Columbia, which use high fractions of hydroelectricity. The model showed the following for a personal car in Canada, covering 20,000 km/year:

- A hybrid has a cost increment of $3000, the cost to reduce a tonne of CO_2 E was $79/tonne.
- For a plug-in hybrid, with 50 km range, with a cost increment of $8000, the reduction cost for a tonne of CO_2 E was *three times more* at $243/tonne.
- For an electric vehicle with a cost increment of $10,000, the cost to reduce a tonne of CO_2 E was $262/tonne.

2.11 MARKET PENETRATION BY VEHICLE NICHE

PHEV development is primarily focused on the personal vehicle, but as battery costs fall consideration is being given to commercial vehicles [41]. Light vehicle PHEV and PEV penetration is not expected to exceed that for hybrids (HEVs), which has leveled at 2.4% of sales per year in the United States, where penetration is the highest of all countries. Canadian penetration averages 1.25% per year. Sales are not uniform by state or province.

Late to start, the commercial fleets are hybridizing rapidly. All 100%, of the delivery vehicles at a FedEx depot in the Bronx, New York, are hybrids, while Purolator [42] is saying that 19% of its step vans will be hybrids by the end of 2011. Commercial fleets are expected to take 70% of the electric vehicles produced before 2015 [43]. Early Smart Grid focus will be on multiple chargers in commercial fleet yards and well dispersed homes.

2.12 VEHICLE ARCHITECTURE, KEY COMPONENTS, CONTROLS, AND COST

Hybrid architectures vary in efficiency and there are other benefits that can accrue from hybridization. Toyota used a series/parallel drive arrangement and Honda used pretransmission architecture in their first passenger car hybrids in the United States in 1999. The "pretransmission" hybrid has its electric motor between the engine and the transmission. Both companies had enhanced their vehicles with other technologies. Toyota had realized that the engine needed to be less powerful when boosted by an electric motor and used the Atkinson cycle rather than the more powerful, but less efficient, Otto cycle. Meanwhile, Honda used an all aluminum bodied vehicle to bring down vehicle weight and fuel consumption. Honda had a battery with half the voltage and capacity of the Toyota battery and used one, rather than two, electric motors. Honda's later hybrid, Civic, also played the engine power reduction card by using a 1.3-liter engine rather than the 1.8-liter engine in the regular vehicle.

Honda's low-cost "pretransmission" parallel approach is now the route that Volvo [44] is following for their heavy duty truck. A family of "post-transmission" hybrids, where the motor is after the transmission, is used for medium duty trucks and is intermediate in cost between "pretransmission" and the series/parallel approach.

There are also "mild hybrids" in both the light and medium duty market. Typically, these do not have electric drive or electric-assisted drive but they tackle the energy losses associated with operating the engine while idle and in the case of the medium duty market can provide auxiliary power to operate, for example, bucket trucks or refrigeration systems. In the light vehicle market the battery used in some GM vehicles is a 36–42 V lead-acid system [45]. In the medium duty market, where more energy is needed, Azure Dynamics has used the same battery (Ni-MH or Li-ion) as in its parallel hybrid [46].

The PHEV is appearing in two forms. Toyota simply adds a bigger battery to its existing hybrid but General Motors has developed the Volt, which is described as an "extended range electric vehicle" (EREV). Though originally conceived as a series hybrid, it is partly a parallel hybrid in which the engine generates electricity which is sent to the electric drive motor or the battery as decided by a controller, while the 16 kWh battery alone can drive the vehicle for a range of 64 km. Control strategies are important as with all these architectures and battery constraints such as depth of discharge, and rates of charge and discharge must be respected. Regenerative braking is always the lowest cost source of energy and leaving space in the battery for it is important. The larger batteries help regenerative braking.

2.13 GRID TO VEHICLE (G2V) CHARGING: LEVELS 1 TO 3

In the 1990s there was remarkable debate at conferences between Ford and General Motors. The Ford Ranger EV was launched with conductive charging while the GM EV1 was launched with inductive charging. As we move to the launch point for the next generation of electric vehicles with lithium-ion batteries, the North American auto industry has come to an agreement on standards for charging electric vehicles at 120 VAC (AC Level 1) and 240 VAC (AC Level 2). The standards are established by the Society of Automotive Engineers (SAE) and are embodied in SAE standard J1772. AC Levels 1 and 2 describe the charging configurations where the conversion from AC to DC power takes place on-board the vehicle. However, today's standards battle swirls around the DC Level 2 configuration where the AC to DC power conversion takes place off-board the vehicle. DC Level 2 is also known informally as fast charging. The Japanese auto industry has developed a standard for off-board power conversion that operates in the 50 kW range and has been dubbed the standard CHAdeMO [47]. The fast charging standard governs the connector between the cord set and the vehicle and the communication protocol between the EV and the fast charging unit during the charging event. Japanese automakers intend to bring the standard to the North American market but are faced with the proposal of another DC Level 2 standard for the connector. Table 2.5 represents the nomenclature for the six EV charging configurations that are currently being considered by the North American auto industry.

In summary, AC Level 1 (AC L1) and Level 2 (AC L2) are the only two standards that are established and in practice in North America. DC Level 2 (DC L2) is in practice in Japan and North America but it is only an established standard in Japan. AC L3, DC L1, and DC L3 are place holders at this time.

TABLE 2.5. EV Charging Configurations Established and Proposed by the North American Auto Industry

	AC					DC		
	V	Phase	A	kW		V	A	kW
AC L1	120	1	12, 16	1.44, 1.92	DC L1	200–450	≤80	≤36
AC L2	240	1	≤80	≤19.2	DC L2	200–450	≤200	≤90
AC L3		1 or 3			DC L3	200–600	≤400	≤240

BC Hydro developed a set of EV infrastructure guidelines, available online [48] and presented here with permission, that provides a more thorough description of the three EV charging configurations in practice in North America and the sections below are taken from it with permission of BC Hydro. Please note that the nomenclature has evolved since the publishing of the EV infrastructure guidelines and *Level 3* in the excerpt below now refers to DC Level 2 as described in Table 2.5.

2.13.1 Level 1: 125 Volt AC

The Level 1 method uses a standard 125 VAC branch circuit that is the lowest common voltage level found in both residential and commercial buildings. Typical amp ratings for these receptacles are 15 or 20 amps. Level 1 charging is an important aspect of the infrastructure because of the widespread availability of these circuits. Consequently, companies that currently provide vehicle conversions to electric and future EV and PHEV suppliers will likely provide a Level 1 Cord Set (125 VAC, 15/20 amp) with the vehicle. With a rating of 15 amps, the actual current draw is limited to 12 amps so the Cord Set will draw approximately 1.4 kW of power. Level 1 charging typically uses a standard 3 prong electrical outlet (NEMA 5-15R/20R). The Cord Set uses a standard 3-prong plug (NEMA 5-15P/20P) with a charge current interrupting device (CCID) located in the power supply cable within 12 inches of the plug. The vehicle connector at the other end of the cord will be the design approved by the Society of Automotive Engineers in their Standard J1772. This connector will properly mate with the vehicle inlet also approved by J1772. The J1772 standard is the subject of a harmonization project with the Canadian Electrical Code Part II Standards.

Level 1 charging at 20 amps is specifically recognized in the CEC Section 86 for dedicated EV charging. (Note that the CEC derates branch circuits to 80% for continuous duty so the usable capacities for the above circuits would be 16 amps.) The dedicated circuit requires the use of NEMA 5-20R for the premises receptacle. Many electrical utilities provide a rate structure that considers on peak and off-peak hours. Home owners may desire to install a timer device in this circuit to control charging to off-peak times.

Because charge times can be prolonged at this level, many EV and PHEV owners will be more interested in Level 2 charging. Some EV providers suggest their Level 1 cord set should be used only during unusual circumstances when the Level 2 EVSE [Electric Vehicle Supply Equipment—an appliance that requires the detection of an EV connected to the attached cord set before allowing power to flow through the cord set is not available, such as when parked overnight at a non-owner's home.

2.13.2 Level 2: Greater than 125 Volt AC or Greater than 20 amps

Level 2 is typically described as the "primary" and "preferred" method for the EVSE both for private and public facilities and specifies a 240 VAC, single phase branch circuit. The J1772 approved connector allows for current as high as 80 amps AC (100 amp rated circuit). However, current levels that high are rare and a more typical rating would be 40 amps AC which allows a maximum current of 32 amps. This provides approximately 7.7 kW.

This level of charge provides the higher voltage that allows a much faster battery charge restoration. The Level 2 method also employs special equipment to provide a higher level of safety required by the CEC and NEC.

The Society of Automotive Engineers (SAE) has been working to standardize the method of coupling for automakers and EVSE suppliers. A standard EV Coupler will be used by EV and PHEV suppliers following the final acceptance of this approved standard. The Coupler and EV Inlet will be the same for both Level 1 and 2 charging. The onboard charger will measure the inlet voltage and determine the available current from the EVSE through the pilot signal and adjust accordingly.

In addition, when connected, the vehicle charger will communicate with the EVSE to identify the circuit rating and adjust the charge to the battery accordingly. Thus an EVSE that is capable of delivering 25 amps will deliver that current even though connected to a 40 amp rated circuit.

2.13.3 Level 3: Charging

Level 3 charging or "Fast Charging" is for commercial and public applications and is intended to perform similar to a commercial gasoline service station in that charge return is rapid. Typically, this would provide a 50% recharge in 10 to 15 minutes. Level 3 typically uses an off-board charger to provide the AC to DC conversion. The vehicle's on-board battery management system controls the off board charger to deliver DC directly to the battery.

This off-board charger is serviced by a three phase circuit at 208, 480 or 600 VAC. The SAE standards committee is working on a Level 3 connector, but has placed the highest priority in getting the Level 1 & 2 connector approved first.

Note: Although not as common, a vehicle manufacturer may choose not to incorporate an on-board charger for Levels 1 and 2, and utilize an off-board DC charger for all power levels. In this case, the plug-in vehicle would only have a DC charge port. Another potential configuration that may be found, particularly with commercial vehicles, is providing 3-phase power directly to the vehicle. This configuration requires dedicated charging equipment that will be non-compatible with typical public infrastructure.

Commercial charging stations in the fork truck industry have proved their value in providing quick charging, by both reducing the need for spare batteries and reducing on-board battery capacity and weight. This approach is now being developed by several international companies for on-road vehicles [49], led by the Japanese auto industry. Demand for fast charging will accelerate if efficiency is increased and costs are lowered.

2.14 GRID IMPACTS

The first point of impact on the utility grid from EV charging is the distribution transformer, the metal canister the size of a garbage can on top of the power pole

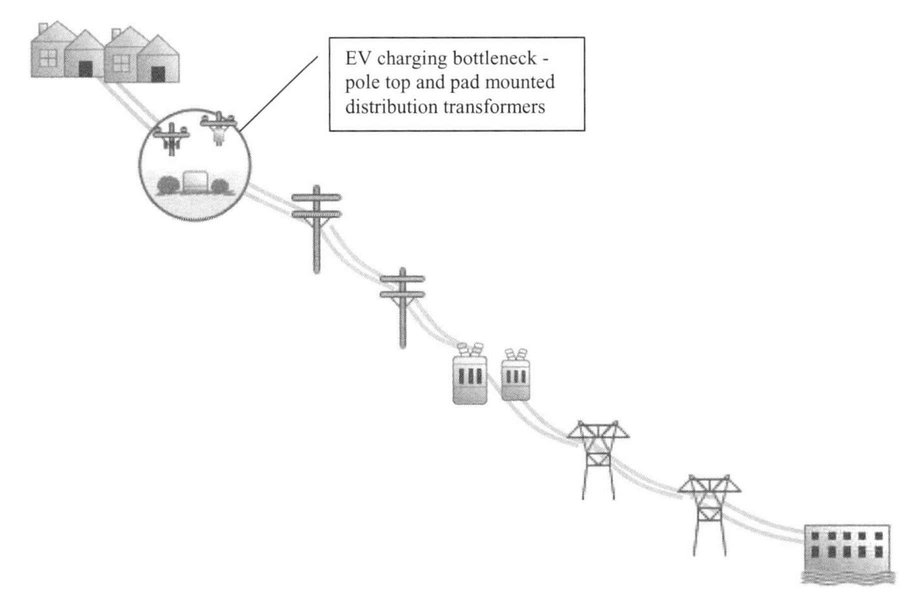

Figure 2.1. Electric utility system.

in the laneway. As EV charging loads are aggregated to the substation, transmission, and generation portions of the electric utility system, the load becomes insignificant in comparison to the existing base load. Even optimistic projections of EV adoption rates appear as gradual load growth beyond the distribution transformers and should be well managed within typical utility planning horizons. Figure 2.1 illustrates the bottleneck for EV charging on the electric utility system.

EV charging at AC L1 (1–2 kW) has negligible impacts on the distribution systems, but charging at AC L2 begins to impinge on the capacity limitations of pole-top distribution transformers. EV owners are expected to conduct the majority of recharging at home, where the personal passenger vehicle is parked the majority of the day. EV recharging is also expected to commence after the end of the daily commute at the beginning of the evening residential peak, the same time when all the other electricity-based home activity commences: cooking, illumination, home heating or cooling, and washing. As described in the Level 2 charging excerpt (Section 2.13.3), a typical AC L2 charging event could be as high as 6–7 kW, doubling the typical household peak load during the evening peak load hours, which can be as broad as 5 P.M. to 10 P.M.. A typical distribution transformer can serve 5, 10, or 15 customers, depending on the size of the transformer. If several households on the same distribution transformer own an EV each, that transformer can become overloaded beyond the expectations of utility distribution planners. The result is a lifecycle derating of the assets. The options are to upgrade transformers where EV clustering is detected or replace assets upon early failure. The problem is that most utilities do not have visibility at the distribution transformer level, never mind the peak consumption of its residential customers. That is, most utilities do not have any automated monitoring of distribution assets downstream of distribution substations.

One solution is advanced metering infrastructure, digital meters with the capability of interval metering. Regardless of how the metering data is collected, autonomously or manually, the utility will be able to detect customers with step function load growths and can plan system upgrades accordingly. Advanced metering solutions can also offer utilities more proactive load management options that could defer system upgrades by shifting nonessential loads to off-peak hours. Demand-side management options range from passive options such as time of use rates to more aggressive options such as demand response or dispatch. Time of use rates simply place a higher value or rate on electricity during peak hours and a lower value during off-peak hours, providing utility customers a time-of-day pricing signal to manage their electricity consumption over the course of a day. Demand dispatch allows the utility to control a customer load by providing the customer financial compensation in return.

The United Kingdom has been successful in using home storage heaters and hot water systems that take advantage of lower priced electricity at night since at least 1970. In North America, the norm is one rate for homeowners and multiple rates for businesses, but some utilities already are installing meters that track use by time of day. For example, some California utilities have customer programs that permit the utility to remotely alter the setting on customer thermostats during peak electricity consumption periods to manage capacity constraints on their system.

2.15 VEHICLE TO GRID (V2G): A FIRST OR SECOND ORDER MATTER?

Using EVs with bidirectional power flow capabilities to shift load from peak to off-peak, also known as "peak shaving," has long been touted as a win–win situation for the utility and EV owners. Utilities could utilize EVs as a distributed energy resource (storage) to provide load during off-peak hours, typically overnight, and provide energy back to the grid during peak hours. The EV owner could benefit financially by using the old stock market adage of buy low and sell high. Utilities practice this type of arbitrage at a much larger scale with facilities such as pumped storage, which typically range from hundreds of megawatts to 1000 MW. It is important to note that storage provides capacity and not energy to a utility system.

In practice, peak shaving or load shifting is a thin margin business. In utility jurisdictions where there is a market for ancillary services, the compensation is much higher than that provided for load shifting. Within the scope of ancillary services the regulation market is the most lucrative. The objective of system regulation is to maintain the operating frequency of 60 Hz, which requires the balancing of load and generation. This requires a quick responding generator or storage facility as the dispatch is in the time scale of minutes at a time rather than the daily charge and discharge cycles required for load shifting. Regulation is more suitable for an EV battery as it avoids deep cycling of the battery, which may conflict with operation of the vehicle, and it should leave the battery at the same state of charge if the utility system is to remain balanced.

Although the concept of EVs providing utility system regulation is technically viable, the questions of scalability and economics still remain to be proved. Once there is a sufficient volume of EVs on the road, an ancillary services aggregator could potentially solicit enough subscriptions from EV owners to negotiate a system regulation contract with an independent system operator (ISO) or a utility could create a tariff that would allow individual EV owners to sell regulations services to the utility. Given that either one of these options were in place, the economic viability of EVs providing regulation services still remains.

Two contrasting utility markets are used to illustrate the question of economic viability for V2G. In British Columbia, the utility BC Hydro is a vertically integrated crown corporation that operates in a regulated environment. PJM [50] is an independent system operator or ISO in a deregulated utility environment that happens to have an ancillary services market.

The following factors void the opportunity for V2G in British Columbia:

- The BC Hydro system is energy constrained and not capacity constrained.
- BC Hydro is more than 90% hydroelectric.
- Hydroelectric generation is a highly responsive generation technology that load follows extremely well in comparison to coal-fired power generation, which comprises a significant portion of the generation mix for PJM.

Alternatively, PJM has an ancillary service market with prices in the $40/MWh range. A joint industry–university study suggested that an EV with a 15 kW V2G setup in the PJM territory could generate nearly $30,000 for its owner on a gross revenue basis discounted at 7% over 10 years [51]. A 15 kW EV charging configuration is twice the charging capacity that is currently offered by first-generation EVs and would require substantial infrastructure upgrades for the customer as well as the utility. These costs have not been factored into the $30,000 revenue stream proposed by the PJM report. Depending on the residential tariff, if the 15 kW setup is not considered normal load growth, the system upgrades would be born by the individual (customer) triggering the upgrade, which could quickly negate the revenue stream suggested in the PJM report.

The United States averages 2.5 cars per household and Canada averages 1.5 cars per household. Let's say conservatively there are two cars per household in North America. A 2.5% market penetration of the fleet on the road by electric vehicles could result in 5% of households requiring intelligent charging and further "opportunity chargers" at places of work and shopping centers. With some personal vehicles carrying more than ten chips already, this is a natural area for vehicle to grid communications. The ability of plug-in electric or hybrid vehicles to provide utility regulation services has been suggested as a reason to pursue electric vehicles. Ford as part of a funded project is ensuring that the vehicle software will be available to return electricity to the grid while protecting the vehicle. The EV charging station (at AC Level 2) is the other component in EV charging where communication capabilities are being included. Therefore, capacity for V2G is being made possible through communication either with the charging station or with the vehicles themselves.

For the full V2G notion to be economically feasible, the following must happen:

- The local utility jurisdiction must have a need for ancillary services.
- An ancillary market should be in place.
- Battery cycle life must rise and costs must fall.

This may happen in due course and vehicle manufacturers are wise to leave the option open in their controllers.

2.16 SECOND LIFE FOR USED VEHICLE BATTERIES GRID-SIDE INSTEAD?

Nissan is projecting that after 10 years, when they are replaced, the Li-ion vehicle batteries will still be able to accept 80% of their original charge capacity. Sold on as certified stationary battery packs they will still have a useful life. Compare hydrogen as an energy storage medium. There are losses: first the loss in a 70% efficient electrolyser and then a loss in compression, where an adiabatic compression energy loss can range up to 14% with hydrogen; but let's say this is 90% efficient. Then a fuel cell is about 55% efficient. This route would return 35% of the electricity applied. The Li-ion battery has low energy loss and high cycle life. More than 90% of the energy applied to a battery could be returned to the grid.

Could the depreciated vehicle batteries be used as stationary storage devices and be competitive? Now a house could have a 24 kWh capacity still capable of high rate discharge. Could it provide load leveling to the grid as envisaged for the V2G strategy? Perhaps, and it could also:

- More rapidly charge a car battery than the Level 2 rate
- Always be available
- Be available for emergency home backup
- Store locally generated energy from a biomass generator or wind turbines
- Be available to the grid without devaluing the car battery

The higher the value of the second use of the battery, the lower the overall cost of the electric vehicle and the greater the market penetration. With DC standards coming into place for all charging levels, here is a driver for the Smart Grid.

2.17 THE CITY AND THE VEHICLE

Frost and Sullivan [52], in a recent study, estimated that about 4.5 billion people will live in cities by 2025, that is, about 60% of the total world population. Rising urban population will lead to rapid expansion of city borders into neighboring suburbs, resulting in the formation of megacities. They say there will be 30 such cities with a population above 10 million by 2025. The link with car growth is complex. Gilbert and Perl [53] in 2008 carefully reviewed the reasons why people use cars in cities and the alternatives that reduce their use. Cairns, and Hietbrink [4] writing in 1981, suggested a vehicle range of 160 km would satisfy 95% of personal travel needs in the United States until late in the 1990s. But vehicle market growth for the last decade has primarily been less in the Americas than in Asia.

Cities in North America are designed around the automobile and need change. There are, however, cities that excel. Schiller and Kenworthy [54] say that "Vancouver has benefited from a Canadian planning tradition which has been influenced by European thinking to a far greater extent than either Seattle or Portland." The city of Vancouver [55] had a population of 578,041 in 2006 [56] in an area 114.67 square kilometers, that is, there were 5041 people/km^2 in the city.

Perhaps as a consequence, car trips have declined in Vancouver as shown in Table 2.6. However, there may not have been a drop in greenhouse gas emissions as heavier personal vehicles increased (see Table 2.4)

2.18 IMPACT OF ELECTRIC DRIVE ON GREENHOUSE GAS EMISSIONS

We can compare hybrids with regular vehicles such as the Ford Escape. It has the same engine size in a hybrid and nonhybrid option. In NRCan Energuide data on combined highway/city driving, the all-wheel drive hybrid saves 28.5% fuel and the two-wheel drive saves 32.9%. The pretransmission Honda Civic saving is 36.6% but the engine size falls from 1.8 to 1.3 L. Today, a post-transmission parallel hybrid medium duty shuttle bus has been independently tested and shows a 30% fuel saving

TABLE 2.6. Reduction in Car Trips in Vancouver from 1992 to 2004

Mode	1992	2004
Auto driver	49%	30%
Auto passenger	13%	9%
Transit	23%	30%
Bike }	15%	3%
Walk }		27%

Source: Translink Trip Diary 2004.

TABLE 2.7. Potential Greenhouse Gas Savings from Hybridizing the Canadian Fleet

GHG by Mode (Mt of CO_2E)	2005	Hybrid Saving	MT Saving
Small cars	22.4	30.0%	6.7
Large cars	23.0	30.0%	6.9
Passenger light trucks	29.0	30.0%	8.7
Freight light trucks	11.5	30.0%	3.4
Medium trucks	11.0	30.0%	3.3
Heavy trucks	37.0	10.0%	3.7
Total	**133.9**		**32.8**

Source: Fitzpatrick [22].

on the New York City composite driving cycle. Volvo presented a 10% fuel saving figure for their Class 8 interurban heavy duty truck.

We can estimate the greenhouse gas savings from hybridization of the vehicle groups discussed in Table 2.4. Let us apply, from hybridization, a 30% fuel saving to the first group, which has combined urban and highway use, a 30% saving to the second group, which drives primarily in cities, and the 10% saving to the third group, that is, interurban heavy trucks. These savings are presented in Table 2.7.

The savings from complete hybridization of the three groups would have resulted in a combined saving of 32.8 Mt of CO_2E in Canada in 2005. Multiply this by ten for the United States. North America could attain a 350 Mt CO_2E reduction. Samaras and Meisterling [37] showed the impact across the United States was similar to that of hybrids because of the mix of power sources until there was a major shift to renewable or nuclear power. And so we can deduce that electrification of road vehicles cannot reduce North American greenhouse gas emissions by much more than 350 Mt CO_2E until lower carbon generation comes into play.

2.19 CONCLUSIONS

The fraction of electric vehicles in some commercial fleets will exceed the fraction of personal vehicles that are electric. The economics of the former will drive a demand for commercial fleet charging infrastructure. Home charged vehicles will also benefit from "opportunity chargers" at places of work and shopping centers. Intelligent communication with the vehicles will be critical to maximize the grid-side benefits of off-peak charging.

As electric vehicles penetrate, there will be a supply of reusable batteries which will be suitable for grid-side storage and renewable energy capture while allowing fast DC charging of vehicles.

The benefits of transferring power back from vehicles to the grid are presently outweighed by the benefits of charging the batteries off-peak. The jurisdictions that will benefit from vehicle to grid transfer will be those that have an ancillary service market with high prices. Vehicle manufacturers are already ensuring software will allow discharge of batteries to the grid in the event that an economic case develops.

The possibility of efficient DC charging of plug-in electric vehicles will be taken into account and perhaps drive the need for a Smart Grid as a home may not need a car to participate as a supply of low cost reusable batteries develops.

For electric vehicles to offer further greenhouse gas reduction benefits over hybrid elcctric vehicles they must be implemented in conjunction with low carbon energy sources and appropriate price signals.

With the North American electric vehicle charging bottleneck being the distribution transformers that service multiple clients, even a small penetration of electric vehicles, if similar to that attained by hybrids, will give more benefit if off-peak price signals are attainable by, for example, advanced metering or separately metered charging circuits.

With electric vehicles arriving for both commercial and consumer use, grids will benefit from being smart enough to charge and communicate with vehicles.

ACKNOWLEDGMENTS

Nigel Fitzpatrick thanks former colleagues, present and past, involved in electric vehicles primarily at Alcan, BC Research, and Azure Dynamics but also in universities and government in Australia, Asia, Europe, and North America over the years. Thanks also to *Advanced Fuel Cell Technology* magazine, the University of Manitoba (PHEV07 conference publisher), and the Engineering Institute of Canada [57] for encouraging recent articles and papers from which material is updated in this chapter.

REFERENCES

[1] E. H. Wakefield, "History of the electric vehicle, hybrid electric vehicles," SAE, 1998.

[2] E. H. Wakefield, "History of the electric automobile, battery-only powered cars," SAE, 1994.

[3] M. L. Poulton, *Alternative Fuels for Road Vehicles*. Computational Mechanics Publications, 1994.

[4] E. J. Cairns and E. H. Hietbrink, "Electrochemical power for transportationm" in Volume 3, *Comprehensive Treatise of Electrochemistry*, edited by J. Bockris, et al. Plenum Press, 1981.

[5] D. W. Kurtz, "The systems role in the development of electric and hybrid electric vehicles," SAE Paper 830222, 1983.

[6] *Popular Science*, May 1991.

[7] Securities Exchange Commission Form 8-K Unique Mobility Inc. June 26, 1997.

[8] H. Pieper, U.S. Patent 913,846, filed 1905 issued March 2, 1909.

[9] V. Wouk, "Two Decades of development and experience with hybrids," EVS 9, Toronto, 1988.

[10] G. G. Harding, B. L. Phillips, and J. E. Hammond, "The Lucas hybrid electric car," SAE 830113, 1983.

[11] Sypher:Mueller International Inc. Report, "BC transit fuel choice study," for Crown Corporation Secretariat, Province of BC, April 1997.

[12] Private communication: R. Geddes (1994) advised NPF that Aisin Seki was assembling a hybrid team in the United States on behalf of Toyota.

[13] W. Edwards, R. Dunlop, D. Wenlo, D. O' Connor, N. Fitzpatrick, and S. Constable, "Alternative and future fuels and energy sources for road vehicles," Report from Levelton Engineering Ltd., December 24, 1999.

[14] "Transportation and climate change: Options for action," Transportation Climate Change Table, published on Transport Canada website, November 1999.

[15] D. V. Ragone, "Review of battery systems for electrically powered vehicles," SAE Paper 680453, Mid-Year Meeting Detroit, 1968.

[16] D. H. Brown, "Hybrid battery system," SAE Paper 710236, January 1971.

[17] M. Shnayerson, *The Car that Could*. Random House, New York, 1996.

[18] R. M. Dell and D. A. J. Rand, "Understanding batteries," Royal Society of Chemistry, 2001.

[19] F. R. Kalhammer, B. M. Kopf, D. H. Swan, V. P. Roan, and M. P. Walsh, "Status and prospects for zero emissions vehicle technology," Report of the ARB Independent Expert Panel, 2007.

[20] "Hybrid vehicles," September 11th–13, Seminar, Windsor, Ontario, 2000.

[21] D. Checkel "Experience with an early PHEV field trial," PHEV2007, Winnipeg, Manitoba, November 1–2nd, 2007.

[22] N. Fitzpatrick, "Electric drive choices for light, medium, and heavy duty vehicles to reduce their climate change emissions in Canada," EIC CCTC2009, Hamilton, May 2009.

[23] W. J. Halliop, J. Stannard, and N. P. Fitzpatrick, "Low cost supercapacitors," Third International Seminar on Double Layer and Similar Energy Storage Devices, December 1993.

[24] D. J. Pavlasek, N. Eryou, N. Dennis, and G. Neufeld, "Design, fabrication and testing of the Marathon C-360 series hybrid vehicle," Report for Marathon Electric Vehicles; filed at the Transport Development Centre Library, Montreal.

[25] D. W. Parish, W. M. Anderson, N. P. Fitzpatrick, and W. B. O'Callaghan, "Demonstration of aluminum–air fuel cells in a road vehicle," Vancouver, BC , SAE paper 891690, August 1989.

[26] S. P. Lapp, G. D. Deuchars, N. P. Fitzpatrick, and R. Rehn, "Demonstration of a 6 kW aluminum/air range extender in a minivan," EVS10, Hong Kong, December 1990.

[27] S. P. Lapp, and J. Dawson, "Testing results and running cost projections for an aluminum–air lead acid battery hybrid electric van," EVS 11, Florence, 1992.

[28] R. G. Weaver, "Air-independent propulsion for submarines: A Canadian perspective," Maritech 90 Victoria, BC, May 1990 with updates January 1991.

[29] G. B. Thornton, "A design tool for the evaluation of air independent propulsion in submarines," MIT Thesis, May 1994.

[30] Ricardo Engineering review, 1st Quarter 2009.

[31] D. O' Connor, "A Model of the lifecycle assessment of transportation fuels with an emphasis on Canada." Available at http://www.ghgenius.ca/.

[32] N. Fitzpatrick, W. Arnold, M. Fairlie, and A. Miller, "The series hybrid pathway; a bridge to hydrogen fuel cells," International Electric Vehicle Symposium, Long Beach, November 2003.

[33] C. E. Thomas, "Comparison of transportation options in a carbon-constrained world, hydrogen, plug-in hybrids, and biofuels," National Hydrogen Association Annual Meeting Sacramento, California, March 31, 2008.

[34] I. Davidson, Y. Abu-Lebdeh, F. Courtel, S. Niketic, H. Duncan, and P. Whitfield, EV2010VÉ, Vancouver, September 13, 2010.

[35] W. M. Carriere, W. F. Hamilton, and L. M. Morecraft, "The future potential of electric and hybrid vehicles," Report to the Congress of the United States, Office of Technology Assessment, 1982.

[36] U.S. Postal Service Press Release, March 27, 2007.

[37] C. Samaras and K. Meisterling, "Life cycle assessment of greenhouse gas emissions from plug-in hybrid vehicles: Implications for policy," *Environ. Sci. Technol.*, 42(9), pp. 3170–3176, 2008.

[38] E. Williams "Plug-in and regular hybrids: A national and regional comparison of costs and CO_2 emissions," Nicholas School of the Environment at Duke University, November 2008.

[39] "A model for lifecycle assessment of transportation fuels." Available at http://www.ghgenius.ca/.

[40] (S&T) 2 Consultants Inc., "GHG emissions from sugar cane ethanol, plug-in jybrids, heavy-duty hybrids and materials review," Report to NRCan, January 30, 2006.

[41] Azure Dynamics presentation in OEM panel EV2010ÉV.

[42] S. Viola, communication to author November 2010.

[43] V. Muralidharan, "Analyst Briefing: EU and NA electric buses, vans and trucks market—depot based delivery vehicles expected to account for 70 percent of overall wlectrification share," Frost and Sullivan , March 31, 2010.

[44] N. Fitzpatrick "Electric drive for climate change, 'The art of the soluble,'" Report on the Electric Vehicle Symposium (EVS 23) Advanced Battery Technology, February 2008, Available at http://www.7ms.com/abt/archive/2008/03/article01.shtml.

[45] PR Newswire, "Delphi's Energen TM 42V belt alternator starter helps improve fuel economy And lower engine emissions," October 23, 2002.

[46] PR Newswire, "Azure Dynamics and Kidron sign supply agreement for low emission electric power (LEEP) systems for pefrigerated trucks." October 10, 2007.

[47] CHAdeMO is the name for the Japanese-based standard for DC fast charging, typically at the 50 kW power level. More information can be found at http://www.chademo.com/.

[48] "Electric vehicle charging infrastructure deployment guidelines, British Columbia," BC Hydro Version 1.0, July 2009. A PDF document can be downloaded at http://www.bchydro.com/ev.

[49] PosiCharge; Enersys, Minit-Charger.

[50] PJM Interconnection is a regional transmission organization (RTO) that coordinates the movement of wholesale electricity in all or parts of Delaware, Illinois, Indiana,

Kentucky, Maryland, Michigan, New Jersey, North Carolina, Ohio, Pennsylvania, Tennessee, Virginia, West Virginia and the District of Columbia. Available at http://pjm.com/about-pjm/who-we-are.aspx.

[51] W. Kempton, Vi. Udo, K. Huber, K. Komara, S. Letendre, S. Baker, D. Brunner, and N. Pearre, "A test of vehicle-to-grid (V2G) for energy storage and frequency regulation in the PJM system," November 2008.

[52] Frost and Sullivan study: "Impact of urbanization and development of megacities on personal mobility and vcehicle technology Planning," November 16 2010.

[53] R. Gilbert and A. Perl, "Transport revolutions, moving people and freight qithout oil," Earthscan, 2008.

[54] P. Schiller and J. Kenworthy, "Prospects for sustainable transportation in the Pacific Northwest: A comparison of Vancouver, Seattle and Portland," *World Transport Policy & Practice*, 5, Number 1, 1999.

[55] The city is the core of the Greater Vancouver Regional District, or Metro Vancouver with a population of 2,116,581 in an area of 2877.36 km^2 and a population density of only 736/km^2.

[56] Census data from Statistics Canada, www.statscan.ca.

[57] N. Fitzpatrick, "Electric drive choices for light, medium, and heavy duty vehicles to reduce their climate change emissions in Canada," EIC CCTC2009, Hamilton, May 2009.

3

AUTONOMOUS DEMAND-SIDE MANAGEMENT

Hamed Mohsenian-Rad and Alberto Leon-Garcia

3.1 INTRODUCTION

The next generation power grid, also known as the *Smart Grid*, is expected to be reliable, scalable, and flexible, in addition to enhancing efficient energy consumption among residential, commercial, and industrial consumers [1–4]. There are active discussions on the right functionalities and architecture needed for a Smart Grid. However, it is widely accepted that a modernized energy infrastructure will be essentially equipped with two-way digital communications capabilities allowing information exchange in both directions between generators/dispatch centers and end users. This will allow more advanced *pricing* and *demand-side management* strategies to be implemented in the future Smart Grid that can provide end users with the opportunity to reduce their electricity expenditure.

In general, demand-side management (DSM) refers to various programs that are implemented by utilities to control energy consumption at the end user side [5].

Smart Grid: Applications, Communications, and Security, First Edition.
Edited by Lars Torsten Berger and Krzysztof Iniewski.

These programs are used to encourage efficient energy consumption in order to *balance supply and demand* without the need for installing new power generation and transmission infrastructures, which are both time consuming and costly. Some of the common traditional demand-side management programs include energy conservation and energy efficiency programs, fuel substitution programs, demand response programs, and residential and commercial load management programs [6–8]. In most cases, a residential load management program seeks to achieve one or both of the following design objectives: *reducing consumption* and *shifting consumption* [9]. The former can be achieved among users by encouraging energy-aware consumption patterns, by constructing more energy efficient buildings, and using energy efficient home appliances. However, there is also a need for practical solutions to shift the high-power household appliances to off-peak hours to reduce the peak-to-average ratio (PAR) in load demand. Note that, in general, the generation capacity of a power network is designed for peak hours in order to prevent service interrupts. Therefore, a high PAR in load demand leads to significant underutilized operation and energy cost inefficiency in the power system [10].

Appropriate load shifting in demand-side management programs is foreseen to become even more crucial as plug-in hybrid electric vehicles (PHEVs) become more popular [9, 11]. Most PHEVs need 0.2–0.3 kWh of charging power for 1 mile of driving [9, 12]. This will introduce a significant new load demand to the existing power generation and distribution systems. In particular, during the charging time, the PHEVs can almost double the average household load and drastically exacerbate the already high PAR in most power networks. Moreover, unbalanced conditions resulting from an increasing number of PHEVs may lead to further degradation of the power quality, voltage problems, and even potential damage to utility and consumer equipment if the system is not reinforced [9, 12].

Most of the current residential load control activities among end users are operated manually. However, recent studies have shown that demand-side management needs to be automatic [9, 13]. That is, it should require minimum involvement of users. In fact, due to the lack of knowledge and the complexity of decision making, it is difficult for most users to follow and properly respond to changes in electricity prices to change their energy consumption in a timely manner [14, 15]. Thus, there is an essential need for *intelligent* home automation systems and technologies to automatically schedule energy consumption on the users' behalf and based on the implemented DSM programs. Understanding the various aspects and challenges of autonomous demand-side management programs is the focus of this chapter.

The rest of this chapter is organized as follows. In Section 3.2 we overview two major classes of demand-side management programs: direct load control and indirect load control. Different classes, key challenges, and technology requirements of distributed and autonomous demand-side management systems are discussed in Section 3.3. An optimal energy consumption scheduling strategy to be implemented in smart meters is introduced in Section 3.4. Price prediction is addressed in Section 3.5. The futuristic model of a demand-side management system where end users have some energy storage and generation capabilities are briefly discussed in Section 3.6. The chapter is concluded in Section 3.7.

3.2 DIRECT AND INDIRECT DEMAND-SIDE MANAGEMENT

There are two general approaches to demand-side management in power generation and distribution systems. One approach is direct load control (DLC) [16–19]. In DLC programs the utility company or an aggregator can remotely control the operations and energy consumption of certain appliances in a household or in an industry. For example, the utility may control lighting, thermal comfort equipment (i.e., heating, ventilating, and air conditioning), refrigerators, and irrigation pumps as described in the field trial outlined in Chapter 14. DLC programs for the purpose of grid stability or reducing the cost of dispatching have also been addressed in Hsu and Su [20] and Ramanathan and Vittal [21]. However, when it comes to load control and home automation systems, users privacy can be a major concern and even a barrier in implementing DLC programs [22, 23]. Therefore, DLC programs have received diminishing attention in recent years.

An alternative for DLC is *indirect load control* via *pricing*. In this approach, no direct energy consumption decision is enforced. Instead, users are *encouraged* to individually and voluntarily manage their load, for example, by reducing their power consumption at peak hours [24–26]. A wide range of pricing models have been proposed over the past few years: *real-time pricing* (RTP), *day-ahead pricing* (DAP), *time-of-use pricing* (TOUP), *critical-peak pricing* (CPP), and so on. In all these variations, the key idea is twofold: first, allowing retail prices to reflect fluctuating wholesale prices to end users such that users pay what electricity is worth at different times of the day; and second, encouraging users to shift high-load household appliances to off-peak hours. For example, in RTP tariffs, the price of electricity varies at different hours of the day. The prices are usually higher in the afternoon, on hot days in summer, and on cold days in the winter [27]. RTP programs have been adopted, for example, by the Illinois Power Company in Chicago [27]. The Ontario Hydro Company in Toronto uses a three-level (on-peak, mid-peak, and off-peak) TOUP electricity pricing tariff. An example daily trend of the real-time prices set by the Illinois Power Company in Chicago on December 15, 2009 is shown in Figure 3.1a. We can see that there are two peak hours (with maximum price of electricity) during the day. One occurs in the morning during the working hours. The other one occurs in the evening due to significant increase in residential load.

Other common pricing models include *energy conservation systems* with *inclining block rates* (IBRs). In IBR pricing, the marginal price increases by the total quantity consumed [28]. That is, beyond a certain *threshold* in the total monthly/daily/hourly residential load, the electricity price will increase to a higher value. This creates incentives for end users to conserve or to distribute their load over different times of the day in order to avoid paying for electricity at higher rates. In addition, IBR helps in load balancing and reducing the PAR [29]. It has been widely adopted in the pricing tariffs by some utility companies since the 1980s. For example, the Southern California Edison, San Diego Gas & Electric, and Pacific Gas & Electric companies currently have two-level residential rate structures where the marginal price in the second level (i.e., the higher block) is 80% or higher than the first

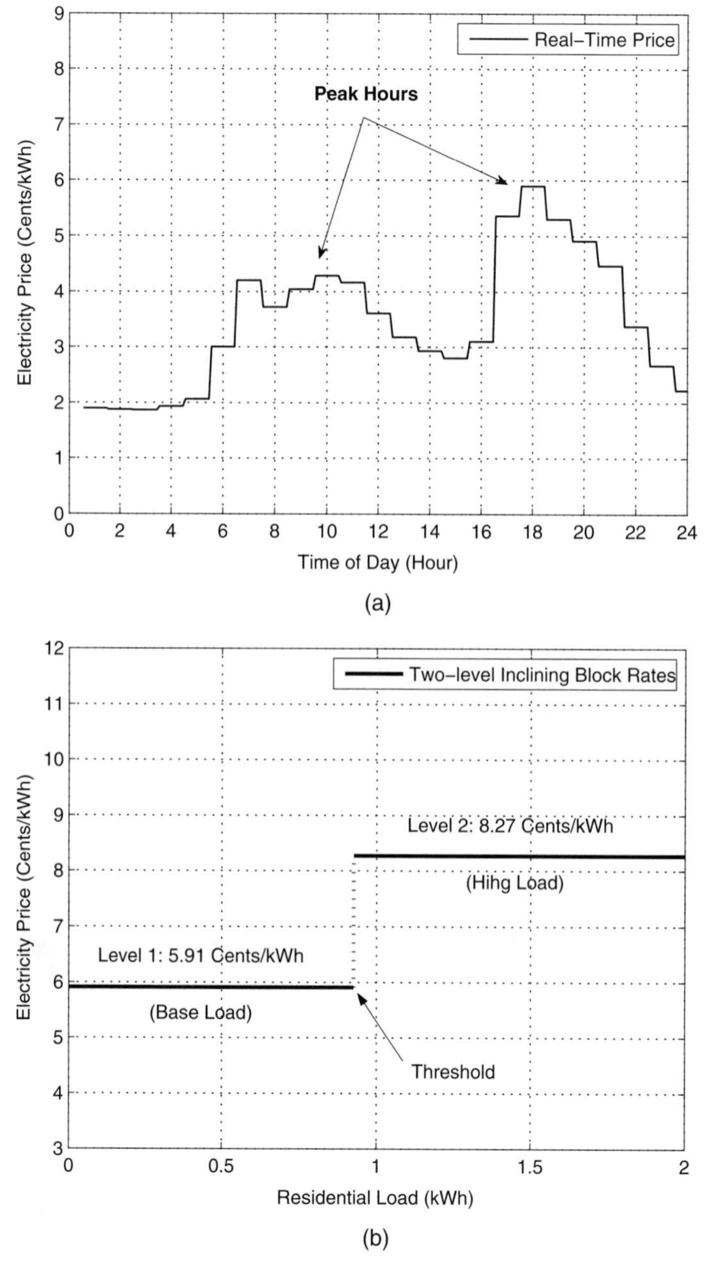

Figure 3.1. Examples of practical (a) real-time pricing and (b) inclining block rates used in North America for demand-side management.

level (i.e., the lower block), depending on the utility [30]. In Canada, the British Columbia Hydro Company currently uses a two-level conservation rate structure with 40% higher prices at the second level [31]. A normalized version of the inclining block rates used by the British Columbia Hydro in December 2009 is shown in Figure 3.1b.

All pricing models mentioned above aim at balancing power supply and demand by reducing or shifting end users' energy consumption. However, the level of success for each of these pricing strategies depends on how end users *respond* to it. Users' reactions to varying electricity prices and their involvements in demand-side management programs will be discussed in detail in the next section.

3.3 AUTONOMOUS DEMAND-SIDE MANAGEMENT

Most demand-side management programs that have been developed over the past three decades focus on managing the direct interactions between the utility company and each end user. This is shown in Figure 3.2a. Under this paradigm, each user communicates with the energy source individually. The users in this setup are sometimes referred to as *price-taking* users [10]. Such users consider the price values as *predetermined* and *fixed*. On the other hand, there are recent studies suggesting that intelligent users can be *price anticipators*. That is, they can take into account the impact of their actions on the prices set by the utility and may try to manipulate the price. Such users can communicate not only with the utility company but also with each other as shown in Figure 3.2b. Due to the recent advancements in Smart Grid technologies, the direct interactions among users can be automatic through message passing over a two-way digital communications infrastructure. While there are recent studies on investigating the behavior of price-anticipating users using tools and techniques from Game Theory, (e.g., see Mohsenian-Rad et al. [9], Caron and

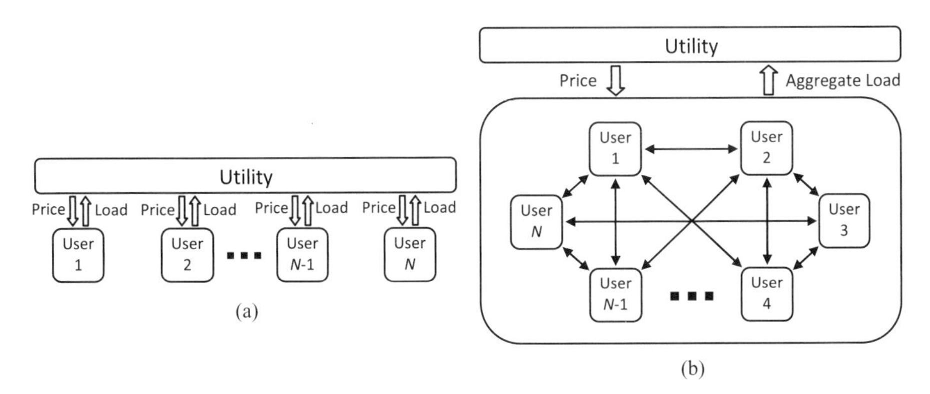

Figure 3.2. Two models for demand-side management: (a) users directly interact with the utility and (b) users may also interact with each other [9, 32, 33].

Kesidis [32], and Ibars et al. [33]), it is also argued that having price anticipating users in power systems may happen only in a futuristic scenario [9].

Recent studies have shown that despite several advantages of various advanced pricing techniques, the lack of knowledge among the end users about how to respond to time-varying prices and the lack of effective home automation systems are the two major barriers for fully utilizing the benefits of these indirect demand-side management programs [14, 15]. In fact, most of the current residential load control activities are operated *manually*. This makes it difficult for users to optimally schedule the operation of their appliances in response to the complex (e.g., time-varying) pricing information they may receive from the utilities in an advanced demand-side management program. For example, the experience of the RTP program in Chicago has shown that although the price values were available via telephone and the Internet, only rarely did households actively check prices as it was difficult for the participants to constantly monitor the hourly prices to respond properly [27]. Another example is the result from a more recent study by The Utility Reform Network (TURN) in San Francisco, which has reported that most users do *not* have time and knowledge to even pursue their *own* interest while they respond to real-time prices [34].

In order to resolve the above problem, recent studies have suggested employing *automated* household energy management strategies. This can be done using different approaches, for example, via Internet Protocol (IP)-based networks, where software on a personal computer is in charge of managing the household's energy consumption [35, 36]. However, as an example to help understand the basic mechanisms in this field, here in this chapter, we study an interesting scenario where automated energy consumption scheduling is done at *smart meters* as part of an *advanced metering system* [37–39]. In this regard, each smart meter is assumed to be equipped with an automated energy consumption scheduling (ECS) function as shown in Figure 3.3 [9, 13]. The ECS function is programmed based on the user's energy consumption needs. Then, it automatically controls the operation of various appliances such as the batteries of plug-in hybrid electric vehicles, washer, dryer, and dishwasher. All smart meters are connected not only to a power line but also to

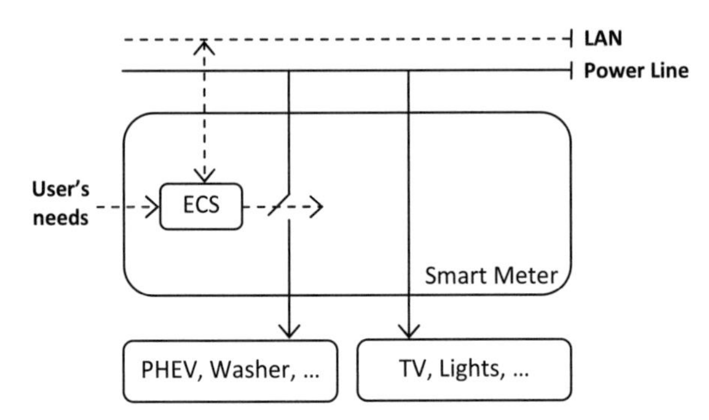

Figure 3.3. The operation of a smart meter with ECS functionality for residential energy consumption management.

a two-way digital communications infrastructure, for example, a local area network (LAN). The pricing information is obtained by each smart meter through the LAN, allowing the ECS function in each smart meter to adequately schedule the household's energy consumption for each appliance to minimize the user's total energy expenses. Note that, as an alternative implementation approach, the ECS function can be implemented as a separate device outside the smart meter as outlined in Chapter 7.

Smart meters and users can be connected to the power grid in different topologies. The most common topology is a tree with an energy source (e.g., a generator or a step-down substation transformer which is connected to the grid) and several users, each one equipped with an ECS function in its smart meter, as shown in Figure 3.4a. The number of users in this system is denoted by N. The energy source is owned or managed by a utility. In this scenario, each user interacts with only a single utility. In other words, users do not have the option of buying electricity from *another*

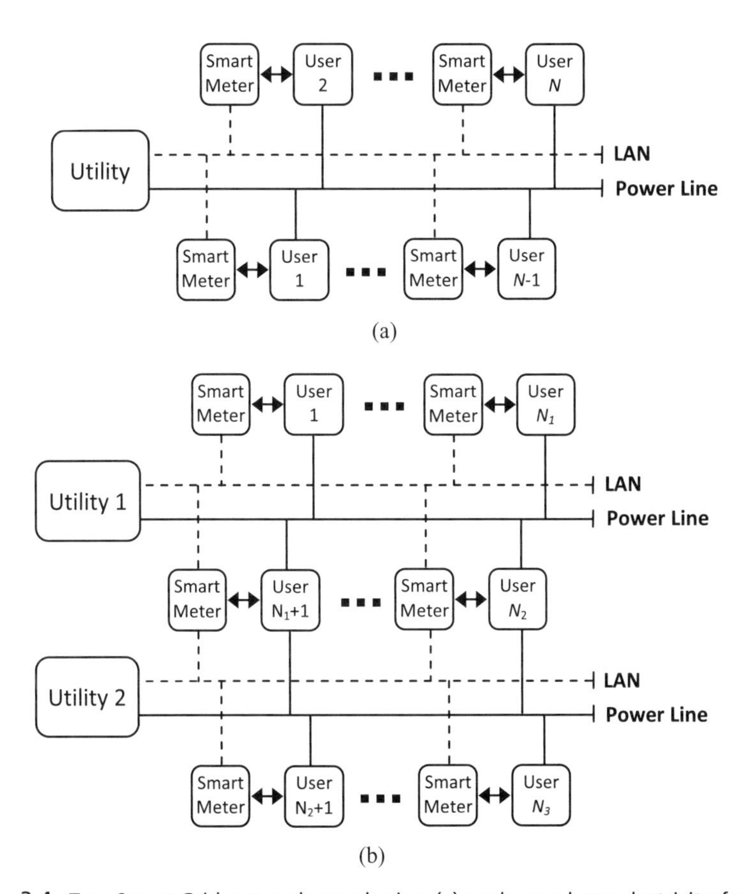

Figure 3.4. Two Smart Grid network topologies: (a) each user buys electricity from one utility (b) users can buy electricity from multiple utilities. Solid lines and dashed lines indicate power and communication infrastructures, respectively.

utility, seeking the lowest prices. This model is the dominant setup for most existing power systems. On the other hand, one may relax this assumption and allow each user to buy electricity from multiple utilities. This will lead to a *deregulated market*. In this setting, users can decide from which utilities they want to buy electricity and how much electricity they want to buy from each. In this case, the system model would look like that in Figure 3.4b. Here, users $N_1 + 1$ to N_2 have the option to buy electricity from either of the two utilities present. In reality, there can be either two separate power lines as shown in this figure, or there can be a single power line with different service agreements for different utility companies. A deregulated market is not the focus of this chapter; however, we believe it is an interesting area to be explored in future research.

Next, we analytically formulate the problem of optimizing energy consumption scheduling to be solved by the ECS function in each smart meter to minimize each user's energy expenses.

3.4 OPTIMAL ENERGY CONSUMPTION SCHEDULING

Consider an end user in the Smart Grid system of Figure 3.4a. Let $\mathcal{A}$ denote the set of all appliances in this user's household that are connected to the grid. For each appliance $a \in \mathcal{A}$, we define an *energy consumption scheduling vector* $\mathbf{x}_a$ as follows:

$$\mathbf{x}_a \triangleq [x_a^1, \ldots, x_a^H] \tag{3.1}$$

where $H \geq 1$ is the *scheduling horizon* that indicates the number of hours *ahead* which are taken into account in the decision making process. For example, $H = 24$ for day-ahead planning. For each upcoming hour $h \in \mathcal{H} \triangleq \{1, \ldots, H\}$, a real-valued scalar $x_a^h \geq 0$ denotes the 1 hour energy consumption level that is scheduled for appliance $a \in \mathcal{A}$. On the other hand, let E_a denote the total energy needed for the operation of appliance $a \in \mathcal{A}$. For example, in the case of a plug-in hybrid electric sedan, in total $E_a = 16$ kWh is needed to charge the battery for a 40-mile driving range [1]. As another example, for a typical front-loading clothes washing machine with warm wash/rinse setting, we have $E_a = 3.6$ kWh per load [40]. Next, assume that for each appliance $a \in \mathcal{A}$, the user indicates α_a, $\beta_a \in \mathcal{H}$ as the *beginning* and *end* of a time interval in which the energy consumption for appliance a can be scheduled, respectively. Clearly, $\alpha_a < \beta_a$. For example, after loading a dishwasher with the dishes used at lunch, the user may select $\alpha_a = 2$ p.m. and $\beta_a = 6$ p.m. for scheduling the energy consumption for the dishwasher as he expects the dishes to be ready to use in the evening. As another example, the user may select $\alpha_a = 10$ p.m. and $\beta_a = 7$ a.m. (the next day) for his PHEV after plugging it in at night. Note that the timing parameters can be *relative*. That is, if, for example, decision making is done at 5:00 p.m., then $\alpha_a = 4$ and $\beta_a = 8$ indicate 9:00 p.m. and 1:00 a.m., respectively. We note that the scheduling parameters need to be entered by the user, for example, using a keypad on the smart meter device, or through computer or a small cell phone using a secure remote access system, such as the one outlined in Chapter 18.

Given the predetermined parameters E_a, α_a, and β_a, in order to provide the needed energy for each appliance $a \in \mathcal{A}$ in times within the interval $[\alpha_a, \beta_a]$, it is required that

$$\sum_{h=\alpha_a}^{\beta_a} x_a^h = E_a \tag{3.2}$$

Furthermore, it is expected that $x_a^h = 0$ for any $h < \alpha_a$ and $h > \beta_a$ as no energy consumption is needed outside the time frame $[\alpha_a, \beta_a]$ for appliance a. We note that the time length $\beta_a - \alpha_a$ needs to be larger than or equal to the time duration required to finish the normal operation of appliance a.

All home appliances have certain *maximum* power levels denoted by $\gamma_a^{\max}$, for each $a \in \mathcal{A}$. For example, a PHEV may be charged only up to $\gamma_a^{\max} = 3.3$ kW per hour [1]. Some appliances may also have *minimum* stand-by power levels $\gamma_a^{\min}$, for each $a \in \mathcal{A}$. Therefore, the following lower and upper bound constraints are needed on the choices of the energy scheduling vector $\mathbf{x}_a$ for each appliance $a \in \mathcal{A}$:

$$\gamma_a^{\min} \le x_a^h \le \gamma_a^{\max}, \quad \forall\, h \in [\alpha_a, \beta_a] \tag{3.3}$$

We can now define a *feasible energy consumption scheduling set* $\mathcal{X}$ for the household of interest as

$$\mathcal{X} = \left\{ \mathbf{x} \,\middle|\, \sum_{h=\alpha_a}^{\beta_a} x_a^h = E_a \quad \text{and} \quad \gamma_a^{\min} \le x_a^h \le \gamma_a^{\max}, \quad \forall h \in [\alpha_a, \beta_a] \quad \text{and} \quad x_a^h = 0, \right.$$
$$\left. \forall h \in \mathcal{H} \setminus [\alpha_a, \beta_a] \right\}, \forall a \in \mathcal{A}$$

where $\mathbf{x} \triangleq (\mathbf{x}_a, \forall a \in \mathcal{A})$ denotes the vector of energy consumption scheduling variables for *all* appliances. An energy consumption schedule is valid if and only if $\mathbf{x} \in \mathcal{X}$. Clearly, the proper choice of $\mathbf{x}$ would depend on the electricity prices and should be determined by the ECS function in the user's corresponding smart meter. The resulting energy consumption schedule set by the ECS function is then applied to all household appliances in the form of on/off commands with specified power levels over a wired or wireless *home area network* as outlined in Chapters 6 and 7.

Let $l^h \triangleq \sum_{a \in \mathcal{A}} x_a^h$ denote the user's *total* hourly energy consumption at each upcoming hour $h \in \mathcal{H}$. Recall that H denotes the scheduling horizon. Let $p_h(l^h)$ denote the hourly price at hour h. The ECS function is expected to select the energy consumption schedules to solve the following problem:

$$\underset{x \in \mathcal{X}}{\text{minimize}} \quad \sum_{h=1}^{H} p^h \left(\sum_{a \in \mathcal{A}} x_a^h \right) \left(\sum_{a \in \mathcal{A}} x_a^h \right) \tag{3.4}$$

That is, the ECS function in the smart meter should schedule energy consumption for all household appliances such that while they are operating within their feasible energy consumption ranges—that is, they finish the job within the required time frame—the total household energy expense is minimized. Of course, the exact schedule depends on the choice of price functions. Here, we focus on a real-time pricing model combined with inclining block rates with three parameters a^h, b^h, $c^h \geq 0$ as follows:

$$p^h(l^h) = \begin{cases} a^h, & \text{if } 0 \leq l^h \leq c^h \\ b^h, & \text{if } l^h > c^h \end{cases} \tag{3.5}$$

Clearly, the price model in Eq. (3.5) is *not* a flat rate structure as the price depends on *time of day* and *total load*. Examples of real-time prices and inclining block rates were already shown in Figure 3.1.

Solving the optimization problem in Eq. (3.4), where the price functions are as in Eq. (3.5), is very difficult in its current form due to the nondifferentiability of inclining block rate price functions. Next, we explain how we can make this optimization problem more tractable. First, consider the illustration of the hourly payment $p^h\ l^h$ in Figure 3.5, where $p^h(l^h)$ is as in Eq. (3.5). Note that in the IBR model, we always have $a^h < b^h$. Therefore, the hourly payment is formed based on two *intersecting* lines:

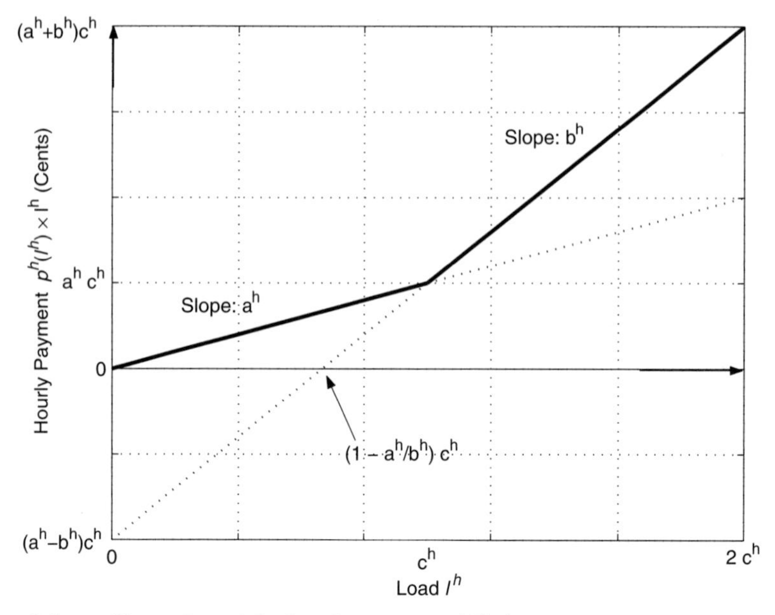

Figure 3.5. An illustration of the hourly payment $p^h(l^h)\ l^h$ with inclining block rates (solid curve). It can be modeled as the intersect of two straight lines (dotted lines). Note that we always have $a^h > b^h$ for all inclining block rates in Eq. (3.5).

$$\text{Payment} = a^h l^h \tag{3.6}$$

and

$$\text{Payment} = b^h l^h + (a^h - b^h)c^h \tag{3.7}$$

Therefore, for each $h \in \mathcal{H}$, we have

$$p^h(l^h)l^h = \max\{a^h l^h, b^h l^h + (a^h - b^h)c^h\} \tag{3.8}$$

The optimization problem in Eq. (3.4) with the model in Eq. (3.5) for the price functions can now be reformulated as

$$\underset{x \in \mathcal{X}}{\text{minimize}} \sum_{h=1}^{H} \max\left\{a^h \sum_{a \in \mathcal{A}} x_a^h, b^h \sum_{a \in \mathcal{A}} x_a^h + (a^h - b^h)c^h\right\} \tag{3.9}$$

However, problem (3.9) is still difficult to solve due to the max term in its objective function. Interestingly, we can introduce an *auxiliary* variable v^h for each $h \in \mathcal{H}$ and also add some new constraints such that we can remove the max terms and eventually rewrite problem (3.9) as

$$\begin{aligned} \underset{\substack{x \in \mathcal{X} \\ v^h, \forall h \in \mathcal{H}}}{\text{minimize}} \quad & \sum_{h=1}^{H} v^h \\ & a^h \sum_{a \in \mathcal{A}} x_a^h \le v^h, \qquad \forall\, h \in \mathcal{H} \\ & b^h \sum_{a \in \mathcal{A}} x_a^h + (a^h - b^h)c^h \le v^h, \quad \forall\, h \in \mathcal{H} \end{aligned} \tag{3.10}$$

We can easily prove by contradiction that problems (3.9) and (3.10) are equivalent [41, p. 130] and have exactly the same optimal solutions in terms of the scheduled energy consumptions. Interestingly, unlike problem (3.9), optimization problem (3.10) is linear and differentiable. Therefore, it can be solved efficiently by using various standard linear programming techniques [42]. In particular, the interior-point method [41, pp. 615–620] can be used to solve optimization problem (3.10) in a polynomial computation time. Therefore, the computational tasks that are required in practice in a smart meter's ECS function to optimally schedule a household's energy consumption are tractable and implementation is not difficult.

Next, we assess the performance of the designed ECS function via computer simulations. We especially want to investigate the resulting users' bill payments and the load PAR. Clearly, each user is interested in reducing its energy expenses while the utility is interested in load balancing with a low PAR. The simulation setting is as follows. We consider a single household with various appliances and assume that it has subscribed for the real-time pricing program adopted by the Illinois Power

Company. We focus on its consumption within a 4-month period from September 1, 2009 to December 31, 2009. In total, this duration includes 122 days. The number of appliances used in this household on each day are assumed to vary from 10 to 25. They include certain appliances with *fixed* consumption schedules, such as refrigerator-freezer (daily usage: 1.32 kWh), electric stove (daily usage: 1.89 kWh for self-cleaning and 2.01 kWh for regular), lighting (daily usage for 10 standard bulbs: 1.00 kW), and heating (daily usage: 7.1 kWh) [40], as well as some appliances with more flexible energy consumption scheduling requirements, such as dishwasher (daily usage: 1.44 kWh), clothes washer (daily usage: 1.49 kWh for energy-star 1.94 kWh for regular), clothes dryer (daily usage: 2.50 kWh), and PHEV (daily usage: 9.9 kWh) [1, 40]. We assume that the scheduling horizon $H = 24$. That is, the user solves optimization problem (3.10) to decide about her consumption for the next 24 hours. In addition, we combine IBR with RTP to increase the prices by 40% at the IBR higher block, where the higher block starts at the hourly consumption of 2.5 kWh.

The trends of daily electricity charges and PAR for a sample residential load based on the *day-ahead* real-time prices adopted by the Illinois Power Company from September 1 to December 31, 2009 are shown in Figures 3.6 and 3.7. From the results in Figure 3.6, by using the designed DSM scheme the user's average daily electricity bill payment decreases by 25% from 108 cents to 81 cents. This is equivalent to reducing the monthly electricity payment from \$32.4 to only \$24.3. On the

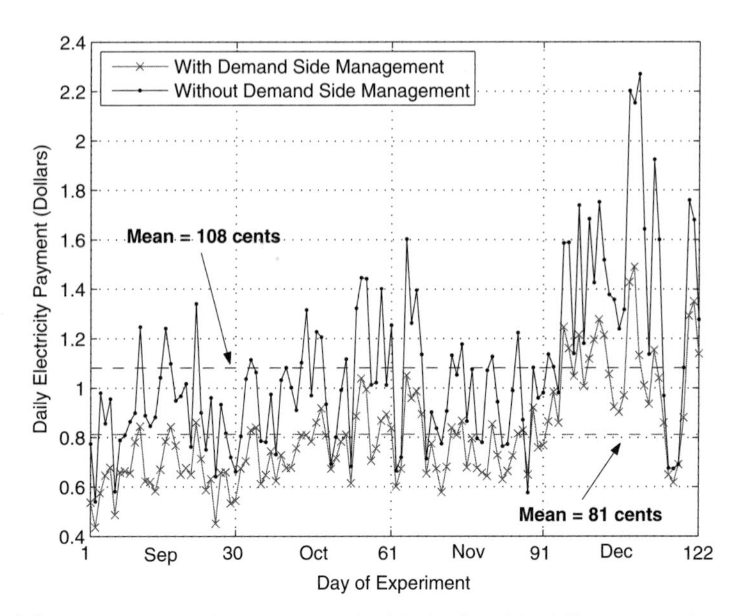

Figure 3.6. Simulation results on the trend of daily electricity bill payments for a sample residential load based on the pricing model adopted by the Illinois Power Company over a 4-month (122 days) experiment from September 1, 2009 to December 31, 2009.

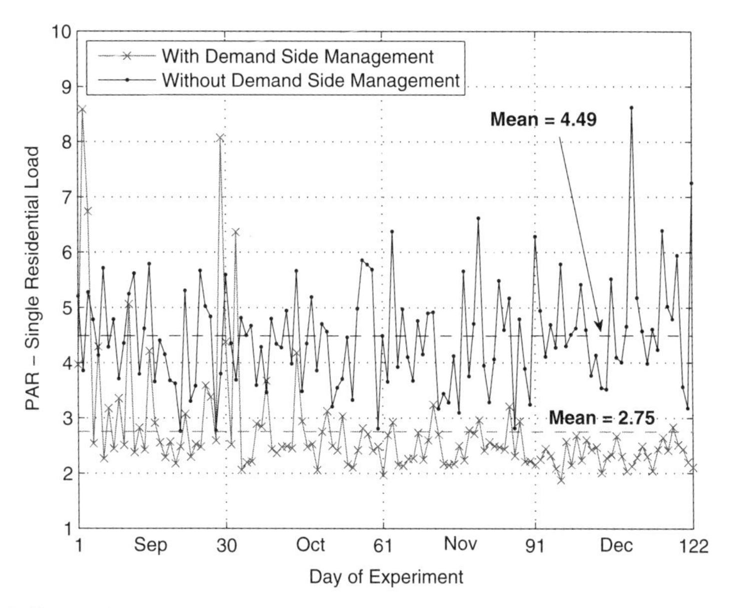

Figure 3.7. Simulation results on the trend of daily peak-to-average ratio in a sample residential load based on the pricing model adopted by the Illinois Power Company from September 1, 2009 to December 31, 2009. Load control is done using the design ECS function in a smart meter.

other hand, from the results in Figure 3.7, the average PAR in daily load is reduced by 38% from 4.49 to 2.75. Thus, the user pays less and produces a more balanced load. Similar trends are observed *almost* every day. Note that the spike in the PAR on September 29th occurred because a major load was scheduled at 4:00 a.m. which happened to have a *very* low price. However, we can reduce PAR by increasing the price at the higher block in IBR. For example, if the price at the higher block is 100% (instead of 40%) higher than that in the lower block, the PAR on September 29th reduces to only 2.3276. The results in this figure suggest that the deployment of the designed DSM structure is beneficial not only for the end user but also for the utility company.

Finally, we would like to add that the deployment of an ECS function can not only help reducing energy expenses via optimal energy consumption scheduling, but can also assist utility companies in their short-term planning process. In fact, one of the main challenges that the utility companies and regional control centers face is the need for predicting the demand load by end users. Clearly, by knowing the upcoming demand, the utility companies can better perform energy dispatching to accurately match supply and demand. Such predictions may only be done statistically in the current electric grid. However, by large deployments of the automatic residential load control strategies that we introduced in this section, the end users will be able to announce their upcoming load back to the utility company through appropriate message passing. More precisely, the end user can send its total *upcoming* daily load $\mathbf{l} \triangleq [l^1, \ldots, l^H]$ as a control message to the utility. Given the expected load from *all* users, the utility company would have an accurate estimation of the load that it needs to provide within the next couple of hours. Therefore, the designed load control structure can not only help users to better respond to real-time prices, but it can also potentially enable the utilities to have a more accurate idea of how much energy is and will be consumed by residential users.

3.5 PRICE PREDICTION

Given the day-ahead prices, the ECS function can make optimal decisions on energy consumption scheduling for several appliances to minimize the household's energy expenses. However, in practice and particularly when real-time pricing tariffs are being used by the utilities, users may know the price values only for the next few (e.g., one or two) hours. This makes decision making very difficult, as in that case the ECS functions need to limit their decision horizon. Alternatively, smart meters can employ efficient *price prediction* schemes such that they can estimate the upcoming price values and use the estimated prices in their decision making process, that is, in formulating and solving optimization problem (3.4).

In general, price parameters may depend on several factors. In particular, they depend on the wholesale market prices, which are not easy to predict themselves. Nevertheless, it is usually expected that the prices are higher during the afternoon, on hot days in summer, and on cold days in winter [27]. Furthermore, one may expect that prices vary depending on working days or weekends. These pieces of information can help in predicting the prices in an RTP environment. While we are interested in accurate predictions, our main focus here is to develop price predictors that have *low* computational complexity and can be implemented in residential smart meter along with the ECS function.

Next, we use the prices adopted by the Illinois Power Company from January 2007 to December 2009 to evaluate different factors that may affect RTP. The results are shown in Figure 3.8. First, we plot the average hourly prices across all days of the year for the years 2007, 2008, and 2009 in Figure 3.8a. We can see that although the trends are partially (not exactly) similar in different years, the exact prices can be drastically different. This can be due to major yearly fluctuations in the wholesale prices, for example, due to changes in the international price of oil. Therefore, making an accurate prediction based on the price values in the previous years does *not* seem feasible. Next, we plot the *correlation* among the hourly prices today with those in the previous days in Figure 3.8b. From the results in this figure, we can see that there is a very high correlation (about 0.84) between the prices today and those yesterday. The correlation decreases as we go further back. However, there is also a noticeable correlation (about 0.67) between the prices today and those on the same day last week. Our analysis further shows that there are *not* major differences in average prices from Monday to Friday, except for slight price reduction on Friday.

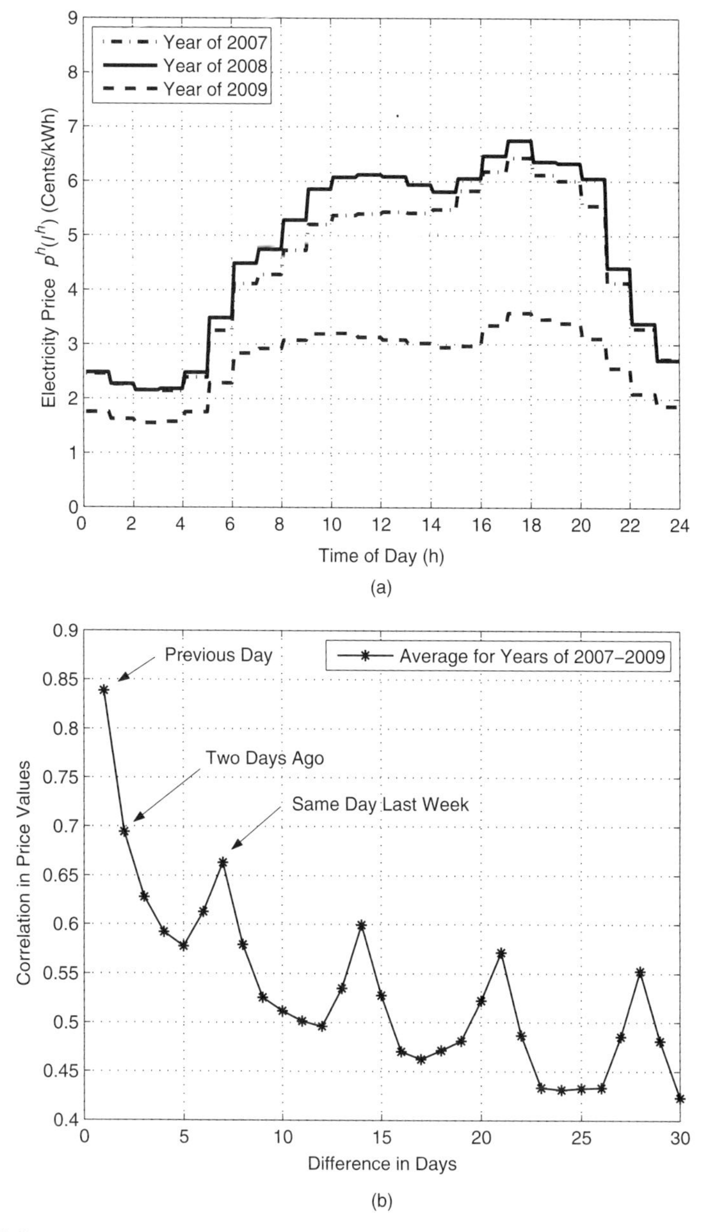

Figure 3.8. Statistical analysis of the real-time prices used by the Illinois Power Company from January 1, 2007 to December 31, 2009: (a) different years and (b) correlation with past prices.

However, the prices on Saturday and Sunday are much less. This suggests that there can be relationships between the upcoming prices and whether the prices are for a working day or for a weekend.

In summary, the observations in Figure 3.8 suggest that an efficient prediction is likely achieved by looking at the prices on *yesterday*, *the day before yesterday*, and *the same day last week*. Let $\hat{a}^h[t]$, $\hat{b}^h[t]$, and $\hat{c}^h[t]$ denote the *predicted* parameters for the upcoming price tariff in Eq. (3.5) for each hour $h \in \mathcal{H}$ on day t. One possible way to predict the price parameter $a^h[h]$ is to use the following simple model [13]:

$$\hat{a}^h[t] = k_1 a^h[t-1] + k_2 a^h[t-2] + k_7 a^h[t-7], \quad \forall\, h \in \mathcal{H} \tag{3.11}$$

Here, $a^h[t-1]$, $a^h[t-2]$, and $a^h[t-7]$ denote the previous values of parameter a^h on yesterday, the day before yesterday, and the same day last week, respectively. In this regard, the expression in Eq. (3.11) formulates a weighted average price predictor filter with coefficients k_1, k_2, and k_7. Similar models can be obtained for predictions of parameters $\hat{b}^h[t]$ and $\hat{c}^h[t]$. However, in practice, the threshold parameter $c^h[t]$ is usually fixed and does not change on a daily basis. It may only change once or twice a year at different seasons. Of course, more complicated price prediction models can also be used. For example, one may include several extra terms in the model in Eq. (3.11), possibly leading to more accurate predictions, but yet at the cost of more computational complexity to obtain the right prediction parameters.

Using minimum squared error prediction, the optimal choices of the prediction filter coefficients at different days of the week are obtained as follows. On Monday, we have $k_1 = 0.355$, $k_2 = 0.465$, and $k_7 = 0.359$. On Tuesday, we have $k_1 = 0.858$, $k_2 = 0$, and $k_7 = 0.126$. On Wednesday, we have $k_1 = 0.837$, $k_2 = 0$, and $k_7 = 0.142$. On Thursday, we have $k_1 = 0.943$, $k_2 = 0$, and $k_7 = 0.050$. On Friday, we have $k_1 = 0.868$, $k_2 = 0$, and $k_7 = 0.092$. On Saturday, we have $k_1 = 0.671$, $k_2 = 0$, and $k_7 = 0.196$. On Sunday, we have $k_1 = 0.719$, $k_2 = 0$, and $k_7 = 0.184$. We can see that the coefficients are significantly different on Monday compared to the other days of the week. This is because, unlike all the other days, there is a low correlation between the prices on Monday and the day before, due to lower prices on weekends. For all other days, the prediction is only based on the prices on the last day and the prices on the same day last week. In fact, for the cases when the prices are highly correlated with those yesterday, there is really no need to know the price values the day before yesterday. Using the coefficients mentioned above, we observed that the prediction error can be as low as only 13% on average. We also note that the proposed prediction structure has minimum computation complexity and its implementation only requires a lookup table for the daily coefficients, a limited memory, and the capability of performing simple arithmetic operations. Price prediction accuracy can further improve by using longer and more computationally complicated price prediction filters, if needed. Given the announced prices by the utility and the predicted prices in longer horizons, the ECS function can follow the same steps explained in Section 3.4 and achieve the best choice for the energy consumption schedule for each appliance.

3.6 MANAGING USER-SIDE STORAGE AND GENERATION

In traditional power systems, the end users were seen as pure energy consumers. In such systems, the electricity is essentially generated in large power plants and users are not expected to make any contribution to power generation. However, there is evidence suggesting that the users in the future Smart Grid will no longer be pure energy consumers. In fact, with the growing interest in using plug-in hybrid electric vehicles, users will be able to use their electric car batteries as small electricity storage units. Such storage capacities can be deployed by users, for example, as a back up power source, or be offered to the grid to contribute in providing various ancillary services [11, 43, 44]. Furthermore, users are expected to also have some power generation capabilities with focus on renewable energy sources. For example, there are noticeable efforts to build efficient and inexpensive *rooftop solar and wind power generation* units for residential customers [45–48]. As distributed power generation and storage become more popular in the future Smart Grid, users will gradually transform from pure consumers into small consumption, storage, and generation units. Having a large number of end users with storage and generation capacities can provide several new opportunities to the grid, for example, by offering ancillary services or integrating renewable energy generation. However, control and management of such a large-scale, complex, and multiplayer system is very difficult and cannot be done by traditional centralized control approaches. Therefore, it is foreseen that there will be a need for developing new and advanced demand-side management programs that focus not only on energy consumption scheduling and load shaping but also on coordinating the aggregated storage and generation capacity of the future end users.

Residential electricity storage and generation can be taken into account in demand-side management by extending the resource management role of smart meters. As one option, one can treat the electricity that is sent back to the grid as *negative* energy consumption. In that case, the ECS function introduced in Section 3.4 can be further expanded to also decide on the amount of power and the timing of selling electricity into the grid. Alternatively, we can equip smart meters with not only ECS functions, but also *energy storage scheduling* (ESS) and *energy generation scheduling* (EGS) functions. The ESS function would decide on when to charge or discharge the PHEVs' batteries and other storage units to participate in frequency regulation or other grid services as outlined in Chapter 2. The decisions are made based on the price of buying and selling electricity, which can change in real-time. The EGS function decides on when to turn on or turn off the standby and renewable generators, for example, when the price of selling electricity is very low. Together, the ECS, ESS, and EGS functions *optimize* the operation of each user in terms of energy consumption, storage, and generation scheduling. This is an interesting area that can be explored further in the future. Interested readers can refer to Han et al. [11], Brooks and Gage [49], Suryanarayanan et al. [50], Saber and Venayagamoorthy [51], and Morren et al. [52] for more details.

3.7 CONCLUSION

With the increase in load demand and given the limited energy resources and the lengthy and costly process of building new power plants, it is essential to improve the utilization of the power grid by reducing the peak-to-average ratio in load demand through deployment of advanced demand-side management strategies. In this chapter, we explored the possibility of developing an autonomous and distributed demand-side management scheme by equipping each smart meter with an intelligent and automated energy consumption scheduling function which schedules energy consumption for each household appliance in response to changes in electricity prices to minimize the household energy expenses. Simulation results show that the proposed design approach can significantly reduce users' electricity bills. In addition, it can reduce the peak-to-average ratio in load demand, making it also attractive to utilities. We also explored the possibility of extending the studied architecture for smart meters to also manage user-side storage and generation capacities, which are foreseen to exist in future Smart Grid systems. This is achieved by further introducing the energy storage scheduling and energy generation scheduling functions.

REFERENCES

[1] A. Ipakchi and F. Albuyeh, "Grid of the future," *IEEE Power Energy Mag.*, pp. 52–62, March 2009.

[2] A. Vojdani, "Smart integration," *IEEE Power Energy Mag.*, pp. 72–79, November 2008.

[3] *The Smart Grid: An Introduction.* U.S. Department of Energy, 2009.

[4] L. H. Tsoukalas and R. Gao, "From smart grids to an energy internet: Assumptions, architecrures, and requirements," *Third International Conference on Electric Utility Deregulation and Restructuring and Power Technologies*, April 2008.

[5] G. M. Masters, *Renewable and Efficient Electric Power Systems.* Wiley, 2004.

[6] B. Ramanathan and V. Vittal, "A framework for evaluation of advanced direct load control with minimum disruption," *IEEE Trans. Power Syst.*, 23(4), pp. 1681–1688, November 2008.

[7] C. W. Gellings and J. H. Chamberlin, *Demand Side Management: Concepts and Methods*, 2nd ed. PennWell Books, 1993.

[8] M. A. A. Pedrasa, T. D. Spooner, and I. F. MaxGill, "Scheduling of demand side resources using binary particle swarm optimization," *IEEE Trans. Power Syst.*, 24(3), pp. 1173–1181, August 2009.

[9] A. H. Mohsenian-Rad, V. Wong, J. Jatskevich, R. Schober, and A. Leon-Garcia, "Autonomous demand-side management based on game-theoretic enegry consumption scheduling for the future smart grid," *IEEE Trans. Smart Grid*, 1, pp. 320–331, December 2010.

[10] P. Samadi, A. H. Mohsenian-Rad, R. Schober, V. Wong, and J. Jatskevich, "Coordination of cloud computing and smart power grids," *IEEE Conference on Smart Grid Communications*, Gaithersburg, MD, October 2010.

[11] S. Han, S. Han, and K. Sezaki, "Development of an optimal vehicle-to-grid aggregator for frequency regulation," *IEEE Trans. Smart Grid*, 1(1), pp. 65–72, June 2010.

[12] A. Ipakchi and F. Albuyeh, "Grid of the future," *IEEE Power Energy Mag.*, pp. 52–62, March 2009.

[13] A. H. Mohsenian-Rad and A. Leon-Garcia, "Optimal residential load control with price prediction in real-time electricity pricing environments," *IEEE Trans. Smart Grid*, 1, pp. 120–133, 2010.

[14] "Demand response program evaluation final report," Quantum Consulting Inc. and Summit Blue Consulting, LLC Working Group 2 Measurement and Evaluation Committee and California Edison Company, April 2005.

[15] M. Ann-Piette, G. Ghatikar, S. Kiliccote, D. Watson, E. Koch, and D. Hennage, "Design and operation of an open, interoperable automated demand response infrastructure for commercial buildings," *J. Computing and Information Sci. Eng.*, 9, pp. 1–9, June 2009.

[16] N. Ruiz, I. Cobelo, and J. Oyarzabal, "A direct load control model for virtual power plant management," *IEEE Trans. Power Syst.*, 24(2), pp. 959–966, May 2009.

[17] A. Gomes, C. H. Antunes, and A. G. Martins, "A multiple objective approach to direct load control using an interactive evolutionary algorithm," *IEEE Trans. Power Syst.*, 22(3), pp. 1004–1011, August 2007.

[18] D. D. Weers and M. A. Shamsedin, "Testing a new direct load control power line communication system," *IEEE Transactions on Power Delivery*, 2(3), pp. 657–660, July 1987.

[19] C. M. Chu, T. L. Jong, and Y. W. Huang, "A direct load control of air-conditioning loads with thermal comfort control," *Proceedings of IEEE PES General Meeting*, San Francisco, CA, June 2005.

[20] Y. Y. Hsu and C. C. Su, "Dispatch of direct load control using dynamic programming," *IEEE Trans. Power Syst.*, 6(3), pp. 1056–1061, August 1991.

[21] B. Ramanathan and V. Vittal, "A small signal stability performance boundary for direct load control," *Proceedings of IEEE PES Power Systems Conference and Exposition*, New York, NY, October 2004.

[22] OpenHAN Task Force of the Utility AMI Working Group, *Home Area Network System Requirements Specification*, August 2008.

[23] P. Hoffert, "Automated homes and offices on the infoway," CulTech Collaborative Research Centre, York University, Toronto, Canada, Technical Report, November 1994.

[24] K. Herter, "Residential implementatin of critical-peak pricing of electricity," *Energy Policy*, 35, pp. 2121–2130, 2007.

[25] C. Triki and A. Violi, "Dynamic pricing of electricty in retail markets," *Q. J. Operations Res.*, 7(1), pp. 21–36, March 2009.

[26] P. Centolella, "The integration of price responsive demand into regional transmission organization (RTO) wholesale power markets and system operations," *Energy*, 35(4), pp. 1568–1574, 2010.

[27] H. Allcott, "Real time pricing and electricity markets," Working Paper, Harvard University, February 2009.

[28] P. Reiss and M. White, "Household electricity demand, revisited," *Rev. Econ. Studies*, 72(3), pp. 853–883, July 2005.

[29] A. H. Mohsenian-Rad, V. Wong, J. Jatskevich, and R. Schober, "Optimal and autonomous incentive-based energy consumption scheduling algorithm for smart grid," *Proceedings of IEEE PES Conference on Innovative Smart Grid Technologies*, Gaithersburg, MD, January 2010.

[30] S. Borenstein, "Equity effects of increasing-block electricity pricing," Center for the Study of Energy Markets, Working Paper 180, November 2008.

[31] BC Hydro, *Electricity Rates*, 2009.

[32] S. Caron and G. Kesidis, "Incentive-based energy consumption scheduling algorithms for the smart grid," *Proceedings of IEEE International Conference on Smart Grid Communications (SmartGridComm'10)*, Gaithersburg, MD, October 2010.

[33] C. Ibars, M. Navarro, and L. Giupponi, "Distributed demand management in smart grid with a congestion game," *Proceedings of IEEE International Conference on Smart Grid Communications (SmartGridComm'10)*, Gaithersburg, MD, October 2010.

[34] M. Toney, "Panel presentation at the IEEE PES conference on innovative smart grid technologies," Gaithersburg, MD, January 2010.

[35] M. Ann-Piette, G. Ghatikar, S. Kiliccote, D. Watson, E. Koch, and D. Hennage, "Design and operation of an open, interoperable automated demand response infrastructure for commercial buildings," *J. Computing and Information Sci. Eng.*, 9, pp. 1–9, June 2009.

[36] S. Tiptipakorn and W. J. Lee, "A residential consumer-centered load control strategy in real-time electricity pricing environment," *Proceedings of North American Power Symposium*, Las Cruces, NM, October 2007.

[37] R. Davies, "Hydro one's smart meter initiative paves way for defining the smart grid of the future," *Proceedings of IEEE Power and Energy Society General Meeting*, Calgary, Alberta, July 2009.

[38] N. C. F. Tse, J. Y. C. Chan, and L. L. Lai, "Development of a smart metering scheme for building smart grid system," *Proceedings of IEEE International Conference on Advances in Power System Control, Operation and Management (APSCOM'09)*, Hong Kong, China, November 2009.

[39] X. Bai, J. Meng, and N. Zhu, "Functional analysis of advanced metering infrastructure in smart grid," *Proceedings of IEEE International Conference on Power System Technology (POWERCON'10)*, Hangzhou, China, October 2010.

[40] Office of Energy Efficiency, Natural Resources Canada, *Energy Consumption of Major Household Appliances Shipped in Canada*, 2005.

[41] S. Boyd and L. Vandenbergher, *Convex Optimization*. Cambridge University Press, 2004.

[42] D. Bertsimas and J. N. Tsitsiklis, *Introduction to Linear Optimization*. Athena Science, 1997.

[43] A. N. Brooks, "Vehicle-to-grid demonstration project: Grid regulation ancillary service with a battery electric vehicle," AC Propulsion Inc., San Dimas, CA, December 2002. Available at http://www.smartgridnews.com/artman/uploads/1/sgnr_2007_12031_001.pdf.

[44] C. Quinn, D. Zimmerle, and T. H. Bradley, "The effect of communication architecture on the availability, reliability, and economics of plug-in hybrid electric vehicle-to-grid ancillary services," *J. Power Sources*, 195(5), pp. 1500–1509, March 2010.

[45] F. Giraud and Z. M. Salameh, "Steady-state performance of a grid-connected rooftop hybrid wind-photovoltaic power system with battery storage," *Proceedings of IEEE Power Engineering Society Winter Meeting*, Columbus, OH, January 2001.

[46] N. Mithraratne, "Roof-top wind turbines for microgeneration in urban houses in New Zealand," *Energy and Buildings*, 41(10), pp. 1013–1018, October 2009.

[47] K. Sedghisigarchi, "Residential solar systems: Technology, net-metering, and financial payback," *Proceedings of IEEE Electrical Power and Energy Conference (EPEC'09)*, Montreal, Quebec, October 2009.

[48] Y. Gurkaynak, Z. Li, and A. Khaligh, "A novel grid-tied, solar powered residential home with plug-in hybrid electric vehicle (PHEV) loads," *Proceedings of IEEE Vehicle Power and Propulsion Conference (VPPC'09)*, Dearborn, MI, 2009.

[49] A. Brooks and T. Gage, "Integration of electric drive vehicles with the electric power grid, a new value stream," *Proceedings of the International Electric Vehicle Symposium and Exhibition*, Berlin, Germany, October 2001.

[50] S. Suryanarayanan, F. Mancilla-David, J. Mitra, and Y. Li, "Achieving the smart grid through customer-driven microgrids supported by energy storage," *Proceedings of IEEE International Conference on Industrial Technology (ICIT'10)*, Vi a del Mar, Chile, March 2010.

[51] A. Y. Saber and G. K. Venayagamoorthy, "Unit commitment with vehicle-to-grid using particle swarm optimization," *Proceedings of IEEE PowerTech*, Bucharest, Romania, July 2009.

[52] J. Morren, J. Meeuwsen, and H. Slootweg, "New network design standards for the grid connection of large concentrations of distributed generation," *Proceedings of the IEEE International Conference on Electricity Distribution*, Prague, Czech Republic, June 2009.

4

POWER ELECTRONICS FOR MONITORING, SIGNALING, AND PROTECTION

Wilsun Xu

4.1 INTRODUCTION

Power electronics technology has gained widespread acceptance in the power industry. Circuits constructed using power electronic components have been adopted to facilitate the transmission and control of electric power. Examples include *high-voltage, direct current* (HVDC) transmission schemes, and *flexible alternating current transmission systems* (FACTSs) that have been introduced in Chapter 1, as well as custom power devices. In all of these applications, power electronics are used to manipulate the electric energy for power transmission and conversion purposes.

Paralleling with these developments, researchers have also utilized power electronics to create controllable disturbances on power system voltage and current waveforms. Such small but discernible disturbances are intended to represent signals and to carry information. They have been used for power line communication, online equipment monitoring, and other information-oriented applications. Although the

Smart Grid: Applications, Communications, and Security, First Edition.
Edited by Lars Torsten Berger and Krzysztof Iniewski.

developments in this field have been sporadic, using power electronics to create signals on high-voltage systems represents a huge opportunity for innovation. All these developments can be categorized into one new area of research and development—power electronic signaling technology—that is, using power electronics to manipulate high-voltage and current waveforms for signal and information oriented applications.

The objective of this chapter* is to survey and review the developments in this fascinating field. Several highly successful power electronic signaling technologies and their applications are illustrated. They are grouped into three areas—power line communication, condition monitoring, and active protection. These examples demonstrate that power electronic signaling technology could be a major source of innovation for the sensing and signaling aspects of the Smart Grid.

4.2 POWER LINE COMMUNICATION

Many power line communication (PLC) systems, discussed in more detail in Chapter 7, use electronic circuits for signal creation. However, due to the low voltage ratings of electronic components, such PLC systems are complicated. With the help of power electronics, it has become possible to "inject" signals into primary power distribution lines directly, resulting in much simpler and robust PLC technologies. A general principle of power electronics based signal generation is as follows: a thyristor or similar device is used to create a waveform disturbance such as a very small but detectable voltage sag. The existence of the sag implies digital "1" and no sag implies digital "0." In this section, several methods of creating signals on medium voltage power lines are reviewed. Some of the methods have become very successful automatic meter reading technologies.

4.2.1 Zero-Crossing Shift Technique

One of the early power line signaling technologies creates a slight shift on the selected zero-crossing points of the carrier voltage waveform [1]. The existence of the shift would represent digital "1" and no shift digital "0." This technique is intended to generate outbound signals propagating from utility substations to customers connected to distribution feeders. The signaling is achieved using a simple circuit shown in Figure 4.1a. The circuit consists of a shunt arm and a series arm, and each arm comprises a pair of antiparallel connected thyristors. When no signaling occurs, gates 3 and 4 are continuously triggered to allow passage of the current

*Most materials in this chapter are adopted from W. Xu and V. Wang,"Power electronic signaling technology—a new class of power electronics Applications," *IEEE Trans. Smart Grid*, 1(3), pp. 332–340, December 2010, with permission of IEEE, ©2010 IEEE.

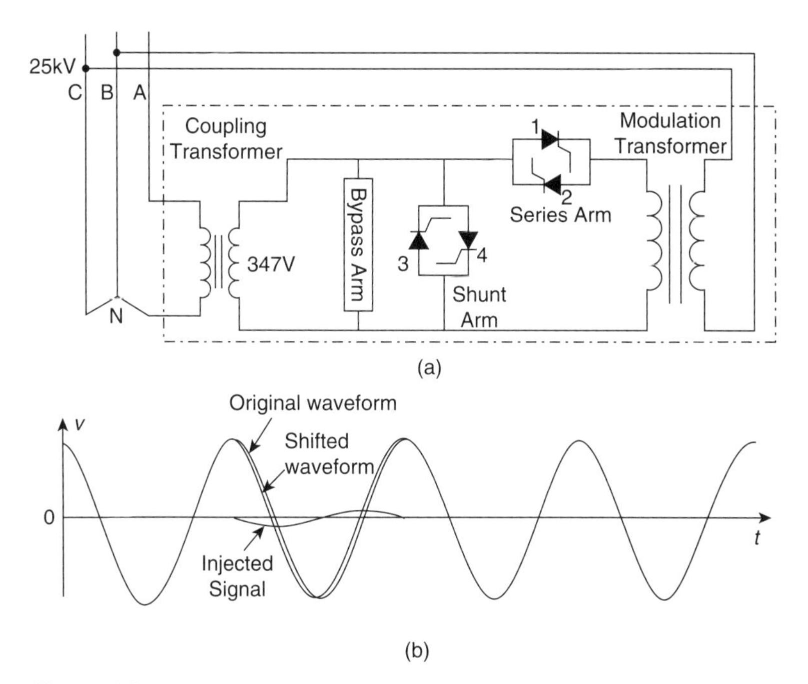

Figure 4.1. Scheme to create a zero-crossing shift disturbance. ©2010 IEEE.

in phase A while gates 1 and 2 are off. In the signaling state, gates 3 and 4 are de-energized but gates 1 and 2 are activated to inject a single-cycle voltage into phase A. The injection voltage is taken from phases B and C. The signal containing voltage waveform is shown in Figure 4.1b. To detect the presence of the signal, one can compare the durations of the half-cycle waveforms. The presence of time difference implies the existence of a signal. One of the main disadvantages of this technology is that the circuit is relatively more complex in comparison with the newer technologies such as the one described next.

4.2.2 Waveform Distortion Technique

Another way to create signals is to cause waveform sags around the zero-crossing point of a voltage waveform. TWACS (two-way automatic communication system) is a representative technology of this scheme. This technology consists of outbound and inbound signaling schemes. The technique of creating outbound signals is surprisingly simple and elegant [2, 3]. As shown in Figure 4.2a, a thyristor (or a pair of thyristors) is shunt connected through a transformer to the MV bus of a substation. The thyristor is triggered ahead of its voltage zero-crossing point and conducts for a short period. During the conduction, which is a controlled short-circuit, a small voltage sag is created near the zero-crossing point. The existence of this sag represents a signal. As it is a voltage disturbance, the sag is experienced by all load points

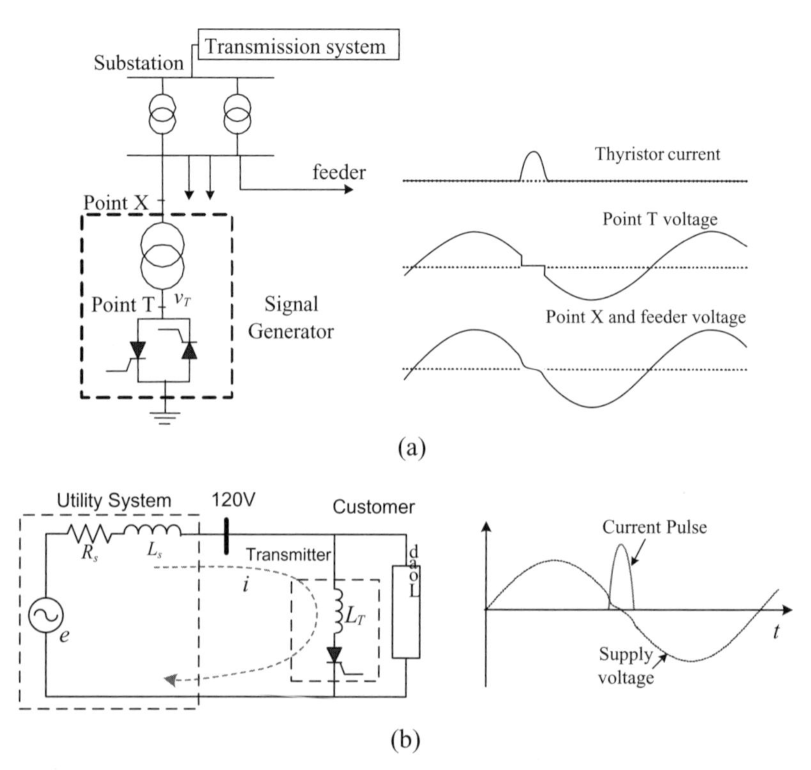

Figure 4.2. Scheme to create a small voltage sag at the zero-crossing point: (a) outbound and (b) inbound signaling. ©2010 IEEE.

downstream of the thyristor location. So information can be sent to various load points this way. A significant feature of this signaling scheme is that the thyristor is fired in such a way that the voltage disturbance exists in only one of two consecutive cycles. The signal is extracted by subtracting two consecutive cycles of voltage waveforms. The difference is the distortion voltage (i.e., the signal) if it exists. This subtraction method is extremely powerful since it filters out the background harmonics in the voltage waveforms. The scheme is therefore essentially immune to (steady-state) background waveform distortions. Only power electronic devices can achieve such a controlled perturbation of voltage waveforms.

The technique of creating an inbound signal that travels from the load points to the substation is also based on the same principle. In this case, a thyristor is connected to the service voltage of a load point (Figure 4.2b). Once triggered near the voltage zero-crossing point, a short-circuit is created and a pulse current is drawn from the supply system. This pulse current can be detected at the substation as information carrier. In typical applications, the pulse current has a peak of 150 A at the 120 V level and a width about 90°. At the 25 kV level, the pulse has a peak of 1.25 A. Although this disturbance is extremely small in comparison with the normal feeder current of 100–200 A, modern signal processing technology has no problem

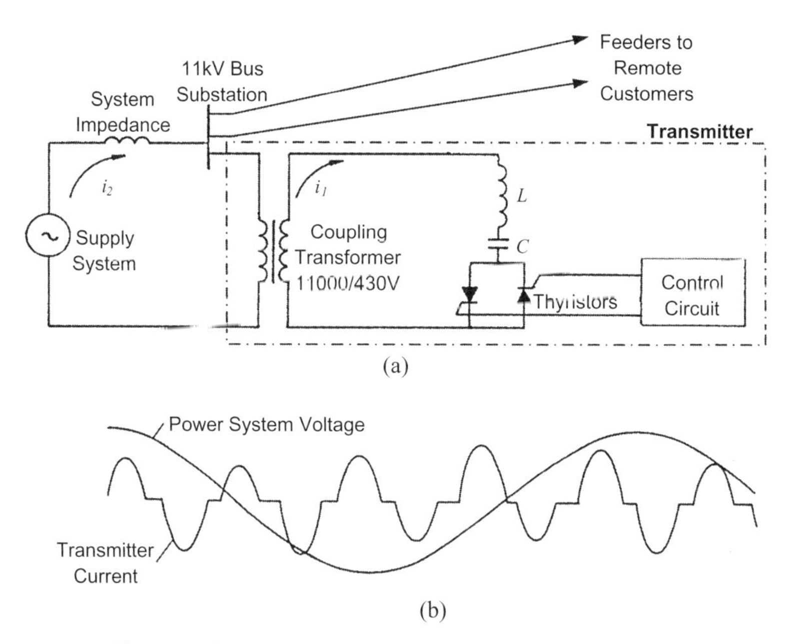

Figure 4.3. Scheme to create a ripple signal. ©2010 IEEE.

to extract it. According to the usage statistics, the TWACS technology seems to be the most widely used PLC technology for automatic meter readings. The main disadvantage of TWACS is its low bit rate since 60 cycles (1 second) can only carry about 30 bits.

4.2.3 Ripple Signaling Technique

The ripple control based signaling scheme is probably the most widely known PLC technique. It has been used for selective load control in distribution networks [4–8]. A ripple signal is essentially a high-frequency waveform superimposed on 60 Hz or 50 Hz carrier waveforms. There are many methods to create such a signal. One of the methods that uses a power electronic device is shown in Figure 4.3.

The signal generator or transmitter consists of an inductor in series with a capacitor and a pair of thyristors [4]. The thyristors are alternately fired at the desired signal frequency. Once a thyristor is fired, a current pulse is drawn from the supply, which goes to zero before the other thyristor is fired, because the L-C resonance frequency is tuned to be 15–25% higher than the signal frequency. The thyristor-off period between thyristor conductions enables the L-C circuit to receive energy from the supply and maintain oscillations. The transmitter oscillatory current (see Figure 4.3b) produces a voltage distortion on the substation secondary bus. This voltage signal propagates along the distribution feeders and can be detected as a load control signal. Galloway [5] shows a ripple generation apparatus using a thyristor-based inverter. Besides thyristors, power transistors such as IGBT and MOSFET have also been used for creating subharmonic ripple signals [6]. In comparison with the TWACS technology, the ripple signaling techniques involve higher frequencies and there is a higher chance to encounter resonance situations that may prevent the signals from traveling further downstream. Note that the ripple signals can also be created using electronic components involving transistors, amplifiers, and coupling transformers. Many ripple signal based PLC technologies are based on such designs [7, 8]. In this chapter, however, we don't include such technologies as power electronic signaling technology since they are essentially electronic circuits and operate at low voltage levels such as 12 V. They cannot couple into the power systems directly in comparison with the power electronic devices. The broadband on power line communication technology [9] belongs to the same category as the non-power-electronics based devices.

4.2.4 Summary

Some of the above signaling techniques have found a number of highly successful applications. Examples are load control, automatic meter reading, and feeder capacitor control. A common feature of these techniques is that the power electronic devices are utilized to create controlled disturbances (i.e., signals). The signal creation methods can be classified into two types, series and shunt. The series type directly inserts a voltage disturbance signal into the main circuit. An example is the zero-crossing shift technique. The shunt type creates a controlled short-circuit which leads to information carrying voltage sags or current pulses. A significant difference between these technologies and electronic based technologies is the direct injection of signals into the power circuit (voltage > 120 V) and, at present, only power electronic circuits can meet this requirement. Potential research directions in this area could include expanding the signal channels, increasing the bit rate, and direct coupling to MV circuits.

4.3 CONDITION MONITORING AND FAULT DETECTION

Equipment condition monitoring has become an area of significant research and development in recent years. One large class of condition monitoring techniques involves injecting some forms of perturbations or disturbances into operating equipment. Since common power equipment deserving monitoring generally operates at high voltages (i.e., above the voltage level acceptable to electronic devices), power electronic circuits have become a natural choice for creating the desired disturbance or signal for active condition monitoring. Similarly, the signals injected by the power electronic circuits can also be used to detect faults in the power system. If a fault exists in a system, the system response to the injected signals will be different from that of a sound system. The fault can therefore be detected.

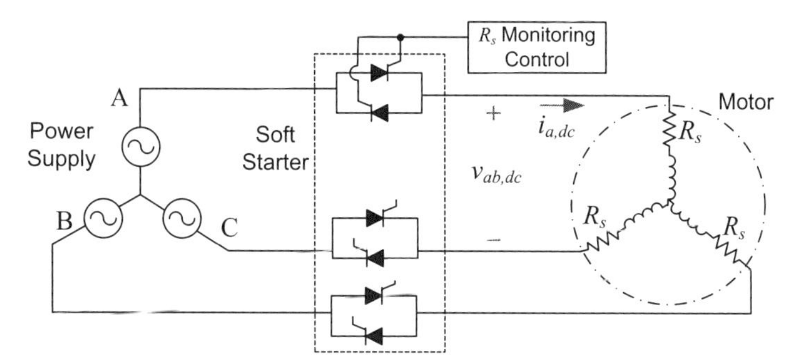

Figure 4.4. Scheme for online motor stator resistance monitoring. ©2010 IEEE.

4.3.1 Online Motor Thermal Protection

One of the representative techniques in the active condition monitoring area is the estimation of motor temperature through the measurement of motor stator resistance R_s. The idea is based on the understanding that $R_s$varies proportionally to stator winding temperature. The best way to estimate the resistance accurately is to estimate its DC value. This requires the injection of a DC current into the motor stator winding.

Zhang et al. [10] propose a method that utilizes the soft starter to inject DC components into the motor windings. As shown in Figure 4.4, phase A of the soft starter, which is a pair of antiparallel thyristors is used for this purpose. This starter stays in the circuit while those in the other two phases are bypassed. A short delay is introduced to the gate drive signal of one of the phase A thyristors. Such a short delay is enough to create a DC component in the motor current. Based on the equivalent DC circuit model of the induction motor, the motor stator resistance R_s can be estimated from the terminal voltages and currents using

$$R_s = \frac{2v_{ab,dc}}{3i_{a,dc}} \tag{4.1}$$

The motor torque pulsation caused by the DC disturbance can be kept within an acceptable level by using a sufficiently small delay angle. Since the thyristor firing can easily be controlled, the DC signal level can be adjusted adaptively to facilitate accurate estimation of R_s. In the above scheme, the signaling circuit is connected in series with the power line. It is more desirable to use a shunt circuit for DC generation, as the failure of a shunt signaling circuit is unlikely to cause the shutdown of the motor under monitoring.

For low voltage motors, power MOSFET has been proposed for DC current injection without using the soft starter [11]. As more and more sophisticated power electronic devices and circuits are becoming available, there are plenty of opportunities to develop alternative low-cost DC injection circuits for the motor resistor monitoring application.

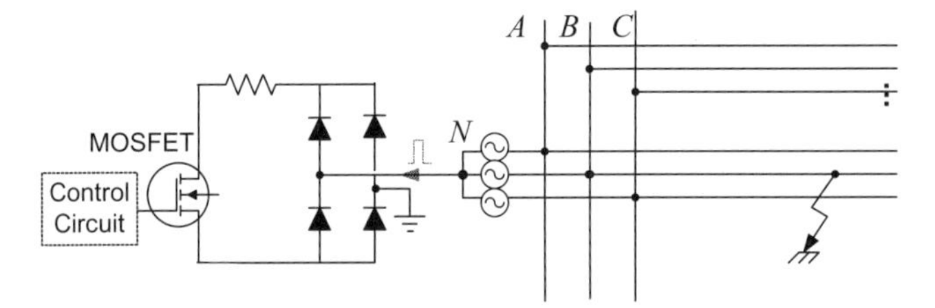

Figure 4.5. A pulsated ground fault detection device. ©2010 IEEE.

4.3.2 Faulted Line Identification in Ungrounded Systems

The power distribution networks in some European and Asian countries, and some industrial power systems in North America, are ungrounded, high-resistance grounded, or resonant grounded. These systems can continue to operate when a single-phase-to-ground fault occurs. Later, when the fault is located, it can be cleared at a convenient time, resulting in minimized operation interruptions. However, identifying the faulted feeder among a number of feeders connected to the same bus is difficult since the fault current is very small. Various techniques have been proposed to solve this problem. Among them, the power electronic based active detection methods seem to be the most promising.

One of the methods works by grounding the distribution system in a pulsating fashion such that a pulsating ground current is produced in the ground fault path [12]. As shown in Figure 4.5, the pulsed grounding circuit comprises a power transistor, such as a power MOSFET, connected in series with a limiting resistor between the neutral and the ground. A multivibrator circuit alternately turns the MOSFET on and off. A diode rectifier allows an AC fault current to flow through the MOSFET when a ground fault exists in the system. This current can be detected by a sensing circuit due to the current's flickering feature. This fault detection circuit is also operable in ungrounded DC circuits, and the rectifier is no longer necessary in this case.

Wang et al. [13] proposed to temporarily convert a noneffectively grounded system into a grounded system using a thyristor-controlled grounding device connected between the neutral and ground (Figure 4.6). A large fault current pulse will thus appear if a ground fault exists in the circuit. This fault current mainly flows through the low-impedance path composed of the faulted line, the fault point, and the ground. By detecting the existence of a current pulse using the CTs installed at the terminal of each outgoing feeder, the faulted line can be identified.

This controlled grounding of system neutral is achieved with a thyristor or a pair of thyristors shown in Figure 4.6. The firing frequency (e.g., one firing per 5 cycles), firing pattern (e.g., one positive pulse followed by two negative pulses), and firing angle can be controlled to create current pulse signatures that are unique and sufficiently strong. The scheme is therefore quite immune to noise and is adaptive to different fault resistances. The use of a neutral transformer makes the scheme workable for different voltage levels. A disadvantage of these schemes is the need to install a signaling device at the neutral point. The neutral point is considered as

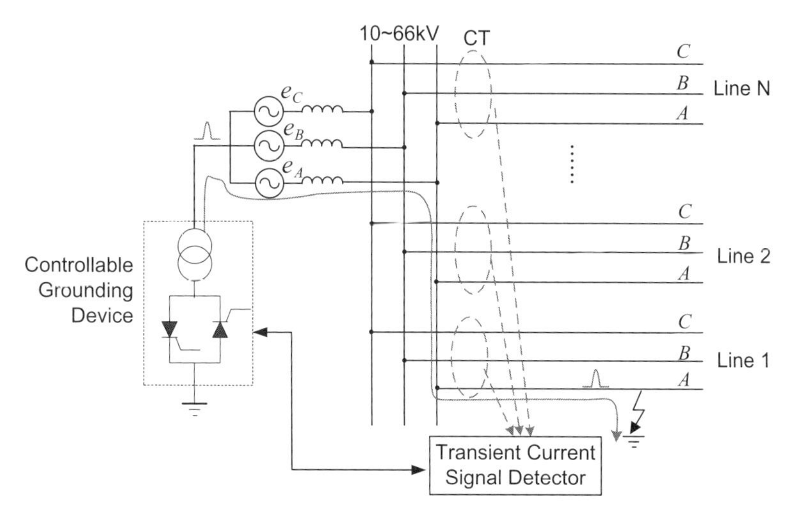

Figure 4.6. A thyristor-controlled grounding scheme. ©2010 IEEE.

belonging to the primary distribution circuit. Approval and installation of the signal generator is more complex. One possible improvement is to use the substation potential transformer as the means for signal injection.

4.3.3 Generator Ground Fault Detection

By far the most prevalent fault that generators are subjected to is a stator short-circuit to ground. For high-resistance grounded generators, the possibility of fault damage due to overcurrent is greatly reduced. Nevertheless, it is recommended that after a ground fault occurs the generator be immediately tripped off, rather than delaying this action until the generator can be shut down more conveniently. The reason is to avoid a high fault current and major damage to the equipment caused by a second ground fault on either a different phase or the same phase at a different position. A difficulty arises because a traditional overcurrent (or overvoltage) generator ground fault protective scheme cannot detect ground faults near the generator neutral.

There is a patent [14] for a proposed power electronics based device that can detect generator stator faults. This device works by injecting a modulated test signal U_v into the generator neutral (see Figure 4.7). With two primary windings controlled to conduct alternately by a pair of triacs, the signal coupling transformer can switch its input between a fundamental voltage v_i and its reverse $-v_i$. Each triac is turned on for 45 ms, and the 15 ms of remaining time serves as a safety redundancy. A large resistor R is connected in series with the signaling transformer secondary winding. When there is no fault, U_v causes no fault current and a zero voltage drop on the resistor R. When a ground fault occurs, a fault current flows through the resistor R, resulting in a voltage drop on R. To detect the fault, the voltage drop across the resistor is multiplied by U_v and then input into a filter-performing

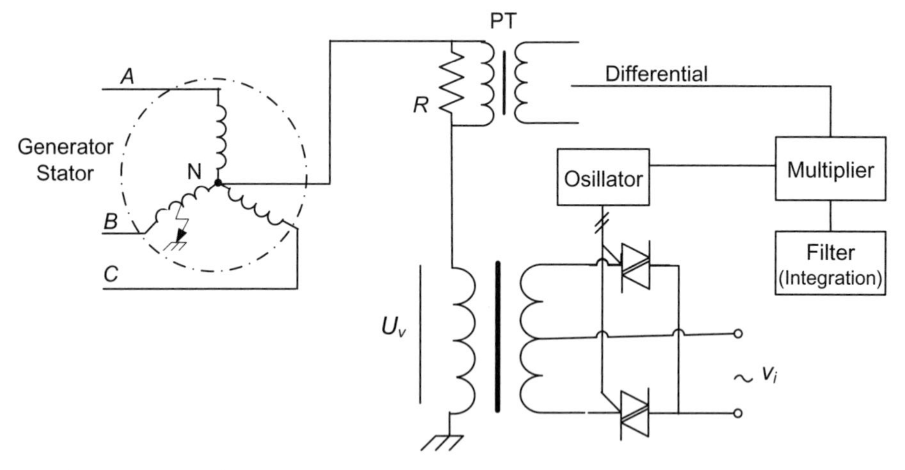

Figure 4.7. A scheme for generator stator ground fault protection. ©2010 IEEE.

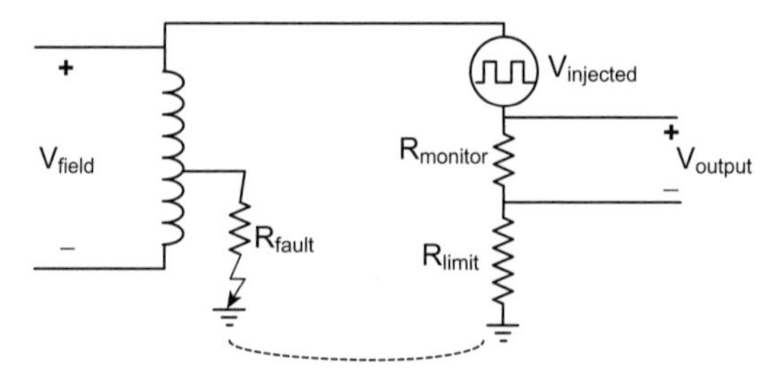

Figure 4.8. A signal injection method for detecting generator rotor ground fault. ©2010 IEEE.

integration. The integration result will be high when a ground fault occurs and will be used to activate a protection device. Otherwise, the integration value will be zero.

The above scheme has several advantages: (1) the fundamental frequency voltage input can be conveniently tapped from the system; (2) the signal generation and the detection method make the scheme immune to random or harmonic noises in the power system; and (3) the scheme does not require an expensive amplifier.

The field circuit of a generator is an ungrounded (typically 600 V) DC system. Timely detection of field winding ground fault is also desirable. Mozina [15] introduced a method that has been widely used in Europe with great success. This method also injects a voltage signal into the system (generator field winding in this case). Since the coupling circuit resistance and the field insulation resistance compose a voltage divider (Figure 4.8), a relay can calculate the field insulation resistance according to the input and return voltage signals. This method is designed for generators with brushes and uses the slip ring for signal injection into the rotor. Although the scheme is implemented using electronic circuits as described by Mozina [15], power electronic switches can also be used in this case.

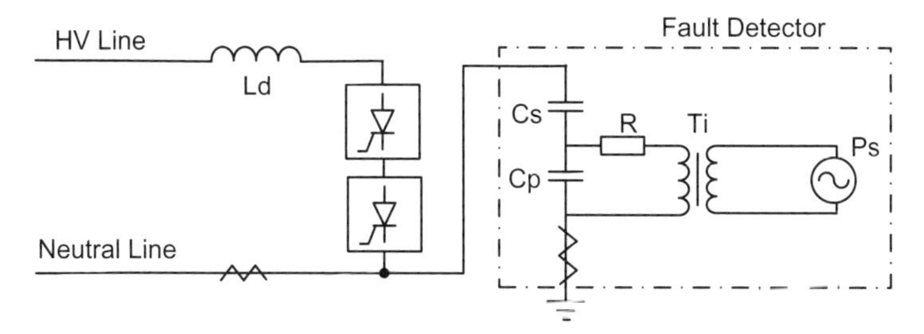

Ps: AC power source; Ti: Insulating transformer; Cp: Coupling capacitor; Cs: Surge capacitor; Ld: Smoothing Reactor;

Figure 4.9. HVDC neutral fault detection system. ©2010 IEEE.

4.3.4 HVDC Neutral Ground Fault Detection

In a bipolar HVDC transmission system with a neutral line, DC current flows through the converters of both poles, and no current flows through the neutral line during normal operation. In this condition, a grounding fault or breaking of the wire occurring on the neutral line does not affect the operation of the HVDC system. However, if this HVDC system with a neutral line fault is transferred from bipolar to monopolar operation due to, for example, a DC line or converter fault, high-level overvoltage might appear and damage equipment in the converter stations. Therefore, a grounding fault or breaking of the neutral line needs to be monitored. Kato et al. [16] introduce one method that has been used successfully in a Japanese HVDC system. This method injects a small-magnitude alternating current (referred to as a "pilot current") into the system through a surge capacitor (see Figure 4.9). This pilot current circulates through the neutral line and the ground in normal condition, and is blocked from flowing into the high-voltage line by the smoothing reactor. The pilot current level is monitored at its sending end and at the other terminal. The pilot current at the sending end increases when a grounding fault occurs or decreases when a breaking of the wire occurs. The pilot current at the other terminal decreases when a grounding fault or breaking of the wire along the neutral occurs. By monitoring the change in the pilot current, fault detection can be performed. The pilot current frequency is chosen as 125 Hz, in view of the signal attenuation and resonance conditions in the system. A dedicated signal generator is required for the interharmonic injection.

4.3.5 Detections of Faults in a De-energized Line

Re-energizing or reclosing to a de-energized overhead distribution feeder safely is a major consideration for a utility's safe work practice. After a feeder is de-energized for an extended period due to events such as repair, maintenance, or storms, there is always the possibility that humans or animals may be in contact with feeder conductors unknowingly. A reclosing action in such a situation can easily lead to fatality. Utility companies are therefore very interested in techniques that can determine if

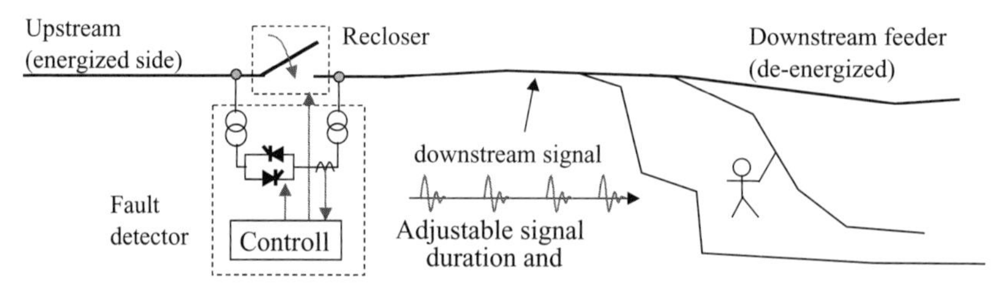

Figure 4.10. Power electronic based method to detect fault in a de-energized distribution line.

a de-energized feeder is clear of short-circuits so that operators can re-energize the feeder with confidence. Compared to detecting faults in an energized system, detecting faults in de-energized systems is more challenging, as it requires the generation and application of a voltage signal to the de-energized feeder. Furthermore, the signal must be high enough to mimic the normal medium voltage stress (e.g., 25 kV) to be experienced by the feeder.

Xun et al. [17] propose using a power electronic signaling technique to help detect faults in a de-energized line. The single line representation of the scheme is shown in Figure 4.10. A thyristor is connected in parallel to a circuit breaker or recloser. A step-down transformer is used to decrease the voltage of the distributed line to a low level for thyristor operation and then a step-up transformer is used to restore the signal back to the system voltage level. When the thyristor is triggered at several degrees before the voltage crosses zero, the energized upstream line is momentarily connected to the de-energized side and then a detection pulse is created in the downstream. The thyristor automatically shuts off when its current drops to zero. Point X in Figure 4.11 is the location for measuring the stimulated voltage and current signals. A significant feature of this technique is that the signal strength is adjustable by changing the thyristor firing angle. Therefore, a low-voltage pulse can be created to satisfy the safety requirement, and a high-voltage pulse can be produced to break down an insulted gap of a high-impedance fault when necessary. Utilizing the thyristor bridge configuration shown in Figure 4.11, the scheme is able to test and detect different kinds of faults such as phase-to-ground or phase-to-phase short-circuits. The actual embodiment of the technique is a low-voltage power electronic device connected to MV feeders through common utility primary-to-secondary service transformers. The device is installed permanently at the recloser or breaker locations and can be operated locally or remotely. As a result, there is no need to replace the existing breaker or recloser.

4.3.6 Summary

In this section, power electronics circuits are applied to inject intentional disturbances for system parameter monitoring. An important concern for these applications is that the injected disturbance must be controllable and must have a negligible impact on

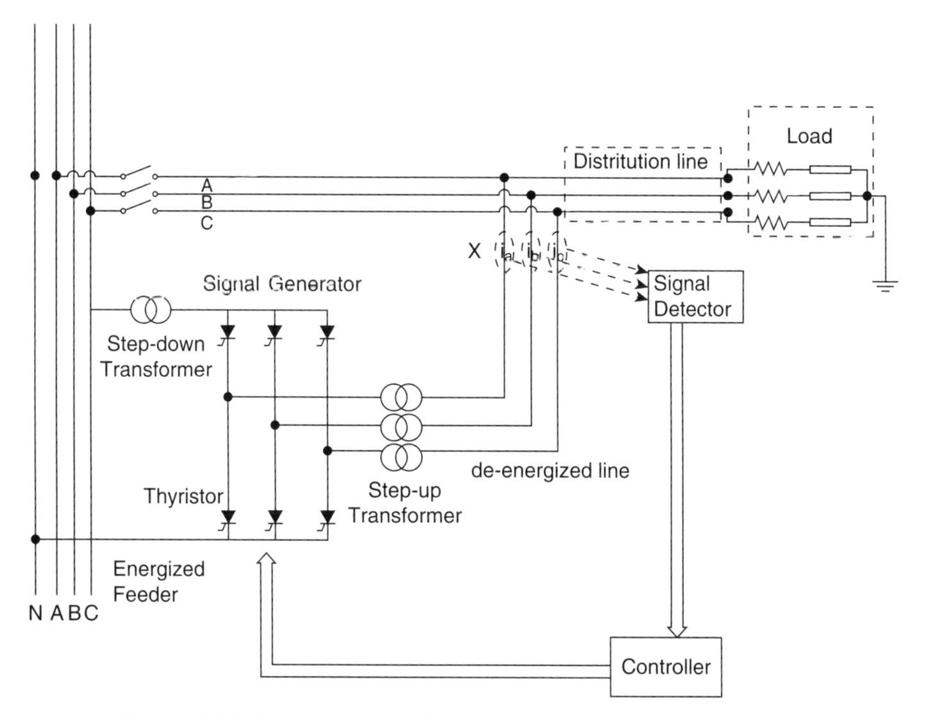

Figure 4.11. Three-phase thyristor bridge based fault testing scheme.

the system operation. The signals often contain unique characteristics. Such characteristic disturbances greatly facilitate signal detection even in a noisy environment. Fault detection seems to be one of the most fertile grounds for power electronic signaling applications. Only power electronic circuits can meet the requirements of creating desired signals on the one hand and operating at higher voltages on the other.

4.4 ACTIVE PROTECTION

Power electronic components or circuits have found their applications in the active protection area as well. The adoption of solid-state switches (e.g., thyristors) as protection devices presents advantages such as higher flexibility and reliability over traditional protection.

4.4.1 Impedance-Based Anti-islanding Protection for Distributed Generators

In recent years, researchers have become interested in online monitoring of the system impedance for the purpose of islanding detection of distributed generators (DGs). A power island forms when a distributed generator becomes isolated from

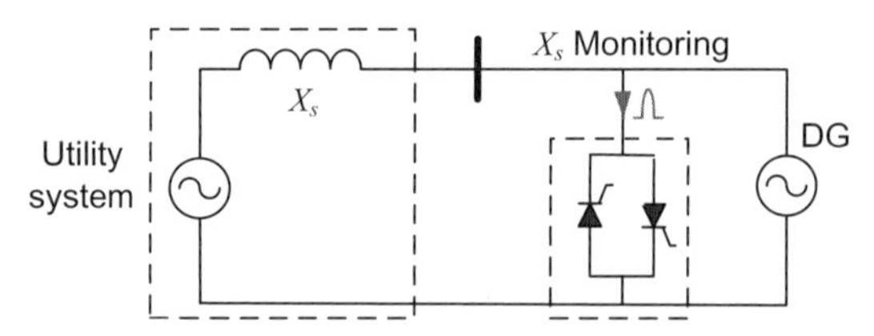

Figure 4.12. The thyristor-based circuit for measuring system impedance. ©2010 IEEE.

the utility grid. An increase in the system impedance at the DG site is regarded as an indication of an islanding condition. According to a European standard, the supply must be isolated within 5 seconds following an impedance increase of 0.5 Ω [18].

A general method for impedance measurement using a power electronic device is proposed by Cooper [19]. The idea is to connect a pair of shunt-connected thyristors at the DG terminal (Figure 4.12). The thyristor is fired near the zero-crossing point of its voltage to create a short-circuit near the voltage zero-crossing point (say, 10° ahead of the voltage zero point). The system impedance can be calculated according to the magnitude of the current pulse drawn by the thyristors. Once the system impedance at the terminals of distributed generators is available, an impedance-based relay can be developed for anti-islanding protection. This method suits all types of DGs. For inverter-based DGs, signals suitable for impedance measurement can easily be created through controlling the operation of the inverter. Several impedance-monitoring-based islanding detection methods have been devised for such DGs [20–22]. These techniques can also be viewed as power electronic signaling techniques. One unsolved issue associated with such monitoring schemes is the interference among the signals when there are multiple DGs and each of them injects similar signals into the system. The interference may corrupt the impedance values estimated at the terminal of each DG.

4.4.2 Power Line Signaling-Based Transfer Trip Scheme

Xu et al. [23] present a power line signaling-based transfer trip scheme for the anti-islanding protection of distributed generators. This scheme broadcasts a signal from a substation to the downstream DG sites using the distribution feeders as the signal paths (Figure 4.13). If the signal detector at a DG site does not sense the signal (caused by the opening of any devices between the substation and the DG) for a certain duration, it is considered as an island condition and the DG can be tripped immediately.

This is a transfer trip protection scheme. However, power lines are used for transmitting the protection signals. As shown in Section 4.2, there are several power line signaling techniques. Xu et al. [23] propose using the waveform distortion technique for DG anti-islanding protection due to its simplicity and response speed. Ropp et al. [24] propose using ripple control signals discussed by Galloway [6]. It is worthwhile to point out that the power line based transfer trip scheme can also be

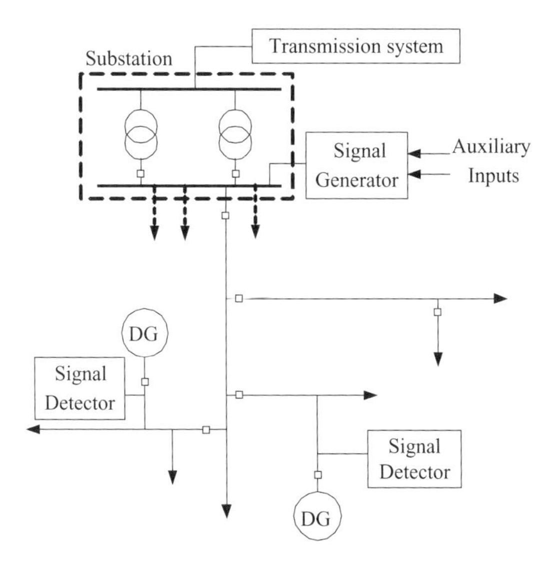

Figure 4.13. A power line signaling-based transfer trip scheme. ©2010 IEEE.

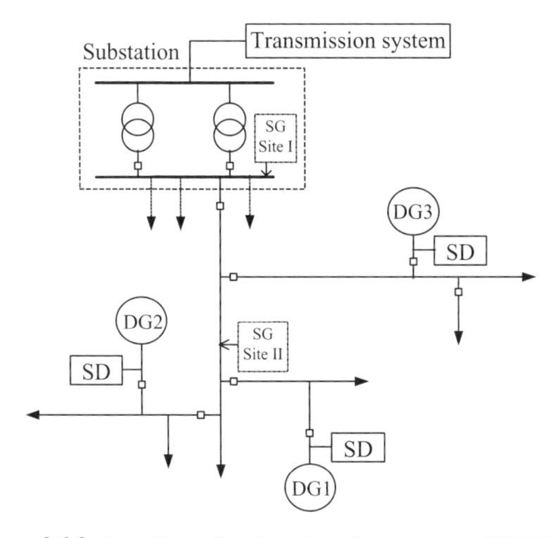

Figure 4.14. Locations for the signal generator. ©2010 IEEE.

used in certain transmission systems. Although the above schemes are simple in concept, their implementation involves many complex signal generation and signaling processing issues. For example, if one wants to detect the opening of one of the phases, signals need to be sent on all three phases. The interference among the three phase signals becomes a problem that must be addressed.

As shown by Wang et al. [25], the above signaling scheme can be further improved. The signal generators (SGs) can be installed at any point downstream. As shown in Figure 4.14, when only DG1 and DG2 exist, the signal generator can be moved from Site 1 to Site 2. In this way, the anti-islanding signal broadcast

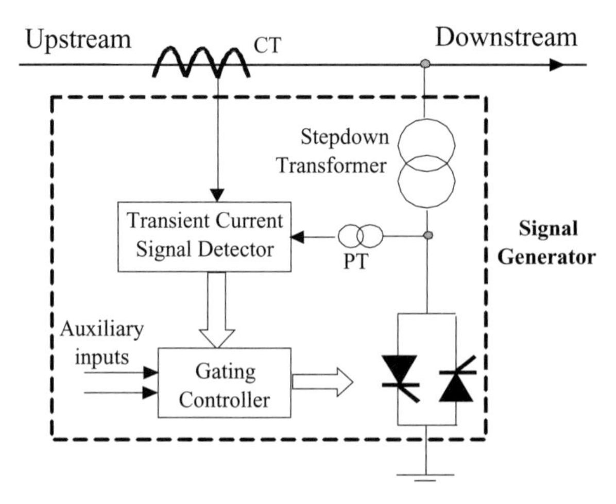

Figure 4.15. Signal generator with current pulse monitoring capability. ©2010 IEEE.

downstream by the signal generator has a smaller coverage area, which still contains DG1 and DG2. This arrangement avoids the need to access the substation, and a smaller signal transformer can be used. This feature makes the power line signaling-based anti-islanding scheme more adaptable to various DG interconnection scenarios. As the SG is moved away from the substation secondary bus, a new problem arises: How can one know if the SG is still electrically connected to the main supply? This condition must be monitored at all times. Once the SG is determined to be isolated from the main supply, it must stop broadcasting immediately and therefore trip all the downstream DGs. The solution is to monitor the transient current signal flowing in the line at the upstream side of the signal generator. This current signal is created as a by-product of the voltage distortion signal generation process. Figure 4.15 shows the architecture of the signal generator in the improved scheme. The same step-down transformer and thyristors as those in the original scheme are used to create the voltage distortion signals. Moreover, a current transducer (CT) is installed at the SG upstream side, monitoring the line current. When the thyristor is fired to create a voltage distortion signal, a transient current signal flows through the CT. If the SG is connected to the main supply, this transient current will have a pulse-like shape. If the SG is disconnected from the main supply, the shape of this current waveform will be different. The transient current waveform can therefore be used for monitoring the SG connection status to the main supply.

4.4.3 PT Ferroresonance Protection

Ferroresonance may occur in ungrounded systems between a primary-side Yg-connected potential transformer (PT) and a line shunt capacitor. Once excited, ferroresonance causes sustained overvoltage and overcurrent and therefore must be suppressed. Wang et al. [26] proposed a new device that can mitigate this phenom-

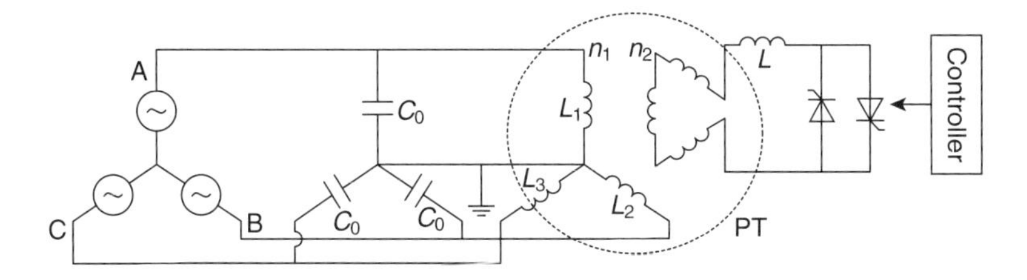

Figure 4.16. Thyristor-based PT ferroresonance suppression. ©2010 IEEE.

enon. This device comprises a pair of antiparallel connected thyristors in series with a small current-limiting inductor and is inserted into the PT's delta tertiary winding (see Figure 4.16). In normal condition, the thyristors are off. When the voltage across the open-delta winding exceeds a preset threshold, indicating that ferroresonance may have occurred, the thyristors will be triggered to conduct. The thyristor current introduces losses to the circuit and will damp the ferroresonance effectively. Since a single-phase-to-ground fault may also cause the voltage across the open-delta winding to exceed the limit, the thyristor is triggered to conduct three times, each time for a short period of 10 ms. If the voltage signal disappears after the three conductions, the scheme considers that ferroresonance has occurred and the thyristors have suppressed the resonance; otherwise the system is regarded to be experiencing a single-phase-to-ground fault and a warning signal will be sent. According to Zhou et al. [26], this device makes use of the fast-switching capability of thyristors and has better performance than the traditional method utilizing a damping resistor.

4.4.4 Summary

Active protection could be an area of significant research and development in the future and power electronic devices are the natural choice for such applications. Although traditional passive protection schemes can satisfy most power system protection requirements nowadays, there are still some challenging protection problems remaining to be solved. Examples are the detection of high-impedance faults and prevention of ferroresonance involving distribution transformers. Active protection schemes could be developed from at least two perspectives. One of them is to use PLC or other PE signals to trigger protection actions. Another is to utilize intentional signature disturbances for active fault detection, which in turn triggers protective action. These perspectives may lead to out-of-the-box and effective solutions to some of the challenging protection problems.

4.5 POWER ELECTRONICS SIGNALING TECHNOLOGY

Examples illustrated in the previous sections help to clarify the characteristics and scopes of the PE signaling technology. The most common characteristics of these applications are the use of power electronic devices to manipulate the voltage and/or current waveforms for *signaling purposes*. These disturbances either propagate in power systems and are used to transmit information over distance, or carry signature equipment responses and are used to monitor system conditions online. The PE signaling technologies reviewed can be classified into the following three types:

- *Communication-Oriented Applications.* In this type of applications, small but discernible signals propagate in the system to transmit information across distances. The presence of a signal represents a binary 1 (or 0), while the absence of a signal represents 0 (or 1). The signals could be used for purposes beyond communications. For example, a PLC-based transfer trip scheme can be developed.
- *Monitoring-Oriented Applications.* In this type, small disturbances with unique characteristic signatures are created in a power system. Information about the system or equipment under monitoring is derived according to the characteristics of the response signals. The characteristics of the signals can be shaped to provide superior antinoise performance. The technologies can monitor continuous parameters such as temperatures or sudden changes such as short/open circuits in a system.
- *Active Protection.* This type of application encompasses two aspects. One aspect is to use actively created signals to develop novel protection schemes. The second aspect is to make the power electronic devices directly involved in the change of equipment operation modes (such as the thyristor-based PT ferroresonance suppression technique). Although traditional passive protection schemes can satisfy most power system protection requirements nowadays, there are still some challenging protection problems remaining to be solved and active protection could be one of the solutions.

Power electronic signaling technology is likely to become an active area of research attracting broader interest from industry and research communities. There are at least three driving forces for this trend:

1. Active monitoring and protection, especially equipment condition monitoring, is gaining increased interest from industry. In many cases, it is impossible to detect the condition change of a piece of equipment by just relying on passively collected waveform data alone. Injecting intentional disturbances for diagnostic purposes is a natural evolution of condition monitoring techniques. This is happening in various fields such as medical diagnosis and seismic imaging and is likely to become a more active area in the power and energy field. Power electronic circuits are probably the only candidate for such applications since they are able to create controllable disturbances at higher voltage levels.
2. Dedicated field tests for system parameter measurement are becoming continuous online applications. One example is the measurement of system impedances. Capacitor switching and other staged tests were traditionally used for

impedance measurement [27]. The technique proposed by Cooper [19] could make online monitoring of system impedance possible. The same approach can also be applied to online monitoring of substation grounding impedances. Similarly, some offline measurement techniques such as cable fault location are migrating into online applications. In all these cases, intentional disturbances are involved. Power electronics are relied upon to make such disturbances continuously available for online measurement.

3. The advancement of signal processing algorithms and hardware makes it highly feasible to detect very small signals in high-voltage and large-current power systems. The signals injected into a power system are often very small for two reasons. One is the power quality concern—the signals shall not disturb the normal operation of power system equipment. The other is a technical constraint—it is very difficult to create signals with good strength in power systems due to high background voltages and currents in the system. The modern signal processing technologies are reducing the above two barriers at an unprecedented rate, making detection of power electronic signals practically and economically achievable.

In terms of future innovations in the field, it is likely that the communication-oriented PE technologies and applications will continue to grow. This is partially due to the fact that powerful signal processing technologies developed by the telecommunication industry are ready for adoption into the PLC schemes. For example, a recent patent [28] has shown that very small capacitor-switching transients can be used as detectable signal carrier for PLC. The resulting signaling rate approaches 10 kb/s. Many active condition monitoring and fault detection schemes will also emerge. A potential research area could be online substation grounding condition monitoring. If the grounding impedance can continuously be estimated, one can use the change of the estimated value over time, instead of a precise measurement value that is very hard to obtain, to evaluate the grounding condition of a substation.

4.6 CONCLUSIONS

Unlike many well-known power electronic applications such as the FACTS devices or HVDC terminals, the applications reviewed in this chapter deploy power electronic devices to manipulate *information* or *signals,* not *electric energy*. From this perspective, we regard these applications as a new class of power electronics technology and refer to them as power electronic signaling technology. The common features of these technologies were presented in this chapter through example applications. This is the time to recognize such an important class of power electronics applications. Three driving forces are likely to accelerate the research and development in the area of power electronic signaling technology. These driving forces are part of the much larger trend of Smart Grid development.

REFERENCES

[1] R. H. Johnston, D. C. Jeffreys, and L. J. Stratton, "Method and apparatus for transmitting interlligence over a carrier wave," U.S. Patent 4106007, August 1978.

[2] S. T. Mak and T. G. Moore, "TWACS, a new viable two-way automatic communication system for distribution networks. Part II: Inbound communication," *IEEE Trans. Power Apparatus Syst.*, 103(8), pp. 2141–2147, August 1984.

[3] S. T. Mak, "A new method of generating TWACS type outbound signals for communication on power distribution networks," *IEEE Trans. Power Apparatus and Syst.*, 103(8), pp. 2134–2140, August 1984.

[4] P. M. Foord, "Bi-directional muti-frequency ripple control system," U.S. Patent 4868539, September 1989.

[5] J. H. Galloway, "Control logic for an inverter ripple controlled power system," U.S. Patent 4215394, July 1980.

[6] P. C. Hunt, "Low frequency bilateral communication over distribution power lines," U.S. Patent 6154488, November 2000.

[7] U. Oehrli, "Method for remote control through a power supply system and apparatus for carrying out the same," U.S. Patent 3986121, October 1976.

[8] R. A. Ausfeld, "Power line communication systems," U.S. Patent 3483546, December 1969.

[9] *IEEE Standard for Broadband over Power Line Hardware*, IEEE Std. 1675, IEEE February 2009.

[10] P. Zhang, B. Lu, and T. G. Habetler, Summary: "A nonintrusive induction motor stator resistance estimation method using a soft-starter," *Proceedings of IEEE International Symposium on Diagnostics for Electric Machines, Power Electronics and Drives*, 2007. SDEMPED, pp. 197–202, September 2007.

[11] S. Lee and T. Habetler, "An online stator winding resistance estimation for temperature monitoring of line-connected induction machines," *IEEE Trans. Ind. Appl.*, 39, pp. 685–694, May/June, 2003.

[12] H. Guzman, "Ground fault detector and locator," U.S. Patent 4,884,034, November 1989.

[13] W. Wang, K. Zhu, P. Zhang, and W. Xu, "Identification of the faulted distribution line using thyristor-controlled-grounding," *IEEE Trans. Power Delivery*, 24(1), pp. 52–60, January 2009.

[14] BBC—Brown Boveri & Cie, "Method and apparatus for detecting ground shorts in electrical systems," Worldwide Patent GB1601235, October 1981.

[15] C. J. Mozina, "Advances in generator field ground protection using digital technology," *Proceedings of IEEE Industrial and Commerical Power Systems Technical Conference*, May 2002, pp. 126–130.

[16] Y. Kato, A. Watanabe, H. Konishi, T. Kawai, Y. Inoue, and H. Irokawa, "Neutral line protection system for HVDC transmission," *IEEE Trans. Power Delivery*, 1(3), pp. 326–331, July 1986.

[17] L. Xun, W. Xu, and Y. W. Li, "A new technique to detect faults in de-energized distribution feeders," *IEEE Trans. Power Delivery*, 26(3), pp. 1893–1901, 2011.

[18] *Photovoltaic Semiconductor Converters Part 1: Utility Interactive Fail Safe Protective Interface for PV-Line Commutated Converters—Design Qualification and Type Approval*, European Standard EN 50330-1, 1999.

[19] C. B. Cooper, "Standby generation—problems and prospective gains from parallel running," *Power System Protection '89*, Singapore, 1989.

[20] L. Asiminoaei, R. Teodorescu, F. Blaabjerg, and U. Borup, "Implementation and test of an online embedded grid impedance estimation technique for PV inverters," *IEEE Trans. Ind. Electron.*, 52(4), pp. 1136–1144, August 2005.

[21] G. Hernández-González and R. Iravani, "Current injection for active islanding detection of electronically-interfaced distributed resources," *IEEE. Trans. Power Delivery*, 21(3), pp. 1698–1705, July 2006.

[22] H. Karimi, A. Yazdani, and R. Iravani, "Negative-sequence current injection for fast islanding detection of a distributed resource unit," *IEEE Trans. Power Electron.*, 23(1), pp. 298–307, January 2008.

[23] W. Xu, G. Zhang, C. Li, W. Wang, and J. Kliber, "A power line signaling based technique for anti-islanding protection of distributed generators: Part I: Scheme and analysis," *IEEE Trans. Power Delivery*, 22(3), pp. 1758–1766, July 2007.

[24] M. Ropp, D. Larson, S. Meendering, D. McMahon, J. Ginn, J. Stevens, W. Bower, S. Gonzalez, K. Fennell, and L. Brusseau, "Discussion of a power line carrier communications-based anti-islanding scheme using a commercial automatic meter reading system," *IEEE 4th World Conference on Photovoltaic Energy Conversion*, 2, pp. 2351–2354, 2006.

[25] W. Wang, J. Kliber, and W. Xu, "Scalable power-line-signaling-based scheme for islanding detection of distributed generators," *IEEE Trans. Power Delivery*, 24(2), pp. 903–909, April 2009.

[26] L. Zhou, Z. Yin, and L. Zheng, "Research on principle of PT resonance in distribution power systems and its suppression," *Trans. China Electrotech. Soc.*, 22(5), pp. 153–158, May 2007.

[27] A. A. Girgis and R. B. McMains, "Frequency domain techniques for modelling distribution or transmission networks using capacitor switching induced transients," *IEEE Trans. Power Delivery*, 4(3), pp. 1882–1890, July 1989.

[28] P. Bertrand, "Method and device for emitting pulses on an electricity distribution network," U.S. Patent 7,078,982 B2, July 18, 2006.

PART II

COMMUNICATIONS

5

INTRODUCTION TO SMART GRID COMMUNICATIONS

Wenbo Shi and Vincent W. S. Wong

Today, the electric power grid is undergoing a significant transition into an intelligent, reliable, and fully automatic grid which is called the Smart Grid. In the Smart Grid vision, a variety of energy services including demand response, load management, distributed generation, real-time pricing, and substation automation can be achieved by incorporating advanced information technologies with the power systems. The key to the Smart Grid vision is the communications network, which serves as the fundamental information infrastructure to provide bidirectional end-to-end data communications in the Smart Grid. Although a myriad of existing communications technologies can be applied to the Smart Grid, new communications protocols and enhancement of existing protocols are necessary to capture the unique characteristics and requirements of the Smart Grid. In this chapter, we first introduce and describe the communications network architecture for the Smart Grid. Then we discuss some potential communications standards and protocols for the construction of communications networks for the Smart Grid. Moreover, we give an overview of the recent standardization activities on communications networks for the Smart Grid.

Smart Grid: Applications, Communications, and Security, First Edition.
Edited by Lars Torsten Berger and Krzysztof Iniewski.

5.1 INTRODUCTION

Currently, modernization and automation of the electric power grid has taken place worldwide to increase energy efficiency, reduce greenhouse gas emissions, and transit to renewable energy. Ultimately, the grid will become smart by incorporating traditional power engineering and advanced information and communications technology to support a variety of energy services and functionalities for the utilities and the customers. Based on a report of the U.S. Department of Energy [1], the Smart Grid is an electricity delivery system enhanced with communications facilities and information technologies to enable more efficient and reliable grid operations with an improved customer service and a cleaner environment. By exploiting the two-way communications capabilities between the utilities and the customers, it becomes possible to replace the current power system with a more intelligent infrastructure. Some of the distinguishing characteristics of the Smart Grid include [2]:

- Increased use of digital information and control technology to improve reliability, security, and efficiency of the electric grid
- Dynamic optimization of grid operations and resources, with full cyber security
- Deployment and integration of distributed resources and generation
- Deployment and integration of demand response
- Deployment of smart technologies for metering, communications concerning grid operations and status, and distributed automation
- Integration of smart appliances and consumer devices

The development of the Smart Grid is still in an early phase. A variety of technological innovations from different fields and disciplines such as power engineering, system control, and communications are required to enable the Smart Grid. Among them, communications is one of the fundamental technologies. One of the key features of the Smart Grid is the integration of modern communications networks with power systems. Communications network is the key information infrastructure to support Smart Grid applications such as demand response, load management, distributed generation, real-time pricing, and substation automation. Unlike existing computer or wireless networks, the Smart Grid is a complex network interconnecting with a large number of heterogeneous devices and systems with various ownerships and management boundaries. Such complexity adds difficulty to constructing the communications network for the Smart Grid, which is essential to exchange information and share resources among different devices and appliances.

Although there are various existing wired and wireless communications standards that can be applied to the Smart Grid, design of the communications network architecture and protocols that can capture the characteristics and meet the specific requirements of the Smart Grid is crucial in order to provide an affordable, reliable, and sustainable supply of electricity. Compared with digital data, electricity has the following characteristics that create challenges for the design of communications networks for the Smart Grid [3]:

- Electricity cannot be effectively stored on a large scale. This is the most distinguishing difference between electricity in the Smart Grid networks and digital data, which can easily be stored in computer networks. There are no mature technologies that can effectively store a large amount of electricity generated but not being used.
- Electricity is mostly generated centrally and consumed locally. The generation of power is controlled centrally through scheduling generation for power stations by the utilities. The customers are fully distributed. Therefore, routing options in the network are limited and long distance transmission is common.
- The quality of service (QoS) is the top priority in the Smart Grid. Unlike the Internet, which uses the best effort delivery service, QoS is crucial in order to satisfy the demand of consumers at any time in the Smart Grid. The Smart Grid should be capable of monitoring and forecasting the consumers' peak demand so that power can be scheduled for generation, transmission, and distribution to meet the demand.

The above features of electricity must be considered carefully in the design of network architectures and protocols for the Smart Grid. The objective of the communications networks in the Smart Grid is to provide bidirectional end-to-end communications between the utilities and the appliances. The typical requirements for communications networks in the Smart Grid can be summarized as follows [4, 5]:

- *Reliability.* The networks must be able to provide reliable communications that coincide with or exceed the reliability of the power grid itself.
- *Scalability.* The networks are expected to last decades and serve an ever increasing number of household appliances.
- *Availability.* Protection mechanisms, redundancy, and fault tolerance with self-healing abilities must exist to guarantee a high availability.
- *Security.* The networks must guarantee end-to-end security including the complete privacy of the networks from unauthorized access as well as confidentiality of communications across the networks.
- *Low Latency.* The latency requirement for some Smart Grid applications are extremely demanding (e.g., 10 ms for teleprotection), which is far beyond telecommunications applications [e.g., 200 ms for voice over IP (VoIP)].
- *Hard QoS.* QoS services must be provided for Smart Grid applications with predictable latency and low error rate.
- *Cost Effectiveness.* The communications networks for the Smart Grid need to provide financial feasibilities and economics. The capital and operational expenditures need to be low.
- *Standards-Based and Interoperability.* Standards on communications networks must be developed to enable interoperability.

The above requirements are some general requirements for communications networks in the Smart Grid. Specific network requirements vary for different applications in terms of bandwidth, latency, security, and priority. For example, teleprotection data, which is critical for power system stability, requires extremely low latency (less than 10 ms), extremely high reliability and security, but low bandwidth. On the other hand, information technology data, which is essential for enterprise operation, requires moderate latency (10–100 ms), only enterprise class reliability and security, but high bandwidth. The Smart Grid communications networks must be able to accommodate data flows of different classes simultaneously.

The Smart Grid includes communications networks from diverse technologies with a hierarchical architecture. In the following sections, we first introduce the network architecture for the Smart Grid. Then we discuss the functionalities of each type of the communications networks in the architecture and give some guidelines in the design of communications networks for the Smart Grid by introducing related communications technologies. Next, we give an overview of the recent standardization activities on the Smart Grid. Conclusions are given at the end of this chapter.

5.2 AN OVERVIEW OF NETWORK ARCHITECTURE

To obtain a general idea of what the Smart Grid consists of, we first introduce the system architecture [6] for the Smart Grid as shown in Figure 5.1. In this architecture, the Smart Grid is divided into four layers: the application layer, the communications layer, the power control layer, and the power system layer from the top to the bottom. Each layer is a set of similar systems that provide services and interfaces to the layer above it and receives services from the layer below it. In this architecture, the top layer is the application layer, which provides Smart Grid applications for customers and utilities based on the information infrastructure. Under the application

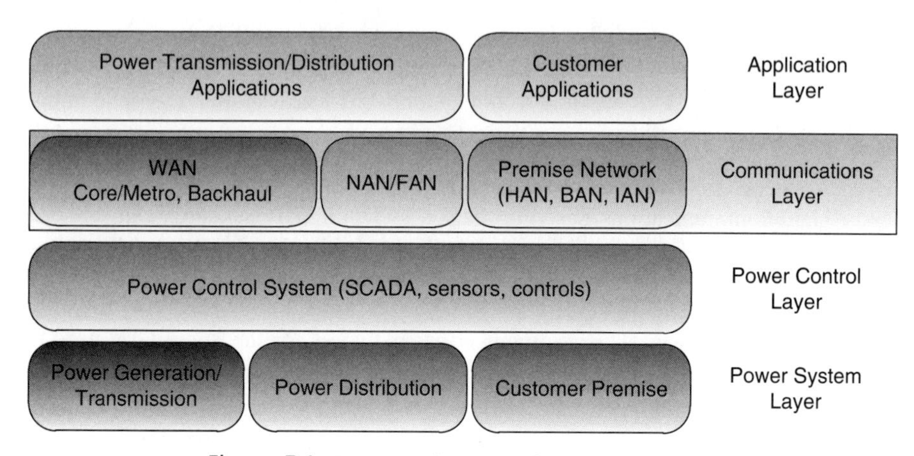

Figure 5.1. System architecture for the Smart Grid.

layer is the communications layer, where a two-way reliable, efficient, and secure information exchange is provided for upper-layer applications in the Smart Grid.

The communications network in the Smart Grid has a hierarchical structure consisting of the premises network [i.e., the home area network (HAN), the building area network (BAN), and the industrial area network (IAN)], neighborhood area network (NAN)/field area network (FAN), and wide area network (WAN) including the backhaul network, core network, and metro network according to their reach and functions in the Smart Grid [7]. The layer under the communications layer is the power control layer, where the power control system such as supervisory control and data acquisition (SCADA) exists to enable functions for monitoring, control, and management in the grid. The bottom layer is the power system layer where the electric power flows via power generation, transmission, and distribution systems. Note that this layered architecture for the Smart Grid is not unique. A different architecture has also been proposed to decompose the technologies needed in the Smart Grid [5]. But communications can always be separated as one layer to provide connectivity in the Smart Grid. In this chapter, we mainly focus on the communications layer of the Smart Grid.

Communications networks in the Smart Grid, according to their reach and characteristics, come in mainly three types: the premises network, NAN/FAN, and WAN, as shown in Figure 5.2. The figure shows the building blocks of an end-to-end communications network for the Smart Grid with multiple network segments and boundaries. In this network architecture, the premises network provides access to appliances in the customer premises. Depending on the specific environment, the premises network can be further divided into HAN, BAN, and IAN. NAN/FAN connects to smart meters, field devices, and distributed resources. WAN provides long distance communications links between the grid and the utility core network. From the premises network to the WAN, network complexity increases due to the increasing network size and coverage area. This trend can be seen in Figure 5.3, showing the communications range and data rate requirement for the three types of network.

According to the National Institute of Standards and Technology (NIST) framework and roadmap for Smart Gird interoperability standards [8], the communications network for the Smart Grid should be primarily Internet Protocol (IP) [9] based. The migration toward IP brings a number of benefits to the Smart Grid including simplified system architecture and control, end-to-end visibility, interoperability with different networks, and support for existing IP applications. IP-based networks provide end-to-end communications based on packet switching. IP standards can serve as the basis for upper-layer applications, which enables applications to be developed without the dependence on a specific data link layer communications protocol. This greatly reduces the complexity for developing upper-layer Smart Grid applications. In the following discussions on different types of networks, we introduce the protocols and standards in the medium access control (MAC) layer and the physical (PHY) layer assuming that IP can serve as a unified interface for upper-layer applications. Furthermore, IP has good network scalability. Any smart meters, household devices, and smart appliances can be connected to the network. IPv6 [10] is

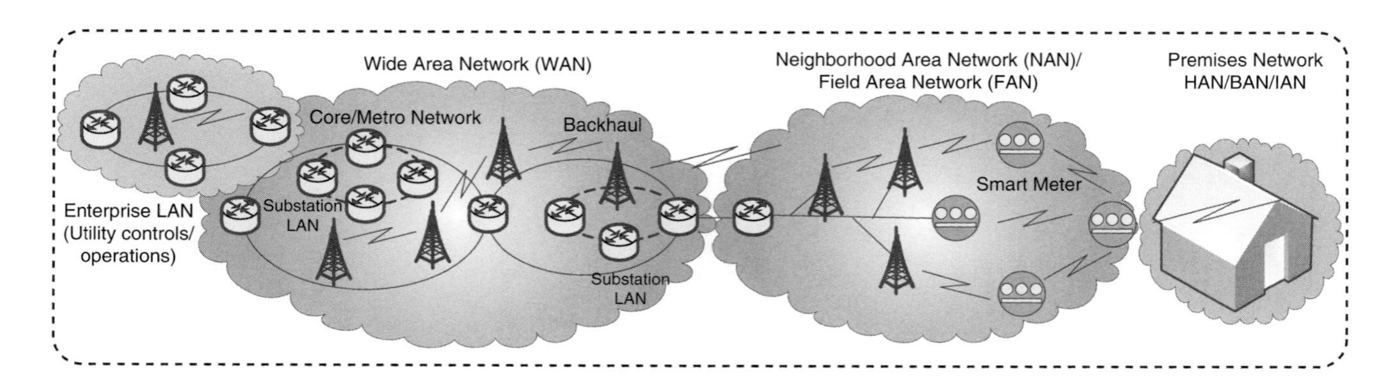

Figure 5.2. Communications network architecture for the Smart Grid.

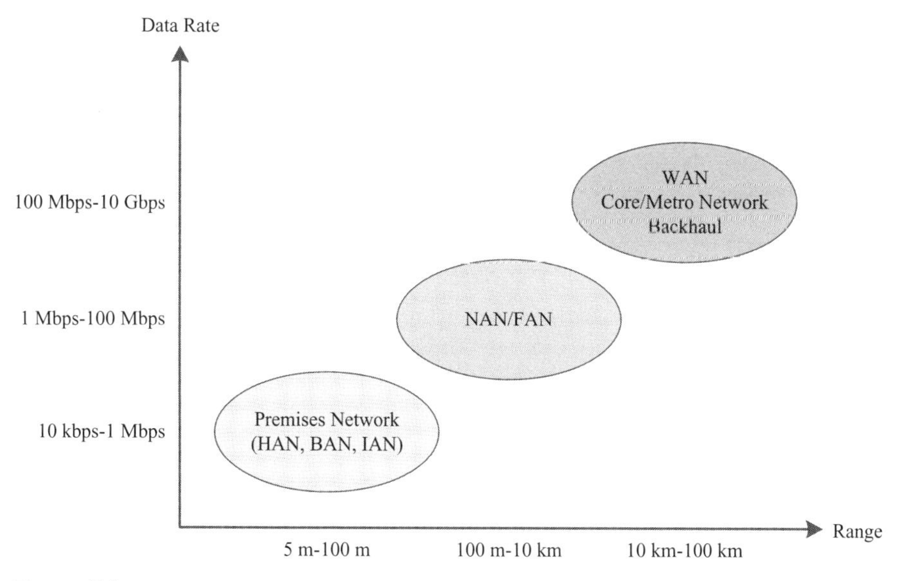

Figure 5.3. Communications range and data rate requirement for different networks in the Smart Grid.

recommended to be used in the Smart Grid, since the addressing would be a problem if the scale of the Smart Grid network expands fast. Although IP is expected to be used in the Smart Grid, it is necessary to assess if the network performances and security meet the requirements of the Smart Grid. For Smart Grid applications, a number of IP protocol suites need to be specified to meet different network requirements.

5.3 PREMISES NETWORK

The premises network is on the customer's end of the network architecture, consisting of interconnected household appliances, electric vehicles, and other electric equipment in the customer premises. It provides communications access to appliances in the customer premises and interfaces them to the Smart Grid. The premises network is of great significance to the Smart Grid, since it is the essential network dedicated to advanced metering infrastructure (AMI) and demand-side management (DSM) to realize energy efficiency, demand response, and direct load control. The premises network can be further classified into HAN, BAN, and IAN depending on residential, commercial, and industrial environment. Therefore, the functions for premises networks vary in different environments. For example, HAN provides communications for household appliances and equipment to enable energy management for in-home uses. BAN and IAN are used for commercial and industrial customers with focus on building automation, heating, ventilating, and air conditioning

(HVAC), and other industrial energy management services. The premises network is connected to the Smart Grid via smart meters, providing energy management abilities for the consumers and the utilities. With the premises network, energy consumers can manage energy from the demand side. The premises network also supports various energy services for the utilities such as prepaid service, user information messaging, real-time pricing and control, load control, and demand response. For these energy management applications, there is no great need for a high bandwidth. The most important network requirements are for low power, low cost, low latency, and secure communications coverage of the premises.

There are many communications standards and protocols to meet the Smart Grid requirements for the premises network. They can be divided into wired and wireless technologies. Wireless communications technologies have several advantages over wired communications in the premises network. Wireless networks are easier to deploy, more flexible, scalable, and portable than wired networks with costly infrastructure. Such characteristics make wireless communications technologies preferred in the premises network. The main challenges for wireless networks are power consumption, reliability, and security. Wired networks can be complementary to wireless networks to assure the coverage of the premises and to increase the network reliability. The wired communications technologies used in the premises network should be optimized for energy management applications. An overview of the existing communications standards across different protocol layers that can be used in the premises network for the Smart Grid is shown in Figure 5.4.

In the following paragraphs, we mainly introduce the features of two communications technologies for the premises network: ZigBee and HomePlug. Note that more information on ZigBee and other wireless standards can be found in Chapter 6. Besides, complementary information on power line communication standards apart from HomePlug can be found in Chapter 7.

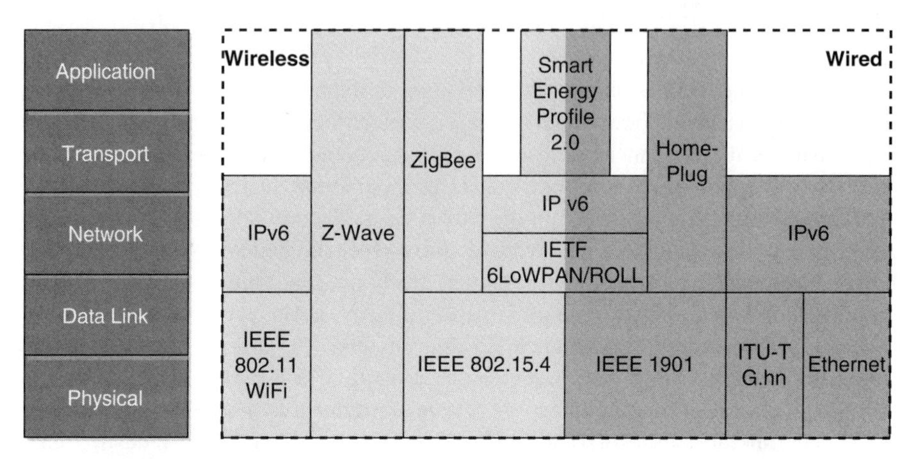

Figure 5.4. Potential communications technologies for the premises network.

ZigBee is a low power, low cost, two-way wireless communications standard for residential home control, commercial building control, and industrial plant management. It is based on the Institute of Electrical and Electronics Engineers (IEEE) 802.15.4 standard [11], which specifies the MAC/PHY layer for low power, low rate wireless personal area networks (WPAN). The IEEE 802.15.4 standard also serves as the basis for the Internet Engineering Task Force (IETF) IPv6 over low power wireless personal area networks (6LoWPAN) [12] and IETF routing over low power and lossy networks (ROLL) [13] to support IPv6. IETF 6LoWPAN is designed to support IPv6 packets to be sent to and received from over IEEE 802.15.4 based networks. IETF ROLL is used to address the routing issue based on the IPv6 architecture in low power and lossy networks such as IEEE 802.15.4, Bluetooth, low power WiFi, wired, or other low power power line communications links. Both IETF 6LoWPAN and IETF ROLL are considered to play an important role in the premises network as the networks in the Smart Grid are expected to be all-IP based. The IEEE 802.15.4 PHY layer operates on the industrial, scientific, and medical (ISM) bands: 868 MHz in Europe, 915 MHz in the United States and Australia, and 2.4 GHz worldwide. Using direct sequence spread spectrum (DSSS), the data rate can be as high as 250 kbit/s in the 2.4 GHz band, 40 kbit/s in the 915 MHz band, and 20 kbit/s in the 868 MHz band. Transmission range for ZigBee is between 10 and 75 meters and up to 1500 meters for ZigBee Pro. The IEEE 802.15.4 MAC layer controls access to the radio channel using the carrier sensed multiple access with collision avoidance (CSMA-CA) mechanism. ZigBee supports multiple network topologies including star, tree, and mesh network topology. It is able to accommodate as many as 65,000 nodes in one network. By adopting the industry standard AES-128 security scheme [14], ZigBee is capable of establishing a secure wireless network.

The ZigBee Alliance has created several public profiles, which define standard interfaces and device definitions to enable interoperability of different products from various manufacturers. The ZigBee smart energy profile [15] offers a wireless solution to establish the premises network for the Smart Grid defining standards for applications including metering, pricing, demand response, and load management in the residential or light commercial environment. The ZigBee Alliance is now developing a new IP-based energy profile called the smart energy profile 2.0 [16] in cooperation with the HomePlug Alliance. The new profile will remove the dependency on the IEEE 802.15.4 standard, which will enable manufacturers to implement any MAC/PHY, such as IEEE 802.15.4 and IEEE 1901 [17], under an IPv6 layer based on IETF 6LoWPAN and IETF ROLL. By merging into IP-based networks, wireless ZigBee devices based on IEEE 802.15.4 can easily establish end-to-end communications with wired HomePlug devices based on IEEE 1901. The integration of wireless ZigBee and wired HomePlug infrastructure will assure the network coverage of a large home or building in the Smart Grid. For more on the ZigBee protocol side see Chapter 10.

HomePlug is a power line communications technology using the power line as the provider of electric current and the carrier of high speed digital data at the same time. The infrastructure for HomePlug is the power lines already installed within the home, which makes no new wiring or cable necessary. Customers can easily establish a high speed power line network by simply plugging adapters into wall outlets within home. The simplicity of HomePlug makes it a low cost, high speed, and easy-to-use network solution for in-home communications. Power line communications is a natural choice for the Smart Grid, since it can reach each AC outlet in the premises so that all electric devices can be connected to achieve smart energy management. The MAC/PHY layer for HomePlug is standardized by the IEEE. The corresponding IEEE 1901 standard for broadband power line communications was approved in September 2010 with a plan for publication in February 2011. Currently, there are two HomePlug specifications: HomePlug AV and HomePlug Green PHY (GP). Both of them are compliant with the IEEE 1901 standard. HomePlug AV is designed for broadband applications such as high definition television (HDTV) and VoIP. The peak data rate can be as high as 100 Mbit/s. HomePlug GP is a low power, highly reliable, cost-optimized power line communications specification targeting Smart Grid connectivity for home energy management to devices such as HVAC, smart meters, appliances, and PHEVs. HomePlug GP is a subset of HomePlug AV specifically designed for the application of Smart Grid. It has a data rate of 10 Mbit/s. Compared with HomePlug AV, HomePlug GP can save up to 75% energy. The ZigBee and HomePlug Smart Energy Liaison is now creating a common application layer to enable interoperability between wireless ZigBee devices and wired HomePlug devices, which can assure coverage in the customer premises. The combination of ZigBee and HomePlug has the advantages of both wireless communications and power line communications, which is one of the promising solutions in the premises network for the Smart Grid.

In addition to ZigBee and HomePlug, there are other competing communications standards that may be used for the premises network. Different communications technologies have different availability, coverage, bandwidth, interoperability, and security characteristics that limit their suitability for certain applications. Thus, the capabilities and weakness of different communications technologies must be assessed for the specific Smart Grid application to check whether they meet the reliability, efficiency, and security requirements of the Smart Grid.

Z-Wave [18] is a low power wireless communications technology designed for home automation. Unlike ZigBee, which is based on the IEEE 802.15.4 MAC/PHY layer, Z-Wave covers a complete protocol stack from the physical layer to the application layer. It has better bandwidth efficiency to support smart home applications such as remote home control, home safety, and energy reservation. ITU-T G.hn [19] is a wired networking standard that conducts communications over power lines, phone lines, and coaxial cables with data rates up to 1 Gbit/s developed by ITU-T. ITU-T G.hn aims to become the universal wired home networking standard. Other commonly used LAN and wireless LAN (WLAN) standards supporting IP such as Ethernet and Wi-Fi/IEEE 802.11 [20] can also be considered for construction of the premises network. A comparison of some of the potential communications technologies for the premises network in the Smart Grid can be found in Table 5.1.

So far, we have focused mainly on the communications protocols in the MAC/PHY layer. In the application layer, there are a few standards that already exist or are being developed for Smart Grid applications. As we have already mentioned,

TABLE 5.1. Comparison of Communications Technologies for the Premises Network

Technology	Media	Data Rate	Range
ZigBee	Radiofrequency	20–250 kbit/s	10 m to 1.5 km
Z-Wave	Radiofrequency	9.6–40 kbit/s	1 m–70 m
Wi-Fi	Radiofrequency	11–248 Mbit/s	30 m–100 m
HomePlug	Power line	14–200 Mbit/s	200 m
ITU-T G.hn	Power line, phone line, coaxial cable	Up to 1 Gbit/s	NA
Ethernet	Twisted pair	10 Mbit/s to 1 Gbit/s	100 m

the ZigBee/HomePlug smart energy prolife being developed specifies potential Smart Grid applications in the residential environment. In the commercial and industrial environment, existing application level standards such as BACnet, LonWorks, and OpenADR can be used. BACnet [21] (see also details in Chapter 10) is a widely used communications standard developed by the American National Standards Institute (ANSI) and the American Society of Heating, Refrigerating, and Air-Conditioning Engineers (ASHRAE) for building system communications and control networks using IP protocols. It is also adopted internationally as a standard of the International Organization for Standardization (ISO) and used in more than 30 countries. LonWorks [22] is a set of local area networking standards over twisted pairs, power lines, optical fibers, and radio frequency channels supported by the LonMark International Users Group. It can be used for various applications including electric meters, street lighting, home automation, and building automation. LonWorks control networking technology was first standardized in the United States by ANSI and the Consumer Electronics Association (CEA). It is now adopted as an international standard approved by ISO and the International Electrotechnical Commission (IEC). OpenADR [23] (see also details in Chapter 10) is a standard developed at the Lawrence Berkeley National Laboratory and primarily used in California. It can be used by the utilities to implement demand response programs, where pricing signals and load control requests can be issued to industrial and commercial customers. New application layer standards are still needed for emerging Smart Grid applications such as electric vehicles charging.

5.4 NEIGHBORHOOD AREA NETWORK

The neighborhood area network (NAN) spans a greater distance than the premises network, providing communications links for smart meters in a neighborhood area. The NAN connects several premises networks within a neighborhood area via smart meters at the customer premises edges. Depending on the role in the utility network, the NAN can also be called FAN or AMI. The NAN becomes FAN if it is connected to field devices such as intelligent electronic devices (IEDs). In this case, the NAN/

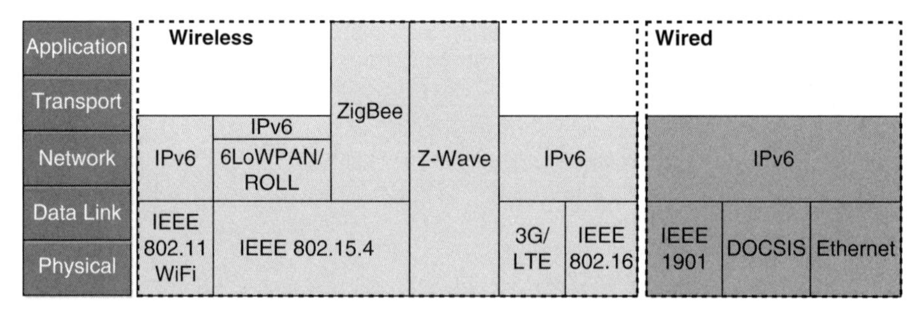

Figure 5.5. Potential communications technologies for NAN/FAN.

FAN is connected to the power distribution network to support distribution automation. Alternatively, the NAN can simply be viewed as a metering network that is part of the AMI providing services such as remote meter reading, control, and detection of unauthorized usage. If distributed energy resources are connected to the NAN, then distributed generation can be implemented. The NAN is connected to a WAN via the backhaul network, where data from many NANs are aggregated and transported between the NANs and the WAN.

The coverage area and bandwidth requirement for the NAN varies for different scenarios, primarily depending on the size of deployment area. Thus, most of the MAC/PHY layer standards for the premises network can be applied to the NAN as well. Some long distance communications technologies can be considered for the NAN if the coverage area is broad. Figure 5.5 shows an overview of the communications technologies that can be used for the NAN/FAN. The choice of technology should be application specific by carefully assessing the network requirements for the Smart Grid. In the following paragraphs, we mainly introduce two sets of wireless communications technologies: the worldwide interoperability for microwave access (WiMAX) and cellular communications technologies including the third generation (3G) and long term evolution (LTE).

WiMAX [24] (see also Chapter 6) is a wireless broadband communications standard originally designed for wireless metropolitan area network (WMAN) based on the IEEE 802.16 standard. It provides long range (around 5 km) and high capacity wireless connections. WiMAX offers flexible broadband links and features low latency (10–50 ms). There are different revisions of the IEEE 802.16 standards of which IEEE 802.16e is currently the most popular implementation. In the PHY layer specified by IEEE 802.16e, several advanced technologies are adopted, such as the scalable orthogonal frequency division multiplexing (OFDM), multiple input multiple output (MIMO), and hybrid automatic repeat request (HARQ) to achieve better frequency reuse, bandwidth efficiency, and larger coverage. The IEEE 802.16e PHY layer can achieve a data rate up to 40 Mbit/s. The frequency band for WiMAX is not uniform. The three licensed spectrum profiles for WiMAX are 2.3 GHz, 2.5 GHz, and 3.5 GHz. In the MAC layer, IEEE 802.16e is connection oriented, which is in contrast to the contention-based CSMA/CA in Wi-Fi/IEEE 802.11. A scheduling algorithm is employed for communications between the subscriber station and the base station. A time slot will be allocated to the subscriber station exclusively if entry to the network is allowed. Other subscribers cannot use the slot once it has been allocated. Such a scheduling algorithm in WiMAX makes QoS possible in the network, which is necessary for Smart Grid applications. The IEEE 802.16e standard supports five QoS classes: unsolicited grant service such as VoIP, real-time polling service such as streaming video/audio, extended real-time polling service such as VoIP with activity detection, non-real-time polling service such as the file transfer protocol (FTP), and best effort service such as the hypertext transfer protocol (HTTP). The priority for the five classes decreases respectively. This QoS classification may not be appropriate for Smart Grid applications. But it is important that WiMAX provides the QoS mechanism for provisioning network resources for different types of applications with a wide range of QoS requirements.

Another set of wireless communications technologies for the NAN are the cellular technologies, which were originally designed for supporting mobile phone services. As cellular technologies evolved, cellular networks today can simultaneously provide voice and data services. Cellular technologies that can be used in the Smart Grid include several standards branded as the 3G technology specified by ITU. The main characteristics that separate 3G technologies from 2G technologies such as global system for mobile communications (GSM) are that 3G can provide data services at a high speed besides conventional mobile phone services. Some of the recently released 3G standards can even provide broadband network access to mobile users with data rate up to several Mbit/s. According to the International Mobile Telecommunications (IMT-2000) specification [25] for 3G technologies, it is expected to provide a minimum speed of 2 Mbit/s for stationary or walking users, and 348 kbit/s for a moving vehicle. Some of the widely used 3G technologies include wideband code division multiple access (WCDMA), high speed packet access (HSPA), and high speed downlink packet access (HSDPA), all of which are based on universal mobile telecommunications systems (UMTS) specified by the 3rd Generation Partnership Project (3GPP), and evolution data optimized (EVDO), which is based on code division multiple access (cdma2000) systems specified by the 3rd Generation Partnership Project 2 (3GPP2). UMTS is the 3G upgrade to GSM networks while cdma2000 is the 3G upgrade to CDMA networks. Note that cdma2000 is the registered trademark of the Telecommunications Industry Association (TIA). In the United States 3G technologies can be used to construct NANs for the Smart Grid using the existing cellular network infrastructure. The widely deployed base stations can cover most of the areas needed. However, to meet the requirements for communications networks in the Smart Grid, optimizations in QoS, reliability, and security are necessary to apply the 3G technologies in the Smart Grid.

Long term evolution (LTE) [26] is the latest cellular technology devolved by 3GPP. It is considered to be the 4G cellular technology to substantially improve end-user throughputs and capacity and reduce latency. LTE has a high spectral efficiency by using advanced techniques such as OFDM, MIMO, and smart antennas. LTE is able to provide peak downlink data rates of at least 100 Mbit/s and uplink data rates of 50 Mbit/s. LTE supports scalable bandwidth from 1.4 MHz to 20 MHz and supports both time division duplexing (TDD) and frequency division duplexing (FDD) on the same platform. Enhanced multicast services and enhanced support for end-to-end QoS in the LTE architecture are also being targeted. One distinguishing feature of LTE is that it is all IP-based. It is scheduled to provide support for IP-based traffic with end-to-end QoS. Voice traffic will be supported mainly as VoIP, enabling better integration with other data services. The support for IP is in line with the requirement for communications networks in the Smart Grid, making LTE one of the promising wireless technologies to be applied in the Smart Grid. Although the initial deployment of LTE has already started in some places, large scale deployment is expected in a few years.

For the NAN in the Smart Grid, we mainly introduced some wireless communications technologies, including WiMAX, 3G and LTE, with more details to be found in Chapter 6. Generally, wireless communications have the advantages of simple infrastructure, good coverage, and flexibility. Once the wireless network is constructed, smart meters can simply be connected to the network. In addition to wireless communications, wired communications technologies such as Ethernet, power line communications, and data over cable service interface specification (DOCSIS) [27] (see Chapter 7 for details) can also be considered to be deployed in the NAN as a complement to wireless communications in order to guarantee robustness and reliability. Ethernet is the most widely used computer networking technology for LANs. It offers a variety of transmission rates ranging from 10 Mbit/s to 100 Gbit/s over twisted-pair cables or optical fibers for different applications. Power line communications standards such as the IEEE 1901 standard are able to provide broadband connectivity over power lines. DOCSIS enables high speed data transmission over an existing cable TV system (CATV) supporting up to at least 160 Mbit/s in the downstream and 120 Mbit/s in the upstream.

In the application layer, functions for the NAN primarily include sensing, monitoring, and controlling the equipment and resources in the power distribution system. There are a number of industrial standards widely used for such purposes. However, they need to be modified for the Smart Grid mainly due to lack of support for IP. The data acquisition, control commands to IEDs, smart meters, and possibly the distributed resources often use a variety of networking technologies and standards. The information exchange of sensing measurements, meter reading, and control signals involves different types of devices. Many different standards in the application layer are used to support these exchanges. To support smart metering, the ANSI C12 protocol suite [28], which standardized the two-way communications with smart meters, can be adopted. These standards cover many aspects of smart meters including electricity metering, watt hour meter sockets, device data tables, meter interfacing to data communications networks, and optical ports. For substation automation, distributed network protocol 3 (DNP3) [29] and IEC 61850 suite [30] can be used to define the communications within substations as well as the communications between the control center and substations. Details on both can be found in Chapter 10. IEC 60870 [31] is a standard used for communications between control centers for message exchange. If distributed resources are connected to the NAN to provide distributed generation, the IEEE 1547 protocol suite [32] can be used for interconnecting distributed resources with electric power systems. This family of standards specifies the interconnections between the utility and distributed generation and storage.

5.5 WIDE AREA NETWORK

The WAN is at the utility's end of the network architecture. It consists of the core/backbone network, the metropolitan area network (MAN), and the backhaul network. The core/backbone network connects the utility backbone and substations commonly using optical fibers, which can provide high capacity communications and minimal latency. If the WAN is owned by the utility company, the term *backbone* is often used, while for service providers, the term *core network* is often used. The MAN refers to the network that connects the backhaul networks within a large city or a metropolitan region. The backhaul network is the link between the WAN and the NAN providing broadband connectivity to the NAN. It is also connected to distribution substation LANs, mobile workforces, automation and monitoring devices in the power transmission and distribution systems including SCADA, remote terminal unit (RTU), phasor measurement unit (PMU), and other sensors.

WAN in the Smart Grid is connected to transmission substation LANs, the utility enterprise LANs, and the public Internet. The transmission substation LAN consists of interconnected transmission substations with protection and control devices in order to provide substation automation. The utility enterprise LANs enable utility controls and operations. It is used for the utility to manage, monitor, and control the information flows from the smart meters, SCADA, substations, and other information flows to control, manage, and supervise the utility's assets, processes, and services such as substation automation, field devices automation, metering, billing, outage management, demand response, and load control. The WAN is interconnected to the public Internet using secure communications, which enables third parties to participate in Smart Grid services. In WAN, the main service is to transport the Smart Grid data reliably and efficiently to distant sites. Therefore, the network devices are mainly switches and routers to transport data at a lower layer to reduce costs. Applications exist at utility control centers and enterprise LANs to monitor, control, and manage the operations and processes in the grid.

Utilities have been operating WANs for various applications such as SCADA, grid monitoring and control, and communications with power plants. These WANs have incorporated a variety of communications technologies over optical fibers, power lines, leased lines, and wireless channels. We mainly focus on optical communications due to its high capacity and predominance in today's WANs. The optical communications standards and protocols in the MAC/PHY layer include multiprotocol label switching (MPLS) [33], MPLS-Transport Profile (MPLS-TP) [34], synchronous optical network (SONET) [35], synchronous digital hierarchy (SDH) [36], optical transport network (OTN) [37], wavelength division multiplexing (WDM) [38], and Metro-Ethernet [39]. These technologies can provide reliable long distance

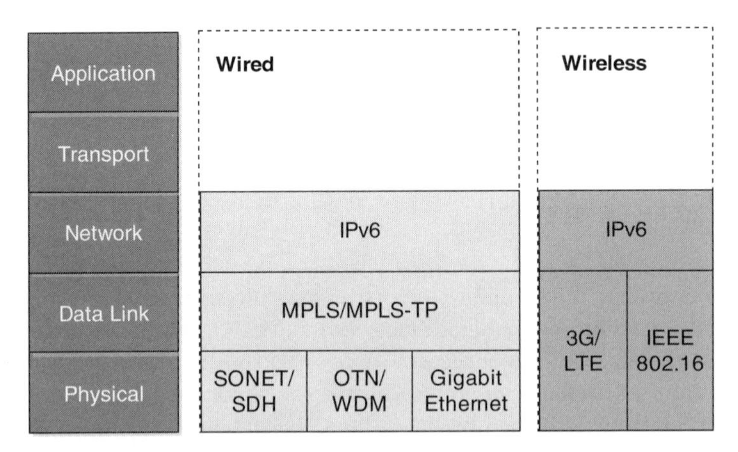

Figure 5.6. Potential communications technologies for WAN.

connectivity up to several thousand kilometers for data transport services in the Smart Grid to support upper-layer applications. In addition to optical communications, wireless technologies such as 3G/LTE and WiMAX sometimes can be a good complement due to the ease of deployment and proven reliability. Figure 5.6 shows an overview of the potential communications technologies for the WAN in the Smart Grid.

MPLS (also discussed in Chapter 9) is a protocol agnostic, connection-oriented, packet-based, data-carrying mechanism using label switching. In MPLS, each packet is assigned with a short fixed-length label. The packet forwarding decision is made using the label. In this way, MPLS is able to create end-to-end virtual connections over any media using any protocols to connect networks over long distances with low latency. It can be used to carry a variety of traffic including IP packets, asynchronous transfer mode (ATM), and Ethernet frames over different physical layer standards such as SONET/SDH, OTN/WDM, and Gigabit Ethernet. MPLS has traffic engineering abilities to explicitly configure the path hop by hop or dynamically set up the path using constrained routing algorithms. The strong traffic engineering abilities of MPLS make it widely used in IP core/backbone networks today. For transport networks, MPLS-TP with transport-oriented extensions in operation administration and maintenance (OAM), survivability, network management, and control plans is now being developed by IETF and ITU-T. MPLS-TP is based on the same architecture of MPLS and it is designed to enable the migration of current time division multiplexing (TDM) based systems such as SONET/SDH toward next generation packet transport networks (PTNs). In the Smart Grid, both MPLS and MPLS-TP are expected to be the main Layer 2 (L2) communications protocol for the construction of a WAN mainly due to their low latency and abilities to support hard QoS and guarantee delay.

SONET/SDH standardized by ANSI and ITU-T, respectively, are standards widely used to provide long distance data transmissions over optical fibers as outlined in more detail in Chapter 8. SONET and SDH are essentially the same. SONET is the standard used in the United States and Canada, while SDH is widely used in the rest of the world. SONET/SDH uses TDM to multiplex different traffic into high speeds and transports them over the networks at a data rate up to 10 Gbit/s. One advantage of SONET/SDH is that it is protocol agnostic and transport oriented, so that different upper-layer traffic such as ATM frames and IP packets can be mapped into the SONET/SDH frames to be transported at high speeds between distant networks. The SONET/SDH network is highly reliable. It supports 50 ms protection, which means even if the optical fiber is cut the transmission path is backed up and can be restored within 50 ms. Furthermore, most of the SONET/SDH networks have a ring topology, which adds high reliability to the whole network. SONET/SDH networks are highly reliable with extensive monitoring and OAM functions required for carrier-class transport. Nowadays, SONET/SDH is still predominant in transport networks.

OTN, standardized within ITU-T, is an emerging transport technology. It specifies the formats for mapping and multiplexing the client signals. It also defines an optical transport module, the corresponding bit rates, and the functionality of the overheads. It supports both coarse and dense WDM to increase the network capacity. WDM is able to multiplex different optical carrier signals on a single optical fiber using separate wavelengths to carry different signals at the same time. It can significantly increase the network capacity. Using WDM, the OTN technology can achieve reliable, cost-effective, versatile, and high-capacity optical networks. Currently, 10 gigabit and 40 gigabit WDM transmission systems are widely deployed in operator networks around the world [40]. The maximum payload capacity in one optical channel, which can be provided by WDM today, is slightly higher than 100 Gbit/s. It is used as the physical layer technology for the emerging 40 gigabit /100 gigabit Ethernet [41].

Metro-Ethernet is a set of networking technologies that connect different LANs in a broad area, usually a metropolitan area based on the Ethernet standard. Ethernet has been the most widespread wired LAN technology for decades. It is also an evolving technology from the original transmission media of coaxial cable to copper, optical fibers, and wireless media. The bandwidth for Ethernet is always increasing from fast Ethernet of 10–100 Mbit/s to Gigabit Ethernet of 1–100 Gbit/s. All of the Ethernet technologies share the same frame formats and can be interconnected through bridging. Therefore, Metro-Ethernet has the natural advantage of interoperability with LANs due to the prevalent use of Ethernet in corporate and residential networks. In addition, an Ethernet interface is much less expensive than a SONET/SDH interface for the same bandwidth. Metro-Ethernet can be implemented using pure Ethernet, Ethernet over SONET/SDH, or Ethernet over MPLS to provide Ethernet connections between distant networks. Pure Ethernet is cheap to deploy but not reliable and scalable. Ethernet over SONET/SDH is applicable when there is existing SONET/SDH infrastructure. Ethernet over MPLS is highly reliable and scalable, and it can run over different physical layer standards such as SONET/SDH, OTN/WDM, and Gigabit Ethernet.

5.6 STANDARDIZATION ACTIVITIES

Today, many utilities have already carried out or are in the process of implementing the Smart Grid technologies in their power transmission, distribution, and customer systems. The value of the Smart Grid realized by the utilities and customers is closely linked to the pace of the development of technologies to enable a reliable, secure, and efficient power grid. As most of the Smart Grid projects start to go far beyond the smart metering, there is an increasing urgent need to establish interoperable standards and protocols for the Smart Grid. Standards are essential to develop, deploy, and operate the Smart Grids worldwide. They play a key role in enabling interoperability of diverse Smart Grid technologies to ensure the success of the intensified worldwide Smart Grid movements. This is of particular importance to the Smart Grid, as it is broad in its scope, and its potential technologies landscape is also very large and complex. In addition, standards promote the application and commercialization of the Smart Grid technology by creating a competitive market for different vendors to compete based on prices and qualities. Lack of standards may impede the commercial implementation of many promising Smart Grid applications and services, such as smart appliances with demand response abilities. In order to establish standards for the Smart Grid, many standardization organizations have taken steps to expedite the development and implementation of the Smart Grid.

NIST, an agency of the U.S. Commerce Department, is assigned "the primary responsibility to coordinate the development of a framework that includes protocols and model standards for information management to achieve interoperability of Smart Grid devices and systems" under the U.S. Energy Independence and Security Act (EISA) of 2007 [2]. NIST published the framework and roadmap for Smart Grid interoperability standards [8] in January 2010 as an output of NIST's efforts to accelerate the development and implementation of key standards to enable the Smart Grid vision. In this framework, a high-level conceptual reference model for the Smart Grid is described and an initial set of 75 standards that can be implemented to support the Smart Grid is identified. Twenty-five of these standards are identified by NIST as the standards for implementation, while the others are subject to further review. Some of the standards related to the communications networks identified for implementation can be found in Table 5.2. As we can see, most of them are upper-layer standards to provide Smart Grid applications for both the utilities and the customers. As shown in Table 5.3, the MAC/PHY layer standards we have discussed in the previous sections are mostly subject to further review due to the specific requirements of the Smart Grid. In the NIST framework, cyber security strategy and requirements are also introduced to address the security issue in the Smart Grid. Moreover, NIST developed 15 priority action plans (PAPs) in collaboration with other standards-setting organizations (SSOs) to develop new standards or revise existing standards, which are urgently needed to address the gaps in the Smart Grid to ensure interoperability, reliability, and security. The PAPs supporting communications in the Smart Grid include guidelines for the use of IP protocol suite in the Smart Grid (PAP01) and guidelines for the use of wireless communications (PAP02) and to harmonize power line carrier standards for appliance communications in the home (PAP15).

TABLE 5.2. Communications Standards Identified for Implementation

Standard	Application
Internet Protocol (IP) Suite	Foundation protocol for delivery of packets in communications networks for the Smart Grid
OpenHAN	Used for HAN to connect to the utility metering system
ZigBee/HomePlug Smart Energy Profile 2.0	Used for HAN in the Smart Grid to support demand-side management
ANSI/ASHRAE 135-2008 ISO 16484-5 BACnet	Used for home automation and building automation
OpenADR	Used for demand response and load control
ANSI/CEA 709 CEA 852.1 LonWorks	Used for home automation and building automation
ANSI C12 Suite	Used for smart meters
IEC 61850 Suite	Used for substation automation
IEEE 1547 Suite	Used for distributed generation and storage
DNP3	Used for substation automation

TABLE 5.3. Some of the Communications Standards Related to Smart Grid

Standard	Application
HomePlug	In-home broadband power line communications
ISO/IEC 12139-1	High speed power line communications protocols
ITU-T G.hn	In-home communications over power lines, phone lines, and coaxial cables
IEEE P1901	Broadband communications over power lines
IEEE P2030	Draft guideline for Smart Grid interoperability of energy technology and information technology operation
IEEE 802 Family	Wired and wireless communications standards developed by the IEEE 802 local area and metropolitan area network standards committee
TIA TR-45/3GPP2 Family	Standards for cdma2000 and high rate packet data systems
3GPP Family	2G, 3G, 4G cellular communications protocols
ETSI GMR-1 3G Family	Satellite-based packet service equivalent to 3GPP standards
ISA SP100	Wireless communications standards for industrial users
Z-Wave	Wireless mesh networking protocol for HAN

The initial PAPs only mark the beginning of an accelerated development and sustained standardization effort from NIST. New PAPs will emerge to cover the larger scope of standardization efforts that will lead to a fully interoperable Smart Grid.

The NIST framework and roadmap for Smart Grid interoperability standards [8] identified the applicable standards and requirements, the gaps in existing standards, and the priority for standardization activities. It only represents an important first step of the NIST's three-phase plan for sustainable development and implementation of Smart Grid related standards. The second phase of the plan was launched in November 2009 with the establishment of the Smart Grid interoperability standards panel (SGIP) to provide a representative, reliable, and responsive forum for stakeholders and SSOs to sustain the ongoing development of Smart Grid interoperability standards. The last phase of the plan is to develop a Smart Grid conformity testing and certification framework to support a rigorous standards conformity and interoperability testing process for the Smart Grid systems and devices with initial steps toward implementation in 2010. The three-phase plan aims to accelerate the Smart Grid standardization activities by continuously developing and promoting related standards in order to realize the Smart Grid in the near future.

IEEE is a major international standardization organization working closely with NIST. There are more than 100 IEEE standards available or under development relating to the Smart Grid in diverse fields including digital information and control, networking, security, sensors, electric metering, power line communications, and systems engineering. Twenty such IEEE standards including those IEEE standards we have discussed in previous sections have been identified in the NIST framework and roadmap for Smart Grid interoperability standards. IEEE is participating in a number of NIST PAPs to address various issues from different aspects of the Smart Grid. IEEE launched its own Smart Grid project named the IEEE standard 2030 guide for Smart Grid interoperability of energy technology and information technology operation with the electric power system (EPS) and end-use applications and loads in May 2009. In the IEEE P2030 work group, there are three task forces with different focuses on power engineering technology, information technology, and communications technology. The IEEE P2030 standard is identified as a standard subject to review by NIST.

IEC is also working closely with NIST to create standards supporting the Smart Grid. There are approximately 10 IEC standards identified by NIST for distribution automation, substation automation, control center automation, metering, and distributed generation. IEC is forming a Strategic Group (SG 3) on the Smart Grid to help IEC develop Smart Grid standards. The SG 3 published the IEC Smart Grid standardization roadmap [42] in June 2010. The roadmap identifies existing standardization and potential gaps in the IEC portfolio related to different aspects of the Smart Grid including communications, security, and planning. IEC SG 3 is also developing a standardization framework that provides guidelines and a set of standards for the industry on Smart Grid projects. Currently, there are 24 IEC Technical Committees (TCs) participating in the Smart Grid efforts from different aspects. For example, TC 57 has created a family of international standards for power system control equipment and systems. These standards include IEC 61850, which is designed for substation automation, and IEC 61970/61968 for energy management system (EMS) (see Chapter 10).

IETF primarily addresses the networking issues of applying IETF standards and protocols such as IP in the Smart Grid. As the leading SSO in the NIST PAP01 (guidelines for the use of IP protocol suite in the Smart Grid), IETF is identifying the key protocols of the IP suite for the Smart Grid networks and the key networking issues in the Smart Grid such as addressing, routing, and security. It is developing an RFC listing of IETF standards applicable for the Smart Grid. The output of the standardization activities from IETF so far includes Internet Protocols for the Smart Grid (Internet draft) [9], the IETF 6LoWPAN protocol [12], and the IETF ROLL protocol [13], all of which are used to support IP in the Smart Grid as required by NIST. In addition, IETF is also an important SSO in the NIST PAP02 (guidelines for the use of wireless communications) to identify key issues, requirements, and guidelines for effectively, safely, and securely employing wireless technologies for various Smart Grid applications.

ITU-T established a Focus Group on the Smart Grid (FG Smart) to identify and study the related standards to familiarize ITU-T and the standardization community with emerging attributes of the Smart Grid in 2010. The objective of ITU-T FG Smart is to investigate documents and information that would be helpful for ITU-T to develop recommendations to support the Smart Grid from a telecommunications perspective. As one of the initial outputs, ITU-T FG Smart has published its repository of activities in a Smart Grid standardization [43], which identifies the Smart Grid related activities of 20 organizations and groups. Within ITU-T, Study Groups (SGs) 5 and 15 are also closely related to the Smart Grid. SG 5 works on environment, climate change, and energy efficiency. SG 15 works on transport and access networking standards including the ITU-T standards we have discussed in previous sections such as ITU-T G.hn, SDH, OTN, and MPLS-TP.

There are some other SSOs actively promoting worldwide standardization activities for the Smart Grid. Such organizations include ANSI, ISO, the Electric Power Research Institute (EPRI), the European Committee for Standardization (CEN), the European Committee for Electrotechnical Standardization (CENELEC), the European Telecommunications Standards Institute (ETSI), and the UCA International Users Group (UCAIUG). Many industrial alliances such as the ZigBee Alliance, the HomePlug Alliance, the Z-Wave Alliance, the Wi-Fi Alliance, and the WiMAX Forum are involved in the standardization for the Smart Grid as well, since their standards and protocols need to be modified or further developed to be applied in the Smart Grid. As the Smart Grid related technologies develop, standardization activities from various organizations and industrial alliances will keep emerging in order to finally implement the Smart Grid.

5.7 CONCLUSIONS

The electric power grid today is undergoing a significant transition to an intelligent, efficient, and fully interconnected Smart Grid. In the Smart Grid vision, customers are active participants involved in the power system by demand response and distributed generation. They are provided with diverse smart energy services in a mature, well-integrated energy wholesale market. Distributed energy resources can be connected to the Smart Grid with plug-and-play convenience, which enables the integration of renewable energies to the grid. Utilities are able to operate and manage the Smart Grid in a simple way using advanced data acquisition techniques and automation systems to guarantee the reliability and efficiency of the power. The Smart Grid is also robust to malicious attacks and natural disasters with the ability of rapid restoration.

To realize the Smart Grid vision, the key part is the communications network, which connects all the devices, systems, and customers for a two-way information and energy flow in the Smart Grid. The communications network is the fundamental infrastructure to provide end-to-end information exchange between and among stakeholders as well as smart devices in the Smart Grid, which is essential to support upper-layer applications such as SCADA systems with specialized protocols. As the Smart Grid is a complex system of systems, the communications network is a complicated network of networks interconnecting a large number of heterogeneous devices and systems with various ownerships and management boundaries. To deal with such a complex network, understanding the network architecture is essential for the construction of communications networks for the Smart Grid.

In this chapter, we first introduced and described the communications network architecture for the Smart Grid. In this architecture, the communications network for the Smart Grid is divided into three types of networks according to their characteristics and functions in the Smart Grid: the premises network (i.e., HAN, BAN, and IAN), which provides network access to the appliances in the customer premises, NAN/FAN, which connects to smart meters, intelligent field devices, and distributed resources; and WAN, which links the grid to the utility core system. Then, we introduced the characteristics and functionalities as well as some potential technologies to construct each type of communications network for the Smart Grid.

Moreover, we gave an overview of the current standardization activities on communications networks for the Smart Grid. Standards play a vital role in the development and implementation of an interoperable Smart Grid. They are essential for the diverse technologies of the Smart Grid. Existing standards may need to be revised to address the gaps for the Smart Grid and new standards may need to be established to enable various Smart Grid applications. Many standardization organizations have realized the urgency to establish standards for the Smart Grid, which leads to the rise of various Smart Grid standardization projects. Among them, NIST is leading the Smart Grid interoperability standardization activities in collaboration with a number of standard organizations such as IEEE, IEC, IETF, ITU-T ANSI, and ISO. Although these standardization activities have started and some of them have already produced important outputs, sustainable development and implementation of Smart Grid standards are necessary toward the realization of an intelligent, efficient, and fully interconnected Smart Grid.

REFERENCES

[1] U.S. Department of Energy, The Smart Grid: An Introduction, 2009.

[2] U.S. Public Law No. 110-140, "Energy Independent and Security Act of 2007," Title XIII, Sec. 1301, 2007.

[3] L. H. Tsoukalas and R. Gao, "From Smart Grids to an energy internet: Assumptions, architectures, and requirements," *Proceeding of 3rd International Conference on Electric Utility Deregulation and Restructuring and Power Technology*, pp. 94–98, April 2008.

[4] Trilliant Inc., The Multi-Tier Smart Grid Architecture, 2010.

[5] E. Darmois, "Smart Grid: A transformational (standards) journey," *Proceedings of IEEE Smart Grid Comm, Gaithersburg, MD*, October 2010.

[6] Sonoma Innovation, *Smart Grid Communications Architectural Framework*, August 2009.

[7] National Institute of Standards and Technology, *The Role of IP in the Smart Grid*, October 2009.

[8] National Institute of Standards and Technology, *Framework and Roadmap for Smart Grid Interoperability Standards*, January. 2010.

[9] F. Baker and D. Meyer, "Internet protocols for the Smart Grid," *IETF Internet draft draft-baker-ietf-core-08*, September 2010.

[10] S. Deering and R. Hinden, Internet protocol, version 6 (IPv6) specification," *IETF RFC 2460*, December 1998.

[11] IEEE Std. 802.15.4-2003, *Wireless Medium Access Control (MAC) and Physical Layer (PHY) Specifications for Low-Rate Wireless Personal Area Networks (WPANs)*, October 2003.

[12] G. Montenegro, N. Kushalnagar, J. Hui, and D. Culler, "Transmission of IPv6 packets over IEEE 802.15.4 networks," *IETF RFC 4944*, September 2007.

[13] J. Martocci, P. De Mil, N. Riou, and W. Vermeylen, "Building automation routing requirements in low-power and lossy networks," *IETF RFC 5867*, June 2010.

[14] U.S. Federal Information Processing Standards Publication 197, *Specification for the Advanced Encryption Standard (AES)*, November 2001.

[15] ZigBee Document 075356r15, "ZigBee smart energy profile specification," December 2008.

[16] ZigBee Alliance and HomePlug Powerline Alliance liaison, "Smart energy 2.0 draft," April 2010.

[17] IEEE Std. 1901–2010, *Draft Standard for Broadband over Power Line Networks: Medium Access Control and Physical Layer Specifications*, September 2010.

[18] Z-Wave Alliance, "Z-Wave controls any or all of your home," 2010.

[19] ITU Rec. G.9960, *Next Generation Home Networking Transceivers*, June 2010.

[20] IEEE Std. 802.15.11, Wireless LAN Medium Access Control (MAC) and Physical Layer (PHY) Specifications, June 2007.

[21] ISO 16484-5:2007, *Building Automation and Control Systems—Part 5: Data Communication Protocol*, 2007.

[22] ISO/IEC DIS 14908-1, *Interconnection of Information Technology Equipment—Control Network Protocol—Part 1: Protocol Stack*, April 2010.

[23] Lawrence Berkeley National Laboratory and Akuacom, "Open automated demand response communications specification," April 2009.

[24] IEEE Std. 802.16e, *Air Interface for Broadband Wireless Access Systems*, May 2009.

[25] ITU Rec. M.687, *International Mobile Telecommunications—2000*, February 1997.

[26] 3GPP Spec. TS 36.201, *Evolved Universal Terrestrial Radio Access (E-UTRA): LTE Physical Layer; General Description*, March 2010.

[27] Cable Television Laboratories Inc., "Data over cable service interface specifications DOCSIS 3.0—Physical Layer Specification," October 2010.

[28] ANSI C12.20-2010, *American National Standard for Electricity Meters—0.2 and 0.5 Accuracy Classes*, August 2010.

[29] IEEE Std. 1815–2010 *IEEE Standard for Electric Power Systems Communications—Distributed Network Protocol (DNP3)*, July 2010.

[30] IEC 61850-5, *Communication Networks and Systems in Substations—Part 5: Communication Requirements for Functions and Device Models*, July 2003.

[31] IEC 60870-5-101, *Telecontrol Equipment and Systems—Part 5-101: Transmission Protocols—Companion Standard for Basic Telecontrol Tasks*, Feb. 2003.

[32] IEEE Std. 1547.2-2008, *IEEE Application Guide for IEEE Std. 1547, IEEE standard for Interconnecting Distributed Resources with Electric Power Systems*, April 2009.

[33] V. Sharma and F. Hellstrand, "Framework for multi-protocol label switching (MPLS)-based recovery," *IETF RFC 3469*, February 2003.

[34] M. Bocci, S. Bryant, D. Frost, L. Levrau, and L. Berger, "A framework for MPLS in transport networks," *IETF RFC 5921*, July 2010.

[35] ANSI T1.105.03-2002(R2007), *Synchronous Optical Network (SONET): Physical Layer Specifications*, 2007.

[36] ITU-T Rec. G.803, *Architecture of Transport Networks Based on the Synchronous Digital Hierarchy (SDH)*, March. 2003.

[37] ITU-T Rec. G.709/Y.1331, *Interfaces for the Optical Transport Network (OTN), December.* 2009.

[38] ITU-T Rec. G.694.1, *Spectral Grids for WDM Applications: DWDM Frequency Grid*, October 2002.

[39] S. Halabi, *Metro Ethernet*. Cisco Press, October 2003.

[40] M. Carroll, J. Roese, and T. Ohara "The operator's view of OTN evolution," *IEEE Commun. Mag.*, September 2008.

[41] The Ethernet Alliance, "40 gigabit Ethernet and 100 gigabit Ethernet technology overview," November 2008.

[42] IEC SG3, "IEC Smart Grid standardization roadmap," June 2010.

[43] ITU-T FG Smart, "Activities in Smart Grid standardization," April 2010.

6

WIRELESS COMMUNICATIONS IN SMART GRIDS

Juan José García Fernández, Lars Torsten Berger,
Ana García Armada, María Julia Fernández-Getino García,
Víctor P. Gil Jiménez, and Troels B. Sørensen

6.1 INTRODUCTION

Wireless technologies are one option contending for deployment in *Smart Grid* (SG) applications. Their counterparts, wireline and fiber optical communications, are discussed in Chapter 7 and Chapter 8 respectively. Sometimes, for example, in difficult to access terrain, they might even be the most cost-effective solution. In other situations they might be deployed in hybrid wireless–wireline SG solutions as, for example, presented in Chapter 14.

Wireless communications employ *radiofrequency* (RF) as the medium for the interconnection of the nodes in the network. The RF spectrum is not only a scarce resource that should be regulated to allow for proper use of it, but it also has different properties depending on the operating frequency, which makes different frequencies interesting for different applications. Figure 6.1 shows the different technologies and the approximate frequency bands they use. As a general rule of thumb, the lower

Smart Grid: Applications, Communications, and Security, First Edition.
Edited by Lars Torsten Berger and Krzysztof Iniewski.

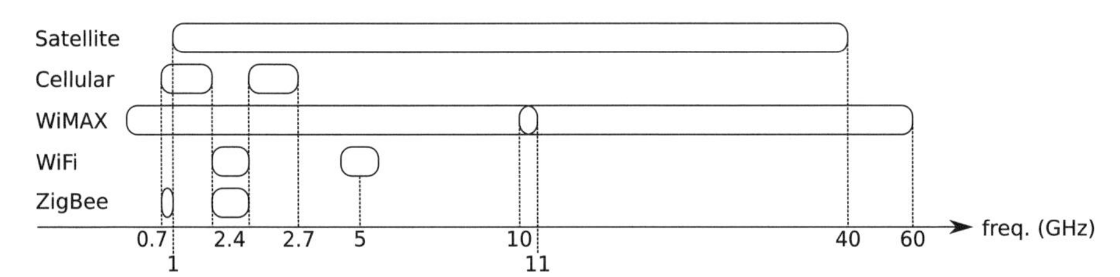

Figure 6.1. RF spectrum.

the frequency the better the obstacle penetration. This makes lower frequency bands appropriate for *non-line-of-sight* (NLOS) communications. Higher frequencies imply higher attenuation both in free space propagation and caused by obstacles, which restricts these bands to *line-of-sight* (LOS) communications. On the other hand, lower frequency bands have less bandwidth available, which impacts directly on the capacity achievable, while higher frequency bands have the potential to employ higher bandwidths.

Different types of wireless technologies also have different availability, time-sensitivity, and security characteristics that may constrain the applications for which they are suitable. Wireless technologies can be used in field environments across the SG including generation plants, transmission systems, substations, distribution systems, and customer premises communications.

The choice of the type of wireless system to use must be made with knowledge of the appropriate applicability of the technology. *Priority Action Plan* (PAP) 2 [1] from the *National Institute of Standards and Technology* (NIST) is focused on identifying the requirements for use of wireless technologies for the SG [2] and delivers a framework for evaluating the strengths and weaknesses of candidate wireless technologies [3] to assist SG design decisions. This framework is summarized in Figure 6.2, where *physical* (PHY) and *medium access control* (MAC) blocks, together with the channel propagation model, are the most important. In addition to the documentation provided, the PAP2 provides some tools that can be used for the modeling and simulation of the candidate technologies, as an input for Task 6 in reference [1].

Figure 6.3 shows a simplified version of the conceptual diagram of the SG from NIST [3]. It contains the different parts of the SG including the communication networks required for the interconnection of the main actors in the SG. The network model depicted in Figure 6.3 was already described in Chapter 5. Three types of networks are identified:

- *Premises network*, located at the customer premises, is responsible for the interconnection of the customer equipment, appliances, meters, and such actors that play a role within the customer's location.
- *Neighborhood area network* (NAN) or *field area network* (FAN) covers a larger area and therefore is used for communication between customer premises and both distribution and transmission parts of the SG.

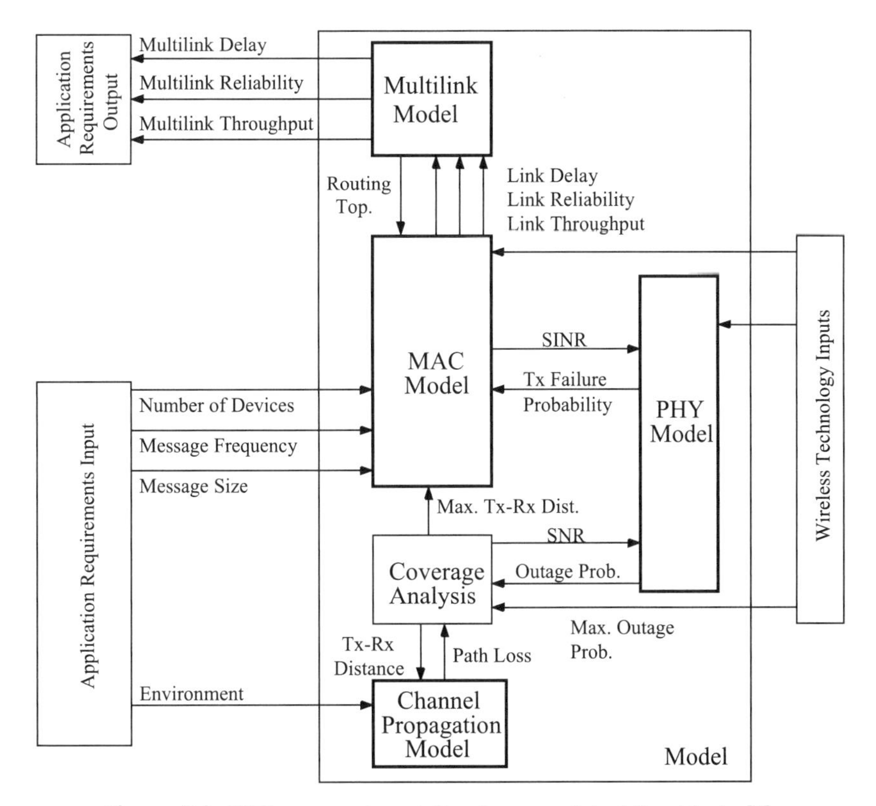

Figure 6.2. NIST proposed modeling framework building blocks [3].

- *Wide area network* (WAN) has the widest range and is used mainly for interconnection within the transmission part of the SG, and to communicate this with the bulk generation domain.

From all the candidate technologies presented by the NIST [2] the following are described in this chapter:

- ZigBee *wireless personal area networks* (WPANs) in Section 6.2
- Wi-Fi *wireless local area networks* (WLANs) in Section 6.3
- WiMAX *wireless metropolitan area networks* (WMANs) in Section 6.4
- Cellular networks in Section 6.5
- Satellite networks in Section 6.6

There are several other technologies that can be employed within the SG, as it can be seen in references [2] and [3] from the NIST and references [4] and [5] from OPENmeter, a project focused on the specification of a comprehensive set of

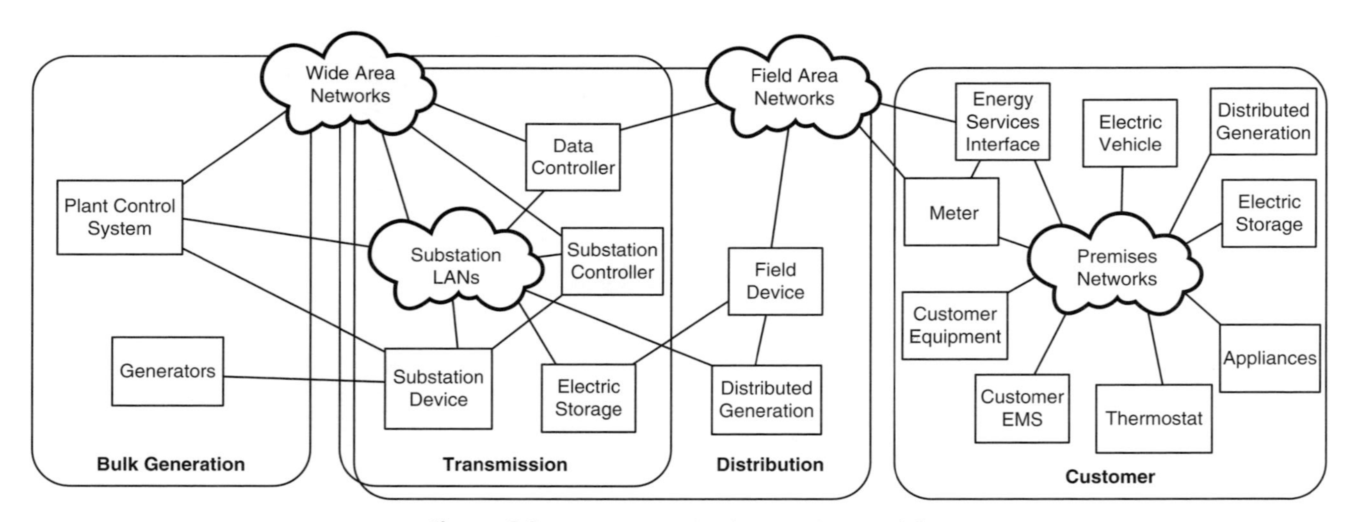

Figure 6.3. SG conceptual reference diagram [3].

standards for *advanced metering infrastructure* (AMI). Some of the alternatives to ZigBee are, for example, Bluetooth [6] and Z-Wave [7]. Nevertheless, for the sake of brevity, only ZigBee is analyzed in this chapter. Notwithstanding, the analysis of ZigBee can provide an overview of the problems that a technology has to deal with in the field of WPANs.

ZigBee, Wi-Fi, WiMAX, and cellular networks are designed to work in a relatively low frequency band, around 2 GHz, which is well suited for the NLOS operation. WiMAX also includes support for higher frequency bands, mainly for LOS applications. Satellite systems cover a wide portion of the spectrum; nevertheless, the long links between satellites and users on the ground make the NLOS communication infrequent even for low frequency bands.

The terms WPAN, WLAN, and WMAN are frequently used in the networks and communications field, while they are not so common in the SG world. They relate to the area covered by the networks and, as such, they do not match perfectly to the network model of the SG. In the SG network model from Figure 6.3 the different types are organized more with regard to the structure of the SG itself, than the geographical area spanned. Nevertheless, the wireless technologies mentioned can be related to some part of the SG network model, although some of them may be applicable for more than one category.

Within the customer premises, the peers that form the *premises network* are located in a confined area, and typically the communications will take place in non-line-of-sight circumstances. Under these conditions the ZigBee and Wi-Fi technologies are good options for covering this network in the SG. WANs potentially have long links which renders technologies like ZigBee and Wi-Fi unsuitable. For these networks, WiMAX and other cellular technologies like *long term evolution* (LTE) are better alternatives. Satellite systems have the unique feature of being the only option that is able to provide global coverage, making them the only solution for connecting remote locations where other technologies simply cannot be deployed. Satellite coverage enforces their use as backup networks especially for WANs, although they require LOS. Finally, for the case of NAN/FAN, the scenarios that can be found are a mixture of the premises networks and WANs and, therefore, more or less any of the technologies covered in this chapter can be used, with the exception of ZigBee, due to its low data rate.

Another advantage that is natural in wireless communications is the ability to connect mobile users. Mobility support is highly dependent on the technology as, for example, satellite communications in high frequency bands require antenna steering, making the system complex and expensive, while for low frequencies the antenna does not necessarily require steering. The NIST [2] provides information about the maximum speed supported by the different technologies. For ZigBee it is about 20 km/h, for Wi-Fi 70 km/h, and for WiMAX 120 km/h, while it exceeds 300 km/h for some cellular systems. The mobility support for satellite systems, as already mentioned, is much more technology dependent.

It is worth noting why in this chapter WiMAX, although similar to other cellular technologies like LTE, is treated separately. Cellular technologies treated in Section 6.5 essentially work in licensed bands, where a license owner is responsible for the deployment and the use of the network. These networks are intended to cover nationwide territory, and they require careful frequency and cell planning to optimize the performance of the network. WiMAX, on the other hand, is not restricted to work in licensed bands. The use of *industrial, scientific, and medical* (ISM) bands allows deployment of custom networks without depending on a third party network operator. Moreover, the specification of WiMAX includes a physical layer that works at high frequencies using single carrier technologies. This can be used to establish high data rate point-to-point links to interconnect distant locations easily.

The chapter is organized as follows. In Section 6.2 ZigBee is introduced and described as the option for short distance and low data rate communications. Section 6.3 treats the WLAN standards, with the IEEE 802.11n as the latest and most advanced of all possible alternatives. In Section 6.4 WiMAX is analyzed. In Section 6.5 cellular systems are presented, both contemporary systems like *Global System for Mobile Communications* (GSM), as well as the latest systems such as LTE. Satellite networks are described in Section 6.6. Finally, in Section 6.7 some conclusions are drawn.

6.2 WIRELESS PERSONAL AREA NETWORKS

Wireless personal area networks (WPANs) address wireless networking of portable and mobile computing devices such as PCs, cell phones, wireless sensors, and other consumer electronics, allowing them to communicate and interoperate with each other. These networks typically require short distance communications, in the range of tens of meters, and focus on long-lasting battery life and low deployment and maintenance complexity. At the same time, these kinds of networks require moderate to low data rates.

Since sensing and communication capabilities are essential for the development of the *Smart Grid* (SG) (see Chapter 10 for more details on sensing protocols), WPANs can play an important role in simplifying sensing and monitoring of several parts of the electric power system.

Among the several working groups that form the 802 Local and Metropolitan Area Network Standards Committee of the *Institute of Electrical and Electronics Engineers* (IEEE) Computer Society, is the 802.15 working group for wireless personal area networks, which focuses on the development of consensus standards for personal area networks or short distance wireless networks. Among all the *task groups* (TGs) that form the 802.15 group, TG 4 was created in 2000 to work on low complexity communications, focusing on battery life more than on achieving high data rates or long distance links. Published in September 2006, the latest standard is IEEE 802.15.4-2006 [8]. In November 2008, a new task group, TG 4g [9], was created to propose an amendment to the IEEE 802.15.4-2006 standard, that intends to provide a global standard for very large scale process control applications including the utility Smart Grid network. The result of this standardization process is a technological basis over which the industry has built ZigBee.

ZigBee has been developed by an industry consortium called ZigBee Alliance [10], as a complete specification covering the physical up to the application layer. The ZigBee Alliance is responsible for ensuring compatibility and interoperability among products, apart from promoting the technology. ZigBee specification [11] is built on top of the *physical* (PHY) and *medium access control* (MAC) layers defined in the standard IEEE 802.15.4-2003, although this was superseded by the publication of IEEE 802.15.4-2006, whose key features are:

- Data rates of 250 kbit/s, 40 kbit/s, and 20 kbit/s
- Support for critical latency devices
- *Carrier sense multiple access-collision avoidance* (CSMA-CA) channel access
- Automatic network establishment by a coordinator
- Power management to ensure low power consumption

Among other applications, the ZigBee Smart Energy profile offers secure, easy-to-use wireless *home area networks* (HANs) for managing energy, offering the possibility to directly communicate with thermostats and other smart appliances. Details are given in Section 6.2.4.

The ZigBee protocol stack is shown in Figure 6.4. Next, the different layers are described.

6.2.1 802.15.4 Physical Layer

The *physical layer* (PHY) transmits and receives data wirelessly and is responsible for the following:

- Activation and deactivation of the radio transceiver.
- *Energy detection* (ED) within the current channel and *link quality indicator* (LQI) for received packets. The LQI measurement is a characterization of the strength and/or quality of a received packet. The measurement may be implemented using receiver ED, a signal-to-noise ratio estimation, or a combination of these methods, as specified in IEEE 802.15.4-2006 [8]. The standard does not dictate how the LQI parameter should be used by upper layers.
- *Clear channel assessment* (CCA) to enable carrier sense multiple access with collision avoidance (CSMA-CA).
- Channel frequency selection.
- Data transmission and reception.

The main PHY parameters are specified in IEEE 802.15.4-2006 [8], 802.15.4c-2009 [9], and 802.15.4d-2009 [13]. These parameters are summarized in Table 6.1.

Another key aspect of the physical layer is the transmitted power, which is not fixed by the standard, and it should conform with local regulations, controlled by the following regional authorities [8]:

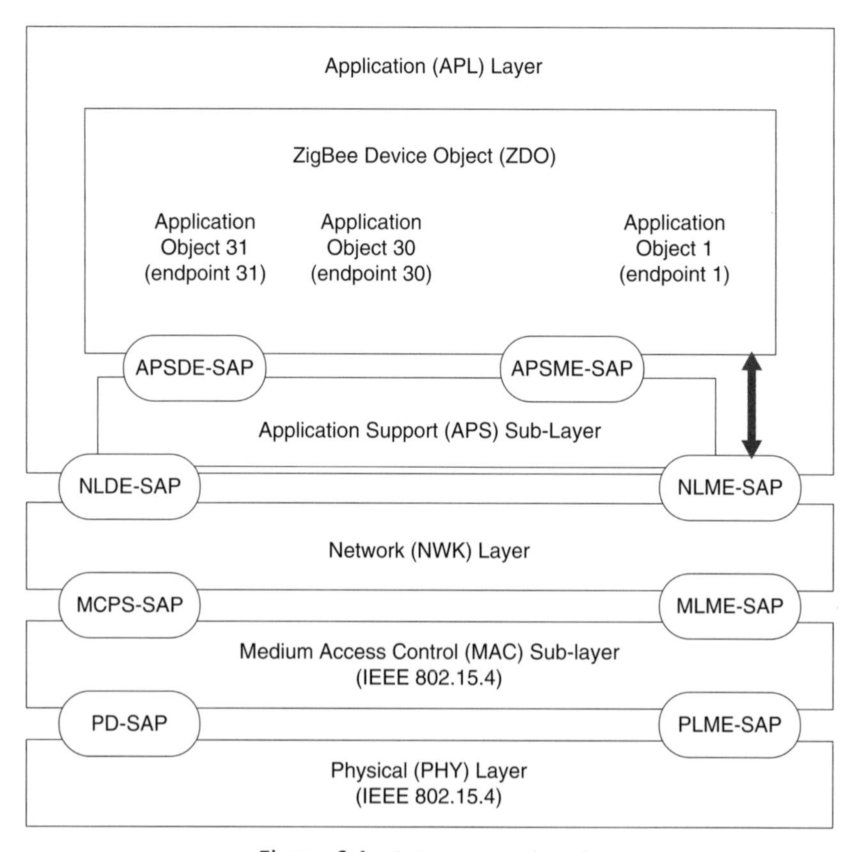

Figure 6.4. ZigBee protocol stack.

TABLE 6.1. The 802.15.4 PHY Parameters

PHY	Frequency (MHz)	Modulation	Spreading	Data Rate (kbit/s)	Rx Sensitivity (dBm)
780	779–787	MPSK	DSSS	250	–85
780	779–787	O-QPSK	DSSS	250	–85
868	868–868.6	BPSK	DSSS	20	–92
868	868–868.6	ASK	PSSS	250	–85
868	868–868.6	O-QPSK	DSSS	100	–85
915	902–928	BPSK	DSSS	40	–92
915	902–928	ASK	PSSS	250	–85
915	902–928	O-QPSK	DSSS	250	–85
950	950–956	BPSK	DSSS	20	–85
950	950–956	GFSK	—	100	–85
2450	2400–2483.5	O-QPSK	DSSS	250	–85

- *European Telecommunications Standards Institute* (ETSI) in Europe
- *Association of Radio Industries and Businesses* (ARIB) in Japan
- *Radio Management Bureau of the Chinese Information Department* in China
- *Federal Communications Commission* (FCC) in the United States
- *Industry Canada* (IC) in Canada

However, IEEE 802.15.4-2006 [8] describes how the low power objective of the standard makes it reasonable to assume that typical transmitted power will be within –3 dBm and 10 dBm. Appendix F in IEEE 802.15.4-2006 [8], 802.15.4c-2009 [12], and 802.15.4d-2009 [13] give more detailed information about regulatory requirements for each of the possible configurations supported by the standard. These figures and those given in the rest of the chapter correspond to the power for the corresponding transmission bandwidth. For the case of 802.15.4, this bandwidth depends on the band, but it is 5 MHz for the 2.4 GHz band (see Section 6.1.2.1 of IEEE 802.15.4-2006 [8]).

Another parameter that is given in IEEE 802.15.4-2006 [8], 802.15.4c-2009 [12], and 802.15.4d-2009 [13] is the receiver sensitivity, defined as the threshold input signal power that yields a specified PER (packet error rate), assuming an interference-free situation. In Table 6.1 the different sensitivity values are given for the different PHY defined by the standard.

The channel and path loss model, dependent on the scenario and application, is also another key aspect to analyze the physical layer performance of the standard. Due to the special characteristics of the technology involved (i.e., short range communications), one common model used is the Hata model [14], as proposed in the NIST PAP 2 [1]. Another channel model used is the 802.11 channel model as adapted in IEEE 802.15.2-2003 [15] and IEEE 802.15.3-2003 [16].

6.2.2 802.15.4 Medium Access Control Sublayer

The *medium access control* (MAC) sublayer handles access to the physical radio channel and is responsible for:

- Generating the network beacon and synchronizing to it
- Supporting WPAN association and disassociation
- Employing the CSMA-CA mechanism for channel access
- Providing frame error-free communications between two peer MAC entities

Network beacons are sent by the network coordinator, described in more detail in Section 6.2.3, and are used to synchronize the attached devices, to identify the WPAN, and to describe the structure of a special type of MAC frame called superframe. Any device wishing to communicate during the contention access period (CAP) between two beacons competes with other devices using a slotted CSMA-CA mechanism. All transactions must be completed by the time of the next network beacon. For low-latency applications or applications requiring guaranteed data bandwidth, the WPAN coordinator may assign portions of the active superframe, called *guaranteed time slots* (GTSs).

Two possibilities exist for the CSMA-CA operation, depending on the use of beacons by the coordinator. Nonbeacon-enabled WPANs use an unslotted CSMA-CA channel access mechanism. Beacon-enabled WPANs use a slotted CSMA-CA channel access mechanism, where the back-off slots are aligned with the start of the beacon transmission. The back-off slots of all devices within one WPAN are aligned to the WPAN coordinator. The general operation of the CSMA-CA algorithm is described in Section 6.3.2.

The MAC establishes the procedure for a device to join or leave a WPAN, called association and disassociation. However, the algorithm for selecting a suitable WPAN with which to associate from the list of WPAN descriptors returned from a channel scan is out of the scope of the standard.

6.2.3 ZigBee Network Layer

The network layer builds upon the IEEE 802.15.4 MAC's features to allow extensibility of coverage. The responsibilities of the network layer include:

- Establishing a network
- Joining and leaving a network
- Configuring a new device
- Assigning addresses to devices joining the network
- Synchronization within a network
- Routing frames to their intended destinations

As defined in standard 802.15.4-2006, devices are one of two types, depending on the functions they can perform within the network. They can be *full function devices* (FDDs) or *reduced function devices* (RFDs). FDDs can operate in any topology, can act as a WPAN coordinator, and can talk to any other device in the network. RFDs, on the other hand, are limited to star topologies, cannot take any coordinator role, and are restricted to communicate with the coordinator of the network.

There are two possible networking topologies—star and peer-to-peer topologies—as described in the 802.15.4-2006 standard [8] and displayed in Figure 6.5. In the star topology the communication is established between each device and a WPAN coordinator that controls the communication in the whole network. In the peer-to-peer topology there is also a coordinator; however, each device can communicate with any other device within reach. Peer-to-peer topologies are more complex and allow mesh networking with self-organizing and self-healing characteristics. Also, the coverage may be extended by multihop communications. Routing in this topology is carried out using a combination of tree routing and on-demand nontree routing. Based on the ZigBee network specification [11], a slightly modified version

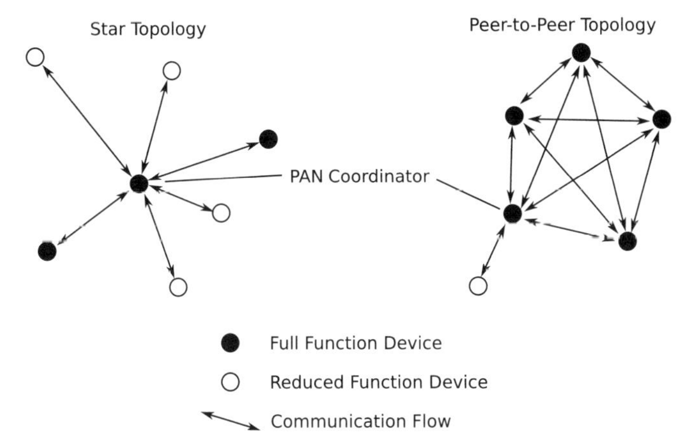

Figure 6.5. ZigBee network topologies.

of the cluster-tree algorithm [17] is used. The on-demand table driven routing is based on the AODV (ad hoc on-demand distance vector) routing algorithm [18] and a simplified version AODVjr (ad hoc on-demand distance vector junior) [19], which guarantees low power consumption.

6.2.4 ZigBee Application Layer

As seen in Figure 6.4 the ZigBee application layer consists of the *application support sublayer* (APS), the *ZigBee device object* (ZDO), and the manufacturer-defined application objects. The APS provides an interface between the network layer and the application layer through a general set of services that are used by both the ZDO and the manufacturer-defined application objects [110].

One of the responsibilities of the ZigBee Alliance is the definition of what are called *application profiles*. These are agreements for messages, message formats, and processing actions that enable developers to create an interoperable, distributed application employing application entities that reside on separate devices. In particular, the ZigBee Smart Energy profile [20] is of interest in the SG context, for it enables users to manage their usage and generation of energy. This is achieved through the information flow between devices such as meters, smart appliances, plug-in electric vehicles, and energy management systems.

The ZigBee Alliance has established a liaison with the HomePlug Powerline Alliance [21], in order to develop a ZigBee Smart Energy profile version 2.0 [22]. This new version will be IP based, and it will include other features like support for MAC/PHY options different from the 802.15.4 (e.g., 802.11, IEEE 1901). As such, this application profile is mostly an application layer specification, where lower layer protocols are not discussed. More on the ZigBee Smart Energy information model may also be found in Chapter 10.

6.3 WIRELESS LOCAL AREA NETWORKS

Local area networks (LANs) are intended to interconnect several devices in a geographical area such as homes, computer laboratories, or office buildings, with a distance on the order of hundreds of meters. Wireless LANs (WLANs) are the wireless alternative to traditional LAN standards such as Token Ring, IEEE 802.5 [23], and Ethernet, IEEE 802.3 [24]. The main wireless alternative to wired solutions is the standard IEEE 802.11 [25], usually known as Wi-Fi, coined by the Wi-Fi Alliance [26].

In 1999 the Wi-Fi Alliance was founded as a global nonprofit industry association of hundreds of companies devoted to the proliferation of Wi-Fi technology. The Alliance adopted the IEEE 802.11 specifications as the technology basis and it launched a program to guarantee interoperability among Wi-Fi products from different manufacturers.

From all the standards created by the 802.11 group, the most widely used are the following:

IEEE 802.11a, approved in 1999, provides up to 54 Mbit/s PHY throughput in the 5 GHz *industrial, scientific and medical* (ISM) band by using *orthogonal frequency division multiplexing* (OFDM) [27].

IEEE 802.11b, approved also in 1999, provides up to 11 Mbit/s PHY throughput in the 2.4 GHz ISM band by using *direct sequence spread spectrum* (DSSS) [28].

IEEE 802.11g, approved in 2003, provides the same data rate and the same technology as IEEE 802.11a but in the 2.4 GHz band by using OFDM.

IEEE 802.11n [29], approved in 2009, provides up to 600 Mbit/s PHY throughput by using multipleantennas and channel bonding.

Standard IEEE 802.11-2007 [30] includes (and supersedes) the first three standards, that is, 802.11a, b, g. It also includes several other documents regarding not only MAC and PHY layers but also *quality of service* (QoS) and security aspects. Of interest in the United States is the standard IEEE 802.11y-2008 [31] that specifies both MAC and PHY layers for operation in the frequency band of 3650–3700 MHz, available under some special licensing conditions, described within the amendment, only in the United States but not in the rest of the world.

The standard [30] supports two network configurations. One, where a device acts as *access point* (AP) and the others are connected to the AP, is called *infrastructure*. The AP usually also interconnects the WLAN to other networks through a wired connection. It is possible to interconnect several APs in order to increase the area covered by a single WLAN and typically all the APs will be connected through some wired network. The other alternative for the network structure is called *ad hoc*, where no AP is present and devices connect directly among them. Within the 802.11 working group, it is Task Group S whose objective is to develop a standard for mesh networking in WLANs [32] that would allow self-configuration mesh networks similar to those in WPANs (see Section 6.2). Although currently in draft version, it is already used in some projects, like the *One Laptop per Child* [33], and is also supported in the Linux and FreeBSD kernels for its use in the interconnection of computers using theses systems.

6.3.1 Wi-Fi Physical Layer (PHY)

The 802.11 standards define several PHY, each of them consisting of two parts:

- A *PHY convergence function* adapts the capabilities of the physical medium to the PHY service provided by the *physical layer convergence procedure* (PLCP) that lays between the PHY and the MAC layer.
- A *PHY layer* adapts to the characteristics of the wireless medium and defines the method of transmitting and receiving data through it.

This separation is used to allow the IEEE 802.11 MAC to operate with minimum dependence on the actual physical medium, and thus there should be a PLCP specification for each PHY defined in the standard.

As the document IEEE 802.11-2007 [30] includes all previous standards, it describes all the possible PHY available. All of these specifications, except the infrared one, are intended to work on the ISM bands made available for such applications by the *International Telecommunication Union, Radiocommunication Sector* (ITU-R) [34] in the Radio Regulations 5.138 and 5.150 [35]. The following PHYs are defined:

Frequency hopping spread spectrum (FHSS) working on the 2.4 GHz band achieves 1 and 2 Mbit/s by using 2GFSK (*gaussian frequency shift key or keying*) and 4GFSK modulation schemes, respectively.

Direct sequence spread spectrum (DSSS) working on the 2.4 GHz band achieves 1 and 2 Mbit/s by using *differential binary phase shift key* (DBPSK) and *differential quadrature phase shift key* (DQPSK) modulation schemes, respectively.

OFDM working on the 5 GHz band achieves from 1.5 to 54 Mbit/s, using a combination of several modulation schemes and *forward error correction* (FEC) codes. Details on all the possibilities can be found in Table 6.2. It can be seen that the standard specifies three possible bandwidths, yielding different options with all the modulation and coding schemes.

Infrared (IR) using this medium for the transmission instead of radiofrequency achieves 1 and 2 Mbit/s by using 16 PPM (*pulse position modulation*) and 4 PPM, respectively. This option is no longer maintained by the working group, and current specification may not include all the features from the standard.

TABLE 6.2. The 802.11 OFDM PHY Parameters for Different Bandwidths (BWs)

Modulation	Coding Rate	20 MHz BW Data Rate (Mbit/s)	10 MHz BW Data Rate (Mbit/s)	5 MHz BW Data Rate (Mbit/s)
BPSK	1/2	6	3	1.5
BPSK	3/4	9	4.5	2.25
QPSK	1/2	12	6	3
QPSK	3/4	18	9	4.5
16-QAM	1/2	24	12	6
16-QAM	3/4	36	18	9
64-QAM	2/3	48	24	12
64-QAM	3/4	54	27	13.5

High rate direct sequence spread spectrum (HRDSSS) increases the rate with respect to regular DSSS by introducing the *complementary code keying* (CCK) modulation scheme. This allows one to achieve 1, 2, 5.5, and 11 Mbit/s.

Extended rate PHY (ERP) takes parts from other PHY specifications to get a set of rates from 1 to 54 Mbit/s using DSSS and HRDSS and applying OFDM in the band of 2.4 GHz. Further more, it can make use of the same parameter options already displayed in Table 6.2.

Due to the operation in unlicensed bands with no exclusive use of the spectrum, coexistence with other systems can be an issue. Maximum transmitted power has to be kept low in order not to interfere with other users or services. This is controlled by regional regulatory authorities. The standard does not pose any restrictions on these terms, but offers information about regional regulation domains in Annex I of IEEE 802.11-2007 [30]. The receiver sensitivity parameter dictates the minimum signal level at the receiver required to obtain a given *frame error rate* (FER) performance. It takes the following values for the different PHY:

- −80 dBm for 1 Mbit/s FHSS and −75 dBm for 2 Mbit/s FHSS.
- −80 dBm for DSSS.
- Table 6.3 gives information about the receiver sensitivity for all possible combinations of modulation, FEC and channel spacing for OFDM PHY.
- −76 dBm for HRDSSS.
- ERP has the same values as the PHY it includes, as stated earlier.

In November 2009 the IEEE 802.11n [29] standard was released. It poses a major step in terms of throughput compared to the previous releases. It consists of modifications of the previous standards. At the same time more advanced technologies are included in order to attain much higher throughput. At the physical layer, the main differences are as follows:

TABLE 6.3. The 802.11 OFDM PHY Rx Sensitivity for Different Bandwidths (BWs)

Modulation	Coding Rate	20 MHz BW Rx Sensitivity (dBm)	10 MHz BW Rx Sensitivity (dBm)	5 MHz BW Rx Sensitivity (dBm)
BPSK	1/2	–82	–85	–88
BPSK	3/4	–81	–84	–87
QPSK	1/2	–79	–82	–85
QPSK	3/4	–77	–80	–83
16-QAM	1/2	–74	–77	–80
16-QAM	3/4	–70	–73	–76
64-QAM	2/3	–66	–69	–72
64-QAM	3/4	–65	–68	–71

- Utilization of multiple antennas both at the transmitter and the receiver, also referred to as *multiple input multiple output* (MIMO), allows one to use up to four spatially separated data streams, *space time block coding* (STBC), and/or beamforming.
- The channel bandwidth is 20 MHz, and it supports the possibility to do channel bonding [36]. This allows the use of two channels to duplicate the throughput using, then, a total bandwidth of 40 MHz.

With these improvements, the rate that can be obtained is up to 600 Mbit/s, when using the full potential of the technology (four spatial streams, 40 MHz bandwidth, and the highest rate FEC code).

The standard supports up to four spatial streams through the use of a maximum number of four transmitting and receiving antennas. In MIMO there exists a trade-off between throughput and diversity. The use of STBC reduces the number of data streams available, reducing with it the throughput achievable but increasing the diversity and therefore the uncoded BER performance at the receiver. In clause 20.6 of IEEE 802.11n-2009 [29] all possible combinations of configuration parameters and rates achievable from 6 Mbit/s to 600 Mbit/s can be found.

The receiver sensitivity required is slightly different from that of previous standards 802.11a,b,g with details available in Table 20–22 of IEEE 802.11n-2009 [29].

The maximum distance will depend not only on the selected configuration but also on the particular scenario where the network is to be deployed. In general, the calculus of coverage and optimal placing of APs for a given area would be done using software tools that take into account all the parameters required, both specified by the standard and constrained by the scenario of deployment. These tools often use geographical data, maps, and environment information along with advanced channel simulation tools. Most of these are commercial tools, but it is possible to find on the Internet some free-of-charge tools that can help with Wi-Fi coverage planning [37–38]. The NIST [1] provides some simulation tools in the form of MATLAB code that can be used to perform some evaluation of 802.11 technologies, as done in Annex A of reference [3].

6.3.2 Wi-Fi Medium Access Control (MAC)

Use of the PLCP introduces an abstraction that allows the MAC sublayer to be common for all the different PHYs described in Section 6.3.1. The tasks of the MAC sublayer include providing reliable data delivery, access control, and protection of data. Nevertheless, the first task it has to fulfill is to manage the mechanism through which users access the shared medium.

The 802.11 MAC sublayer specifies a different mechanism through which the users get access to the wireless medium. The simplest is the *carrier sense multiple access–collision avoidance* (CSMA-CA) that is also used in other wireless technologies, like 802.15.4 described in Section 6.2.2. On top of this *distributed coordination function* (DCF), the standard builds more advanced mechanisms of contention that allow handing of QoS, for instance, the *hybrid coordination function* (HCF), which relies on different user priorities to provide differentiated and distributed access to the wireless medium, or the use of a centralized QoS-aware coordinator. Details on this complex coordination mechanism can be found in the specification [30]. Next, the characteristics of the underlaying DCF are described.

In a wireless medium it is not possible to detect the collisions when two users are accessing the shared medium concurrently. This is why the CSMA-CA protocol is designed to reduce the collision probability among multiple users accessing the medium at the point where collisions would most likely occur, which is just after the medium becomes idle following a busy medium. This is the situation that requires a random back-off procedure to resolve medium contention conflicts. The status of the medium, idle or busy, should be obtained using physical and virtual mechanisms to sense the medium.

The need for the virtual mechanism is due to something known as the *hidden node problem*, where, for example, two stations can communicate with an access point but they cannot communicate with each other. In this case sensing the medium will be useless. Instead, a slight modification called RTS/CTS (*ready to send, clear to send*) can be used immediately after checking the status of the medium and prior to data transmission. This mechanism assumes that the stations send small messages called RTS prior to sending application data, and the destination of those messages replies back with a CTS marking that it is OK to start the communication. All other stations that receive the CTS, apart from the one that initiated the communication, should refrain from starting communication for a time that is informed in the RTS/CTS messages. This leads to a slight throughput loss, but significantly reduces the probability of collisions especially in hidden node situations.

Figure 6.6 shows graphically the operation of the possible medium access procedures, that is, with and without RTS/CTS.

Additionally, the MAC sublayer can provide some security within the communications in terms of authentication and data encryption. The IEEE 802.11 standard defines two alternatives—*wired equivalent privacy* (WEP) and *Wi-Fi protected access* (WPA). As specified in Section 8 of IEEE 802.11-2007 [30], WEP security has been deprecated because it fails to meet the security goals, although it has been

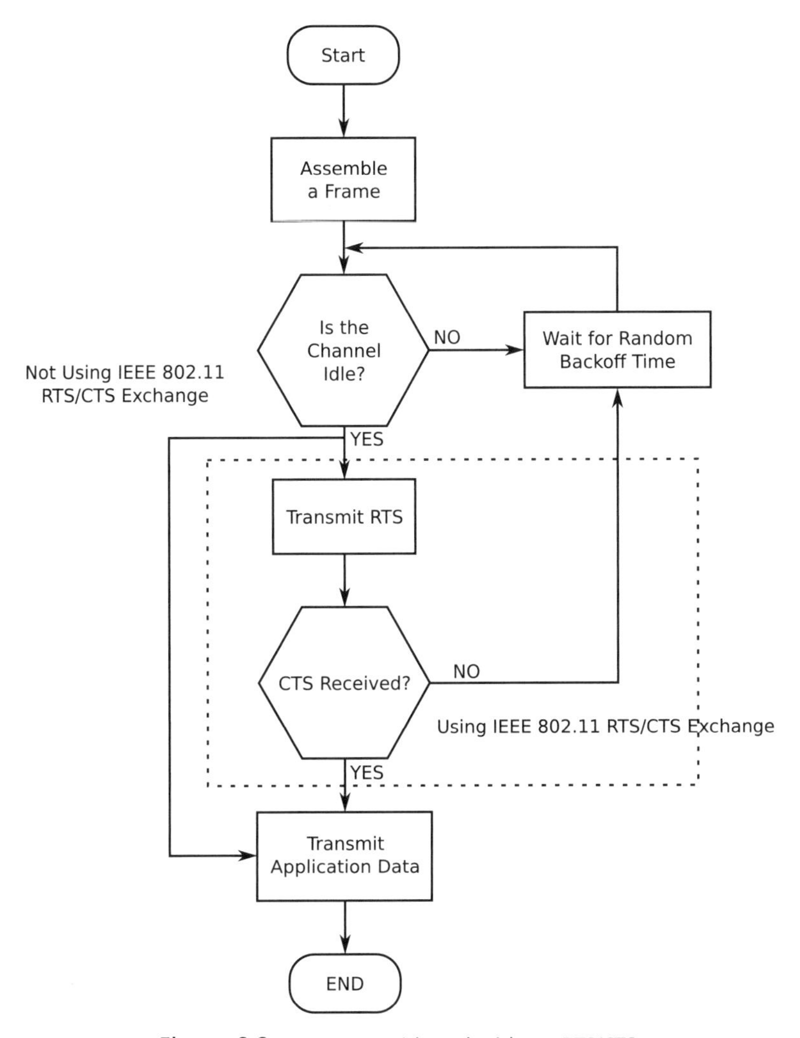

Figure 6.6. CSMA/CA with and without RTS/CTS.

included in the standard for backward compatibility reasons. On the other hand, WPA security involves a secure association of users to the network and the protocols required to guarantee data confidentiality and integrity. These meet the *Smart Grid* security requirements, as mentioned in Chapter 5, and discussed in more detail in Chapter 11, that is, end-to-end confidentiality of the communications and complete privacy of the network from unauthorized access.

Finally, note that IEEE 802.11n [29] includes the possibility to aggregate several MAC frames into one, so that the overhead is reduced and the throughput performance is increased.

6.4 WIRELESS METROPOLITAN AREA NETWORKS

When communication distance reaches the order of kilometers, the standards 802.15.4 (Section 6.2) and 802.11 (Section 6.3) might be an option, but they are not the best possible technologies to be used.

Metropolitan area networks (MANs) is the common name used for networks that cover areas such as small towns, campuses, neighborhoods, or even wider areas. Within the framework of the *Smart Grid* (SG) these kinds of networks are called *neighborhood area network* (NAN), *field area network* (FAN), or *wide area network* (WAN). These networks can be used to communicate, for example, with premises networks or power generation plants placed in hard to access geographical zones (mountains, forests, etc.). In such situations, wireless technology is possibly the best and most affordable option available.

Within the IEEE 802 LAN/MAN standards committee is the IEEE 802.16 Working Group on Broadband Wireless Access Standards [39], which fosters the development of the 802.16 Wireless MAN standard [40].

In 2001 the WiMAX Forum [41] was formed as a nonprofit organization to certify and promote the compatibility and interoperability of broadband wireless products based on the harmonized IEEE 802.16 standard. The name WiMAX stands for *worldwide interoperability for microwave access*. In South Korea a data communication service called WiBRO [42] exists, which is also based on the access technology of the standard IEEE 802.16e called Mobile-WiMAX that was the first approach of the standard to deal with mobility issues and was consolidated in the current IEEE 802.16-2009 standard.

The WiMAX Forum's current efforts are on completing WiMAX Release 2. It is based on the standard IEEE 802.16m [43]. WiMAX 2 delivers higher system capacity with peak data rates of more than 300 Mbit/s, lower latency, and increased *voice over IP* (VoIP) capacity, meeting the *International Telecommunications Union* (ITU) requirements for 4G or "IMT-Advanced" [41].

6.4.1 The 802.16 Physical Layer

IEEE 802.16 presents several options as air interface for the physical layer. They can be selected depending on the application and frequency range to be used. There are two main propagation situations to be differentiated: *line of sight* (LOS) and *non-line of sight* (NLOS). The main difference in these cases is the working frequency. The standard covers two frequency ranges:

- Frequencies between 10 and 66 GHz for LOS applications
- Frequencies below 11 GHz, both licensed or unlicensed, for NLOS applications

The 10–66 GHz band provides a physical environment where, due to the short wavelength, line-of-sight (LOS) is required and multipath is negligible. In the

10–66 GHz band, channel bandwidths of 25 or 28 MHz are typical. With raw data rates in excess of 120 Mbit/s, this environment is well suited for *point-to-multipoint* (PMP) access serving applications from *small office/home office* (SOHO) through medium to large office applications. The PHY defined for this case is called WirelessMAN-SC and is based on single carrier communications. In order to allow for flexible spectrum usage, both *time division duplexing* (TDD) and *frequency division duplexing* (FDD) configurations are supported. Both cases use a burst transmission format. Their framing mechanism supports adaptive burst profiling in which transmission parameters, including the modulation and coding schemes, may be adjusted individually to each subscriber station on a frame-by-frame basis. The FDD case supports full-duplex as well as half-duplex subscriber stations.

Frequencies below 11 GHz provide a physical environment where due to the longer wavelength, LOS might not be necessary and multipath may be significant. The ability to support near-LOS and NLOS scenarios requires additional PHY functionality, such as the support of advanced power management techniques, interference mitigation/coexistence, and multiple antennas. In this range there are some license exempt bands, for example, ISM frequencies where additional interference and coexistence constraints are introduced. At the same time regulatory constraints limit the allowed radiated power. There are two PHYs defined for these applications—WirelessMAN-OFDM (*orthogonal frequency division multiplexing*) and WirelessMAN-OFDMA (*orthogonal frequency division multiple access*). The use of OFDM with its cyclic prefix allows for the mitigation of multipath effects in an easy way. Additionally, the use of OFDM enables a high flexibility and scalability for the WirelessMAN-OFDMA PHY as both subcarriers and time slots can be assigned to different users to meet the particular requirement of each user. The main difference between WirelessMAN-OFDM and WirelessMAN-OFDMA physical specifications is the configurability of both options. Table 6.4 shows the differences in the number of subcarriers (*fast Fourier Transform*—FFT—size), *multiple input single output* (MISO) number of transmit antennas, *forward error correction* (FEC) options, and minimum bandwidth allowed.

TABLE 6.4. The IEEE 802.16 OFDM PHY Versus OFDMA PHY

Parameter	OFDM	OFDMA
FFT size	256	128, 512, 1024, 2048
MISO number of transmit antennas	2	2, 4
FEC options	Concatenated Reed–Solomon–convolutional, block turbo coding, convolutional turbo codes	Convolutional, block turbo coding, convolutional turbo codes, zero-tailed convolutional coding, *low density parity check* (LDPC) code
Minimum bandwidth (MHz)	1.25	1

Both PHYs share other parameters such as the length of the cyclic prefix used, to handle different channel response durations, which can be selected from the set $\{\frac{1}{4}, \frac{1}{8}, \frac{1}{16}, \frac{1}{32}\}$ as a fraction of the OFDM symbol time. Modulation schemes are also the same and can be selected among QPSK (*quadrature phase shift key*), 16-QAM (*quadrature amplitude modulation*), and 64-QAM.

Appendix B of IEEE 802.16-2009 [40] provides information about raw data rates achievable with WirelessMAN-OFDM and WirelessMAN-OFDMA for different modulation and coding schemes and typical bandwidths. These data rates range from 2 Mbit/s to more than 70 Mbit/s.

The WiMAX Forum provides useful information about recommended fading and mobility channel models as well as path loss models for different scenarios (suburban macrocell, urban macrocell, urban microcell, indoor picocell) [44]. In addition to this, the *National Institute of Standards and Technology* (NIST) in Annex E of reference [3] offers an example of how to evaluate the applicability of this technology for a particular scenario, that is, the backhauling of data aggregator points.

6.4.2 The 802.16 Medium Access Control Layer

The *downlink* (DL) of the 802.16 standard operates on a PMP basis, where a centralized base station that can be arranged into independent sectors transmits to all the users located in the coverage area. In this configuration the BS needs to coordinate with no other transmitter except for the overall time division duplexing that may divide time into *uplink* (UL) and DL periods.

The UL, on the other hand, is shared among all the users in a given sector. Within each sector users adhere to a transmission protocol that controls contention between users and enables the service to be tailored to the delay and bandwidth requirements of each user application. This is accomplished through different types of UL scheduling mechanisms. These are implemented using unsolicited bandwidth grants, polling, and contention procedures. Mechanisms are defined in the protocol to allow vendors to optimize system performance by using different combinations of these bandwidth allocation techniques while maintaining consistent interoperability definitions.

The MAC in opposition to the CSMA-CA mechanism introduced in Section 6.3.2 is connection oriented. All data communications are performed within the context of a transport connection. This allows one to associate varying levels of *quality of service* (QoS) to different subscribers. The different QoS parameters (data rate, peak rate, delay, jitter, etc.) are defined by what is called a service flow. Each of the service flows is associated with one transport connection.

Optionally, for the WirelessMAN-OFDMA PHY it is possible to use a hybrid version of the ARQ (*automatic repeat request*) error control method called HARQ on a per-connection basis. It can be used to mitigate the effect of channel and interference fluctuation. HARQ provides performance improvement due to *signal-to-noise* (SNR) gain and time diversity achieved by combining previously erroneously decoded packets and retransmitted packets. The two HARQ mechanisms supported by the standard are compared and explained by Frenger et al. [45].

6.5 CELLULAR NETWORKS

Cellular communication systems are mainly characterized by the concept of channel reuse by which system capacity and coverage can be provided over a wide geographic service area by a suitable planning of radio cell sizes, configuration of cells, and handling of system resources within and between the cells. Cellular systems offer and maintain predetermined levels of service to subscribers, including the ability to handle roaming of mobile subscribers within the service area. Cellular systems are interesting for *machine-to-machine* (M2M) communications, first of all, because of the already immense existing infrastructure provided by a multitude of cellular systems and standards. M2M communications can benefit from their highly optimized communications infrastructure, where quality of service is a central part of the service offering and, as part of this, the widespread coverage provided by cellular systems. On the other hand, M2M communications are of interest also to cellular operators in order to diversify their networks with a relatively low maintenance subscriber base.

This section gives a brief overview of selected cellular systems, particularly those that are of interest for M2M communications in general and *Smart Grid* (SG) communications in particular. Included here are the *Global System for Mobile Communications* (GSM) first standardized by *European Telecommunications Standards Institute* (ETSI) in 1990 [46] and later extended in *3rd Generation Partnership Project* (3GPP) releases 1997 and 1999 with *General Packet Radio Service* (GPRS) [47] and *Enhanced Data Rates for GSM Evolution* (EDGE) (series 44 and 45 of 3GPP specifications [48]), respectively. The *Universal Mobile Telecommunications System* (UMTS) first introduced with 3GPP release 4 (or release 1999) [49] and later enhanced with *High Speed Downlink Packet Access* (HSDPA) in release 5 [50] and *High Speed Uplink Packet Access* (HSUPA) in release 7 [51]. *Long-Term Evolution* (LTE) was introduced in 3GPP release 8 in 2008 [52]. And finally, the CDMA2000 1xEV-DO (*evolution data-optimized*) system was first introduced with *Telecommunications Industry Association*-856 (TIA-856) revision 0 in 2000 [53] and later Enhanced EV-DO in revision A [54] and revision B [55]. For a general overview of the CDMA2000 standards see reference [56].

Cellular system performance estimates, specifically looking at SG meter reading applications, are also presented in the description [3] for LTE, HSPA, and CDMA2000 in Annex B to Annex D, respectively.

6.5.1 Cellular Systems

One of the very first references to treat the cellular concept is the classical paper by MacDonald [57]. The concept described therein, relating to the reuse of frequency spectrum assigned to different cells, governs all modern cellular systems. In the chase for spectrum efficiency the spectrum reuse factor has gone from a small reuse factor to almost full reuse (i.e., where channels and frequencies are reused between neighboring cells). This has been possible in part due to a significant technology evolution.

The latest evolution of the third generation of cellular communications, commonly accepted as being the 3GPP LTE, is expected to come in widespread use across Europe in the 2010–2011 time frame and has already begun service in some areas of Europe and the United States. This will lead to a situation where several concurrent cellular systems are operating side-by-side from second generation GSM over third generation UMTS and its evolution *High Speed Packet Access* (HSPA) to the latest technology evolution which brings broadband communications to cellular (e.g., LTE). While these systems have been adopted in many places around the globe there are a considerable number of other regional and prevalent standards. Primarily, there has been a parallel development of cellular generations for the United States from cdmaOne (IS-95) to CDMA2000 with its EV-DO extensions until the merge with LTE in the latest release. Both UMTS and CDMA2000 are systems within the *International Telecommunications Union* (ITU) vision of *International Mobile Telecommunications* (IMT-2000) [58]. The UMTS system actually covers a number of air interface standards of which *wideband code division multiple access* (WCDMA) is the most common form; hence, UMTS is often synonymous with WCDMA. These cellular systems are of interest for M2M communications already or are likely to be when the M2M subscriber base and applications evolve.

There are other cellular standards relevant for M2M communications, notably the (mobile) *worldwide interoperability for microwave access* (WiMAX) described in Section 6.3. WiMAX is similar to LTE in many aspects including characteristics and performance [59] and might be considered a cellular technology, but it has some particular characteristics that make it different. WiMAX is designed to work also using unlicensed bands, and it includes support for high speed point-to-point links using high frequency single carrier communications. This can make WiMAX a good option in some scenarios. Another option that is available for facilitating the wide area coverage is the use of *television white spaces* for opportunistic transmissions that do not interfere with licensed users [60], that is, the portion of the VHF and UHF spectrum that is not being used at some location at a particular moment.

6.5.2 Applicability to Machine-to-Machine Communications

There are many important arguments for using cellular systems for M2M communications, notably the coherent radio coverage achieved within cells, the availability in terms of system deployment (local, regional, national, continental coverage), and, not to forget, ease of use. GSM/GPRS modems are readily available even for hobbyists (e.g., m2mtec in Germany [61]), which allow one to send and receive SMS text messages through a modem-like interface. While the demands on real-time communications in today's M2M communications might be quite limited, the prospect and hence the need to consider broadband wireless comes from the need to serve hundreds of thousands of devices with ever increasing demands for updating rates and real-time data availability.

The need for instant information for SG applications can be important for stability reasons and this is where the advances made in wireless broadband for minimizing air interface, core network, and connection setup delays become particularly important. From a capacity point of view, contemporary systems, such as GSM or GPRS, could be overloaded in providing low delay services for frequent M2M communications to and from many devices. Another consideration is that some systems may not be able to handle frequent and low payload traffic very efficiently due to issues with bandwidth scalability and signaling overhead.

Cellular is already in use for smart meter applications in which the base station communicates with a dedicated smart meter appliance installed at the customer premises [62], which is capable of higher output power and antenna efficiency than a handheld mobile device. Another typical alternative is a smart meter concentration point (usually a transformer station) where dedicated equipment concentrates and passes the power line communication (see Chapter 7) from many smart meters into a cellular connection. To name an example, EverBlu technology [63] uses a wireless mesh network for meter data collection. Typical cellular solutions in use are based on GSM/GPRS and CDMA2000.

The integration of a cellular modem to facilitate a direct cellular connection to devices has yet to be seen. At the moment, this solution is both too expensive and too power consuming. One of the trends in the cellular evolution that might change this, besides the price erosion and continued evolution of low power electronics, is the use of *discontinuous reception and transmission* schemes (DRX/DTX). Such schemes will allow terminals to sleep for most of the time. The LTE cellular system, for example, has very effective signaling structures and DRX/DTX mechanisms that will allow a terminal to wake up only when absolutely necessary and thereby bring down the average power consumption.

6.5.3 Cellular Characteristics

Table 6.5 shows some main characteristics of the considered systems. Although not mentioned in the table, all the systems employ *time division multiplexing* (TDM) on top of their basic multiple-access technique in order to share the channel capacity between different users. The operating frequency range (band) has not been tabulated since it mostly depends on the regional division as set by the ITU [64] and further national regulations. Also, the latest *World Radiocommunication Conference* (WRC) frequency band allocations, and their implementation in national regulations, are becoming increasingly technology neutral. This leads to the concept of refarming, where spectrum once reserved for a specific system is being converted into use for a new system, for example, LTE being deployed in 900 MHz GSM spectrum. Similarly, there is the concept of *digital dividend* where spectrum in the broadcast bands becomes available due to the digitalization of television broadcast. Therefore, generally, the systems can be deployed in many bands across the UHF range from 300 to 3000 MHz. Some complication to this picture is that despite the technology neutral band assignments, legacy terminal capability in some cases could limit the possible

TABLE 6.5. Selected Characteristics of Second (2G) and Third (3G) Generation Systems of Relevance to Machine-to-Machine Communications

Standard	2G GPRS	3G WCDMA	3.5G HSDPA	3.9G LTE	3G CDMA2K 1xEV-DO	3G (Nx) EV-DO
Bandwidth (MHz)	0.2	5	5	1.25–20	1.25	(Nx) 1.25
Duplex	FDD	FDD TDD	FDD TDD	FDD TDD	FDD	FDD
Multiple access	FDMA TDMA	CDMA	CDMA	OFDMA SC-FDMA	CDMA	CDMA
Frame time (ms)	4.615	10	2	1	20	10
RTT (ms)	600	150	50–100	10–30	150	150

uses. For instance, although in theory it is possible to operate HSPA in the 1800 MHz band, legacy terminals in Europe do not support HSPA in this band and for this reason it is more likely to be refarmed for LTE. The operating frequency is especially important for coverage and, hence, lower frequencies will likely be more beneficial for M2M communications because of the relatively low data rate requirements but high demands for coverage, for example, to reach utility meters inside buildings or buried deep in utility wells, which requires good building penetration.

There is a clear trend in the evolution of basic parameters, more bandwidth, shorter frame time, and, partly due to this, less delay. The code division multiple access (CDMA) principle is used in many third generation systems, in which transmissions are separated by modulation codes. In the latest evolution, LTE has adopted the *orthogonal frequency division multiple access* (OFDMA) principle for the direction from the base station to the terminal [65–66]. OFDMA is a spectrally efficient access principle that can deal with the degradation from wideband frequency-selective fading in a simple manner. The technology selected for the reverse direction is based on single-carrier FDMA (SC-FDMA) due to the better (battery) power efficiency at the terminal side. The multiple access principle in both cases can be said to reintroduce some of the frequency and time division multiple access (FDMA, TDMA) principles from GSM, but just with a more flexible resource allocation, higher granularity, and faster allocation based on the basic 1 ms time frame interval. The separation of the two transmission directions in LTE can be based on either frequency division duplexing (FDD) or time division duplexing (TDD), but as in the case of wideband CDMA (WCDMA) the most common version is expected to be FDD. The bandwidth supported by LTE is scalable up to 20 MHz with possible subdivisions into smaller bandwidths of 1.25, 2.5, 5, 10, and 15 MHz for operators having less spectrum available.

Figure 6.7 shows the peak data rate capability of the different systems including their evolution up to the latest addition of LTE. One thing to note is that the peak data rates are obtained in different system bandwidths ranging from 200 kHz for

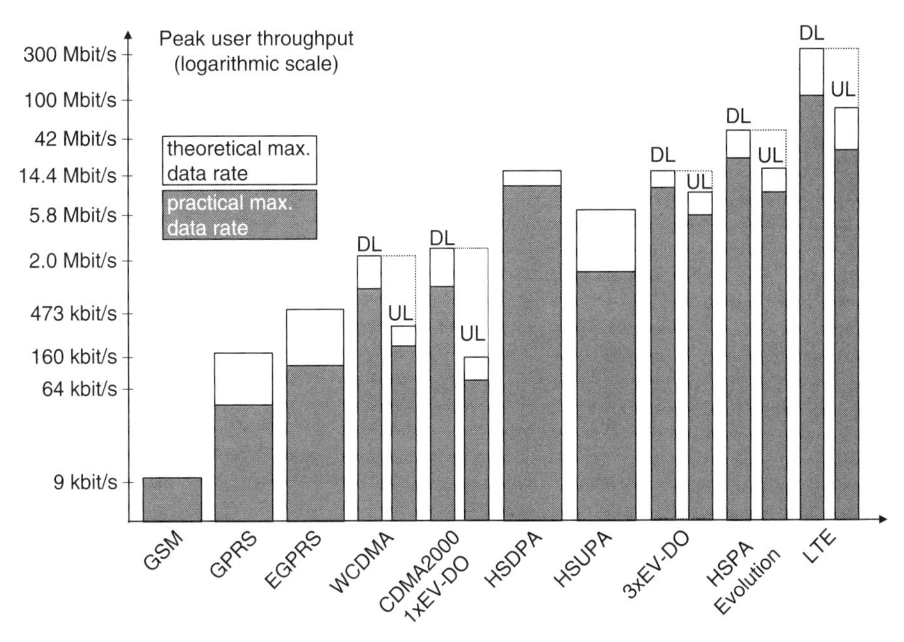

Figure 6.7. Peak data rates of selected 2G and 3G systems.

GSM technologies, 1.25 MHz for CDMA2000 technologies, 5 MHz for WCDMA, 3.75 MHz for 3xEV-DO, and 20 MHZ for LTE. Peak data rates are theoretical and mostly cannot be achieved in practice due to protocol overhead and practical equipment capabilities. For example, in the case of HSDPA (HSDPA for downlink direction from the base station to the terminal and similarly HSUPA for the opposite uplink direction) a maximum of 14.4 Mbit/s is possible with release 5, whereas in practice the maximum is more likely to be around 11–12 Mbit/s. Technological evolution such as the possibility to support phase coherent demodulation of high order signal constellations, the use of multiple antennas at both the base station and the terminal through *multiple input multiple output* (MIMO) technology, and the possibility to operate over several simultaneous carriers has steadily increased the peak data rates. Both NxEV-DO revision B [67] and HSPA [68] support multicarrier operation. HSPA, for example, boosts the peak data rate in the downlink to 42 Mbit/s by combining 64-ary *quadrature amplitude modulation* (QAM) with dual-carrier operation or, alternatively, with the use 64-QAM and MIMO. A major difference between these two combinations is that dual-carrier operation doubles the experienced user throughput throughout the network, or allows serving a larger number of terminals, whereas MIMO mainly benefits the data rate of users close to the base station.

Speaking of average user data rates, they are naturally much lower since, first of all, users in the cell must share the available air interface capacity between them and, second, most users will have much worse cell conditions (channel and interference) than what is required to obtain the peak data rate. Again, using HSDPA as an example, average user data rates are often quoted in the range of 0.5–1.5 Mbit/s.

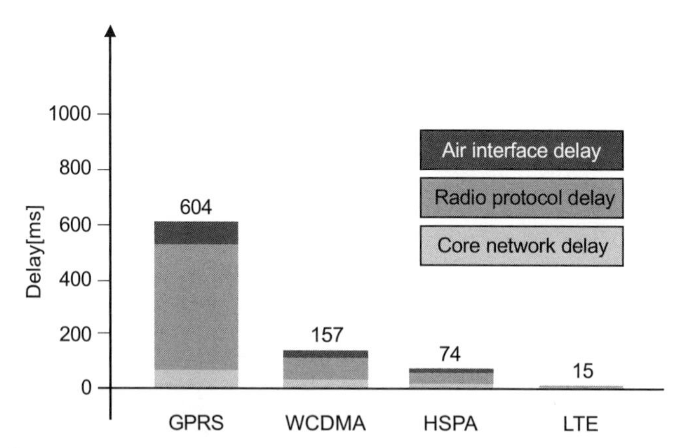

Figure 6.8. Round-trip time for transmission of a small data packet.

Nevertheless, the information about average data rates is not usually available and they are highly dependent on the scenario. Moreover, the *National Institute of Standards and Technology* (NIST) proposes as a metric the peak data rate in its NIST Priority Action Plan 2 guidelines [3].

The air interface delay, which is determined by the peak data rate, is a part of the calculation of round-trip time (RTT) (the time it takes for a short package to go from a core network server to the terminal device and back again, or vice versa) and, hence, the expected response time in M2M communications. As can be seen in Figure 6.8, the air interface delay is only a small part of the total delay budget. Comparing the different generations of cellular systems, it can be seen that the largest reduction has happened to the delay associated with the radio protocol. This is the result of reducing the basic frame timing and concentrating more of the radio protocol close to the air interface, that is, in the base station.

If the terminal is not already connected to the radio network, the setup delay from terminal sleep to active (data) mode is also of importance. When a terminal wakes from sleep mode it first needs to set up an active connection with the base station—a procedure that delays the actual transfer of data compared to the case where the terminal already had an active connection. While such a connection procedure takes several seconds for GPRS, that number has been reduced considerably with LTE due to new protocol structures and procedures, especially given the capability to keep the device in data active mode with the DRX/DTX feature. DRX is used to keep the terminal in sleep except for certain predefined instants where it wakes up to check for data. In this manner, the terminal can save power and avoid the long setup delay.

6.6 SATELLITE COMMUNICATIONS

Over half a century has passed since Sir Arthur C. Clark published the first major geostationary satellite communications article in 1945 [69]. Nowadays, satellite communications are used on a daily basis, be it to receive television channels, to find a location via the *Global Positioning System* (GPS) [70], or even for browsing the Internet [71].

Being still the only technology able to deliver global coverage, satellites also play an important role in upcoming *Smart Grid* (SG). For example, weather satellites are used to predict electricity production of wind farms and solar parks, as well as electricity demand for air conditioners [72]. Further more, GPS time signals are used to calculate the *Coordinated Universal Time* (UTC), which is used in synchronization and time-stamping operations [73]. Moreover, utilities use GPS and satellite imaging services to feed information into their enterprise *geographic information system* (GIS) [74] used for grid management, fault localization, and repair and maintenance operations. More futuristic lines of thinking even deal with problems of worldwide energy transmission via satellite microwave links and solar energy harvesting in outer space [75].

Leaving these satellite technologies on the sideline, the rest of the section focuses on satellite communication aspects. Usually, remote grid elements like solar parks, offshore wind farms, hydro storage, or mobile field dispatch units are connected via a satellite link to a satellite teleport (satellite hub), which is linked to the *public switched telephony network* (PSTN), a private network, or the Internet. Examples are Iberdrola Renovable wind farms connected to a control center in Toledo, Spain [76], or ASTRA2Connect services connected to the Italian *Gestore dei Servizi Energetici* (GSE) for telemetry and video surveillance applications [77]. Further more, satellite links can serve as a redundant backup path in case of terrestrial link disruption [78].

6.6.1 Satellite Orbits

The nature of the satellite communications link is strongly influenced by the satellite's orbit. Important orbits for communication purposes are the *low Earth orbit* (LEO) between around 160 and 2000 km of altitude, the *geostationary orbit* (GEO) at 35,786 km altitude above the equator, the *medium Earth orbit* (MEO) filling the space between LEO and GEO, as well as the *highly eccentric orbits* (HEO) with apogee and perigee around 40,000 km and 1000 km, respectively. A brief comparison of the different orbits and some of their properties can be found in Table 6.6.

Satellites placed in LEO travel with an approximate speed of 27,400 km/h (8 km/s), making one complete revolution around the Earth in about 90 minutes. The main advantages resulting from the close proximity to Earth are lower launching cost per satellite, lower free-space signal loss, and lower propagation delay. According to the ETSI [79], the LEO one-way connection latency is on the order of 20–60 ms assuming a simple link from an Earth station to the satellite and back to another Earth station. This latency corresponds to approximately half the round-trip time. Further more, the lower free-space loss allows the deployment of small and low gain antennas at the ground station (user terminal) side, making LEO the preferred orbit for mobile satellite systems. Nevertheless, from Earth LEO satellites can only be seen within a radius of around 1000–2000 km around the subsatellite point. Hence, several dozen LEO satellites, jointly referred to as a constellation, are

TABLE 6.6. Orbits and Related Properties[a]

	LEO	MEO	GEO	HEO
Altitude (km)	160–2000	2000–35,000	35,786	40,000 (apogee) 1000 (perigee)
Orbital period (h)	1.5–2	5–10	24	12
Visibility duration	15–20 min/pass	2–8 h/pass	Permanent	8–11 h/pass (apogee)
Number of satellites for global coverage	48–66	10–15	3	5–12
Distance-dependent attenuation	Low, suitable for handheld terminals without antenna pointing	Medium, less favorable for handheld terminals	High, difficult or not suitable for handheld terminals	High, difficult or not suitable for handheld terminals
Example system	Iridium, Iridium NEXT, Globalstar	GPS, Galileo, 03b	Hispasat, Eutelsat, Intelsat, Inmarsat, Wild-Blue, SES/Astra, Thuraya, ACeS	SIRIUS, ESA Iris (study)

[a]Based on reference [79] and Table 6.1 by the ITU [86].

required to deliver continuous coverage and handovers between satellites are necessary. Further more, a high density of Earth stations or else *intersatellite links* (ISLs) are required to assure that a bidirectional communication path can be established between a user terminal and the core network. Examples of LEO mobile communications satellite systems are Globalstar (altitude 1414 km), Iridium (altitude 780 km) and the upcoming Iridium NEXT with 40, 66, and 66 active satellites per constellation, respectively. See Section 6.6.7 for details on these systems.

Popular MEO system representatives are the navigation satellite systems GPS at 26,559.7 km [80] and Galileo at 23,222 km [70]. Further more, the upcoming *Other 3 billion* (O3b) constellation at 8063 km with the mission "to make the Internet accessible and affordable to those who remain cut off from the information highway" [81] plans to use MEO. The main advantages of MEO constellations are that only around six satellites are required to communicate around the Earth and that the simple one-way connection latency is on the order of 80–120 ms [79], which is considerably lower than for GEO satellites.

GEO satellites travel at the Earth's angle velocity and appear stationary with respect to an Earth-based reference point. Thus, they are often used in combination with relatively inexpensive fixed dish antennas on the ground. Due to its altitude, a single GEO satellite can see one-third of the Earth's surface, excluding polar regions, which means that a constellation of three satellites suffices to deliver close to whole Earth coverage. However, due to the high altitude (almost one-tenth of the way to the moon), satellite launch costs are high and the typical simple one-way connection latency is on the order of 280 ms [79], which can be undesirable for some communication purposes. Nevertheless, GEO is the most popular satellite orbit, especially suited to serve fixed or nomadic users. Example systems are the broadcasting systems Hispasat, Eutelsat, Asiasat, and Intelsat Americas 5. Further more, two-way GEO communication services are offered, for example, by Inmarsat *Broadband Global Area Network* (BGAN), Thuraya 3, Astra (Astra2Connect), and Hispasat (AMAZONAS). Details on these systems can be found in Sections 6.6.6 and 6.6.7.

Seen from a point at relatively high latitude (e.g., northern Europe), a GEO satellite appears very low above the horizon. Thus, in high-build urban or mountainous areas it is possible that no line of sight path can be established. Further more, the low elevation above the horizon can lead to undesirable multipath effects that are attributed to reflections from the Earth's surface. To cover these problematic high latitude locations, the HEO can deliver alternative solutions. A satellite in this orbit can be seen for approximately 8 h so that a constellation of three satellites suffices to deliver constant coverage of a selected area. The one-way propagation latency is on the order of 200–310 ms [79] and is therewith comparable to the GEO latency. The Soviet-Russian broadcasting system Molniya, deployed in 1967, is possibly the most well-known HEO system representative. Currently, the HEO is used, for example, for digital radio broadcasting by the U.S. SIRIUS XM Radio [82] and is being examined by the *European Space Agency* (ESA) within the Iris project to deliver ATM services in high latitudes [83].

6.6.2 Satellite Regulations

The main areas in which regulation is required are orbital location allocation of GEO satellites, frequency spectrum allocation, and the maximum allowable effective isotropically radiated power. On the international scale, regulation is handled by the Radiocommunication Sector of the *International Telecommunication Union* (ITU-R) [34]. In terms of space radiocommunications services the ITU distinguishes:

- *Fixed satellite service* (FSS)
- *Mobile satellite service* (MSS)
- *Broadcasting satellite service* (BSS)
- *Earth exploration satellite service* (EES)
- *Space research service* (SRS)
- *Space operation service* (SOS)
- *Radiodetermination satellite service* (RSS)
- *Intersatellite service* (ISS)
- *Amateur satellite service* (ASS)

In light of SG communications, the first two (i.e., FSS and MSS) are of main importance while BSS and ISS are of secondary importance.

The ITU's spectrum management is region specific. For this purpose the Earth is subdivided into Region 1 (Europe, Africa, the former Soviet Union states, and the Middle East), Region 2 (North America and South America), and Region 3 (Asia Pacific). It is assumed that services in different regions do not interfere significantly with each other due to geographic separation. International transregion allocations, however, apply for satellite systems covering more than one region.

6.6.3 Frequency Bands and Propagation Effects

IEEE letter band designations [84] are commonly used when referring to satellite frequency bands. These designations are displayed in Table 6.7 together with space radiocommunications services (limited to FSS, MSS, BSS, ISS) and example systems that operate within the bands.

While the L, S, and C bands are crowded with various partly terrestrial services (see Figure 6.2), only recently has extensive satellite use been made of the Ku, K, and Ka bands, the X band being reserved for government and military operations. The initial reluctance to use the higher Ku, K, and Ka bands was due to the increased rain attenuation experienced at frequencies above 10 GHz. Dealing with this *rain fading* requires either a significant fade margin (around 12 dB in Ku band and up to 40 dB in Ka band [85]) or a link adaptive system that is able to lower the data rate accordingly. Further options are to deploy frequency diversity, for example, switching to L band during the rain event, or site diversity—that is, avoiding the rain link using a different satellite beam or a different earth station. More detailed treatments on rain fade and other propagation effects as well as mitigation techniques can be found, for example, in ITU [86], Ippolito and Louis [87], and Maral and Bousgust [85].

TABLE 6.7. Satellite Frequency Bands, Services, and Example Systems

Band	Frequencies (GHz)	Service	Example System
L	1–2	MSS	Inmarsat BGAN, ACeS, Globalstar, Thuraya, Iridium (subscriber links)
S	2–4	BSS, MSS	Globalstar
C	4–8	FSS	Intelsat
X	8–12	FSS	Government, military use, deep space exploration
Ku	12–18	FSS, BSS	Eutelsat, Astra, Hispasat
K	18–26.5	FSS, BSS, ISS	Iridium (feeder links and satellite cross-links)
Ka	26.5–40	FSS, MSS	O3b

6.6.4 Satellite Technology and Topology Considerations

Over the years satellite technology has advanced significantly, making current satellite broadband solutions in many regions a cost-effective alternative to terrestrial services [71]. Especially worth mentioning in this respect are advances of satellite antenna technology, that started with broadcast beams covering entire continents, over area coverage beams illuminating countries or regions of interest, leading to advanced spot beams that cover a large amount of small cells within a region, operating in many aspects similar to 2G terrestrial cellular systems with frequency reuse [88–89]. Further more, shifting away from the link budget fade margin provisioning paradigm toward varying data rates using link adaptation schemes with the ability to adjust modulation constellations and forward error correction protection has led to a much more efficient utilization of the scarce bandwidth resources. Instead of operating simple active reflectors in space, also referred to as bent pipe or transparent satellites, the aforementioned technologies require *onboard processing* (OBP), which has become possible due to significant advances in digital signal processing.

Looking at the overall satellite network, two topologies can be found—star topology, where remote satellite terminals communicate via the satellite with a central hub, and mesh topology, where remote terminals are able to communicate directly with each other via the satellite [85]. Hybrid networks mixing star and mesh structures are also possible. With respect to star networks, the link from a user terminal over the satellite to the central hub is called the return channel. The link from the central hub over the satellite to one or several user terminals is called the forward channel. Usually, a star topology can be realized with cheaper user terminal equipment, that is, *very small aperture terminals* (VSATs), taking advantage of more advanced and expensive equipment at the hub. Nevertheless, to connect two remote satellite terminals implies going through the hub. This requires two satellite hops and, hence, doubles the propagation delay. Instead, the ability of the mesh topology to directly connect two remote user terminals via the satellite demands more advanced and, hence, more expensive hardware. Nevertheless, the propagation delay is approximately half compared to a similar link realized within the star topology. Independent of star or mesh topology, the link from any user terminal to the satellite is commonly referred to as the uplink, while the link from the satellite to a user terminal is commonly referred to as the downlink.

Further more, an important topological criterion is whether satellites are able to communicate via *intersatellite links* (ISLs) or if they are limited to establish a communication link between two points on Earth. Especially for LEO satellite constellations, which due to their low altitude have a relatively small coverage area, the capability of establishing ISLs can be important. Considering a hypothetical case where a link is to be established from one side of the Earth to the other, a LEO system with ISLs, (e.g., IRIDIUM) (see Section 6.6.7) can relay the signal over several ISLs before reaching its destination. Without ISLs a tighter net of ground stations, like in the Globalstar system (see Section 6.6.7), is required over which multiple satellite–Earth hops could be established to reach the destination.

6.6.5 Satellite Communication Standards

Satellite systems frequently used proprietary communications solutions. Nevertheless, the need to drive development time and equipment prices down has led to the development of standardized solutions with the aim of enabling mass market production and multivendor product interoperability. *Fixed satellite services* (FSSs) are possibly the most important for SG deployments. However, dispatch units and field crews require *mobile satellite services* (MSSs). Furthermore, there are situations where MSS terminals are statically deployed for M2M communication, as is already happening for many terrestrial 2G GSM terminals (see Section 6.5). From a standardization standpoint the ETSI possibly draws the most complete picture of FSS and MSS standards. To achieve this ETSI not only developed standards but also endorsed some standards that had initially been developed by other organizations such as the *Telecommunications Industry Association* (TIA).

The ETSI Technical Committee *Satellite Earth Stations and Systems* (SES) has defined a *broadband satellite multimedia* (BSM) standards family [90], which, although not bound to be stationary, largely coincide with fixed satellite services. The BSM protocol architecture defines several satellite PHY/MAC air interfaces. Each air interface links to a common satellite independent–service access point (SI-SAP). This means that the same BSM higher layer protocols (e.g., handling IP traffic) can be shared independent of the air interface. A list of available BSM air interfaces is found in Table 6.8.

A three-letter designation is used to distinguish the air interfaces [91]. The first letter indicates whether a satellite operates *transparently* (T), also known as *bent pipe*, or in a *regenerative* (R) manner, also known as *onboard processing* (OBP). The second letter shows whether the return link is established via the *satellite* (S) or over a terrestrial component, which would be called a *hybrid* (H) solution. The third designator refers to the *star* (S) or the *mesh* (M) topology as introduced in

TABLE 6.8. Broadband Satellite Multimedia (BSM) Air Interface Standards Overview

	TSS-A	TSS-B/IPoS	RSM-A	RSM-B
ETSI No.	TS 102 352 TS 102 402	TS 102 354	TS 102 188-x TS 102 189-x	TS 102 429-x
Technology	DVB-S/ DVB-RCS	DVB-S/ DVB-RCS	Uplink: FDMA–TDMA, OQPSK, Reed–Solomon–Hamming Downlink: TDM, QPSK, Reed–Solomon–convolutional	DVB-S/DVB-RCS
Examples		Hispersat-1D	Hughes Network Systems, Spaceway	Hispersat AMERHIS, AMAZONAS satellites

Section 6.6.4. The abbreviation TSS, hence, means *transparent satellite star*. The abbreviation RSM means *regenerative satellite mesh*. The mesh topology requires onboard processing. It is thus not found in connection with purely transparent satellites. Further more, the topology of a regenerative satellite star is seen as a subcase of the more general mesh topology and is therefore also not explicitly mentioned. Note that TSS-B and RSM-A are explicitly treated by the U.S. National Institute of Standards and Technology (NIST) in their SG standards considerations [2].

Most BSM air interfaces use the *digital video broadcasting over satellite* (DVB-S) and the *digital video broadcasting return channel satellite* (DVB-RCS) standard. DVB-S [92] is deployed in transmissions to user terminals. It uses *quaternary phase shift keying* (QPSK) and concatenated *forward error correction* (FEC). The inner code is a convolutional code adjustable in rate to the transponder bandwidth. The outer code is a shortened Reed–Solomon code. Besides, in situations with very poor carrier to interference plus noise ratios, it is possible to opt for *binary phase shift keying* (BPSK) instead of QPSK [93]. DVB-RCS [94] is used for the signal from the user terminals to the satellite. Terminal access is managed with a *multifrequency time division multiple access* (MF-TDMA) scheme. A time slot as well as a frequency and a bandwidth are allocated to a requesting terminal, there-by allowing flexible "bandwidth on demand" allocation. Further more, DVB-RSC uses QPSK modulation. The deployed FEC can either be a turbo code or a concatenation of convolutional and Reed–Solomon code.

Both standards have undergone a major evolution. The second generation of DVB-S, referred to as DVB-S2, employs link *adaptive coding and modulation* (ACM) from 4 to 32 *amplitude phase shift keying* (APSK) paired with 1/4 up to 9/10 *low density parity check* (LDPC) coding [95] and is therewith able to cope with, for example, cloud coverage and rain fade without the need for a fixed fade margin. This leads to an approximately 30% average throughput increase compared to its predecessor. Also, the DVB-RCS has evolved. Since a revision in 2004 DVB-RCS integrates the DVB-S2 standard for the forward link transmission. The latest revision now also caters to terminal mobility and is sometimes referred to as DVB-RCS+M [96–97]. This leads to mobile satellite services whose main ETSI standards are outlined in Table 6.9.

Some early systems deployed different interface types of the *GEO mobile radio* (GMR) family, which is based on the popular terrestrial *second generation* (2G) cellular GSM standard and its packed data evolutions GPRS/EDGE (see Section 6.5). The interfaces termed GMR-1 and GMR-2 have undergone an evolution with different releases. Albeit, GMR-2 has undergone much less evolution than GMR-1 and is up to the present date used to deliver primarily voice, fax, and short message services, for example, over the *ASIA Cellular Satellite* (ACeS) constellation [89]. GMR-1, on the other hand, has significantly evolved from its first release to the current third release. Release 1 used circuit switched connections. Release 2 added packed data services comparable to terrestrial GPRS and is also referred to as GMR-1 packet mode services (GMPRS-1). With release 3, GMR-1 became a 3G standard delivering 3GPP packet data services based on GSM/EDGE. GMR-1 3G is also explicitly treated by NIST in its SG wireless standards considerations [2].

TABLE 6.9. Mobile Satellite Services (MSSs) Standards Overview

	GMR-1	GMR-2	S-UMTS-A SW-CDMA	S-UMTS-SL	S-UMTS-G
ETSI No.	TS 101 376-x	TS 101 377-x	TS 101 851-1-2 TS 101 851-2 TS 101 851-2-2 TS 101 851-3 TS 101 851-3 TS 101 851-3 TS 101 851-3-2	TS 102 744	TS 101851-1-1 TS 101851-2-1 TS 101851-3-1 TS 101851-4
Technology	Based on GSM/ GPRS FDMA/ TDMA	Based on GSM FDMA/ TDMA	Based on UTRA FDD WCDMA	(π/4) QPSK, 16-QAM FDMA/ TDMA	Compatiblity with UTRA FDD WCDMA
Examples	Inmarsat (Hughes), Thuraya	ACeS/ Inmarsat (Lockheed)	No system[a]	Inmarsat BGAN	No system[a]

[a]As of June 2009.

Turning to the *Universal Mobile Telephony System* (UMTS), the satellite component, referred to as S-UMTS, has long been talked about. Interesting reports can be found in references [79], [98], and [99]. As in BSM, different S-UMTS families exist. Families A and G are specified in the various parts of ETSI TS 101 851. Especially Family G seems attractive as its air interface is fully compatible with the terrestrial part of 3GPP UTRA FDD WCDMA, currently already in widespread terrestrial use. Related research activities are presented in Boudreau et al. [100], Maitin et al. [101], and Azizan et al. [102]. Further more, a detailed comparison of the S-UMTS families A and G as well as the South Korean proposal known as SAT-CDMA or family C can be found in reference [103]. In terms of deployed S-UMTS systems especially, the Inmarsat *Broadband Global Area Network* (BGAN) is worth mentioning. It stems from the proprietary *Inmarsat Air Interface-2* (IAI-2). As an SL family it has currently been assigned the designation ETSI TS 102 744 [104] and is discussed in more detail in Section 6.6.7. It is worth noting that Inmarsat BGAN is currently also the only S-UMTS standard treated in the NIST SG standardization process [2].

The following two sections turn the view from standards to actual systems that in some cases use standards, but in many cases also still use proprietary communication solutions. Fixed satellite systems are discussed first in Section 6.6.6, followed by a discussion of mobile satellite systems in Section 6.6.7.

6.6.6 Fixed Satellite Systems

Astra2Connect [105] by *Société Européenne des Satellites* (SES) [106] offers bidirectional broadband Internet and M2M services in Europe and parts of Africa using the geostationary satellites ASTRA 1E and 3E. The digital video broadcasting satellite second generation standard (DVB-S2) is used on the forward link. On the reverse link the signals traveling from many user terminals to the satellite are managed using *multifrequency time division multiple access* (MF-TDMA) as used in DVB-RCS. However, the modulation and coding scheme follows SATMODE, a CENELEC standard [107–108] with constant envelope *Gaussian minimum shift keying* (GMSK) and turbo coding. When targeting the consumer market the joint forward and return link solution is also marketed as *Sat3Play*, as it allows the seamless integration of the three services *voice over IP* (VoIP), digital video, and broadband Internet access. The Sat3Play system emerged from a *European Space Agency* (ESA) project with the same name [109]. Sat3Play equipment is produced by the Belgium based Newtec. It mitigates long GEO satellite link latencies through TCP-acceleration and HTTP prefetching implemented in Newtecs' TelliNet software suite [110]. Commercially available Sat3Play currently provides up to 4 Mbit/s download and 320 kbit/s upload capacity. An example of Astra2Connectet SG deployment is reported by Payer [77].

Another main European satellite operator, Eutelsat, supplies through its subsidiary Skylogic broadband IP (Internet protocol) and TV satellite communication services for small- and medium-sized businesses, great industrial conglomerates, and the Public Administration [111]. The coverage area includes Europe, South America, Africa, and the Middle East. Skylogic makes use of the geostationary Eutelsat fleet including satellites such as the ATLANTIC BIRDs, W3A, W6, EUROBIRD 3, SESAT 2, and TELSTAR 12. The main technological platform is called Eutelsat D-Star [112]. D-Star is provided in cooperation with the satellite equipment manufacturer ViaSat and uses ViaSat's LinkStar architecture. The latest product, LinkStarS2 [113], is compliant with DVB-S/DVB-S2 and DVB-RCS. As the names *D-Star* and *LinkStar* indicate, the satellite network architecture is star-shaped. Nevertheless, ViaSat's product line LinkWayS2 is also able to support mesh networks. Alongside its D-Star service, Eutelsat introduced in 2007 a broadband access service called Eutelsat Tooway. It is tailored for the needs of small business and consumers in Europe and along the Mediterranean. Using the satellites HOT BIRD 6, EUROBIRD 3, and Ka-Sat, Tooway provides download speeds of 3.6 Mbit/s and upload speeds of 512 kbit/s. The Tooway service operates in the Ka and Ku bands and has also been developed in close cooperation with ViaSat. The underlying technology is known as SurfBeam [114]. It uses an adapted version of the *Data Over Cable Service Interface Specification* (DOCSIS) [115] (see also Chapter 7) and can therefore benefit from mature mass production of its digital signal processing components.

Since 2005, SurfBeam technology has also been used by the service provider WildBlue (since 2009 owned by ViaSat) to deliver broadband Internet access to over 400,000 homes in North America. The geostationary satellites delivering the WildBlue service are Telesat's Anik F2 and WildBlue-1 [116].

Another satellite operator, the Spanish Hispasat, was founded in 1989. It focuses with its fleet primarily on Europe as well as North America and South America, and has grown over the years to become a main global satellite operators [117]. Especially on the Iberian Peninsula, many photovoltaic parks, as well as wind farms make use of Hispasat's broadband services through intermediation of companies like

Neo-Sky, an Iberdrola subsidiary [118]. Cornerstones of Hispasat's broadband activities are currently the GEO satellites Hisparsat-1D, as well as AMAZONAS, working in the Ku band with DVB-RCS/DVB-S technology [117]. Especially noteworthy is the Hispasat AMAZONAS/AMERHIS. It was developed in collaboration with ESA and the Spanish government's *Center for the Development of Industrial Technology* (CDTI) and it is the first regenerative mesh system based on DVB-RCS/DVB-S [119].

Finally, O3b, short for *Other 3 billion*, is an ambitious MEO satellite project by O3b Networks [81], aiming to bring fiber-like coverage primarily to underdeveloped and emerging markets. The project attracted capital from SES, Google, Allen-and-Company, Liberty Global, HSBC Principal Investments, and Northbridge Venture Partners. O3b Networks is aiming to launch the first eight satellites of an eventually 16 satellite constellation into MEO in 2013, with commercial services supposed to commence shortly afterward. The satellite fleet, built by Thales Alenia Space of France, will follow the equator (equatorial orbit), resulting in around the Earth broadband connectivity within 45 degrees of latitude north and south. Ground equipment will be provided by ViaSat and other partners such as Advantech, Gilat, iDirect, and MDA [81]. Communication Technology will be based on DVB-S2, DVB-RCS. Although it sounds thrilling to provide fiber-like Internet services to a large part of the world's poorest population, there also exist unsuccessful histories of similar satellite projects like the Microsoft-backed Teledesic project or the Alcatel Space project Sky Bridge, which were never put into operation due to financial difficulties [120].

6.6.7 Mobile Satellite Systems

Iridium is the first and, up to the present day, only mobile satellite system that provides coverage over the entire globe including the polar regions. It uses a constellation of 66 LEO satellites with onboard processing capabilities that are interconnected via ISLs. Services are limited to voice, low rate data, fax, paging, messaging, and supplementary services delivered mainly to small handheld terminals operating in dual mode with terrestrial GSM. Iridium operates in the L band and uses QPSK modulation and FDMA/TDMA-TDD for the uplink and downlink. Technical details can be found in reference [121]. Important service contractors include the U.S. Department of Defense and other government agencies. In 1999 the Iridium consortium went through a bankruptcy process but emerged with new strength and is now planning to launch an IP-based second generation system, Iridium NEXT, for 2014, with mobile user data rates of up to 10 Mbit/s and nomadic rates up to 30 Mbit/s [122–123].

For a long time Globalstar was Iridium's main rival. The constellation consists of 48 LEO satellites, offering round the Earth coverage up to ±70° of latitude. The satellites do not use ISLs. A connection is usually established from a user terminal over the satellite to a gateway, which connects either to the PSTN or to the Internet.

Due to a lack of gateways in Africa and Asia, no coverage is obtained in these regions despite the fact that they lie well within the ±70° of latitude. In all other regions the provided services are voice, fax, paging, and short messages up to a data rate of 9.6 kbit/s. Many Globalstar terminals are handheld, operating in dual mode together with a terrestrial GSM component. As a multiple access strategy the satellite channels use DS-CDMA, together with QPSK modulation. The uplink takes place in the L band, while the downlink is performed in the S band [89, 124].

Another satellite operator, Inmarsat, has a longstanding reputation in delivering mobile and nomadic services. In 1979 it started as the *International Maritime Satellite Organization* serving primarily large vessels and other maritime users. Over the years it extended its service portfolio to aeronautical and land users and since 1994 has been called International Mobile Satellite Organization. Currently, it has 11 GEO satellites in operation and has teamed up with ACeS to provide handheld voice and low data rate services over parts of Asia using ACeS's Garuda 1 satellite [125]. The latest launched satellites (2005 to 2008) were three Inmarsat-4 providing 3G UMTS/BGAN services while the aging constellations Inmarsat-2 and Inmarsat-3 are still delivering low rate data and voice services. The BGAN system is 3GPP compliant and delivers services including voice, Internet, and video conferencing at user data rates up to 492 kbit/s. The user links are operated in the L band, while the feeder links make use of the C band [104]. The launch of a new GEO satellite Inmarsat-XL/Alphassat is scheduled for 2012. It has been developed in cooperation with ESA and is supposed to boost packed data rates as well as to provide single hop mesh connectivity [126].

Finally, the Thuraya fleet [127–128] of three GEO satellites uses the GMR-1 standard and delivers voice and low rate data services to Europe, North and Central Africa, the Middle East, Central Asia, and the Indian subcontinent. Commercial operation started in 2001 with Thuraya 1, now out of service. Thuraya 2 and 3 followed in 2003 and 2008, respectively. Services are operated in the L band with feeder links in Ku and C bands. User data rates are now reaching 444 kbit/s.

6.7 CONCLUSIONS

Information flow in the Smart Grid (SG) is likely to consist, mainly, of low control data payloads so that the ability to handle these payloads efficiently is a key parameter for the candidate technologies. Low overhead protocols are suitable for such communications as this is a design parameter in the long term evolution (LTE) or in the 802.11n standard.

It is also important to consider the ability to handle different quality of service (QoS) levels for the information to be transmitted. QoS is already an integral part of, for example, IEEE 802.16/WiMAX and IEEE 802.11/Wi-Fi.

One of the main objectives of the SG is to achieve a more efficient management of energy generation and consumption. Therefore, the energy efficiency of the communication technology is of great importance and has specifically been considered in IEEE 802.15.4/ZigBee or in the discontinuous reception and transmission schemes of LTE.

Mobility is a natural characteristic of the wireless networks, except for the case of fixed wireless networks (e.g., fixed satellite services or point-to-point single carrier WiMAX links). Mobility support of the different wireless technologies is provided in the NIST document [2]. Except for the case of ZigBee with a modest 20 km/h, the remaining technologies support higher speeds. Cellular technologies allowing speeds of over 300 km/h guarantee that for almost any mobile application there would be at least one wireless technology candidate to be used.

Satellite communications have the unique ability to provide data connectivity at nearly any place on Earth. Hence, they are predestined to be used within the Smart Grid in remote sites as a "low cost" alternative, as well as in many other sites as redundancy/diversity communications solution.

In summary, wireless technologies can be used both for normal operation or as backup option within the SG network. There is no single technology appropriate for all the SG applications. SG communication solutions will have to be custom made, adjusting to the scenario—that is, user density, propagation characteristics, data throughput, latency requirements, and so on. And they are likely to consist of heterogeneous networks, potentially also including wireline communications as introduced in Chapter 7 and as implemented in a real world case study, as described in Chapter 14.

ACKNOWLEDGMENTS

This work has been partly funded by the Spanish national projects Multi-Adaptive (TEC2008-06327-C03-02), COMONSENS (CSD2008-00010), MAMBO4 (CCG-UC3M/TIC-4620), and by the European Regional Development Fund within the ERDF Operational Programme of the Valencian Community 2007-2013 (grant number IMIDTA/2011/905).

REFERENCES

[1] National Institute of Standards and Technology (NIST), "PAP02: Wireless Communications for the Smart Grid." Available at: http://collaborate.nist.gov/twiki-sggrid/bin/view/SmartGrid/PAP02Wireless (accessed January 2012).

[2] "National Institute of Standards and Technology (NIST)—Priority Action Plan 02 (PAP02)—Objective 2," Available at: http://collaborate.nist.gov/twiki-sggrid/bin/view/SmartGrid/PAP02Objective2 (accessed January 2012).

[3] "National Institute of Standards and Technology (NIST)—Priority Action Plan 02 (PAP02)—Objective 3—Guidelines for Assessing Wireless Standards for Smart Grid Applications." Available at: http://collaborate.nist.gov/twiki-sggrid/bin/view/SmartGrid/PAP02Objective3 (accessed January 2012).

[4] OPENmeter, "Description of State-of-the-Art Wireless Access Technologies." Available at: http://www.openmeter.com/files/deliverables/OPEN-Meter%20WP2%20D2.1%20 part3%20v1.0.pdf (accessed January 2012).

[5] OPENmeter, "Open Public Extended Network Metering," Available at: http://www.openmeter.com/ (accessed January 2012).

[6] "Bluetooth Special Interest Group Home Page." Available at: https://www.bluetooth.org (accessed January 2012).

[7] "Z-Wave Home Page." Available at: http://www.z-wave.com/ (accessed January 2012).

[8] 802.15.4-2006, "Local and Metropolitan Area Networks—Specific Requirements Part 15.4: Wireless Medium Access Control (MAC) and Physical Layer (PHY) Specifications for Low-Rate Wireless Personal Area Networks (WPANs)," IEEE Standard for Information Technology—Telecommunications and Information Exchange Between Systems, September 2006.

[9] IEEE, "IEEE 802.15 WPAN Task Group 4g (TG4g) Smart Utility Networks." Available at: http://www.ieee802.org/15/pub/TG4g.html (accessed January 2012).

[10] "ZigBee Alliance." Available at: http://www.zigbee.org/ (accessed January 2012).

[11] "ZigBee Specification," ZigBee Document 053474r17, ZigBee Alliance, January 2010.

[12] 802.15.4c-2009, "Local and Metropolitan Area Networks Specific Requirements Part 15.4: Wireless Medium Access Control (MAC) and Physical Layer (PHY) Specifications for Low-Rate Wireless Personal Area Networks (WPANs)—Amendment 2: Alternative Physical Layer Extension to Support One or More of the Chinese 314–316 MHz, 430–434 MHz, and 779–787 MHz Bands," IEEE Standard for Information Technology—Telecommunications and information exchange Between Systems, April 2009.

[13] 802.15.4d-2009, "Local and Metropolitan Area Networks Specific Requirements Part 15.4: Wireless Medium Access Control (MAC) and Physical Layer (PHY) Specifications for Low-Rate Wireless Personal Area Networks (WPANs)—Amendment 3: Alternative Physical Layer Extension to Support the Japanese 950 MHz Bands," IEEE Standard for Information Technology—Telecommunications and information exchange Between Systems, April 2009.

[14] M. Hata, "Empirical formula for propagation loss in land mobile radio services," *IEEE Trans. Vehicular Technol.* 29(3), pp. 317–325, August 1980.

[15] 802.15.2-2003, "Local and Metropolitan Area Networks Specific Requirements Part 15.2: Coexistence of Wireless Personal Area Networks with Other Wireless Devices Operating in Unlicensed Frequency Bands." IEEE Recommended Practice for Information Technology—Telecommunications and information exchange Between Systems.

[16] 802.15.3-2003, "Local and Metropolitan Area Networks Specific Requirements Part 15.3: Wireless Medium Access Control (MAC) and Physical Layer (PHY) Specifications for High Rate Wireless Personal Area Networks (WPANs)," IEEE Standard for Information Technology—Telecommunications and information exchange Between Systems.

[17] J. Sun, Z. Wang, H. Wang, and X. Zhang, "Research on routing protocols based on ZigBee network," *Third International Conference on Intelligent Information Hiding and Multimedia Signal Processing, 2007. IIHMSP 2007*, 1, pp. 639–642, November 2007.

[18] C. E. Perkins, E. M. Belding-Royer, and S. Das, "Ad hoc on demand distance vector (AODV) routing," *IETF REFC* 3561, July 2003.

[19] I. D. Chakeres, and L. Klein-Berndt, "AODVjr, AODV simplified." *ACM Mobile Computing and Commun. Rev.*, 6(3), pp. 100–101, 2002.

[20] "ZigBee Smart Energy Profile," ZigBee Document 075356r15, ZigBee Alliance, December 2008.

[21] "HomePlug Powerline Alliance." Available at: http://www.homeplug.org/ (accessed January 2012).

[22] "Smart Energy Profile 2.0 Technical Requirements Document, Draft," ZigBee Document 105553, ZigBee Alliance and Home-Plug Powerline Alliance Liason, April 2010.

[23] IEEE, "IEEE 802.5 Web Site." Available at: http://www.ieee802.org/5/ (accessed January 2012).

[24] "IEEE 802.3 Ethernet Working Group." Available at: http://www.ieee802.org/3/ (accessed January 2012).

[25] "IEEE 802.11 Wireless Local Area Networks." Available at: http://www.ieee802.org/11/ (accessed January 2012).

[26] "Wi-Fi Alliance." Available at: http://www.wi-fi.org (accessed January 2012).

[27] R. van Nee and R. Prasad, *OFDM for Wireless Multimedia Communications*, ser. Universal personal communication. Artech House Publishers, 2000.

[28] T. Rappaport, *Wireless Communications: Principles and Practice*, 2nd ed. Prentice Hall PTR, 2001.

[29] 802.11n-2009, "Local and Metropolitan Area Networks—Specific Requirements Part 11: Wireless LAN Medium Access Control (MAC) and Physical Layer (PHY) Specifications Amendment 5: Enhancements for Higher Throughput," IEEE Standard for Information Technology—Telecommunications and Information Exchange Between Systems, October 2009.

[30] 802.11-2007, "Local and Metropolitan Area Networks—Specific Requirements Part 11: Wireless LAN Medium Access Control (MAC) and Physical Layer (PHY) Specifications," IEEE Standard for Information Technology—Telecommunications and Information Exchange Between Systems, June 2007.

[31] 802.11y 2008, "Local and Metropolitan Area Networks—Specific Requirements Part 11: Wireless LAN Medium Access Control (MAC) and Physical Layer (PHY) Specifications Amendment 3: 3650–3700 MHz Operation in USA," IEEE Standard for Information Technology—Telecommunications and Information Exchange Between Systems, November 2008.

[32] IEEE, "IEEE P802.11—Task Group S." Available at: http://www.ieee802.org/11/Reports/tgs_update.htm (accessed January 2012).

[33] "One Laptop per Child." Available at: http://one.laptop.org (accessed January 2012).

[34] International Telecommunication Union (ITU), "About ITU—Radiocommunication Sector." Available at: http://www.itu.int/net/about/itu-r.aspx (accessed January 2012).

[35] ITU-R, "ITU-R Terrestrial FAQs." Available at: http://www.itu.int/ITU-R/terrestrial/faq/index.html#g013 (accessed January 2012).

[36] X. Liang, K. Yamamoto, and S. Yoshida, "Performance comparison between channel-bonding and multi-channel CSMA." *Wireless Communications and Networking Conference, 2007. WCNC 2007. IEEE*, March 2007, pp. 406–410.

[37] AirTight Networks, "802.11n WLAN coverage estimator." Available at: http://www.airtightnetworks.com/home/solutions/80211n/80211n-wlan-coverage-estimator.html (accessed January 2012).

[38] Aerohive, "Aerohive Wi-Fi Planner." Available at: www.aerohive.com/planner/ (accessed January 2012).

[39] IEEE, "The IEEE 802.16 Working Group on Broadband Wireless Access Standards," Available at: http://www.ieee802.org/16/ (accessed January 2012).

[40] 802.16-2009, "Part 16: Air Interface for Broadband Wireless Access Systems," IEEE Standard for Local and Metropolitan Area Networks, May 2009.

[41] "WiMAX Forum." Available at: http://www.wimaxforum.org/ (accessed January 2012).

[42] "WiBro." Available at: http://www.wibro.or.kr (accessed January 2012).

[43] IEEE, "IEEE Std 802.16m," Available at: http://www.ieee802.org/16/pubs/80216m.html (accessed January 2012).

[44] "Requirements and Recommendations for WiMAX Forum™ Mobility Profiles," WiMAX Forum™, Document NO. WMF-T21-001-R010v01, September 2005.

[45] P. Frenger, S. Parkvall, and E. Dahlman, "Performance comparison of HARQ with Chase combining and incremental redundancy for HSDPA," *Vehicular Technology Conference, 2001. VTC 2001 Fall. IEEE VTS 54th*, vol. 3, October 2001, pp. 1829–1833.

[46] 3rd Generation Partnership Project (3GPP), "General Description of a GSM Public Land Mobile Network (PLMN)," Tech. Rep. TS 01.02 ver. 6.0.1.

[47] 3GPP, "General Packet Radio Service (GPRS); Service Description; Stage 2 (Release 1998)," Tech. Rep. TS 03.60 ver. 7.9.0.

[48] 3rd Generation Partnership Project, "Overview of the 3GPP Specification Numbering." Available at: http://www.3gpp.org/specification-numbering (accessed January 2012).

[49] 3rd Generation Partnership Project (3GPP), "UE Radio Access Capabilities (Release 1999)," Tech. Rep. TS 25.306 ver. 3.10.0.

[50] 3GPP, "High Speed Downlink Packet Access (HSDPA); Overall Description; Stage 2 (Release 5)," Tech. Rep. TS 25.308 ver. 5.7.0.

[51] 3GPP, "Enhanced Uplink; Overall Description; Stage 2 (Release 7)," Tech, Rep. TS 25.319 ver. 7.8.0.

[52] 3GPP, "Evolved Universal Terrestrial Radio Access (E-UTRA) and Evolved Universal Terrestrial Radio Access Network (E-UTRAN); Overall Description; Stage 2 (Release 8)," Tech. Rep. TS 36.300 ver. 8.12.0.

[53] 3rd Generation Partnership Project 2, "Physical Layer Standard for cdma2000 Spread Spectrum Systems," Tech. Rep. C.S0002-0 ver. 1.0, July 1999.

[54] 3rd Generation Partnership Project 2, "cdma2000 High Rate Packet Data Air Interface Specification," Tech. Rep. C.S0024-A ver. 1.0, March 2004.

[55] 3rd Generation Partnership Project 2, "cdma2000 High Rate Packet Data Air Interface Specification," Tech. Rep. C.S0024-B ver. 1.0, April 2006.

[56] 3rd Generation Partnership Project 2, "Introduction to cdma2000 Standards for Spread Spectrum Systems (Revision E)," Tech. Rep. C.S0001-E ver. 2.0, June 2010.

[57] V. H. MacDonald, "The cellular concept," *Bell Syst. Tech. J.*, 58(1), pp. 15–43, January 1979.

[58] H. Holma and A. Toskala, *WCDMA for UMTS; Radio Access for Third Generation Mobile Communications*. John Wiley & Sons, 2004.

[59] C. Ball, T. Hindelang, I. Kambourov, and S. Eder, "Spectral efficiency assessment and radio performance comparison between LTE and WiMAX," *IEEE International*

Symposium on Personal, Indoor and Mobile Radio Communicatios (PIMRC), Cannes, France, 2008.

[60] O. Fatemieh, R. Chandra, and C. A. Gunter, "Low cost and secure smart meter communications using the TV white spaces," *IEEE International Symposium on Resilient Control Systems (ISRCS'10)*, Idaho Falls, ID, August 2010.

[61] m2mtec Germany, "m2mtec Germany Home Page." Available at: http://www.m2mtec.de/ (accessed January 2012).

[62] smartgridnews.com, "Smart Grid Technology: Cellular Emerges as Viable Communications Choices," reporting on the Texas–New Mexico Power (INMP) trial, May 2010.

[63] Itron, "EverBlu, wireles fixed data collection system." Available at: https://www.actaris.com/html/medias/water/EverBlu_tech_pb_EN_07_09.pdf (acessed January 2012).

[64] Article 5 of The Radio Regulations, "Smart Grid Technology: Cellular Emerges as Viable Communications Choices," International Telecommunication Union, 2008.

[65] "LTE Part I: Core network," *IEEE Commun. Mag.*, 47, February 2009.

[66] "LTE Part II: Radio access," *IEEE Commun. Mag.*, 47, April 2009.

[67] R. Attar, D. Ghosh, C. Lott, M. Fan, P. Black, R. Rezaiifar, and P. Agashe, "Evolution of cdma2000 cellular networks: multicarrier EVDO." *IEEE Commun. Mag.*, 44, pp. 46–53, March 2006.

[68] J. Bergman, D. Gerstenberger, F. Gunnarsson, and S. Ström, "Continued HSPA evolution of mobile broadband," *Ericsson Review: A Cost-Effective Way of Migrating Mobile Broadband Access*, No. 1, 2009.

[69] A. C. Clarke, "Extraterrestrial relays," *Wireless World*, 51, pp. 305–308, October 1945.

[70] A. Leick. *GPS Satellite Surveying*, 3rd ed. John Wiley & Sons, 2004.

[71] G. L. Fong and K. Nour, "Broadband and the role of satellite services," Frost and Sullivan, White Paper, 2004.

[72] S. Brusch, S. Lehner, and J. Schulz-Stellenfleth, "Synergetic use of radar and optical satellite images to support severe storm prediction for offshore wind farming," *IEEE J. Selected Topics in Appl. Earth Observations and Remote Sensing*, 1(1), pp. 57–66, 2008.

[73] B. Heinert, "GPS-based Time Synchronization Improves Plant Reliability," online article on the Power-Gen Worldwide portal. Available at: http://www.powergen worldwide.com/index/display/articledisplay/183732/articles/power-engineering/volume-107/issue-8/features/gps-based-time-synchronization-improves-plant-reliability.html (accessed January 2012).

[74] P. A. Longley, M. F. Goodchild, D. J. Maguire, and D. W. Rhind, *Geographic Information Systems and Science*, 2nd ed. John Wiley & Sons, 2005.

[75] N. Komerath, "The space power grid: synergy between space, energy and security policies," *Atlanta Conference on Science and Innovation Policy*, October 2009, pp. 1–7, 2010.005, IEEE 5367831.

[76] Control Engineering Europe, "Satellite Systems Control Spanish Wind Farms," online article. Available at: http://www.controlengeurope.com/article.aspx?ArticleID=28275 (accessed January 2012).

[77] M. Payer, "SES ASTRA Supplies ASTRA2Connect for Energy Monitoring in Italy," ENP Newswire on-line article. Available at: http://www.allbusiness.com/energy-utilities/utilities-industry-natural-gas/13913818-1.html (accessed January 2012).

[78] Hughes Networks, "Guarding Against Network Failures—How to Ensure the 'Always-on' Business," White Paper. Available at: http://www.hughes.com/HNS%20Library%20Presentations%20and%20White%20Papers/HN-AccessContinuity_H38054_HR.pdf (accessed January 2012).

[79] European Telecommunication Standards Institute (ETSI), Technical Committee Satellite Earth Stations and Systems (SES), "Satellite Component of UMTS/IMT-2000; General Aspects and Principles." Technical Report, TR 101 865 V1.2.1, September 2002.

[80] M. Richharia and L. D. Westbrook, *Satellite Systems for Personal Applications: Concepts and Technology*. John Wiley & Sons, 2010.

[81] O3b Networks, "Home Page." Available at: http://www.o3bnetworks.com/ (accessed January 2012).

[82] Sirius Satellite Radio, "Cooperate Overview." Available at: http://www.siriusxm.com (accessed January 2012).

[83] European Space Agency (ESA), "HEO for ATM," online article. Available at: http://telecom.esa.int/telecom/www/object/index.cfm?fobjectid=29925 (accessed January 2012).

[84] Institute of Electrical and Electronics Engineers, "IEEE Standard Letter Designations for Radar-Frequency Bands," IEEE, Tech. Rep. IEEE Std. 521–2002, 2003.

[85] G. Maral and M. Bousquet. *Satellite Communications Systems—Systems, Techniques and Technologies*. 5th ed. John Wiley & Sons, 2009.

[86] I T U *Handbook on Satellite Communications*, 3rd ed. John Wiley & Sons, 2002.

[87] L. J. Ippolito Jr, *Satellite Communications Systems Engineering—Atmospheric Effects, Satellite Link Design and System Performance*, Wiley Series on Wireless Communications and Mobile Computing. John Wiley & Sons, Chichester, UK, 2008.

[88] S. D. Ilcev, *Global Mobile Satellite Communications—For Maritime, Land and Aeronautical Applications*. Springer, Dordrecht, The Netherlands, 2005.

[89] P. Chini, G. Giambene, and S. Kota, "A survey on mobile satellite systems," *Int. J. Satellite Commun.*, 28(1), pp. 29–57, August 2009.

[90] European Telecommunication Standards Institute (ETSI), Technical Committee Satellite Earth Stations and Systems (SES), "Broadband Satellite Multimedia (BSM); Services and Architectures," Technical Report, TR 101 984 V1.2.1, December 2007.

[91] ETSI, "Broadband Satellite Multimedia (BSM); Overview of BSM Families," Technical Report, TR 102 187 V1.1.1, May 2003.

[92] European Telecommunications Standard Institute (ETSI), Technical Committee BROADCAST, "DVB-S, Framing Structure, Channel Coding and Modulation for 11/12 GHz Satellite Services," European Standard, EN 300 421 V1.1.2, August 1997.

[93] European Telecommunication Standards Institute (ETSI), Technical Committee BROADCAST, "DVB-S, Implementation of Binary Phase Shift Keying (BPSK) Modulation in DVB Satellite Transmission Systems," Technical Report, TR 101 198 V1.1.1, September 1997.

[94] European Telecommunications Standard Institute (ETSI), Technical Committee BROADCAST, "Digital Video Broadcasting (DVB): Interaction Channel for Satellite Distribution Systems, DVB-RCS Standard," European Standard, EN 301 790 V1.4.1, April 2005.

[95] ETSI, "Digital Video Broadcasting (DVB); Second Generation Framing Structure, Channel Coding and Modulation Systems for Broadcasting, Interactive Services, News

Gathering and Other Broadband Satellite Applications (DVB-S2)," European Standard EN 302 307 V1.2.1, August 2009.

[96] ETSI, "Digital Video Broadcasting (DVB): Interaction Channel for Satellite Distribution Systems, DVB-RCS Standard," European Standard EN 301 790 V1.5.1, May 2009.

[97] H. Skinnemoen, "Special Issue on DVB-RCS+M," *International Journal of Satellite Communications and Networking*, vol. 28, no. 3–4, pp. 113–231, May–August 2010.

[98] European Telecommunication Standards Institute (ETSI), Technical Committee Satellite Earth Stations and Systems (SES), "Satellite Component of UMTS/IMT-2000; Evaluation of the W-CDMA UTRA FDD as a Satellite Radio Interface," Technical Specification, TR 102 058 V1.1.1, November 2004.

[99] ETSI, "Satellite Component of UMTS/IMT-2000; W-CDMA Radio Interface for Multimedia Broadcast/Multicast Service (MBMS)," Technical Specification, TR 102 277 V1.2.1, February 2007.

[100] D. Boudreau, G. Caire, G. E. Corazza, R. De Gaudenzi, G. Gallinaro, M. Luglio, R. Lyons, J. Romero-García, A. Vernucci, and H. Widmer, "Wide-band CDMA for the UMTS/IMT-2000 satellite component," *IEEE Trans. Vehicular Technol.*, 51(2), March 2002.

[101] B. Martin, N. Chuberre, H. J. Lee, G. E. Corazza, and A. Vanelli, "IMT-2000 satellite radio interface for UMTS communications over mobile satellite systems," *Advanced Satellite Mobile Systems (ASMS)*, Bologna, Italy, August 2008, pp. 167–172.

[102] A. Azizan, A. U. Quddus, and B. G. Evans, "Satellite high speed downlink packet access physical layer performance analysis," *Advanced Satellite Mobile Systems (ASMS)*, Bologna, Italy, August 2008, pp. 156–161.

[103] European Telecommunication Standards Institute (ETSI), Technical Committee Satellite Earth Stations and Systems (SES), "Satellite Component of UMTS/IMT-2000; Considerations on Possible Harmonization Between A, C and G Family Satellite Radio Interface Features," Technical Specification, TR 102 278 V1.1.1, August 2008.

[104] A. Howell, "Standards and the New Economy—Broadband Global Area Networks," presentation. Available at: http://www.cambridgewireless.co.uk/docs/Standards%20and%20theNew%20Economy_Alan%20Howell%20RTT%20standards%20-%20BGAN_V4_approved.pdf (accessed January 2012).

[105] SES-ASTRA, "Astra2Connect." Available at: http://www.astra2connect.com/ (accessed January 2012).

[106] "SES-ASTRA." Available at: http://www.ses-astra.com/business/en/corporate/index.php (accessed January 2012).

[107] European Committee for Electrotechnical Standardization (CENELEC), "Functional Receiver Specification of Satellite Digital Interactive Television with a Low Data Rate Return Channel Via Satellite—Modem Layer Specification," European Standard EN 50478, February 2007.

[108] European Space Agency (ESA), "SATMODE." Available at: http://telecom.esa.int/telecom/www/object/index.cfm?fobjectid=11843 (accessed January 2012).

[109] ESA, "Sat3Play." Available at: http://telecom.esa.int/telecom/www/object/index.cfm?fobjectid=28634 (accessed January 2012).

[110] Newtec, "TL100, TelliNet V2.6, Tellitec Product Family." brochure. Available at: http://www.newtec.eu/fileadmin/user_upload/product_leaflets/Tellitec/TelliNet_TL100_R3.pdf (accessed January 2012).

[111] Skylogic, "Home Page." Available at: http://www.skylogic.it (accessed January 2012).

[112] T. Lohrey, "Satellite broadband internet everywhere for enterprises, organisations, governmental institutions and consumers," *13th International Seminar/Workshop on Direct and Inverse Problems of Electromagnetic and Acoustic Wave Theory*, September 2008, pp. 15–18.

[113] ViaSat, "LinkSarS2 and LinkWayS2 TDMA Systems," product brochure. Available at: http://www.viasat.com/files/assets/web/datasheets/linkstar_linkway_tdma_system_13_lores.pdf (accessed January 2012).

[114] ViaSat, "SurfBeam Broadband System." Available at: http://www.viasat.com/broadband-satellite-networks/surfbeam (accessed January 2012).

[115] CableLabs, "DOCSIS Specifications." Available at: http://www.cablelabs.com/cablemodem/specifications/ (accessed January 2012).

[116] WildBlue, "How It Works." Available at: http://www.wildblue.com/overview/how-it-works (accessed January 2012).

[117] Hispasat. Available at: http://www.hispasat.com (accessed January 2012).

[118] J. E. Puente, "Instalación de Sistemas de Automatización y de Datos—Parques Fotovoltaicos: Gestion Remota y Video Supervision via Satelite," talk organized by the University of Vigo within the framework of Instalación de Sistemas de Automatización y Datos (ISAD'0910). Available at: http://tv.uvigo.es/uploads/material/Video/4930/Conferencia_ISAD__18_de_Enero_de_2010_.pdf (accessed January 2012).

[119] A. Yun and J. Prat, "Amerhis: DVB-RCS Meets Mesh Connectivity," White Paper. Available at: http://satlabs.org/pdf/WhitePaper_RCS-AmerHisv3.pdf (accessed January 2012).

[120] M. Williamson, "Connecting the other three billion," *Eng. Technol.*, 4(4), pp. 70–73, February 2009.

[121] Iridium Satellite LLC, "Implementation Manual for IRIDIUM Satellite Communications Service," Tech. Rep. Draft v1.0, February 2006.

[122] D. Thoma, "Iridium NEXT generation satellite system and application to CNS," *ICNS Conference*, Herndon, US, May 2007. Presentation available at: http://acast.grc.nasa.gov/wp-content/uploads/icns/2007/Session_G/06-Thoma.pdf. (accessed January 2012).

[123] Iridium Satellite LLC, Available at: http://www.iridium.com (accessed January 2012).

[124] Globalstar. Available at: http://www.globalstar.com (accessed January 2012).

[125] ASIA Cellular Satellite (ACeS), "ACeS Goes Global with Inmarsat," online news article. Available at: http://www.acesinternational.com/corporate/index.php?fuseaction=News.read&type=1&id=04092006170000 (accessed January 2012).

[126] M. Vilaca, A. Franchi, and G. Huggins, "The Inmarsat XL payload—a proposal for alphasat," 3rd Advanced Satellite Mobile Systems Conference (ASMS), Herrsching am Ammersee, Germany, May 2006.

[127] Boing Satellite Systems, "Thuraya-2, 3—Completed Systems for Mobile Communication." Available at: http://www.boeing.com/defense-space/space/bss/factsheets/geomobile/thuraya2_3/thuraya2_3.html (accessed January 2012).

[128] Thuraya. Available at: http://www.thuraya.com/ (accessed January 2012).

7

WIRELINE COMMUNICATIONS IN SMART GRIDS

Lars Torsten Berger

7.1 INTRODUCTION

The material within this chapter is partly based on Berger and Moreno-Rodríguez [1] and Berger [2].

There is a long list of complementary and sometimes competing wireline specifications and standards that can be used in Smart Grid deployments. Industry adoption and large-scale customer roll-outs are still in their infancies and it is hard to make an accurate prediction of the "winners" and "losers." The technologies discussed throughout this chapter are the telephone line standards *ADSL/VDSL*, the coaxial cable standard *DOCSIS*, the power line standards *G3*, *PRIME*, and *IEEE 1901* and *IEEE 1901.2*, as well as the multiwire standards *ITU-T G.hn* and *ITU-T G.hnem*, which have different areas of applications within the *Smart Grid* (SG). To structure these areas, the U.S. *National Institute of Standards and Technology* (NIST) devised a domain-based conceptual model [3]. Each domain contains actors that

Smart Grid: Applications, Communications, and Security, First Edition.
Edited by Lars Torsten Berger and Krzysztof Iniewski.

TABLE 7.1. Domains and Actors in the Smart Grid Conceptual Model

Domain	Actors in the Domain
Customers	The end users of electricity. May also generate, store, and manage the use of energy. Traditionally, three customer types are discussed, each with its own domain: residential, commercial, and industrial.
Markets	The operators and participants in electricity markets.
Service providers	The organizations providing services to electrical customers and utilities.
Operations	The managers of the movement of electricity.
Bulk generation	The generators of electricity in bulk quantities. May also store energy for later distribution.
Transmission	The carriers of bulk electricity over long distances. May also store and generate electricity.
Distribution	The distributors of electricity to and from customers. May also store and generate electricity.

Source: Based on Table 3-1 in NIST [3].

with the help of communications might act over domain boarders. The definitions of domains and actor are reproduced in Table 7.1. The interconnections between domains are displayed in Figure 7.1.

Considering customer *Access*, it can be said that many households in the developed world already have a connection to the Internet. This general Internet Access might be reused in some situations, for example, to connect the customer to third party energy *service providers*, to *operations*, and to *markets*. At the end of 2008, 65% of the worlds, Access customers were using a *digital subscriber line* (DSL) modem connected to their telephone line, while 21% were using a cable modem linking them with a cable network operator [4]. Delivering Internet Access over the electrical grid, also known as *broadband over power line* (BPL), is less frequent (less then 1% at the end of 2008) but certainly on the rise especially in rural areas [5], and in developing countries with a poorly developed fixed line telephone infrastructure, not to mention *coaxial cable* (coax). While it is believed that globally DSL and coax services, as discussed in more detail in Sections 7.2 and 7.3, will dominate the Internet traffic oriented Access market for the next couple of years, early *automated meter reading* (AMR) systems, frequently used *ultra narrowband power line communication* (UNB-PLC) technology like Turtle [6] and TWACS [7, 8] to gain access and in parts control over the energy meters within private homes. UNB-PLC systems are usually designed to communicate over long distances with their signals passing through LV/MV-transformers (*low voltage/medium voltage*). This helps to keep the amount of required modems and repeaters to a minimum. Drawbacks are low data rates on the order of 0.001 bit/s and 60 bit/s for Turtle and TWACS, respectively, and sometimes limitations to unidirectional communications. These UNB-PLC technologies are mentioned here as they are among the pioneers in the AMR and *distribution automation* field. Power electronics related to TWACS transmission and reception have been presented in Chapter 4. However, in the light of many

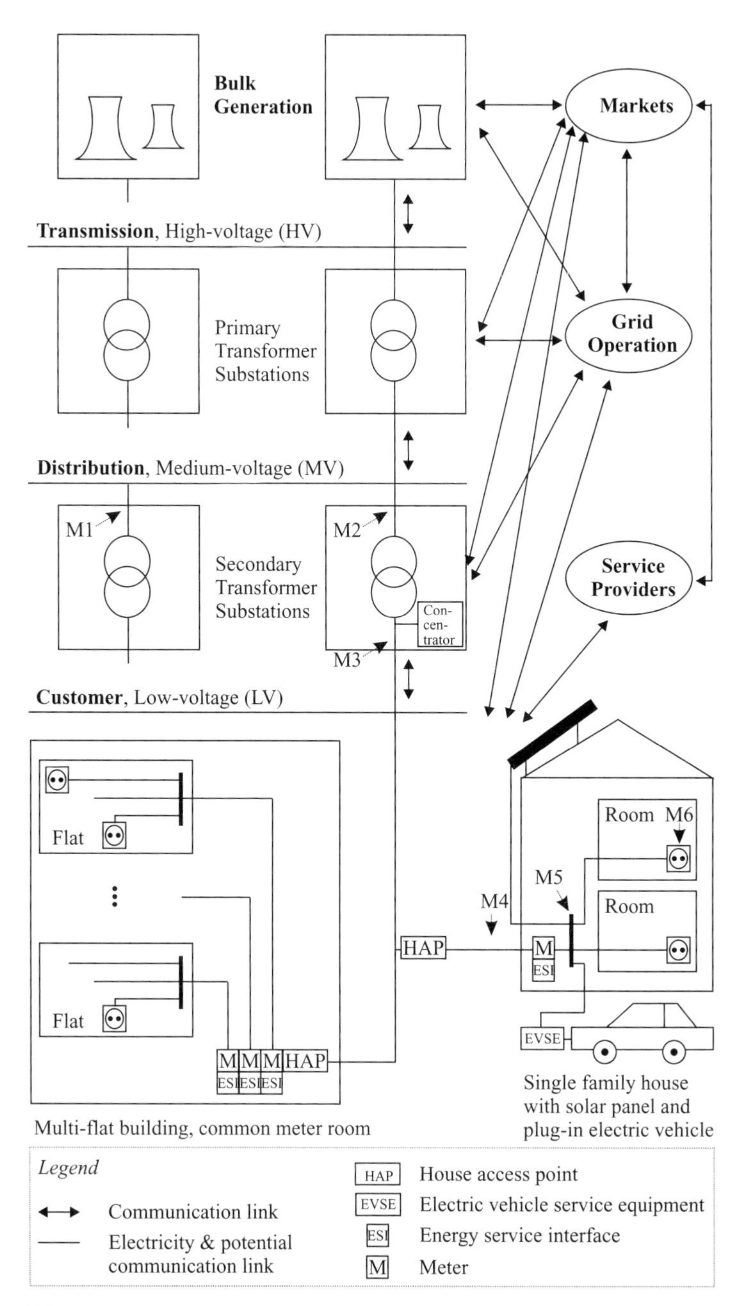

Figure 7.1. Smart Grid domains and their electrical as well as communication interconnections.

upcoming Smart Grid deployments, there are much higher requirements on the communication infrastructure, for example, to support *demand response*, *distributed generation*, and *demand-side management* applications as discussed in Chapters 1 and Chapter 3. Due to the reliability requirements of these services, it is frequently considered that instead of a general third party Internet Access a dedicated Access in the framework of *advanced metering infrastructure* (AMI) will be used. A whole wealth of material on AMI requirements and architectures is freely available, for example, from the ongoing European *OPEN meter* project [9]. To cope with increased AMI requirements, Section 7.4 on PLC in general will leave UNB-PLC solutions on the sideline in favor of more recent *narrowband PLC* (NB-PLC) technologies, such as *Powerline Related Intelligent Metering Evolution* (PRIME) [10] and G3 [11, 12]. NB-PLC bidirectional data rates are on the order of a hundred kbit/s, while partly preserving the advantage to communicate over long ranges and through transformers. PRIME and G3 are also being examined in the move to develop the ITU/IEEE NB-PLC standards, *ITU-T G.hnem* and *IEEE 1901.2*.

Turning to *In-Home* scenarios, also referred to as premises networks, or *home area networks* (HANs), it is noted that in industrial and office buildings, as well as in some newer private homes, Ethernet connections are used over preinstalled twisted pair wiring. However, the majority of private homes do not have an ubiquitous twisted pair nor a coaxial cable supplying every room. Many homes are limited to a handful of coax and telephone outlets in a low subset of rooms. This is why Internet connectivity is frequently provided using wireless networks as discussed in Chapter 6. Nevertheless, for reliable high data rate applications power line technologies are becoming more and more attractive as power sockets are available in every room and current *broadband power line communication* (BB-PLC) technologies provide data rates of more than 200 Mbit/s [13], making it easy to fulfill the users' home entertainment needs including *high definition television* (HDTV) support. Furthermore, NB-PLC solutions have been used for a long time for home automation applications [14], and it is believed that some existing automation systems, like BACnet [15], KNX [16], and LON [17], are being integrated into upcoming Smart Home concepts [18]. It is further anticipated that to support demand response and distributed generation schemes (see Chapter 1), In-Home communication links will have to be established between plug-in electric vehicles (PEVs) and their charging station, as well as communications between smart appliances such as heaters, air conditioners, and washers. Upcoming SG services in the home include granular control of smart appliances, the ability to remotely manage electrical devices, and the display of consumption data to better inform consumers, and thus motivate them to conserve power. As most of the In-Home SG appliances will be connected to the mains network, PLC solutions seem to be especially attractive to support these communication needs.

Moving from the in-building scenarios to the transmission and distribution grid and its related *supervisory control and data acquisition* (SCADA) substations, it can be said that there exists a large variety of custom SCADA communication solutions, in part based on (satellite) radio links and leased telephone lines, as well as fiber optical SONET/SDH connections. Wireless and optical communication solutions are discussed in Chapters 6 and 8, respectively. From a wireline point of view, phone line connections still play an important role. Nevertheless, the benefit of the full control over the transmission medium may in some cases be so attractive to utilities that they opt for power line solutions, which are presented in detail in Section 7.4.

The domains of the NIST conceptual model are also reflected in the *"IEEE P2030 Draft Guide for Smart Grid Interoperability of Energy Technology and Information Technology Operation with Electric Power System (EPS), and End-User Applications and Loads"* [19]. There they form three *interoperability architectural perspectives* (IAPs), one dealing with *power systems* (PS-IAP), one dealing with *communication technology* (CT-IAP), and, finally, one dealing with *information technology* (IT-IAP). The IEEE Guide provides a detailed framework on how to map *power system perspective interfaces* to *communication technology interfaces* and related *information technology interfaces*. However, the Guide is to a large extent communication technology agnostic, leaving room for integration of newer communication technologies that in the past have developed at a much faster pace than their power system counterparts. It becomes clear that a requirements and applicability assessment of communication technology has to be performed on a case-by-case basis. In this line, the remainder of this chapter does not suggest certain wireline communications solutions for certain SG applications, but rather provides some details on existing wireline communication technology options.

7.2 PHONE LINE TECHNOLOGY

Over the last two decades an extended variety of systems have been developed under the denomination of *digital subscriber line* (DSL). Commonly, DSL systems are designed to be used over twisted pair telephone cabling, which originally had been installed to carry telephone signals, that is, the *public switched telephone network* (PSTN) carrying *plain old telephone services* (POTS). Coexistence is possible as telephony signals use frequencies between 300 and 3400 Hz, while DSL services currently operate from around 25 kHz to 30 MHz (the exact operating frequencies depend on the DSL system, country-specific regulations, and the selected band plan). Lowpass filters, also referred to as *splitters*, are used on the cable to the fixed-line telephone to avoid audible DSL interference. Similar options for coexistence (bundling) with *integrated services digital network* (ISDN) and *time compression multiplexed ISDN* (TCM-ISDN) exist.

7.2.1 DSL Overview

Although there is a long list of DSL systems—for example, HDSL (*high bit rate DSL*), IDSL (*integrated services digital network DSL*), and SDSL (*symmetric DSL*)—the following focuses on the more recent systems ADSL (*asymmetric DSL*) and VDSL (*very high bit rate DSL*), and their evolutions as summarized in Table

TABLE 7.2. Recent DSL Standards

Name	ITU Standard	Approved	Up Peak Rate (Mbit/s)	Down Peak Rate (Mbit/s)
ADSL	G.992.1&2	1999	0.8	7
ADSL2	G.992.3/4	2002	1	8
ADSL2+	G.992.5	2003	1	24
VDSL	G.993.1	2004	15	55
VDSL2	G.993.2	2005	100	100

7.2. A detailed overview of *International Telecommunications Union* (ITU) recommendations related to DSL technology can be found in ITU publications [20–21]. The standards from the ITU G Series can be downloaded [22].

The ADSL standard was approved by the ITU in 1999, reaching 5 million subscribers in 2000 [24]. Standard evolution led to a second version, *ADSL2*, approved in 2002. Among others, improvements were made in the areas of data rate and supported range, as well as with regard to fault diagnosis and power saving operations. For example, keeping the range constant at around 5 km, ADSL2 approximately increases the rate by 50 kbit/s. Alternatively, by keeping the data rate constant, ADSL2 can reach customers 183 m further away [24]. With the need to support even higher data rates ADSL2+ was approved in 2003. It doubles the spectrum used for downlink transmissions, that is, going from 0.14–1.1 MHz in ADSL2 to 0.14–2.2 MHz in ADSL2+. ADSL2+ can therefore double downlink data rates. Nevertheless, since signal attenuation increases with frequency, and the extra spectrum was made available at higher frequencies, better downlink performance can only be obtained up to distances of around 2500 m [24]. In 2004 the *very high bit rate DSL* (VDSL) standard was ratified. Using frequencies up to 12 MHz, VDSL is able to boost peak data rates to 15 Mbit/s in the uplink and 55 Mbit/s in the downlink [23]. Aiming for even higher rates, the ITU ratified VDSL2 in 2005. By extending its operating frequency up to 30 MHz, VDSL2 is able to deliver, for example, a symmetric solution of 100 Mbit/s in either direction [23].

7.2.2 DSL Scenarios

Basic structures of DSL Access deployments are presented in Figure 7.2. Figure 7.2a presents an entire copper-based option, where around 15 primary cables with around 1500 copper pairs each are leaving the telecommunication operator's *central office* (CO). These primary cables are subsequently broken down at various *distribution points* (DPs) until they finally reach a larger building or individual customer premises.

In Figure 7.2b an optical fiber link establishes the connection between the service provider's *central office* (CO) and the *digital subscriber line access multiplexer*

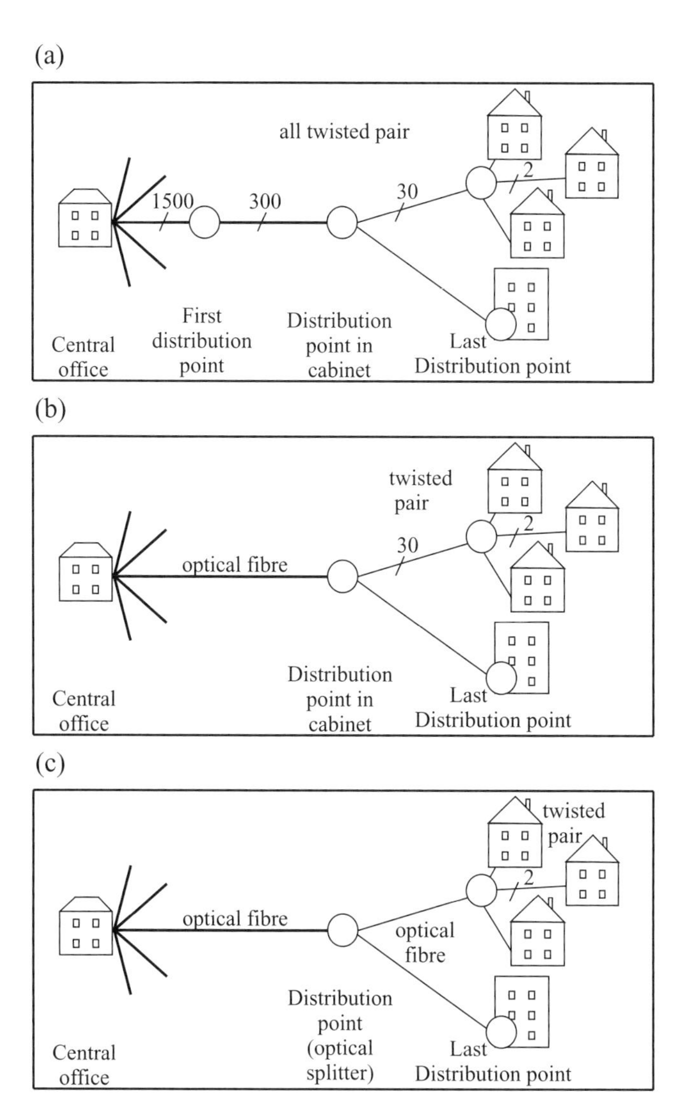

Figure 7.2. DSL Access deployment scenarios: (a) based on Magesacher et al. [25], (b) and (c) based on Lazauskaite [26].

(DSLAM) installed in a street cabinet up to several kilometers away from the customer. From there, VDSL2/ADSL2+ links run over the telephony cabling infrastructure to DSL modems at the customers' premises. The modems are sometimes also referred to as *customer premises equipment* (CPE) and the whole scenario is sometimes referred to as *fiber-to-the-node* (FTTN).

The lines carrying the DSL signals usually consist of *unshielded twisted pair* (UTP) whose diameter frequently complies with the *American Wire Gauge* designations 24 AWG and 26 AWG, that is, 0.511 mm and 0.405 mm, respectively [27].

Especially looking at communication performance, ANSI/TIA/EIA-568-B.1 defines the categories (*Cat 3*, *Cat 5e*, *Cat 6*) [28], while ISO/IEC 11801 defines the classes (*Classes C*, *D*, *E*, and *F*) [29]. Cat3/Class C and Cat 5e/Class D cabling are frequently encountered in DSL applications and undergo test procedures up to 16 MHz and 100 MHz, respectively. The cable can be described in terms of basic parameters like resistance, inductance, capacitance, and admittance, also called the *RLCG parameters*. Analytical and numerical RLCG transmission line models exist and are used to calculate the propagation constant, and the characteristic impedance, and in the sequel the voltages and currents along the line [30]. However, in practice lines are often much less uniform as predicted by these models [31]. *Bridged taps* (i.e., open ended stub lines) and line discontinuities lead to reflections and multipath, whose basic principles are visualized on the example of the power line channel in Section 7.4.1. Hence, safety margins are frequently applied when planning DSL networks.

Wire pairs are usually collected into *binders*. Proximity between pairs leads to crosstalk, which is one of the main performance impairments [32]. To reduce crosstalk different twist rates, also called *pitches*, are used to avoid conductors of different pairs repeatedly lying next to each other. Nevertheless, crosstalk cannot be fully avoided, and one usually distinguishes between the *near-end crosstalk* (NEXT) and *far-end crosstalk* (FEXT). In NEXT a victim DSL receiver experiences interference from a close by DSL transmitter. In FEXT a victim receiver experiences interference from a transmitter located at the other end of the link [33]. Other impairments include interference from radio transmitters and impulsive noise, for example, due to control voltages to elevators, or ringing of phones on lines sharing the same binder [33], or, of course, thermal background noises. Even *power line communications* (PLC) can lead to interference if DSL cabling and PLC cabling are run less than 2 cm apart [34]. The cable binders may be treated as *multiple input multiple output* (MIMO) communication channels [32, 35] with the aim of applying multiuser coordination and interference mitigation techniques, also referred to as *dynamic spectrum management* (DSM) Level 3 [25, 36]. However, at present, real world installations mainly attack line length, also referred to as *loop length*, to combat signal attenuation that increases with length and frequency. To get a better feeling for this effect, consider the signal and noise *power spectral densities* (PSDs) plotted in Figure 7.3. The PSDs have been derived by Magesacher et al. [25] based on various regulatory, cable property, and crosstalk assumptions. Considering the difference between the signal PSDs and the noise plus ingress PSDs, one can obtain a feeling for the *signal-to-noise plus interference ratio* (SNIR) that could be obtained at a certain frequency and for a certain loop length.

Especially, to leverage sufficient benefit from the higher frequencies used by ADSL2+ and VDSL2, operators try to minimize the distance between DSLAM and CPE. Such a situation is depicted in Figure 7.2c, where first an optical splitter is used to bring the fiber link close to the customer. Then a DSLAM is installed in a street cabinet close to individual houses, that is, less than 300 m from the house, or even inside the basement of multidwelling units, sometimes referred to as *fiber to the cabinet (or curb)* (FTTC) and *fiber to the building* (FTTB), respectively. For more on fiber optical communication see Chapter 8.

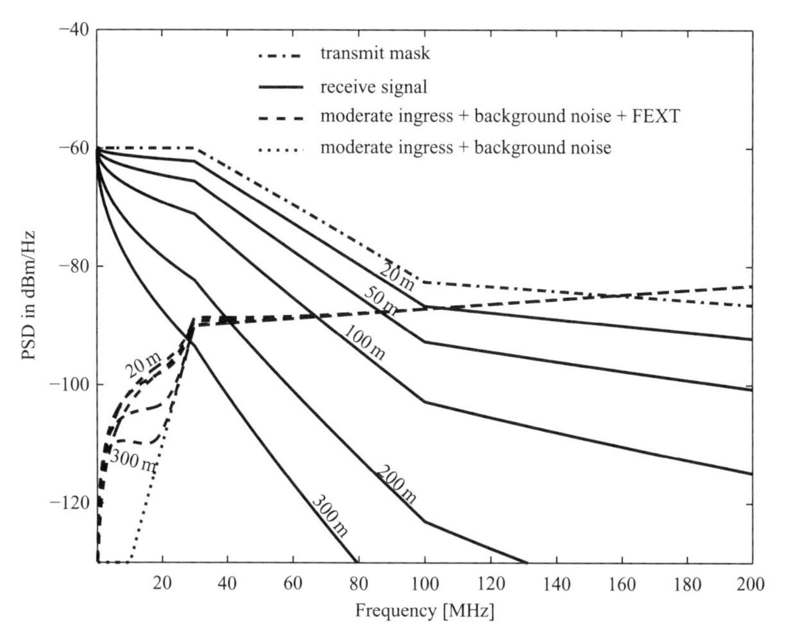

Figure 7.3. Transmit and receive power spectral densities (PSDs). The uppermost line (dash–dot) is the transmit PSD mask, followed by receive PSDs (solid lines) for loop lengths 20 m, 50 m, 100 m, 200 m, and 300 m (AWG24/0.5 mm). Transmission is impaired by moderate ingress (–110 dBm/Hz), FEXT from one equal-length crosstalker, and background noise (–130 dBm/Hz). From Magesacher et al. [25], © 2010 Wiley.

7.2.3 ADSL2+ and VDSL2

ADSL2+ as well as VDSL2 are based on a modulation called *discrete multitone* (DMT). DMT is a form of *orthogonal frequency division multiplexing* (OFDM) [37] reaching down to base band. Generally, OFDM has achieved widespread adoption throughout current wireline as well as wireless communication systems. It provides a high flexibility in terms of communication over frequency selective channels, flexible frequency assignments, notching (PSD shaping) options, and interference avoidance possibilities. Furthermore, OFDM possesses high spectrum efficiency, and its signal processing techniques are well understood. Often *quadrature amplitude modulation* (QAM) is used to assign bits to the OFDM subcarriers, which in the context of DSL systems are usually referred to as *tones*. Specifically, in ADSL2+/VDSL2 odd and even constellations are used. Up to 15 bits can be loaded onto a tone, for example, as a function of the tone's SNIR. Up to 4096 tones are available in VDSL2, which is eight times the number of tones available in ADSL2+. A combination of Reed–Solomon and Wei's Trellis coded modulation is used for *forward error correction* (FEC) [38–39]. VDSL2 has been designed with ADSL, ADSL2, ADSL2+, and VDSL backward compatibility in mind. This means that operators can, for example, upgrade their ADSL line card inside the DSLAM to VDSL2 without the need to immediately upgrade all CPEs. Furthermore, VDSL2 defines eight profiles—*8a/b/c/d*, *12a/b*, *17a*, and *30a*—operating over 8, 12, 17, and 30 MHz, respectively. To claim compliance with the VDSL2 standard only the implementation of one profile is required, offering the option of reduced complexity implementations targeting specific applications. Among others, the profiles contain specifications on the *maximum* and *minimum aggregate transmit power* in the downstream and upstream directions, the subcarrier spacing (4.3125 or 8.625 kHz), the *minimum bidirectional net data rate capability* (MBDC), and whether to support *upstream band zero* (US0) or not. US0 lies in the kilohertz range, where signal attenuation is usually low. It therefore allows longer reach deployments. For the high bandwidth profiles 12b, 17a, and 30a, that are intended for high data rates in short loops, the support of US0 is optional. In addition to the profiles, several frequency band plans are defined for North America (Annex A), Europe (Annex B), and Japan (Annex C). As an example, the band plan for Japan is presented in Figure 7.4.

With the many configuration options it is a difficult task to present a single throughput performance estimate. Nevertheless, to get at least a feeling for potential aggregated throughput performance, Figure 7.5 presents a comparison of ADSL2+ 2.2 MHz, VDSL 12 MHz, and VDSL2 30 MHz performance adapted from Lindecke [23].

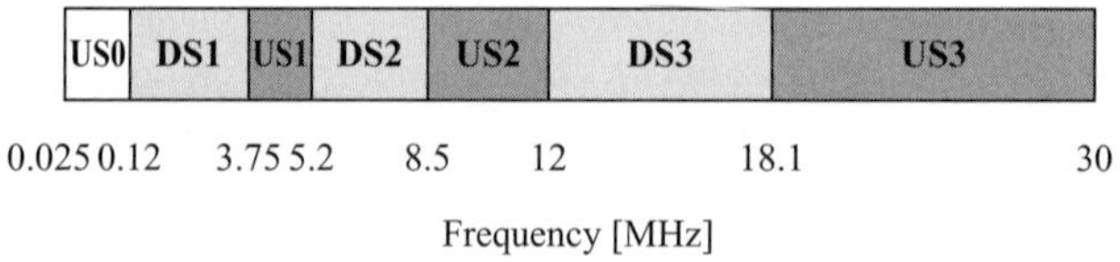

US0	DS1	US1	DS2	US2	DS3	US3

0.025 0.12 3.75 5.2 8.5 12 18.1 30

Frequency [MHz]

Figure 7.4. VDSL2 band plan for Japan (DS for *downstream*, US for *upstream*).

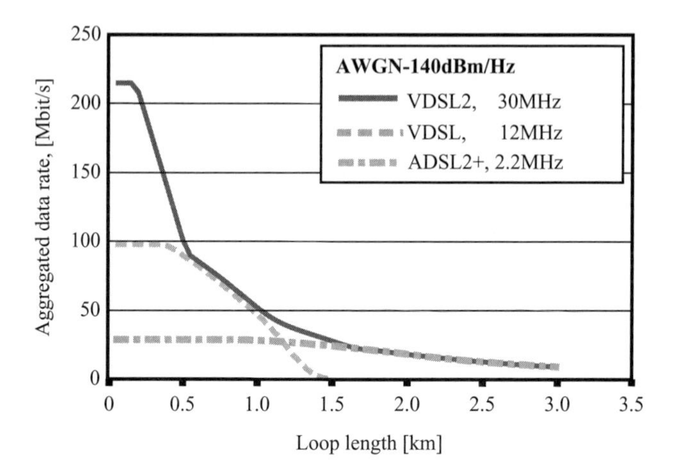

Figure 7.5. ADSL2+, VDSL, and VDSL2+ aggregated throughput performance example based on Lindecke [23].

It can be seen that VDSL usability is limited to short local loops, where it is able to provide a clear throughput advantage over ADSL2+. Beyond around 1.2 km, however, ADSL2+'s range extension capabilities allow a graceful throughput decline while VDSL becomes unusable. VDSL2 is able to deliver more than 200 Mbit/s aggregated throughput close to the transmitter, but then deteriorates sharply to around 100 Mbit/s at 0.5 km. Finally, for loop length over 1.5 km, the throughputs obtained by VDSL2 and ADSL2+ are roughly the same.

In any case the data rates achieved by today's DSL technologies are more than sufficient to cover most Smart Grid needs. Looking at stability and the ability to resist impairments like impulsive noise, ADSL2+ introduced a parameter called *impulsive noise protection* (IMP), which also forms part of VDSL2, with values ranging from 2 to 16. The parameter influences FEC coding and interleaving depth and allows one to correct a burst of up to 16 OFDM/DMT symbol erasures. Furthermore, to cater for services with different latency and noise protection requirements ADSL2+/VDSL2 have dual latency options, where two data streams with different FEC/interleaving settings can be multiplexed. With the additional ability to use Ethernet in the last mile and manufacturers announcing support for IPv6 [40] in their DSL products [41], existing DSL connections become an interesting option for many Smart Grid situations and have, for example, been used in Smart Grid field trials within the DISPOWER project or within the SmartGridCity pilot in Bolder, Colorado [42–43].

7.3 COAXIAL CABLE TECHNOLOGIES

In the context of Smart Grid, *coaxial cable* (coax) technologies can play a role providing Internet Access, which is dealt with in more detail in Sections 7.3.1 and 7.3.2. On the other hand, coax can be used to connect different devices within the home. One specification covering coax In-Home deployment was developed by the *Multimedia over Coax Alliance* (MoCA) [44], which is, however, not discussed further in this chapter in favor of the more general ITU-T G.hn multiwire standards discussed in detail in Section 7.4.4.

Coax Access networks are generally much younger than their telephone network counterparts. Initially delivering *community antenna television* (CATV), cable operators have long shifted their activities to bidirectional voice, video, and data services over *hybrid-fiber-coax* (HFC) networks. The main difference to a DSL Access network is possibly the shared nature. While in DSL deployments individual twisted pair cables serve individual customers—albeit with some negative crosstalk effects—the coax Access network was primarily designed for broadcast applications; that is, the same television and radio channels were delivered to all the users. In general, good coax propagation properties lead to relatively high SNIRs and in the sequel to relatively high shared peak data rates, which goes well in line with broadcast applications or with bursty Internet traffic generated by a few users. However, to increment bidirectional and especially unicast centered data rates, cable network

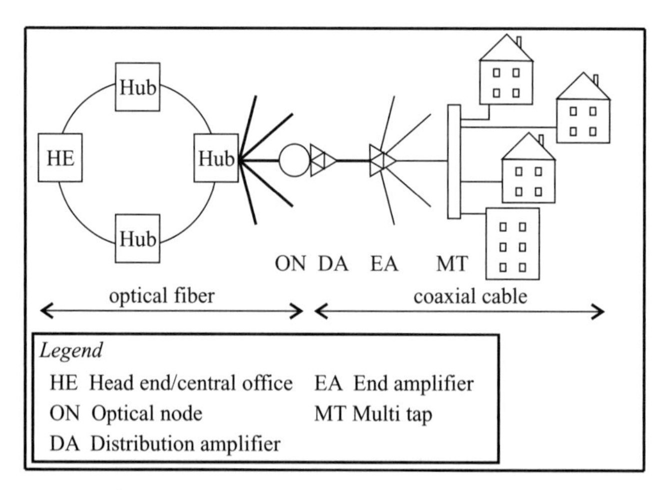

Figure 7.6. Hybrid-fiber-coax Access network.

operators are also bringing the fiber closer to the end customers, also known as *deep fiber* deployments. Different from DSL, the intention is not so much to increase the available SNIR, but to sectionalize the shared medium, so that the common bandwidth is shared among fewer users, in consequence augmenting each user's share.

7.3.1 Coax Scenarios

Looking at real world deployment scenarios across Europe, in the framework of the EU funded ReDeSign project [45], it was found that the *tree-and-branch*, *star*, and *hybrid* scenarios occur with a probability of 47%, 29%, and 18%, respectively. A common HFC deployment is shown in Figure 7.6. A *head end* (HE), in analogy to a central office in the Telco world, is connected via a bidirectional single fiber ring to various *hubs*. Together these constitute the core network with a fiber ring length usually below 100 km. From the hubs, fiber with narrow as well as broadcast capabilities and return channel runs to *optical nodes* (ONs). These nodes establish the interconnection between fiber and coax segment. On average, 1074 cable modems are served from an optical node. The modems are connected via segments of coaxial cable and cascades of *distribution*, *trunk*, and *end amplifiers*. Hence, when modeling the coax propagation channel, one not only has to deal with coaxial cable propagation but also with the representation of passive multitap splitters, as well as the nonlinear characteristics of active amplifier elements. On average, 24 customers are passed per end amplifier, and 95% of all customers are within 30 km from a cable operator's fiber hub. In general, most scenarios comply well with the requirements set forth in IEC standards [46–47], with the average upstream SNIR of the coax network at 30 dB and the downstream SNIR at 36 dB. Details on channel models as well as reference scenarios can be found in documents of ReDeSign Partners

TABLE 7.3. DOCSIS Standard Overview

Name	ITU Standard	Approved	Down Rate (Mbit/s)	Up Rate (Mbit/s)
(Euro)DOCSIS 1.1	J.112 (Annex B)	2001	42.88 (55.62)	10.24
(Euro)DOCSIS 2.0	J.122 (Annex F)	2007	42.88 (55.62)	30.72
(Euro)DOCSIS 3.0	J.222.0 to 3, J.210 (Annex A)	2007	$m \cdot 42.88$ ($m \cdot 55.62$)	$n \cdot 30.72$

[48–49], respectively. Interesting channel and noise characterization results on cable networks as well as on customer home wiring may also be found in a report by Cable Television Laboratories (CableLabs) [50].

7.3.2 Data Over Cable Service Interface Specification (DOCSIS)

Leaving radio and television broadcasts on the sideline, the transmission technology dominating in the Access segment is DOCSIS (*data over cable service interface specification*). Originally developed within CableLabs [51], DOCSIS was standardized by ITU in the J.112 and J122 family as outlined in Table 7.3. The documents [51–52] are available.

Communication takes place between the *cable modem termination system* (CMTS), usually located at the cable operators head end, and the various *cable modems* (CM), at the customer's premises. This implies that CMs cannot directly communicate with each other. The downstream bandwidths are oriented on those used for television channel broadcast, that is, 6 MHz in North America and 8 MHz in Europe. The system adapted to the 8 MHz downlink channel is referred to as *EuroDOCSIS*. Reflecting this downlink dichotomy, the ITU standard [53] and the CEA standard [54] apply to DOCSIS, while the ETSI standard [55], *digital video broadcasting cable* (DVB-C), applies to EuroDOCSIS. Looking at *electromagnetic compatibility* (EMC) requirements, FCC regulations [56–57] are relevant to North America and ETSI [58] and CENELEC standards [59–65] are relevant to the European Union.

In terms of technology EuroDOCSIS uses equalized single carrier modulation with QAM constellation mapping and concatenated Trellis/Reed–Solomon forward error correction [39]. Multiple access is realized via time division (i.e., TDMA). DOCSIS 2.0 additionally adds the option of *synchronous code division multiple access* (SCDMA). The maximum uplink bit rates are 10.24 Mbit/s and 30.72 Mbit/s for DOCSIS 1.1 and DOCSIS 2.0, respectively [66]. Maximum downlink bit rates per channel are identical between DOCSIS 1.1 and DOCSIS 2.0. However, downlink bit rates differ between DOCSIS and EuroDOCSIS, amounting to 42.88 Mbit/s and 55.62 Mbit/s, respectively.

DOCSIS 3.0, whose main specifications are available online [67], is backward compatible. It allows upstream as well as downstream channel bonding, which assuming, for example, four channels, allows uplink data rates of 122.88 Mbit/s, and downlink data rates of 171.52 Mbit/s and 222.48 Mbit/s for DOCSIS and EuroDOCSIS, respectively.

An additional important step forward was made in terms of security. While DOCSIS 1.1 and DOCSIS 2.0 support the 56-bit *data encryption standard* (DES) [68], DOCSIS 3.0 additionally supports the more secure 128-bit *advanced encryption standard* (AES) [69]. Furthermore, DOCSIS 3.0 was designed to support IPv6 [40], which is believed to play an essential role for Smart Grid applications. Just recently, IPv6 was also specified for DOCSIS 2.0 [70].

Especially with respect to Smart Grid/Smart Home applications, CabelLabs issued the specification "*Security, Monitoring, and Automation Specification* (SMA)" [71–72], which is currently in the NIST *Smart Grid Interoperability Panel* (SGIP) review process [5, 73]. The SMA reference architecture can be broken down into four *domains*. The *Home* domain includes all sensors, actuators, monitoring, and control devices within a user's home. Over various wireless and wireline communication protocols, they interface with the *SMA Gateway*. The SMA Gateway could be integrated within a cable modem. Using, for instance, a DOCSIS Access deployment the SMA Gateway interfaces with the *Operator Domain* through the *Access Domain*. The operator has control over the SMA Gateway and uses it to query and configure SMA devices within a user's home. An example of a device within the Operator Domain would be an *event server*, that in case of alarms could forward them to the so called *Central Station Domain*, which in the sequel would perform tasks like notifying the home owner or contacting emergency or security personnel. Although the SMA architecture is not bound to use DOCSIS cable technology, it is at least implicitly suggested. Similar architectures, that in most cases can be mapped to the NIST *conceptual model*, are also presented by other industrial stakeholders, as in the power line case discussed in the following.

7.4 POWER LINE TECHNOLOGY

The idea of using power lines also for communication purposes has been around since the beginning of the last century [74–75]. It is now broadly referred to as *power line communications* (PLC). The obvious advantage is the widespread availability of electrical infrastructure, so that theoretically deployment costs are confined to connecting modems to the existing electrical grid. Power line technologies can be grouped into *narrowband PLC* (NB-PLC), operating usually below 500 kHz, and *broadband PLC* (BB-PLC), operating usually at frequencies above 1.8 MHz. These are discussed in Sections 7.4.3 and 7.4.4, respectively. Nevertheless, the following starts out with an introduction to PLC scenarios including channel and noise aspects in Section 7.4.1, as well as *electromagnetic compatibility* (EMC) issues in Section 7.4.2. The subsections are in parts based on Berger and Moreno-Rodríguez [1] and

Berger [2]. Freely available complementary reading on PLC state-of-the-art can also be found in material by the OPEN meter Consortium [76].

7.4.1 PLC Scenarios, Channel, and Noise Aspects

Before diving into channel and noise aspects, let's briefly revisit the different scenarios from Figure 7.1.

High voltage (*HV*) lines, with voltages in the range from 110 to 380 kV, are used for nationwide or even international power transfer and consist of long overhead lines with little or no branches. This makes them acceptable waveguides with less attenuation per line length compared to their medium and low voltage counterparts. However, their potential for broadband SG communication services has up to the present day been limited. Time varying high voltage arcing and corona noise with noise power fluctuations on the order of several tens of decibels, as well as the practicalities and costs of coupling communication signals in and out of these lines have been an issue. Furthermore, there is fierce competition from fiber optical links. In some cases, fiber links might even be spliced together with the ground conductor of the HV system [77–78]. Nevertheless, several successful trials using HV lines have been reported [79–82].

Medium voltage (*MV*) lines, with voltages in the range from 10 to 30 kV, are connected to the HV lines via primary transformer substations. The MV lines are used for power distribution between cities, towns, and larger industrial customers. They can be realized as overhead or underground lines. Furthermore, they exhibit a low level of branches and directly connect to *intelligent electronic devices* (IEDs) such as reclosers, sectionalizers, capacitor banks, and phasor measurement units. IED monitoring and control requires only relatively low data rates and NB-PLC can provide economically competitive communication solutions for these tasks. MV related studies and trials can be found in Wouters et al. [83], Benato and Caldon [84], and Cataliotti et al. [85].

Low voltage (*LV*) lines, with voltages in the range from 110 to 400 V, are connected to the MV lines via secondary transformer substations. A communication signal on an MV line can pass through the secondary transformer onto the LV line, however, with a heavy attenuation on the order of 55–75 dB [86]. Hence, a special coupling device (inductive, capacitive) or a PLC repeater is frequently required if one wants to establish a high data rate communications path. As indicated in Figure 7.1, the LV lines lead directly or over street cabinets to the end customers' premises. Note that considerable regional topology differences exist. For example, in the United States one smaller secondary transformer on a utility pole might service a single house or a small number of houses. In Europe, however, it is more common that up to 100 households get served from a single secondary transformer substation. Furthermore, as pointed out by Rubinstein et al. [87] significant differences exist between building types. They may be categorized as *multiflat buildings with riser*, *multiflat buildings with common meter room*, *single-family houses*, and *high-rise buildings*. Their different electrical wiring topologies influence signal attenuation as well as interference between neighboring PLC networks [88]. In most cases the electrical grid enters the customers' premises over a *house access point* (HAP) followed by an *electricity meter* (M) and a distribution board (fuse box). From the distribution board the LV lines run up to the different power sockets in every room. Lines may also run to an *electric vehicle service equipment* (EVSE) and to a photovoltaic installation as indicated in Figure 7.1.

Similarly, as for the DSL and coax scenarios discussed previously, hybrid fiber optical–power line Access scenarios are common and are discussed in Chapter 8. The power line channel and noise situations heavily depend on the scenario and, hence, span a very large range. Generally, it can be said that the PLC channel exhibits frequency selective multipath fading and a low pass behavior. Furthermore, *alternating current* (AC) related cyclic short-term variations and abrupt long-term variations can be observed.

To understand the effects that lead to frequency selective fading consider, for example, the stub line schematic in Figure 7.7 adapted from Zimmermann and Dostert [89]; see also Berger and Moreno-Rodríguez [1]. An impedance matched transmitter is placed at *A*. *B* marks the point of a branch, also called an electrical T-junction. An impedance matched receiver is placed at *C*. A parallel load is connected at *D*. Transmissions and reflections lead to a situation where a PLC signal travels in the form of a direct wave from *A* over *B* to *C* as indicated by the dotted line. Another signal travels from *A* over *B* to *D*, bounces back to *B*, and reaches *C*, as indicated by the dashed line. All further signals travel from *A* to *B*, and undergo multiple bounces between *B* and *D* before they finally reach *C*, as indicated by the dash-dot line. The result is a classical multipath situation, where frequency selective fading is caused by in-phase and anti-phase combinations of the arriving signal components. The corresponding transfer function can readily be derived in closed form as an infinite impulse response filter [1].

One important parameter capturing the frequency selectivity characteristics is the *root mean square* (rms) *delay spread* (DS). For example, designing OFDM systems

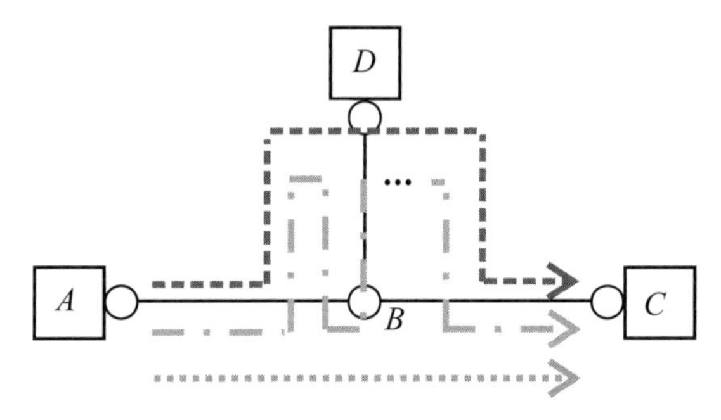

Figure 7.7. Multipath propagation in stub line.

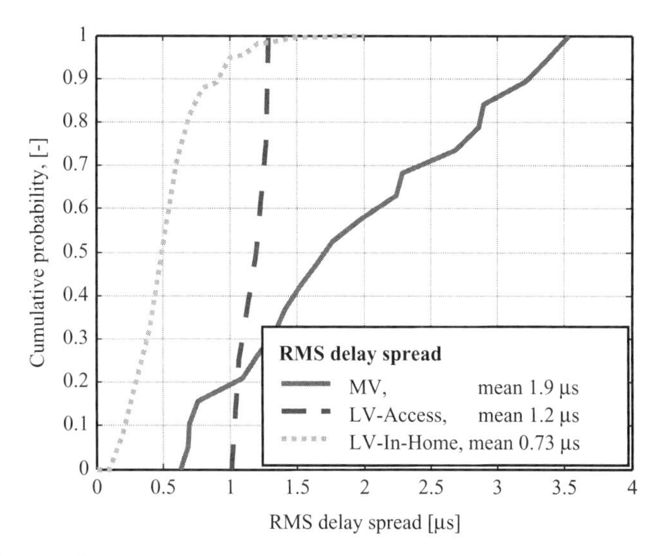

Figure 7.8. RMS delay spread statistics. From Berger [2], © 2010 Wiley.

[37], the guard interval might be chosen as 2 to 3 times the rms DS to deliver good system performance [90]. MV, LV-Access and LV-In-Home DS statistics extracted from Meier et al. [86] and Galli [90] are presented in Figure 7.8. The displayed rms DS statistics correspond to a band from 1 MHz up to 30 MHz, and give a good indication which order of rms DS to expect in the different scenarios.

Besides multipath fading, the PLC channel exhibits time variation due to loads and/or line segments being connected or disconnected [91]. Furthermore, through synchronizing channel measurements with the electrical grid AC mains-cycle, Cañete and co-workers were able to show that the In-Home channel changes in a cyclostationary manner [92–94].

Until now the low pass behavior of PLC channels has not been considered. It results from dielectric losses in the insulation between the conductors and is more pronounced in long cable segments such as outdoor underground cabling. Transfer function measurements on different cable types and for different lengths, can be found in Zimmermann and Dostert [95] and Babic et al. [96]. Using a large set of field trials, lowpass mean gain models are derived by Meier et al. [86]. Over the range from 1 to 30 MHz the mean gain in decibels is approximated by linear models. Consider again the PLC scenarios from Figure 7.1. The mean gain from the secondary transformer to the HAP, labeled M3 and M4, is [86]

$$\bar{g}_{\text{LV-Access}} = -(a_1 fd + a_2 f + a_3 d + a_4) \tag{7.1}$$

f is frequency (MHz), d is distance in meters, and the coefficients a_1 to a_4 are 0.0034 dB/(MHz m), 1.0893 dB/MHz, 0.1295 dB/m, and 17.3481 dB, respectively.

The mean gain model in decibels for MV lines, as well as for LV-In-Home situations is given by [86]

$$\overline{g}_{\text{MV or LV-In-Home}} = -(b_1 f + b_2) \tag{7.2}$$

For the LV-In-Home situation, the mean gain is given from the mains distribution board to a socket in a room, labeled M5 and M6 in Figure 7.1. The coefficients are $b_1 = 0.596$ dB/MHz and $b_2 = 45.325$ dB. The MV gain describes the channel between two primary transformers on the MV side, indicated by M1 and M2 in Figure 7.1. Its coefficients are $b_1 = 1.77$ dB/MHz and $b_2 = 37.9$ dB. In both situations the model is not distance-dependent. For the MV situation this is due to the fact that not enough measurement results were available to construct a distance-dependent model. Hence, in this case, the model is limited to situations where the distance between M1 and M2 is around 510 m. Nevertheless, correction factors are proposed by Meier et al. [86] to determine the mean gain at other distances. For the LV-In-Home situation the model is not distance-dependent either, since "distance" in an In-Home situation is a hard to define term. Power line networks in such situations usually exhibit a large amount of branches, and a detailed floor plan to determine cable length cannot always be obtained.

Using these linear models the mean gains for the three cases are plotted in Figure 7.9. A distance of 300 m is used in the LV-Access graph. It can be seen that the lowpass behavior is less pronounced in the In-Home case. It can further be seen that in the MV and the LV-Access situation the attenuation drastically increases with frequency. This goes well in line with the findings by Liu et al. [97] and is one of the reasons why BB-PLC Access networks are frequently operated in the lower frequency range, for example, between 1 and 10 MHz, while the BB-PLC In-Home networks might operate at frequencies above 10 MHz.

Power line noises can be grouped based on temporal as well as spectral characteristics. Following, for example, Zimmermann and Dostert [98] and Babic et al.

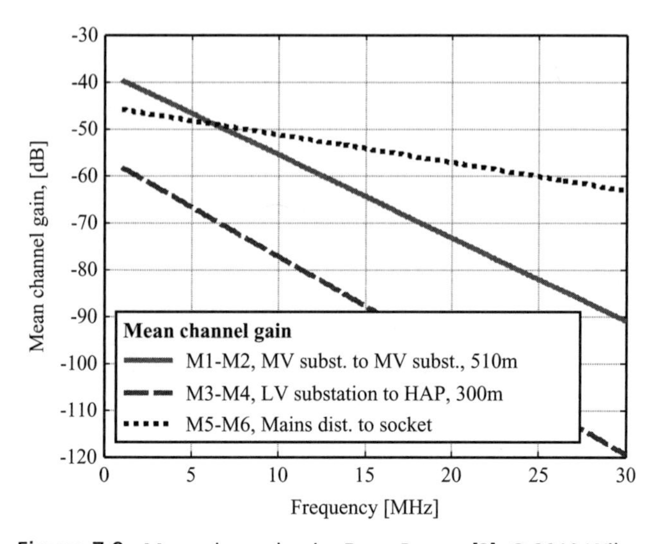

Figure 7.9. Mean channel gain. From Berger [2], © 2010 Wiley.

[96], one can distinguish *colored background noise*, *narrowband noise*, *periodic impulsive noise asynchronous to the AC frequency*, *periodic impulsive noise synchronous to the AC frequency*, and *aperiodic impulsive noise*. In Babic et al. [96] all these noises are modeled directly at the receiver using a superposition of spectrally filtered *additive white Gaussian noise* (AWGN), modulated sinusoidal signals, and Markov processes. Instead of modeling the noise directly at the receiver, Cañete and co-workers proposed to model the noise at its origin and to filter it by the channel transfer function [92, 99].

A statistical approach to average colored background noise modeling is presented by Meier et al. [86] based on a large amount of noise measurements in MV as well as LV-Access and LV-In-Home situations. Although a lot of the details get lost by averaging, the results can still deliver some interesting rules of thumb when one wants to determine a likely average noise level. One general finding is that the mean noise power falls off exponentially with frequency. Derived from Meier et al. [86], the mean noise PSD (in dBm/Hz) is given by

$$\bar{P}_{\mathrm{N}} = c_1 e^{(-c_2 f)} + c_3 - 10\log_{10}(30000) \tag{7.3}$$

where the last term normalizes out the 30 kHz bandwidth used in the noise measurement process. The coefficients c_1 to c_3 are given in Table 7.4. The resulting noise models correspond to the measurement points M1 to M6 in Figure 7.1 and are plotted in Figure 7.10.

Details on tx limits will be discussed in Section 7.4.2. For simplicity, assume for now that a power line signal with $\bar{P}_{\mathrm{S}} = -55$ dBm/Hz may be injected. Using the gain and noise models from Eqs. (7.1) to (7.3) the mean *signal-to-noise ratio* (SNR) can then be approximated as

$$\overline{\mathrm{SNR}} = \bar{g} + \bar{P}_{\mathrm{S}} - \bar{P}_{\mathrm{N}} \tag{7.4}$$

The mean SNRs for the various connections between the measurement points M1 to M6 are plotted in Figure 7.11. One should note that although the channel gain between two measurement points is symmetric, the noise at the measurement points differs. Hence, five different graphs are produced.

TABLE 7.4. Mean Noise Model Coefficients

Location	c_1 (dB)	c_2 (1/MHz)	c_3 (dBm/Hz)
M1&2, secondary transformer, MV	37	0.17	−105
M3, secondary transformer, LV	24.613	0.105	−116.721
M4, house access point, LV	29.282	0.12	−114.941
M5, main distribution board, LV	39.794	0.07	−118.076
M6, socket in private home, LV	17.327	0.074	−115.172

Source: Berger [2], © 2010 Wiley.

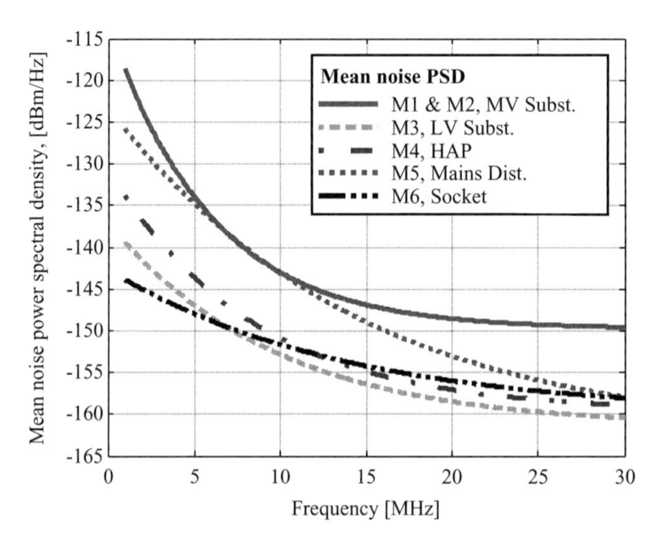

Figure 7.10. Mean noise power spectral densities. From Berger [2], © 2010 Wiley.

It can be seen that especially the lower part of the spectrum, up to 10 MHz, is very well suited for Access and Backhaul applications. Furthermore, for In-Home applications the entire spectrum from 1 to 30 MHz promises high mean SNRs on the order of 40 dB, which goes also well in line with the findings in Schwager et al. [100]. Further interesting results for the frequency range up to 100 MHz are available in Tlich et al. [101].

In general, the results show that there is a high potential for PLC if the estimated mean SNRs can be exploited in PLC modems. However, the presented results have to be handled with care. One should bear in mind that the mean SNR models from Meier et al. [86] exhibit a significant standard deviation. Furthermore, effects due to frequency selectivity, narrowband interference, and impulsive noise, as well as time variation are not reflected in Figure 7.11. Whether the estimated mean SNRs translate into high PLC data rates depends, among other factors, on the PLC modem's signal processing algorithms and its component quality.

7.4.2 PLC Electromagnetic Compatibility Regulations

Power line cables were not designed to carry communication signals and, hence, give rise to conducted emission, as well as radiated emission that may interfere, for example, with amateur radio or television broadcast receivers. When looking at power line *electromagnetic compatibility* (EMC) regulations, one may distinguish between regulations for NB-PLC and BB-PLC.

The NB-PLC regulations deal with the spectrum from 3 kHz up to around 500 kHz. Important NB-PLC regulations are listed in Table 7.5. Being a subset of all other bands, the CENELEC (*European Committee for Electrotechnical Standard-*

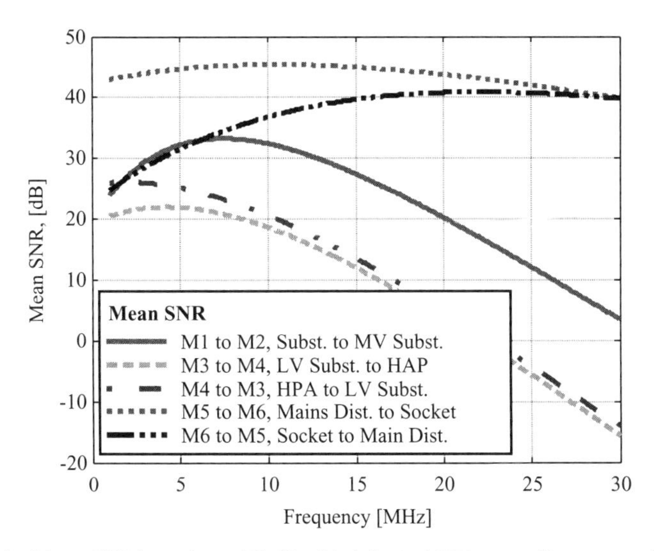

Figure 7.11. Mean SNR based on −55 dBm/Hz injected PSD, as well as mean channel gain and mean noise power spectral density models. From Berger [2], © 2010 Wiley.

TABLE 7.5. Important Regulations Related to NB-PLC

States	Frequencies	Institution	Regulation
European Union	3–148.5 kHz	Comité Européen de Normalisation Électrotechnique (CENELEC)	[103]
United States	10–490 kHz	Federal Communications Commission (FCC)	[56]
Japan	10–450 kHz	Association of Radio Industries and Businesses (ARIB)	[104]

ization) bands are the only ones available on a global basis. Four CENELEC bands are defined as *A* (3–95 kHz), *B* (95–125 kHz), *C* (125–140 kHz), and *D* (140–148.5 kHz). Besides specifying transmission limits and their measurement procedures, the CENELEC standard also mandates that the A-band may only be used by energy suppliers and their licensees, while the other bands may be used by consumers. Furthermore, devices operating in C band have to comply with a *carrier sense medium access/collision avoidance* (CSMA/CA) protocol that allows a maximum channel holding period of 1 s, a minimum duration between channel uses by the same device of 125 ms, and a minimum time of 85 ms before declaring the channel idle. In the United States, there are currently ongoing efforts to specify the band from 9 to 534 kHz for NB-PLC operations with a mandatory CSMA/CA protocol compliant with CENELEC EN 50065-1 [102]. The advantages are that equipment manufacturers would easily be able to adapt their NB-PLC products to the EU and U.S. markets and to many other markets that follow those standards.

Turning the view to BB-PLC, one may again distinguish two frequency ranges, that is, 1 to 30 MHz, where conducted emission is at the focus of regulation, and 30 to 100 MHz, where the focus shifts to radiated emission.

The *Comité International Spécial des Perturbations Radioélectriques* (CISPR), founded in 1934 and now part of the *International Electrotechnical Commission* (IEC), is currently making efforts to regulate BB-PLC generated interference. The CISPR/I/PT PLT working group, in charge of PLC standardization, has not yet been able to agree on a standardization proposal [105–106]. It could, however, be that an amendment to the existing CISPR 22 standard [107] will in the future regulate PLC emissions. A special PLC amendment to CISPR 22 could include the following:

- An emission measurement procedure at the mains-communications port while no communication takes place
- A second emission measurement procedure at the mains-communications port when normal PLC communication takes place
- A general cap on the injected PSD (e.g., –55 dBm/Hz)
- Permanent notching of certain parts of the radio spectrum (e.g., related to amateur radio bands)
- A procedure for adaptive notching, meaning that the PLC equipment senses the presence of radio services, and notches the affected frequencies for its own operation
- A procedure of adaptive transmit power management, meaning that the transmitting equipment limits its transmit power as a function of channel attenuation and noise to a level below the allowed maximum, that is just sufficient to achieve the required data rate

Once an amended CISPR 22 standard is in place, there is a good chance that it will become part of European Union legislation. Responsible here is CENELEC, which through liaison groups maintains a close collaboration with CISPR.

In the United States the *Federal Communications Commission* (FCC) is in charge of regulating electromagnetic emissions. In general, all digital equipment has to comply with the FCC part 15 standard (47 CFR §15) [56]. Specifically, Access PLC systems over medium and low voltage power lines, and for a frequency range from 1.705 to 80 MHz, are treated in the standard's Section G. Conducted emission limits are explicitly not applicable but radiated emission limits are imposed through a transmit power spectral density mask. Additionally, PLC systems have to be able to notch certain frequencies that might be used by other services. Furthermore, the FCC defines excluded bands where no PLC signal shall be injected, as well as geographical exclusion zones close to which no Access PLC systems may be deployed. In addition, procedures in which service providers inform about prospective PLC Access deployments as well as complaint handling procedures are a requirement.

More details on PLC EMC regulations as well as conducted and radiated interference measurement results can be found in Razafferson et al. [108] and the descriptions by OPEN meter Consortium [76]. Besides, the *IEEE Standard for Power Line*

Communication Equipment—Electromagnetic Compatibility (EMC) Requirements—Testing and Measurement Methods [109] was recently released, intending to provide an internationally recognized EMC measurement and testing methodology. It endorses among others CISPER 22 and FCC part 15 as normative references, but does not establish any emission limits itself. Looking at the developments in CISPR 22, as well as at FCC part 15, it becomes clear that next generation PLC equipment has to be highly configurable to apply a power spectral density shaping mask, as well as adaptive notching. For a rough rule of thumb one may resort to a −55 dBm/Hz and a −85 dBm/Hz tx power limit for frequencies below and above 30 MHz, respectively.

7.4.3 Narrowband PLC

Narrowband power line communication (NB-PLC) systems usually operate in the frequency range from 3 to 500 kHz, that is, the CENELEC/ARIB/FCC bands (see Section 7.4.2). They can be subdivided into *low data rate* (LDR) and *high data rate* (HDR) systems. LDR systems have throughputs of a few kilobit per second and usually are based on single carrier technology. Example standards, listed in the *NIST SGIP Catalog of Standards* [73] are ISO/IEC 14908-3 (LON, ANSI/EIA 709) [17, 110] and ISO/IEC 14543-3-5 (KNX) [16]. These standards span all layers of the *open systems interconnection* (OSI) model and can be used over power lines as well as over other media such as twisted pair and in some cases even wirelessly. Their main area of application is building automation. In this respect a further popular protocol is BACnet (ISO 16484-5 [15]). More on the higher layers of BACnet and LonWorks can be found in Chapter 10. The following, however, will focus on the *physical layer* (PHY) of high data rate (up to 500 kbit/s) NB-PLC. As in many other communication systems, OFDM [37] has emerged as the modulation scheme of choice for HDR NB-PLC. Example HDR NB-PLC systems are PRIME [111] and G3 [13].

PRIME, for *Powerline Related Intelligent Metering Evolution*, was developed within the PRIME Alliance, with its steering committee chaired by the Spanish utility heavy weight Iberdrola [12]. The PRIME system uses a total of 96 ODFM subcarriers over the frequencies from 42 to 89 kHz, that is, within the CENELEC A band. Furthermore, it deploys differential *binary*, *quaternary*, and *eight phase shift keying* (BPSK, QPSK, and 8PSK), and an optional 1/2-rate convolutional code. Therefore, it is able to achieve a PHY peak data rate of 128.6 kbit/s [112]. The OFDM symbol interval is 2240 μs including a 192 μs cyclic prefix, which—considering the rms delay spreads in Figure 7.8—easily suffices to deal with any possible power line delay spread. Furthermore, to deal with unpredictable impulsive noise PRIME offers the option to implement *automatic retransmission request* (ARQ), based on the *selective repeat* mechanism [113]. Turning to the system architecture, PRIME is forming subnetworks, where each subnetwork has one *base node* and several *service nodes*. The base node is the "master" that manages the subnetwork's resources and connections using a periodically sent *beacon* signal. The base node is further responsible for PLC channel access arbitration. A contention-free and a contention-based access mechanism exists, whose usage time and duration are decided by the base node. Within the contention-free *time division multiplex* (TDM) channel access period, the base node assigns the channel to only one node at a time. The contention-based access uses CSMA/CA [111–112]. To assure *privacy*, *authentication*, and *data integrity* a *Security Profile 1* is defined that uses 128-bit AES encryption [69]. The specification also defines a *Security Profile 0* equivalent to no encryption and leaves room for definition of two further security profiles in future releases. To interface MAC and *application layer*, PRIME defines a *convergence layer* (CL) between the two. The CL can be split into a *common part convergence sublayer* (CPCS) and a *service specific convergence sublayer* (SSCS). The CPCS performs tasks of data segmentation and reassembling and is adjusted to the specific application. Three SSCSs are currently defined: The "*NULL Convergence Sublayer* provides the MAC layer with a transparent way to the application, being as simple as possible and minimizing the overhead. It is intended for applications that do not need any special convergence capability. The *IPv4 Convergence Layer* provides an efficient method for transferring IPv4 packets over the PRIME network." Finally, the "*IEC 61334-4-32 Convergence Layer* supports the same primitives as the IEC 61334-4-32 standard" [114], making it easy, for example, to support advanced metering applications that make use of the standardized data models of IEC 62056-62 [115]. PRIME could therefore also be used to replace the aging PHY and MAC layer of the single carrier power line standard IEC 61334-5-1 [116], also known as S-FSK, for *spread frequency shift keying*.

The other OFDM based HDR NB-PLC specification—G3-PLC [117–119], was published in August 2009. It can be configured to operate in the internationally accepted bands from 10 to 490 kHz (FCC, CENELEC, ARIB). Using differential BPSK, QPSK, and 8PSK for constellation mapping, and concatenated convolutional Reed–Solomon forward error correction coding [39], it reaches PHY peak data rates close to 300 kbit/s. Peak and typical data rates for various frequency bands have been reported by Razazian et al. and are for convenience also listed in Table 7.6. The MAC layer is based on IEEE 802.15.4-2006, described in more detail in Chapter 6. 6LoWPAN [120] is used to adapt the IEEE 802.15.4-2006 MAC (see Chapter 6) to IPv6 [40]. This allows the application layer to comply with ANSI C12.19/C12.22 [121] and IEC 62056-61/62 (DLMS/COSEM) [122], or to run standard Internet services.

TABLE 7.6. G3-PLC Data Rates

Frequency Band	Peak Rate (kbit/s)	Typical Rate (kbit/s)
CENELEC (36–91 kHz)	46	44
FCC (150–487.5 kHz)	234	187
FCC (10–487.5 kHz)	298	225

Source: Based on Razazian et al. [12].

The specifications PRIME and G3-PLC form the baseline in the ongoing NB-PLC standardization processes within IEEE P1901.2 and ITU-T G.hnem (also referred to as ITU G.9955/G.9956: Narrow-band OFDM power line communication transceivers). It is to be decided whether the NB-PLC standard IEEE 1901.2 shall contain two optional interoperability profiles for G3-PLC (CENELEC A) and PRIME (CENELEC A), and whether an IEEE 1901.2 compliant device shall implement at least one of the two PHY/MACs [123]. With the aim to obtain widespread industry and consumer acceptance, the NB-PLC standardization groups within IEEE and ITU-T agreed to work on a joint NB-PLC standard [124], which is expected by November 2012. The standard is supposed to specifically consider SG use cases like support of *pricing awareness*, *load control*, and *demand response*. Use cases as well as resulting requirements may be found in Su and Singletary [125]. Implementers that cannot wait for the joint NB-PLC standard might consider that, due to relatively low signal processing complexity, NB-PLC modems can be implemented on a *digital signal processor* (DSP). This allows for upgradeability via software updates, and can especially be an advantage for early stage customer premises deployments where a hardware update would be prohibitively expensive. An upgradeable DSP solution that supports PRIME as well as G3-PLC is, for example, offered by Texas Instruments [126].

Cases where NB-PLC and BB-PLC systems operate on the same physical medium will become more frequent with increasing PLC system deployments. Nevertheless, coexistence of the two should be straightforward as they are using different frequency ranges. Interconnections could then be established by OSI Layer 3 bridges as suggested, for example, by Oksman and Egan [127].

7.4.4 Broadband PLC

In the last decade, BB-PLC chips by semiconductor vendors, such as Intellon [128] (in 2009 acquired by Atheros), DS2 (in 2010 acquired by Marvell) [129], and Panasonic [130] came to market and can operate in the band from around 1 to 30 MHz. The chips are mainly based on three consortia backed specifications developed within the frameworks of the *HomePlug Powerline Alliance* (HomePlug) [131], the *Universal Powerline Association* (UPA) [132], and the *High Definition Power Line Communications* (HD-PLC) [133]. Related products allow data rates around 200 Mbit/s and are not interoperable. However, to make PLC systems a broad success, an internationally adopted BP-PLC standard became essential. The *International Telecommunications Union* (ITU) as well as the *Institute of Electrical and Electronics Engineers* (IEEE) commenced work on such next generation standards, namely, *ITU-T G.hn* and *IEEE P1901*. ITU-T G.hn is not only applicable to power lines but also to phone lines and coaxial cables, thus for the first time defining a single standard for all major wireline communications media. At the end of 2008, the PHY layer and the overall architecture were defined in ITU-T Recommendation G.9960 [134]. The *Data Link Layer* (DLL) Recommendation G.9961 [135] was approved in June 2010. In addition, the *HomeGrid Forum* was founded to promote the ITU-T G.hn standard and to address certification and interoperability issues [136]. Simultaneously, IEEE P1901 [137] was working on the "*Draft Standard for Broadband over Power Line Networks: Medium Access Control and Physical Layer Specifications*" [138]. It covers the aspects of Access and In-Home, as well as coexistence of Access–In-Home and In-Home–In-Home networks.

IEEE 1901 uses the band from 2 MHz up to 50 MHz with services above 30 MHz being optional. ITU-T G.hn (G.9960/G9961) operates from 2 MHz up to 100 MHz using bandwidth scalability, with three distinct and interoperable bands defined as 2–25, 2–50, and 2–100 MHz. The architectures defined by IEEE 1901 and ITU-T G.hn (G9960/G9961) are similar in several aspects. In ITU-T G.hn one refers to a subnetwork as *domain*. Operation and communication is organized by the *domain master*, who communicates with various *nodes*. Similarly, the subnetwork in IEEE 1901 is referred to as *basic service set* (BSS). The equivalent to the domain master is the *BSS manager*, which connects to so-called *stations*. These basic network components with their system specific terminology are summarized in Table 7.7.

While the general concepts are similar, one should note that ITU-T G.hn defines a PHY/DLL used for operation over any wireline medium. Primarily, the OFDM parameters are adjusted to account for different medium dependent channel and noise characteristics. On the contrary, IEEE 1901 defines two disparate PHY/MAC technologies based on HomePlug AV and HD-PLC. In consequence, a special coexistence mechanism has to be used when operating IEEE 1901 devices from the two camps on the same power line. This coexistence mechanism is standardized within IEEE 1901 as *intersystem protocol* (ISP) (see also Galli et al. [139]). A nearly identical mechanism was standardized by ITU-T in G.9972 [140], also known as G.cx. Technical contributions to ITU and IEEE from members of the NIST *Priority Action Plan 15* (PAP15) assured the alignment of both standards. As a result it is likely that the NIST SGIP will mandate "that all BB-PLC technologies operating over power

TABLE 7.7. Synopsis of Terms Used in the BB-PLC Standards ITU-T G.hn and IEEE 1901

Network Item	ITU-T G.hn Term	IEEE 1901 Term
Subnetwork	Domain	Basic service set (BSS)
Transceiver	Node	Station (STA)
Subnetwork controller	Domain master	BSS manager
Access control schedule	*Media access plan* (MAP)	Beacon
Schedule time frame	MAC cycle	Beacon interval
Access methods	CSMA/CA, TDMA, STXOP (*shared transmission opportunities*)	CSMA/CA, TDMA
Relaying transceivers	Relay (Layer 2)	Repeater (Layer 2)
Network controller proxy	Relay (assigned as a proxy)	Proxy BSS manager
Transceiver that cannot be be reached directly	Hidden node	Hidden station

lines (either the ones currently listed in the list of recommended standards or the ones that may be added in the future) include in their implementation either Recommendation ITU-T G.9972 or ISP as specified in IEEE 1901" [141]. ISP/G.cx provides coexistence by splitting time equally among systems present on the line. IEEE 1901 additionally specifies the optional *coexistence protocol* (CXP) that provides coexistence among multiple technologies based on a first-come-first-serve basis.

The coexistence mechanism is used for the case were disparate networks would otherwise be interfering with each other. Another likely scenario is that same-technology networks exist in close proximity, with the risk of so-called *neighboring network* interference. To deal with neighboring network interference, G.hn uses different preamble-symbol-seeds in each network. Therefore, G.hn networks are able to coexist and communicate simultaneously (i.e., not using time division). Instead, link adaptation procedures adjust the throughput to cope with degraded SNIR. In many cases the throughput will be throttled only slightly, allowing G.hn networks to coexist nearly unimpeded. On the other hand, IEEE 1901 relies on a CSMA/CA access strategy, which may lead to an increased number of collisions. As a countermeasure, IEEE 1901 introduces a *coordinated mode* that allows neighboring networks to allocate times over the shared medium for specific communications. This coordinated *time division multiple access* (TDMA) mode enables traffic to get through unimpeded, albeit at the price of time division (orthogonal throughput sharing).

For the sake of brevity the following focuses on ITU-T G.hn (G.9960, G9961, G.9972) and its usage in the Smart Grid context. It is envisioned that G.hn nodes in the future will be embedded into *Smart Grid home area network* (SGH) devices. SGH nodes will typically make use of the G.hn *low complexity* profile, operating in the frequency range from 2 to 25 MHz. This allows for reduced component cost and power consumption. Examples of SGH nodes could be heating and air conditioning appliances, as well as plug-in electric vehicles (PEVs) and electric vehicle supply equipment (EVSE) as indicated in Figure 7.12. Together they form a multidomain HAN.

The SGHs interact with the *utility's access network* (UAN) and its *advanced metering infrastructure* (AMI) through an *energy service interface* (ESI). The AMI domain comprises *AMI meters* (AM), *AMI submeters* (ASM), and an AMI *head end* (HE). The HE is a local hub (concentrator) that controls all meters downstream from it and interfaces to the utility's wide area/backhaul network upstream from it. Each AMI HE supports up to 250 AM and/or ASM nodes, forming an AMI domain (in dense urban areas 150–200 meters are a frequently encountered maximum [78]). Furthermore, a network supports up to 16 AMI domains, delivering support for up to $16 \cdot 250 = 4000$ AMI devices. The ability to support 16 domains with 250 nodes each is a general property of G.hn not limited to Smart Grid/AMI applications. Domains may be formed over any kind of wiring. The nodes within a domain are grouped into SGH and non-SG nodes. For security reasons, non-SG nodes are logically separated from SGH nodes using a secure upper-layer protocol.

In every domain there is a domain master that coordinates operation of all nodes. G.hn nodes of different domains communicate with each other via *interdomain*

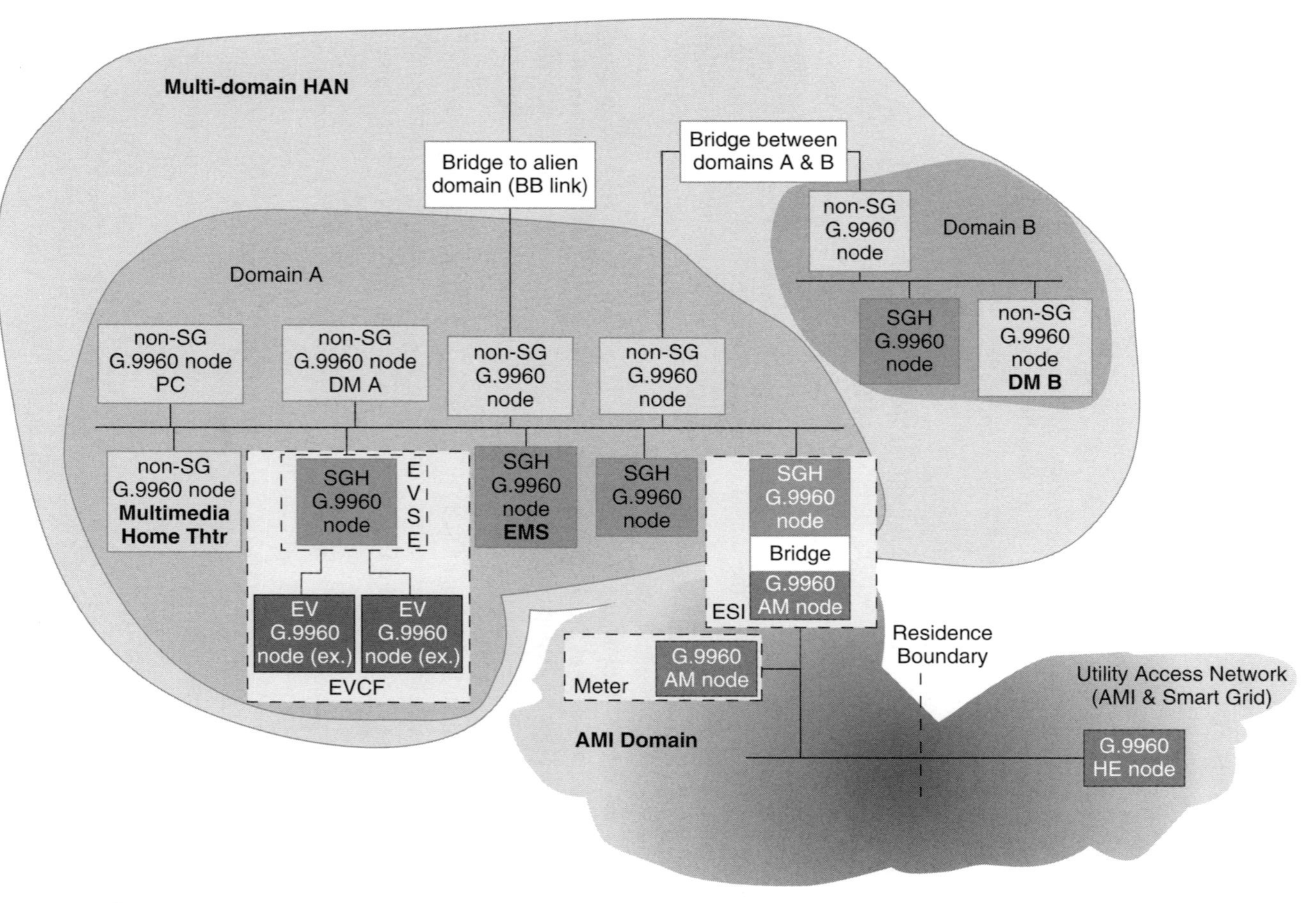

Figure 7.12. Smart Grid HAN implementation based on G.9960. From Oksman and Egan [127], © 2010 ITU.

bridges (IDBs). IDBs are simple data communications bridges on OSI Layer 3 and above, enabling a node in one domain to pass data to a node in another domain. In a multidomain situation, a *global master* (GM) provides coordination of resources, priorities, and operational characteristics between G.hn domains. Besides, G.hn domains can be bridged to alien (non-G.hn) domains, for example, to IEEE 1901, IEEE 1901.2, NB-PLC, wireless technologies, and so on. For example, besides the UAN/AMI connection through the ESI, the HAN might be connected to the outside world via a DSL or cable modem gateway communicating with the G.hn HAN via an alien domain bridge.

Let's now turn our attention to the *electric vehicle charging facility* (EVCF) in Figure 7.12. It consists of the *electric vehicle supply equipment* (EVSE) and the *electric vehicle* (EV) itself. The G.hn nodes inside EVs are expected to use the low complexity profile or even a reduced version of it. The required EV node data rate is not expected to exceed 5 Mbit/s in transmission and 20 Mbit/s in reception. Generally, data exchanged between EVSE and EV nodes can be of three types: recharging/discharging management, EV maintenance, and multimedia data. The first type may only require low throughput, while the second and third types may require relatively high data throughput with downstream dominance. A zoom on a simple EVCF is provided in Figure 7.13.

Figure 7.13 shows the usage of a SAE J1772 (IEC 62196-2 Type 1) charge connector favored in North America [142–143]. Alternative options are the IEC 62196-2 Type 2 or 3 connectors dominating in Europe [143]. More on charging procedures for example, can be found in Chapters 2 and 15. Independent of the type of connector, once plugged in a G.hn EV node tries to connect to the domain master. In the sequel it runs through registration and authentication procedures. Remote authorization is supported through a trusted channel to the service provider (established from the EVSE via the ESI and the UAN). Through this trusted channel the service provider validates the EV's identity credentials, authorizes its access in the EVSE, and authorizes charging the EV at a given charge rate, at a given time, and for a given duration. The EVSE may charge the EV using AC or DC voltage. The single EVSE–EV link scenario may be extended to support up to four EVs connected in parallel.

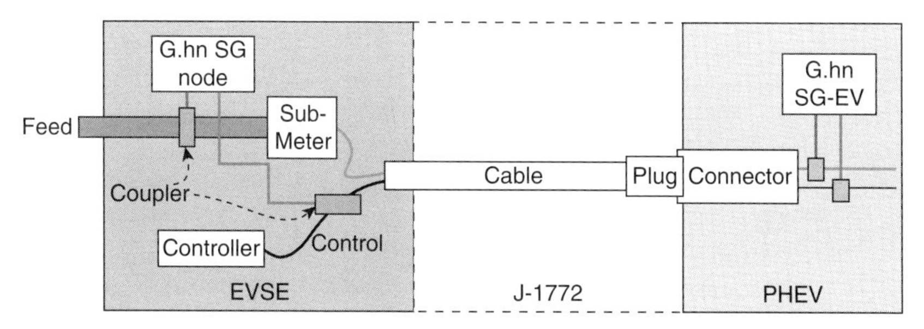

Figure 7.13. Simple electric vehicle charging facility (EVCF). An electric vehicle supply equipment (EVSE) links to an electric vehicle (EV) using a J-1772 cable. From Oksman and Egan [127], © 2010 ITU.

7.5 CONCLUSIONS

There are various wireline communication technologies available to fulfill Smart Grid communication needs in a plurality of scenarios, a subset of which were discussed in this chapter. A matching of communication requirements and communication technology has to be performed on a case-by-case basis. It is likely that the presented wireline technologies, DSL, coax/DOCSIS, as well as narrowband and broadband power line communications, all have a role in specific Smart Grid applications. Moreover, wireline technologies are likely to be complemented by wireless technologies as discussed in Chapter 6. In this respect a study on a "hybrid wireless-PLC broadband Smart Grid solution" can be found , for example, in Chapter 14.

ACKNOWLEDGMENT

This work has been supported by the European Regional Development Fund within the ERDF Operational Program of the Valencian Community 2007-2013 under grant number IMIDTA/2011/905.

REFERENCES

[1] L. T. Berger and G. Moreno-Rodríguez, "Power line communication channel modelling through concatenated IIR-filter elements," *Academy Publisher J. of Commun.*, 4(1), pp. 41–51, February 2009.

[2] L. T. Berger, "Broadband powerline communications," Chapter 10 in *Convergence of Wireless, Wireline, and Photonics Next Generation Networks*, K. Iniewski, Ed. John Wiley & Sons, 2010, pp. 289–316.

[3] National Institute of Standards and Technology (NIST), U.S. Department of Commerce, "NIST Framework and Roadmap for Smart Grid Interoperability Standards," NIST Draft Publication, Release 1.0, January 2010.

[4] Broadband Forum, "Next Generation Broadband Access," White Paper MR-185, Issue: 1 August 2009. Available at: http://www.broadband-forum.org/marketing/download/mktgdocs/NextGenAccessWhitePaper.pdf (accessed February 2011).

[5] Wikipedia, "List of Broadband Over Power Line Deployments—Wikipedia, The Free Encyclopedia." Available at: http://en.wikipedia.org/w/index.php?title=List_of_broadband_over_power_line_deployments&oldid=386780074 (accessed November 2010).

[6] D. Nordell, "Communication systems for distribution automation," *IEEE Transmission and Distribution Conference and Exposition*, Bogota, Colombia, April 2008.

[7] S. Mak and D. Reed, "TWACS, a new viable two-way automatic communication system for distribution networks. Part I: Outbound communication," *IEEE Trans. Power App. Syst.*, 101(8), pp. 2941–2949, August 1982.

[8] S. Mak and T. Moore, "TWACS, a new viable two-way automatic communication system for distribution networks. Part II: Inbound communication," *IEEE Trans. Power App. Syst.*, 103(8), pp. 2141–2147, August 1984.

[9] The OPEN meter Consortium, "Open Public Extended Network Metering," Available at: http://www.openmeter.com/ (accessed April 2011).

[10] PRIME Alliance, "Powerline Related Intelligent Metering Evolution (PRIME)." Available at: http://www.prime-alliance.org/Technology/PRIMETechnology/tabid/57/Default.aspx (accessed December 2010).

[11] Électricité Réseau Distribution France, "G3-PLC: Open Standard for Smart Grid Implementation." Availabel at: http://www.maxim-ic.com/products/powerline/g3-plc/ (accessed December 2010).

[12] K. Razazian, M. Umari, A. Kamalizad, V. Loginov, and M. Navid, "G3-PLC specification for powerline communication: Overview, system simulation and field trial results," *IEEE International Symposium on Power Line Communications and Its Applications (ISPLC)*, Rio de Janeiro, Brazil, March 2010.

[13] P. Siohan, A. Zeddam, G. Avril, P. Pagani, S. Person, M. Le Bot, E. Chevreau, O. Isson, F. Onado, X. Mongaboure, F. Pecile, A. Tonello, S. D'Alessador, S. Drakul, M. Vuksic, J.-Y. Baudais, A. Maiga, and J.-F. Herald, "State of the Art, Application Scenario and Specific Requirements for PLC," OMEGA, 2008, European Union Project Deliverable D3.1, v1.0, IST Integrated Project No ICT-213311. Available at: http://www.ict-omega.eu/publications/deliverables.html (accessed December 2010).

[14] K. Dostert, *Powerline Communications*, Prentice Hall, 2001.

[15] International Organization for Standardization, "Building Automation and Control Systems—Part 5: Data Communication Protocol," 2010, International Standard ISO 16484-5.

[16] International Organization for Standardization, "Information Technology—Home Electronic System (HES) Architecture—Part 3–5: Media and Media Dependent Layers—Powerline for Network Based Control of HES Class 1," May 2007, International Standard ISO/IEC 14543-3-5, first edition.

[17] International Organization for Standardization, "Interconnection of Information Technology Equipment—Control Network Protocol—Part 3: Power Line Channel Specification," January 2011, International Standard ISO/IEC 14908-3, Revision 11.

[18] DKE German Commission for Electrical, Electronic & Information Technologies of DIN and VDE, "The German Roadmap E-Energy / Smart Grid," April 2010. Available at: http://www.e-energy.de/documents/DKE_Roadmap_SmartGrid_230410_Engllish.pdf (accessed April 2011).

[19] IEEE 2030 Smart Grid Interoperability Working Group, "IEEE Draft Guide for Smart Grid Interoperability of Energy Technology and Information Technology Operation with the Electric Power System (EPS), and End-Use Applications and Loads," IEEE Standard Draft, IEEE P2030/D5.0,2-23-2011, Febuary 2011.

[20] International Telecommunications Union (ITU), Telecommunication Standardization Sector, "Digital Sections and Digital Line System—Access Networks Overview of Digital Subscriber Line (DSL)," ITU-T Recommendation G.995.1, Series G: Transmission Systems and Media, Digital Systems and Networks, February 2001. Available at: http://www.itu.int/rec/dologin_pub.asp?lang=e&id=T-REC-G.995.1-200102-I!!PDF-E&type=items (accessed February 2011).

[21] International Telecommunications Union (ITU), Telecommunication Standardization Sector, "Digital Sections and Digital Line System—Access Networks Overview of Digital Subscriber Line (DSL), Amendment 1," ITU-T Recommendation G.995.1, Series G: Transmission Systems and Media, Digital Systems and Networks, November 2001. Available at http://www.itu.int/rec/dologin_pub.asp?lang=e&id=T-REC-G.995.1-200111-I!Amd1!PDF-E&type=items (accessed February 2011).

[22] International Telecommunication Union (ITU), "Transmission Systems and Media, Digital Systems and Networks—G Series." Available at: http://www.itu.int/rec/T-REC-G/e (accessed Feburary 2011).

[23] S. Lindecke, "DSL technology and deployment—VDSL2," DSL Forum presentation, April 2005. Available at: http://kambing.ui.ac.id/onnopurbo/library/library-ref-eng/ref-eng-3/physical/adsl/VDSL2_Tutorial-2005.ppt (accessed February 2011).

[24] Aware, Inc., "ADSL2 and ADSL2+—The New ADSL Standards," White Paper, revision 3, 2006. Available at: http://www.aware.com/dsl/whitepapers/WP_ADSL2_Plus_Rev3_0505.pdf (accessed Febuary 2011).

[25] T. Magesacher, P. Ödling, M. Berg, S. Höst, E. Areizaga, P. O. Börjesson, and E. Jacob, "Paving the road to Gbit/s broadband access with copper," Chapter 7 in *Convergence of Mobile and Stationary Next-Generation Networks*, K. Iniewski, Ed. John Wiley & Sons, 2010.

[26] V. Lazauskaite, "Developments of Next Generation Networks (NGN): Country Case Studies," Report, International Telecommunications Union (ITU), Geneva, Switzerland, 2009. Available at: http://www.itu.int/ITU-D/treg/Documentation/ITU-NGN09.pdf (accessed February 2011).

[27] American Society for Testing and Materials (ASTM) International, "Standard Specification for Standard Nominal Diameters and Cross-sectional Areas of AWG Sizes of Solid Round Wires Used as Electrical Conductors," 2008, ASTM Standard B 258-02.

[28] American National Standards Institute/Electronic Industries Association (ANSI/EIA), "Commercial Building Telecommunications Cabling Standard Part 1: General Requirements," May 2001, TIA/EIA-568-B.1.

[29] International Organization for Standardization, "Information Technology—Generic Cabling for Customer Premises," 2002, International Standard ISO/IEC 11801.

[30] D. Rauschmayer, *ADSL/VDSL Principles*. Macmillan Technical Publishing, 1999.

[31] J. Cook, R. Kirkby, M. Booth, K. Foster, D. Clarke, and G. Young, "The noise and crosstalk environment for ADSL and VDSL systems," *IEEE Commun. Mag.*, 37(5), pp. 73–78, May 1999.

[32] G. Ginis and J. M. Cioffi, "Vectored transmission for digital subscriber line systems," *IEEE J. Selected Areas in Commun.*, 20(5), pp. 1085–1104, June 2002.

[33] T. Starr, J. M. Cioffi, and P. J. Silverman, *Understanding Digital Subscriber Line Technology*. Prentice Hall, 1999.

[34] F. Moulin, P. Péron, and A. Zeddam, "PLC and VDSL2 coexistence," *IEEE International Symposium on Power Line Communications and Its Applications (ISPLC)*, Rio de Janeiro, Brazil, March 2010, pp. 207–212.

[35] W. Xu, C. Schroeder, and P. A. Hoeher, "A stochastic MIMO model for far-end crosstalk in VDSL cable binders," *International Conference on Communications (ICC)*, Dresden, Germany, June 2009.

[36] B. Lee, J. M. Cioffi, S. Jagannathan, and M. Mohseni, "Gigabit DSL," *IEEE Trans. Commun.*, 55(9), pp. 1689–1692, September 2007.

[37] R. van Nee and R. Prasad, *OFDM for Wireless Multimedia Communications*, Universal personal communication. Artech House Publishers, 2000.

[38] L.-F. Wei, "Trellis-coded modulation with multidimensional constellations," *IEEE Trans. Information Theory*, IT-33(4), pp. 483–501, 1987.

[39] T. K. Monn, *Error Correction Coding, Mathematical Methods and Algorithms*. John Wiley & Sons, 2005.

[40] S. Deering and R. Hinden, "Internet Protocol, Version 6 (IPv6) Specification," RFC 2460, December 1998. Available at: http://tools.ietf.org/html/rfc2460 (accessed February 2011).

[41] E. Chen, "ZyXEL Introduces New Service-on-Demand Solutions at the Broadband World Forum 2010," ZyXEL press release, October 2010. Available at: http://www.zyxel.com/news/press_room_20101206_613572.shtml (accessed November 2010).

[42] T. Degner, J. Schmid, and P. Strauss, "DISPOWER—Distributed Generation with High Penetration of Renewable Energy Sources," Tech. Rep., DISPOWER Consortium, Kassel, Germany, 2006. Available at: http://www.iset.uni-kassel.de/dispower_static/documents/fpr.pdf (accessed February 2011).

[43] K. Oravez and J. Moore, "CURRENT, Qwest to Integrate DSL into Smart Grid—Use of DSL Network Reduces Costs, Speeds Smart Grid Deployment," press release, June 2009. Available at: http://news.qwest.com/index.php?s=43&item=1893 (accessed February 2011).

[44] Multimedia over Coax Alliance, "The Standard for Home Entertainment Networking," Home page. Available at: http://www.mocalliance.org/ (accessed Febuary 2011).

[45] ReDeSign Partners, "Research for Development of Future Interactive Generations of Hybrid Fibre Coax Networks (ReDeSign)," Web page, 2008, FP7-ICT-217014. Available at: http://www.ict-redesign.eu/index.php?id=28 (accessed February 2011).

[46] International Electrotechnical Commission, "Cabled Networks for Television Signals, Sound Signals and Interactive Services—Methods of Measurement and System Performance," International Standard IEC 60728-1, 2001.

[47] International Electrotechnical Commission, "Cabled Networks for Television Signals, Sound Signals and Interactive Services—System Performance of Return Path," International Standard IEC 60728-10, 2005.

[48] ReDeSign Partners, "Research for Development of Future Interactive Generations of Hybrid Fibre Coax Networks—HFC Channel Model," European Union Project Deliverable FP7-ICT-217014, December 2008, Version 1.0. Available at http://www.ict-redesign.eu/fileadmin/documents/ReDeSign-D08_Report_on_Cable_Channel_Model.pdf (accessed February 2011).

[49] ReDeSign Partners, "Research for Development of Future Interactive Generations of Hybrid Fibre Coax Networks—Reference Architectures Report," European Union Project Deliverable FP7-ICT-217014, October 2008, Version 1.0. Available at: (http://www.ict-redesign.eu/fileadmin/documents/ReDeSign-D06_Reference_architectures_report.pdf (accessed February 2011).

[50] Cable Television Laboratories, Inc., "Digital Transmission Characterization of Cable Television Systems," Tech. Rep., November 1994. Available at: http://www.cablelabs.com/downloads/digital_transmission.pdf (accessed February 2011).

[51] Cable Television Laboratories, Inc., "Revolutionizing Cable Technology," Home page. Available at: http://www.cablelabs.com/ (accessed February 2011).

[52] International Telecommunication Union (ITU), "Cable Networks and Transmission of Television, Sound Programme and Other Multimedia Signals—J Series," Web portal. Available at: http://www.itu.int/rec/T-REC-J (accessed Feburary 2011).

[53] International Telecommunications Union (ITU), "Series J: Cable Networks and Transmission of Television, Sound Programme and Other Multimedia Signals—Digital Transmission of Television Signals—Digital Multi-programme Systems for Television, Sound and Data Services for Cable Distribution," December 2007. Available at: http://www.itu.int/rec/dologin_pub.asp?lang=e&id=T-REC-J.83-200712-I!!PDF-E&type=items (accessed February 2011).

[54] Consumer Electronics Association (CEA), "CEA Standard: Cable Television Channel Identification Plan," July 2003.

[55] European Telecommunication Standards Institute (ETSI), "Digital Video Broadcasting (DVB); Framing Structure, Channel Coding and Modulation for Cable Systems," April 1998.

[56] FCC, "Title 47 of the Code of Federal Regulations (CFR)," Tech. Rep. 47 CFR §15, Federal Communications Commission, July 2008. Available at: http://www.fcc.gov/oet/info/rules/part15/PART15_07-10-08.pdf (accessed February 2009).

[57] FCC, "Title 47 of the Code of Federal Regulations (CFR)," Tech. Rep. 47 CFR §76, Federal Communications Commission, October 2005.

[58] European Telecommunication Standards Institute (ETSI), "Electrical Safety; Classification of Interfaces for Equipment to Be Connected to Telecommunication Networks," November 1998.

[59] European Committee for Electrotechnical Standardization (CENELEC), "Cable Networks for Television Signals, Sound Signals and Interactive Services—Part 1: Safety Requirements," European Standard EN 50083-1, 2002.

[60] European Committee for Electrotechnical Standardization (CENELEC), "Cable Networks for Television Signals, Sound Signals and Interactive Services—Part 2: Electromagnetic Compatibility for Equipment," European Standard EN 50083-2, 2005.

[61] European Committee for Electrotechnical Standardization (CENELEC), "Cable Networks for Television Signals, Sound Signals and Interactive Services—Part 7: System Performance," European Standard EN 50083-7, April 1996.

[62] European Committee for Electrotechnical Standardization (CENELEC), "Cable Networks for Television Signals, Sound Signals and Interactive Services—Part 10: System Performance for Return Paths," European Standard EN 50083-10, March 2002.

[63] European Committee for Electrotechnical Standardization (CENELEC), "Information Technology Equipment—Safety—Part 1: General Requirements," European Standard EN 60950-1, December 2001.

[64] European Committee for Electrotechnical Standardization (CENELEC), "Electromagnetic Compatibility (EMC)—Part 6-1: Generic Standards—Immunity for Residential, Commercial and Light-Industrial Environments," European Standard EN 61000-6-1, October 2001.

[65] European Committee for Electrotechnical Standardization (CENELEC), "Electromagnetic Compatibility (EMC)—Part 6-3: Generic Standards—Emission Standard for Residential, Commercial and Light-Industrial Environments," European Standard EN 61000-6-3, 2003.

[66] M. Kar and G. White, "DOCSIS 2.0—the next revolution in cable modem technology," *International Conference on Consumer Electronics*, August 2002, pp. 152–153.

[67] Cable Television Laboratories, Inc., "DOCSIS 3.0 Interface," DOCSIS Specifications. Available at: http://www.cablelabs.com/cablemodem/specifications/specifications30.html (accessed Febuary 2011).

[68] National Institute of Standards and Technology (NIST), U.S. Department of Commerce, "Recommendation for the Triple Data Encryption Algorithm (TDEA) Block Cipher," NIST Special Publication 800-67, May 2008, Version 1.1. Available at: http://csrc.nist.gov/publications/nistpubs/800-67/SP800-67.pdf (accessed February 2011).

[69] National Institute of Standards and Technology (NIST), U.S. Department of Commerce, "Specification for the Advanced Encryption Standard (AES)," Federal Information Processing Standards Publication 197, November 2001.

[70] Cable Television Laboratories, Inc., "Data-Over-Cable Service Interface Specifications—DOCSIS 2.0 + IPv6 Cable Modem Specification," Tech. Rep. CM-SP-DOCSIS2.0-IPv6-I03-110210, February 2011. Available at: http://www.cablelabs.com/specifications/CM-SP-DOCSIS2.0-IPv6-I03-110210.pdf (accessed February 2011).

[71] Cable Television Laboratories, Inc., "PacketCable Security, Monitoring, and Automation Specification," Specification PKT-SP-SMA-I01-081121, November 2008.

[72] Cable Television Laboratories, Inc., "PacketCable—Security, Monitoring, and Automation Architecture Framework Technical Report," Tech. Rep. PKT-TR-SMA-ARCH-V01-081121, November 2008.

[73] M. Burns, "NIST SGIP Catalog of Standards," Web page, October 2010. Available at: http://collaborate.nist.gov/twiki-sggrid/bin/view/SmartGrid/SGIPCatalogOfStandards (accessed Febuary 2011).

[74] K. Dostert, "Telecommunications over the power distribution grid—possibilities and limitations," *International Symposium on Power Line Comms and Its Applications (ISPLC)*, Essen, Germany, April 1997, pp. 1–9.

[75] P. A. Brown, "Power line communications—past present and future," *International Symposium on Power Line Communicatons and Its Applications (ISPLC)*, September 1999, pp. 1–8.

[76] The OPEN meter Consortium, "Description of Current State-of-the-Art of Technology and Protocols Description of State-of-the-Art of PLC-Based Access Technology," European Union Project Deliverable FP7-ICT-2226369, March 2009, D 2.1 Part 2, Version 2.3. Available at: http://www.openmeter.com/files/deliverables/OPEN-Meter%20WP2%20D2.1%20part2%20v2.3.pdf (accessed April 2011).

[77] G. Held, *Understanding Broadband Over Power Line*. CRC Press, 2006.

[78] P. Sobotka, R. Taylor, and K. Iniewski, "Broadband over power line communications: Home networking, broadband access, and smart power grids," Chapter 8 in *Internet Networks: Wired, Wireless, and Optical Technologies*, K. Iniewski, Ed., Devices, Circuits, and Systems. CRC Press, 2009.

[79] R. Pighi and R. Raheli, "On multicarrier signal transmission for high voltage power lines," *IEEE International Symposium on Power Line Communications and Its Applications (ISPLC)*, Vancouver, Canada, April 2005.

[80] D. Hyun and Y. Lee, "A study on the compound communication network over the high voltage power line for distribution automation system," *International Conference on Information Security and Assurance (ISA)*, Busan, Korea, April 2008, pp. 410–414.

[81] R. Aquilu, I. G. J. Pijoan, and G. Sanchez, "High-voltage multicarrier spread-spectrum system field test," *IEEE Trans. on Power Delivery*, 24(3), pp. 1112–1121, July 2009.

[82] N. Strandberg and N. Sadan, "HV-BPL Phase 2 Field Test Report," Tech. Rep. DOE/NETL-2009/1388, U.S. Department of Energy, 2009. Available at: http://www.netl.doe.gov/smartgrid/referenceshelf/reports/HV-BPL_Final_Report.pdf (accessed December 2010).

[83] P. Wouters, P. van der Wielen, J. Veen, P. Wagenaars, and E. Steennis, "Effect of cable load impedance on coupling schemes for MV power line communication," *IEEE Trans. Power Delivery*, 20(2), pp. 638–645, April 2005.

[84] R. Benato and R. Caldon, "Application of PLC for the control and the protection of future distribution networks," *IEEE International Symposium on Power Line Communications and Its Applications (ISPLC)*, Pisa, Italy, March 2007.

[85] A. Cataliotti, A. Daidone, and G. Tiné, "Power line communication in medium voltage systems: Characterization of MV cables," *IEEE Trans. Power Delivery*, 23(4), pp. 1896–1902, October 2008.

[86] P. Meier, M. Bittner, H. Widmer, J.-L. Bermudez, A. Vukicevic, M. Rubinstein, F. Rachidi, M. Babic, and J. Simon Miravalles, "Pathloss as a Function of Frequency, Distance and Network Topology for Various LV and MV European Powerline Networks," Project Deliverable, EC/IST FP6 Project No 507667 D5v0.9, The OPERA Consortium, April 2005.

[87] A. Rubinstein, F. Rachidi, M. Rubinstein, A. Vukicevic, K. Sheshyekani, W. Bäschelin, and C. Rodríguez-Morcillo, "EMC Guidelines," IST Integrated Project Deliverable D9v1.1, The OPERA Consortium, October 2008, IST Integrated Project No. 026920.

[88] A. Vukicevic, *Electromagnetic Compatibility of Power Line Communication Systems*, Dissertation, École Polytechnique Fédérale de Lausanne, Lausanne, Switzerland, June 2008, No. 4094.

[89] M. Zimmermann and K. Dostert, "A multi-path signal propagation model for the power line channel in the high frequency range," *International Symposium on Power Line Communications and Its Applications (ISPLC)*, Lancaster, UK, April 1999, pp. 45–51.

[90] S. Galli, "A simplified model for the indoor power line channel," *IEEE International Symposium on Power Line Communications and Its Applications (ISPLC)*, Dresden, Germany, March 2009, pp. 13–19.

[91] F. J. Cañete Corripio, L. Díez del Río, and J. T. Entrambasaguas Muñoz, "A time variant model for indoor power-line Channels," *International Symposium on Power Line Communications (ISPLC)*, Malmö, Sweden, March 2001, pp. 85–90.

[92] F. J. Cañete, L. Díez, J. A. Cortés, and J. T. Entrambasaguas, "Broadband modelling of indoor power-line channels," *IEEE Trans. Consumer Electronics*, 48(1), pp. 175–183, February 2002.

[93] J. A. Cortés, F. J. Cañete, L. Díez, and J. T. Entrambasaguas, "Characterization of the cyclic short-time variation of indoor power-line channels response," *International Symposium on Power Line Communications and Its Applications (ISPLC)*, Vancouver, Canada, April 2005, pp. 326–330.

[94] F. J. Cañete Corripio, J. A. Cortés Arrabal, L. Díez del Río, and J. T. Entrambasaguas Muñoz, "Analysis of the cyclic short-term variation of indoor power line channels," *IEEE J. Selected Areas in Commun.*, 24(7), pp. 1327–1338, July 2006.

[95] M. Zimmermann and K. Dostert, "A multipath model for the powerline channel," *IEEE Trans. Commun.*, 50(4), pp. 553–559, April 2002.

[96] M. Babic, M. Hagenau, K. Dostert, and J. Bausch, "Theoretical Postulation of PLC Channel Model," IST Integrated Project Deliverable D4v2.0, The OPERA Consortium, March 2005.

[97] H. Liu, J. Song, B. Zhao, and X. Li, "Channel study for medium-voltage power networks," *IEEE International Symposium on Power Line Communications (ISPLC)*, Orlando, FL, March 2006, pp. 245–250.

[98] M. Zimmermann and K. Dostert, "An analysis of the broadband noise scenario in power-line networks," *International Symposium on Power Line Communications and Its Applications (ISPLC)*, Limerick, Ireland, April 2000, pp. 131–138.

[99] F. J. Cañete, J. A. Cortés, L. Díez, and J. T. Entrambasaguas, "Modeling and evaluation of the indoor power line transmission medium," *IEEE Commun. Mag.*, 41(4), pp. 41–47, April 2003.

[100] A. Schwager, L. Stadelmeier, and M. Zumkeller, "Potential of broadband power line home networking," *Second IEEE Consumer Communications and Networking Conference*, January 2005, pp. 359–363.

[101] M. Tlich, P. Pagani, G. Avril, F. Gauthier, A. Zeddam, A. Kartit, O. Isson, A. Tonello, F. Pecile, S. D' Alessandro, T. Zheng, M. Biondi, G. Mijic, K. Kriznar, J.-Y. Baudais, and A. Maiga, "PLC Channel Characterization and Modelling," European Union Project Deliverable D3.3 v1.0, IST Integrated Project No ICT-213311, OMEGA, December 2008. Available at: http://www.ict-omega.eu/publications/deliverables.html (accessed December 2010).

[102] National Institute of Standards and Technology (NIST), Priority Action Plan 15 (PAP15): Harmonize Power Line Carrier Standards for Appliance Communications in the Home, "Coexistence of Narrow Band Power Line Communication Technologies in the Unlicensed FCC Band," April 2010. Available at: http://collaborate.nist.gov/twiki-sggrid/pub/SmartGrid/PAP15PLCForLowBitRates/NB_PLC_coexistence_paper_rev3.doc (accessed December 2010).

[103] European Committee for Electrotechnical Standardization (CENELEC), "Signalling on Low-Voltage Electrical Installations in the Frequency Range 3 kHz to 148.5 kHz—Part 1: General Requirements, Frequency Bands and Electromagnetic Disturbances," Standard EN 50065-1, September 2010.

[104] Association of Radio Industries and Businesses (ARIB), "Power Line Communication Equipment (10 kHz–450 kHz)," November 2002, STD-T84, Ver. 1.0, (in Japanese). Available at: http://www.arib.or.jp/english/html/overview/doc/1-STD-T84v1_0.pdf (accessed December 2010).

[105] W. Baeschlin, S. Arroyo, L. Feltin, M. Koch, B. Wirth, J.-P. Faure, and M. Heina, "First Report on the Status of PLC Standardisation Activities in CISPR," European Union Project Deliverable D30, OPERA Consortium, February 2008, IST Integrated Project No. 026920.

[106] W. Baeschlin, L. Feltin, J.-P. Faure, M. Rindchen, and H. Hirsch, "Second Report on the Status of PLC Standardisation Activities in CISPR," European Union Project Deliverable D31, OPERA Consortium, January 2009, IST Integrated Project No. 026920.

[107] Comité International Spécial des Perturbations Radioélectriques, "Information Technology Equipment; Radio Disturbance Characteristics; Limits and Methods of Measurement," International Standard Norme CISPR 22, Edition 6.0, ICS CISPR, September 2008.

[108] R. Razafferson, P. Pagani, A. Zeddam, B. Praho, M. Tlich, J.-Y. Baudais, A. Maiga, O. Isson, G. Mijic, K. Kriznar, and S. Drakul, "Report on Electro Magnetic Compatibility of Power Line Communications," European Union Project Deliverable D3.3 v3.0, IST Integrated Project No. ICT-213311, OMEGA, April 2010. Available at: http://www.ict-omega.eu/publications/deliverables.html (accessed December 2010).

[109] Institute of Electrical and Electronics Engineers, "IEEE Standard for Power Line Communication Equipment—Electromagnetic Compatibility (EMC) Requirements—Testing and Measurement Methods," January 2011.

[110] American National Standards Institute/Electronic Industries Association (ANSI/EIA), "Control Network Power Line (PL) Channel Specification," September 2006, ANSI/CEA-709.2-A.

[111] PRIME Alliance, "Draft Standard for Powerline Intelligent Metering Evolution," 2010. Available at: http://www.prime-alliance.org/portals/0/specs/PRIME-Spec_v1%20 3%20E_201005.pdf (accessed February 2011).

[112] I. Berganza, A. Sendin, and J. Arriola, "PRIME: Powerline Intelligent Metering Evolution," *CIRED Seminar 2008: SmartGrids for Distribution*, pp. 1–3. CIRED, Frankfurt, Germany, June 2008.

[113] A. S. Tannenbaum, *Computer Networks*, 4th ed. Prentice Hall International, 2003.

[114] International Electrotechnical Commission (IEC), "Distribution Automation Using Distribution Line Carrier Systems—Part 4: Data Communication Protocols—Section 32: Data Link Layer—Logical Link Control (LLC)," November 1997.

[115] International Electrotechnical Commission (IEC), "Electricity Metering—Data Exchange for Meter Reading, Tariff and Load Control—Part 62: Interface Classes," Standard IEC 62056-62, Ed. 2, November 2006.

[116] International Electrotechnical Commission (IEC), "Distribution Automation Using Distribution Line Carrier Systems—Part 5-1: Lower Layer Profiles—The Spread Frequency Shift Keying (S-FSK) Profile," Standard IEC 61334–5-1, Ed. 2.0, 2001.

[117] Electricité Réseau Distribution France (ERDF), "G3-PLC Physical Layer Specification," August 2009. Available at: http://www.maxim-ic.com/products/powerline/pdfs/G3-PLC-Physical-Layer-Specification.pdf (accessed February 2011).

[118] Electricité Réseau Distribution France (ERDF), "G3-PLC MAC Layer Specification," August 2009. Available at: http://www.maxim-ic.com/products/powerline/pdfs/G3-PLC-MAC-Layer-Specification.pdf (accessed February 2011).

[119] Electricité Réseau Distribution France (ERDF), "G3-PLC Profile Specification," August 2009. Available at: http://www.maxim-ic.com/products/powerline/pdfs/G3-PLC-Profile-Specification.pdf (accessed February 2011).

[120] Z. Shelby and C. Bormann, *6LoWPAN: The Wireless Embedded Internet*. John Wiley & Sons, 2009.

[121] American National Standards Institute (ANSI), "Utility Industry End Device Data Tables," ANSI Standard C12.19, 2008.

[122] International Electrotechnical Commission (IEC), "Electricity Metering—Data Exchange for Meter Reading, Tariff and Load Control—Part 62: Interface Classes," November 2006, International Standard IEC 62056-62, 2nd edition.

[123] J. LeClare, "IEEE 1901.2 PAP15 Update," Presentation to NIST PAP-15 SGIP Plenary Meeting, Chicago, IL, USA, at Grid Interop, December 2010. Available at: http://collaborate.nist.gov/twiki-sggrid/pub/SmartGrid/PAP15PLCForLowBitRates/PAP15_IEEE_1901_2_Update.ppt (accessed February 2011).

[124] D. Su, "Announcement of Joint IEEE ITU-T Work on NB Standard," August 2010. Available at: http://collaborate.nist.gov/twiki-sggrid/pub/SmartGrid/PAP15PLC ForLowBitRates/Announcement_of_joint_IEEE_ITU-T_work_on_NB_standard.doc (accessed February 2011).

[125] D. Su and B. A. Singletary, "Smart Appliance PLC Requirements—PAP-15 Recommendations to SSOs Developing Interoperable Low-Frequency Narrowband PLC for Smart Appliances," January 2011. Availabe at: http://collaborate.nist.gov/twiki-sggrid/pub/SmartGrid/PAP15PLCForLowBitRates/Narrow_Band_Power_Line_Protocol_Coexistence_Standard_Requirement.doc (accessed February 2011).

[126] O. Monnier, "TI Delivers Flexible Power Line Communications Solutions," White Paper, September 2010. Available at: http://focus.ti.com/lit/wp/slyy026/slyy026.pdf (accessed February 2011).

[127] V. Oksman and J. Egan, "Applications of ITU-T G.9960, ITU-T G.9961 Transceivers for Smart Grid Applications: Advanced Metering Infrastructure, Energy Management in the Home and Electric Vehicles," ITU-T technical paper, Telecommunication Standardization Sector of ITU; Series G: Transmission Systems and Media, Digital Systems and Networks, June 2010. Available at: http://www.itu.int/dms_pub/itu-t/opb/tut/T-TUT-HOME-2010-MSW-E.doc (accessed February 2011).

[128] Intellon, "About Us," Home page. Available at: [http://www.intellon.com/ (accessed September 2009).

[129] DS2, Design of Systems on Silicon, "Your World. Connected," Web page. Available at: http://www.ds2.es/ (accessed Febuary 2011).

[130] Panasonic Communications Co., Ltd., "Ideas for Life," Web page. Available at: http://panasonic.net/corporate/segments/pcc/ (accessed Febuary 2011).

[131] HomePlug Powerline Alliance, "About Us," Web page. Available at: http://www.homeplug.org/home (accessed Febuary 2011).

[132] Universal Powerline Association (UPA), Web page. Available at: http://www.upaplc.org/ (accessed September 2009).

[133] High Definition Power Line Communication Alliance (HD-PLC), "About Us," Web page. Available at: http://www.hd-plc.org (accessed Febuary 2011).

[134] International Telecommunications Union (ITU), "ITU-T Recommendation G.9960, Unified High-Speed Wire-line Based Home Networking Transceivers—Foundation," August 2009.

[135] International Telecommunications Union (ITU), "ITU-T Recommendation G.9961, Data Link Layer (DLL) for Unified High-Speed Wire-line Based Home Networking Transceivers," June 2010.

[136] HomeGrid Forum, "For Any Wire, Anywhere in Your Home," Web page. Available at: http://www.homegridforum.org/ (accessed Febuary 2011).

[137] Institute of Electrical and Electronic Engineers (IEEE), Standards Association, Working Group P1901, "IEEE Standard for Broadband over Power Line Networks: Medium Access Control and Physical Layer Specifications," Web page. Available at: http://grouper.ieee.org/groups/1901/ (accessed Febuary 2011).

[138] Institute of Electrical and Electronics Engineers (IEEE), Standards Association, Working Group P1901, "IEEE P1901 Draft Standard for Broadband over Power Line Networks: Medium Access Control and Physical Layer Specifications," July 2009.

[139] S. Galli, A. Kurobe, and M. Ohura, "The inter-PHY protocol (IPP): A simple coexistence protocol for shared media," *IEEE International Symposium on Power Line*

Communications and Its Applications (ISPLC), Dresden, Germany, March 2009, pp. 194–200.

[140] International Telecommunications Union (ITU), "ITU-T Recommendation G.9972, Coexistence Mechanism for Wireline Home Networking Transceivers," June 2010.

[141] D. Su and S. Galli, "PAP 15 Recommendations to SGIP on Broadband PLC Coexistence," December 2010. Available at: http://collaborate.nist.gov/twiki-sggrid/pub/SmartGrid/PAP15PLCForLowBitRates/PAP15_-_Recommendation_to_SGIP_BB_CX_-_Final_-_APPROVED_2010-12-02.pdf (accessed February 2011).

[142] Society of Automotive Engineers (SAE), "SAE Electric Vehicle Conductive Charge Coupler," January 2010, Standard SAE J1772.

[143] International Electrotechnical Commission (IEC), "Plugs, Socket-Outlets, Vehicle Connectors and Vehicle Inlets—Conductive Charging of Electric Vehicles—Part 2: Dimensional Compatibility and Interchangeability Requirements for A.C. Pin and Contact-Tube Accessories," CDV—Committee Draft for Voting, December 2010, IEC 62196-2 Ed. 1.

8

OPTICAL COMMUNICATIONS IN SMART GRIDS

Kris Iniewski

8.1 INTRODUCTION

A Smart Grid network will need to integrate electronic meters in either a wireless or wireline fashion into the two-way communications network. Once the meter network is installed and tied into the grid system, real-time load monitoring and new aggregated power loads can be viewed and managed instantly across the grid: at the substation, feeder, neighborhood, and even at the household level. Tracking usage, managing peak loads, sending real-time market signals to the consumers, and verifying conservation programs are now all critical in advanced metering infrastructure (AMI) related benefits as discussed elsewhere in this book.

One of the leading technologies to employ for high bandwidth communication access is optical fiber, which, though expensive to install, is by far the most sophisticated technology, capable of handling the largest bandwidths [1, 2]. Optical networks are expanding at a dramatic rate to support the explosive growth of

Smart Grid: Applications, Communications, and Security, First Edition.
Edited by Lars Torsten Berger and Krzysztof Iniewski.

bandwidth-rich Internet video applications along with traditional voice and data services. Fiber-optic-based networks are ideal for this task because they can carry information further and in greater density than previous copper-based transmission systems. In particular, by using dense wavelength division multiplexing (DWDM), optical fibers can carry upwards of 100 wavelength channels of information simultaneously, with each wavelength channel operating at high bandwidth (2.5, 10, or 40 Gbit/s). Fiber-optic communications has evolved from point-to-point transmission where information is transmitted between two nodes in the network, to intelligent fully meshed optical networks, where individual channels are added, dropped, and routed at individual nodes in the network. Thus, optical network architectures not only provide transmission capacities to higher transport levels, such as inter-router connectivity in an IP-centric infrastructure, but also provide the intelligence required for efficient routing and fast failure restoration. This is possible due to the emergence of optical network elements that have the intelligence required to efficiently manage such networks.

While optical communication offers extremely high level of bandwidth, the fundamental question that has to be answered is whether such high bandwidth is required in low data rate monitoring in the Smart Grid environment. Assume for a second that a data rate per monitoring point of only 10 kbit/s is required for all applications, billing, demand-side management, safety and supply continuity, and generation and distribution management. If we consider that a utility company serves some 1 million households, the aggregate bandwidth required is 10 Gbit/s. If we consider that some significant portion of all households in the future will install some form of generation in the form of wind turbines or solar cells, the required bandwidth to manage these installations will be even larger. Based on this simple argument, optical fiber is clearly required at some aggregation points in the Smart Grid network. While some high-bandwidth wireless services, like WiMAX, can probably be sufficient in some towns, larger cities will need to have optical fiber infrastructure installed by utility companies.

Utility companies require a great degree of visibility and control within the distribution network as data collection and real-time analysis become a more fundamental part of their business model. Fiber networks play an essential role in supporting the information exchange requirements between customers, the distribution network, and the data centers carrying out the real-time analysis of the data. The aggregated volume of data is unlikely to be supported without a fiber infrastructure in some parts of the distribution network.

8.2 PASSIVE OPTICAL NETWORKS (PONs)

Optical fiber can clearly deliver the highest possible bandwidth and can be connected directly to the home (FTTH), to the curb (FTTC), or to the building (FTTB). Whether it is FTTH, FTTC, or FFTB, the benefits of fiber to the node (FTTN) technology are therefore not in debate from the technological point of view. What is critical, however, is exactly how the optical wiring is delivered within a given premises. A passive optical network (PON) is typically the most effective way of delivering the split optical signal to individual users. PON technology utilizes optical splitters to enable a single optical fiber to serve multiple premises, with all users sharing available bandwidth.

PON technology is gaining worldwide acceptance, in particular, in Asia and the developing countries, with massive service roll-outs in United States as well. It offers significant cost advantage versus FTTH (fiber-to-the-home). It has been demonstrated in applications of five homes per one transformer that the sustainable bandwidth of 15–27 Mbit/s per home can be achieved. For the case of 100–150 homes the data throughput drops to 1–4 Mbit/s, which is still an impressive number compared to xDSL. A further advantage of PON architecture is that the access can be upgraded to FTTH at a later time, should there be a demand for even higher bandwidth services. While such high bandwidth numbers are not required for Smart Grid application, the technology offers the possibility of combining Internet-based services with electricity metering needs.

A PON network requires no intermediate electronics or powering between the head end (HE) and the subscriber's home and consequently is operationally simple. The main benefit of this type of architecture is that historically it has been significantly cheaper than other options, requiring lower power lasers and fewer fibers, reducing the capital (CapEx) and operational expense (OpEx) for service providers. The drawbacks of a PON network are significant, however: splitters reduce the distance possible between active electronics and make it so that more users are impacted by a single point of failure in the network. Also, as more users are added to the system, the bandwidth available for each given user decreases.

A typical PON architecture is shown in Figure 8.1. OLT represents an optical line termination unit. The access node offers the last mile access via a PON interface. Traffic is aggregated using passive optical splitter. Typical downstream bandwidth

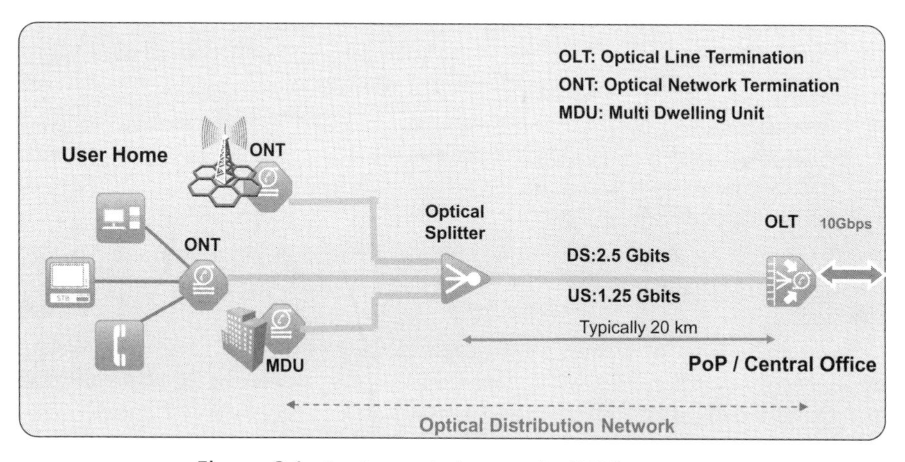

Figure 8.1. Passive optical networks (PON) concept.

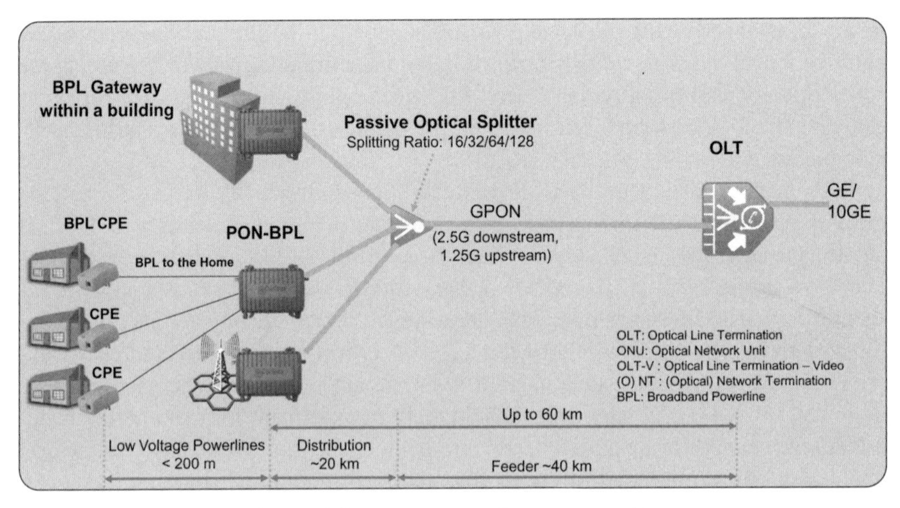

Figure 8.2. BPL-PON implementation [3].

is 2.5 Gbit/s while the upstream bandwidth can be as high as 1.25 Gbit/s. PON networks can easily be connected to other wireline or wireless networks as shown in Figure 8.2 for the case of a broadband power line (BPL) network connecting to optical fiber.

This BPL-PON solution is based on a combination of two technologies. At the central office, a PON concentrator (OLT) is used to aggregate all fiber coming from the field toward the packet core. The distribution to the customer is done at three levels:

- *Aggregation.* Fiber distribution infrastructure is based on point-to-multipoint technology and PON (or point-to-point technology such as Ethernet) from the central office to selected poles at each area cell (determined by the low-voltage transformer and its serving area). A BPL gateway with optical termination capabilities (ONT) terminates the fiber. Distances can be over 20 km.
- *Access.* The BPL gateway is collocated with the low-voltage transformers and connected to 20–50 households, handling the traffic over the power line. Actual distance varies depending on the specific energy grid with typical distances of a few hundred meters.
- *Home.* At the customer environment, the subscribers connect to the network using a home plug.

The combination of gigabit passive optical networks (GPONs) and broadband over power line (BPL) enables the telecommunication carriers and the power utilities to deploy a FTT-BPL (fiber-to-the-BPL) broadband access network, using low-voltage power lines for “last mile” access to the home, instead of leasing copper pairs.

The fiber distribution network provides downstream rates of 2.5 Gbit/s; this can be shared between 16 or 32 BPL gateways (depending on the end user bandwidth requirements), BPL chip sets provide downstream rates of up to 200 Mbit/s, which can be shared between the end users. The combination of GPONs plus BPL is a cost-effective way of meeting the bandwidth requirements for broadband access services such as Internet access, voice over IP, and IPTV.

8.3 WAVE LENGH DIVISION MULTIPLEXING (WDM)

WDM stands for wavelength division Multiplexing and refers to a technology that combines multiple wavelengths on one single-mode fiber. The ability to multiplex wavelengths is of critical importance for creating high-bandwidth optical pipes to transport data. In a WDM system the optical signals of different frequencies are launched into the inputs of a wavelength multiplexer (MUX). At the output of the wavelength multiplexer all wavelengths are effectively combined and coupled into a single-mode fiber. At the end of the transmission link the optical channels are separated again by means of a wavelength demultiplexer (DE-MUX) and thus arrive at the different outputs.

Each WDM wavelength requires its own laser source and light detector. The particular choice of wavelength used is governed by standards established by the ITU organization. As a result, the WDM equipment from different vendors has a chance to interoperate. The so-called krypton line at 193.10 THz was selected to be the reference line. The 100 GHz number was selected as a channel spacing, which is the term for the distance between two neighboring WDM lines. An example of a WDM signal is shown in Figure 8.3 for the case of 16 channels in a 1530.33–

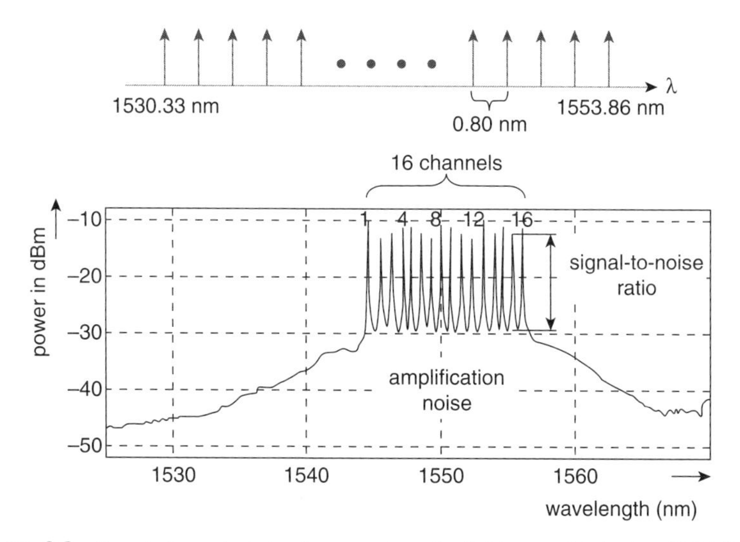

Figure 8.3. Signal characteristics for wavelength division multiplexing (WDM) [2].

1553.86 nm optical window. Note that wavelengths are 0.8 nm apart, which, converted to frequency domain, gives the 100 GHz number.

The ITU has specified six transmission bands for fiber optic transmission:

- O band (original band)—1260 to 1310 nm
- E band (extended band)—1360 to 1460 nm
- S band (short band)—1460 to 1530 nm
- C band (conventional band)—1530 to 1565 nm
- L band (long band)—1565 to 1625 nm
- U band (ultra band)—1625 to 1675 nm

A seventh band, not defined by the ITU, runs around 850 nm and is used in private networks.

Depending on the number of wavelengths desired to be sent on a single fiber, WDM can be classified either as *Coarse* WDM, for 8 or less wavelengths, or *dense* WDM , for systems with 8 or more wavelengths. Coarse WDM (CWDM) equipment is cheaper, as the spacing between wavelengths is large but has limited scalability potential. Dense WDM (DWDM) equipment is much more expensive to build, as very precise and stable separation between wavelengths is required, but its capacity is very scalable. CWDM lasers, on the other hand, are not subjected to the same stringent requirements. Wavelength spacing in CWDM systems is much larger, so laser stability and its small linewidth are not huge concerns. As a result, much cheaper components can be used to build cost-effective CWDM equipment. For Smart Grid applications CWDM is typically sufficient.

With small modifications at the conceptual level, but fairly major growth in a practical sense, the WDM regenerator can be transformed into an optical add/drop multiplexer (OADM), shown in Figure 8.4. An OADM can also serve as a WDM terminal at network entry points. OADMs and WDM terminals are always needed in WDM systems in order to launch and terminate WDM signals. They have to provide the ability to add and drop electrical signals as required. Some of these network elements also have additional grooming capability to convert some low bandwidth signals into one larger bandwidth signal in order to not waste the capacity of the WDM system. Note that from the WDM system capacity point of view, one channel is used whether it carries a 10 Gbit/s electrical signal or only a 1 Mbit/s electrical signal, so it is advantageous to groom electrical signals to the largest denominator prior to WDM transmission.

In the past, telecommunication companies have used OADMs with fixed functionality just to add or drop wavelengths at a node. Because these devices have generally used fixed-wavelength add/drop filters, they turn wavelength reconfiguration into a very extensive manual task, requiring a large effort to implement network changes if requested. Recently introduced remotely reconfigurable OADMs, called ROADMs, solve that problem.

In ROADMs, add/drop wavelength functionality can be programmed from a remote location. To change network configuration, a service provider no longer has

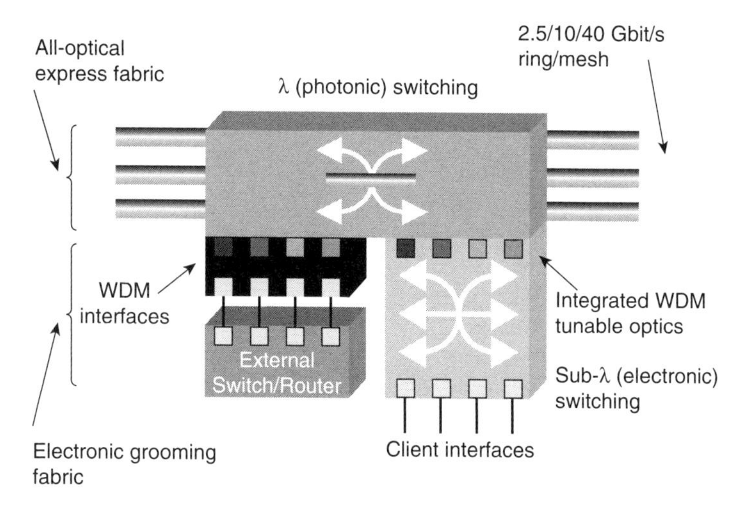

Figure 8.4. Optical add/drop multiplexer (OADM) [2].

to send their service workers to manually perform the change. As a result, the great added value that ROADMs bring is their introduction of increased flexibility and speed by providing remote reconfigurability, simpler service provisioning, and faster fault detection. The benefits for telecommunication companies are obvious, and include simplified, remote network management, fast service delivery, and significant savings in network operating costs.

In the past, adding, dropping, and routing wavelength channels at individual nodes in the network were done by demultiplexing the wavelength channels at each node, manually rearranging the wavelengths using an optical patch panel, and then multiplexing them again into the desired fiber output for transmission to the next node. Needless to say this was time consuming and, far worse, very prone to human error. Over the years, manual switchboards were replaced by automated switches across all types of technologies. For example, in the last few decades digital cross-connects (DCSs) and SONET/SDH add/drop multiplexers (ADMs) have completely replaced the practice of manually rearranging individual circuits using back-to-back channel banks. Similarly, in today's optical networks, optical cross-connects (OXCs) and reconfigurable add/drop multiplexers (ROADMs) are two of the main network elements that were introduced to quickly and efficiently route the optical connections to their desired destinations. These elements have considerably simplified some of these manual operations and in the future they are expected to eliminate them completely.

Over the last decade, the initial focus was on large port-count OXCs, and following a wave of timely technological breakthroughs, in the early 2000s optical network equipment vendors were announcing a variety of optical switching systems capable of exchanging and redirecting several terabits of information per second. The dimensions of the proposed switches were colossal, ranging from a few tens to several thousand ports with each single port capable of carrying millions of voice calls, or thousands of video streams. These switches were opaque (with either an electronic switch fabric or an optical switch fabric but with electrical interfaces at the edges of the switch) or transparent, with either a large port-count switch or several smaller port-count switches (one per wavelength). In the last few years the focus has shifted from these large port-count OXCs to ROADMs, which are smaller network elements that can be used efficiently to direct wavelength channels in an optical network. These ROADMs device are particularly well suited to Smart Grid applications.

8.4 SONET/SDH

Synchronous Optical Network (SONET) is a standard for optical telecommunications transport established by ANSI in the 1980s. Its European/Japanese counterpart—Synchronous Digital Hierarchy (SDH)—was formulated by the ITU standard body. The differences between the two standards are reasonably small for our purposes, so in this chapter, for the most part, we will only use the term SONET, for simplicity.

SONET uses time domain multiplexing (TDM) as a way to multiplex lower bandwidth signals into one high capacity signal that can be sent over optical fiber. Being a networking standard, it provides a set of specifications to determine frame formats, bit rates, and optical conditions in order to ensure multivendor interoperability. It also provides fast restoration schemes, meaning that in the case of a fiber cut or some other network failure it has the capability to restore the service, typically within 50 ms. Finally, SONET has extensive built-in features for operations, administration, maintenance & provisioning (OAM&P) as schematically illustrated in Figure 8.5. These OAM&P features are used to provision connections in point-to-point and ring topologies, to detect defects, and to isolate network failures.

SONET line signals can be optical or electrical. Electrical signals are used for short distances, for example, as interconnections of networking equipment on the same site or between racks of the networking gear. Electrical signals are sent on copper wires, while optical signals used for longer distances require optical fibers. We will use the term STS (synchronous transport signal) for electrical signals and OC (optical carrier) for optical signals.

SONET, as its name implies, is synchronous. That means a highly stable reference point or clock is provided for every piece of SONET equipment through an elaborate clock synchronization scheme. Since a very precise clock is available, there is no need to align the data streams or synchronize clocks. SONET provides extensive monitoring and error recovery functions, and allows for integrated network OAM&P. Substantial overhead information is provided in SONET to allow for quicker troubleshooting and the detection of failures before they degrade to serious levels that can bring the entire network down.

Recently, SONET/SDH networks incorporated mechanisms for mapping generic data transport protocols into the synchronous frames. This was motivated by the fact that whereas the growth of the voice traffic was smooth since the introduction of the

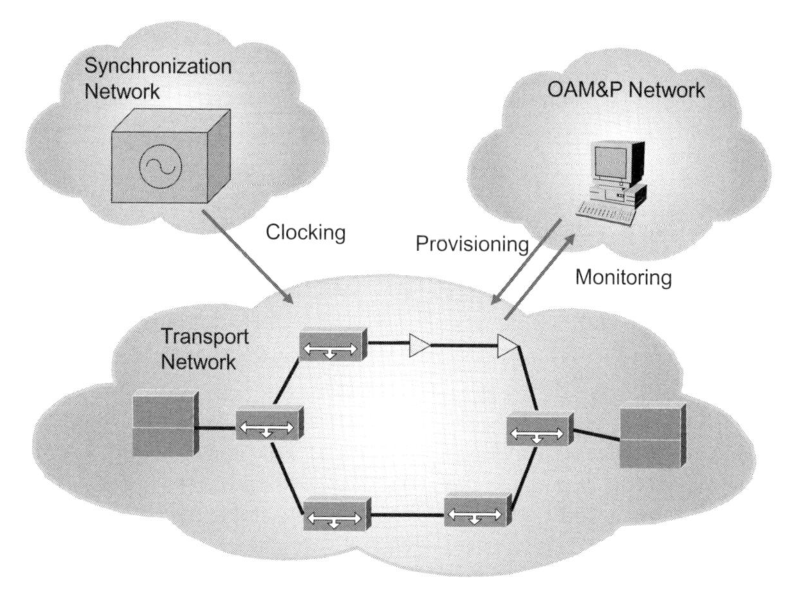

Figure 8.5. Schematic representation of SONET/SDH networks [2].

first SONET optical networks, the Internet traffic was increasing very rapidly. Although SONET networks can be combined with the wavelength division multiplexing (WDM) technology to scale up their transmission capacity due to the aggregation of access traffic, the variety of services and types of traffic presented in today's networks require a more efficient usage of the bandwidth than that provided by TDM circuits.

The limited number of protocols that could be mapped into SONET frames did not reflect the emergence of new technologies for transporting data in the metro segment, especially 100 Mbit/s Ethernet and Gigabit Ethernet. In order to define an efficient and interoperable mapping of generic protocols in SONET networks, the International Telecommunication Union (ITU) defined the generic framing procedure (GFP). To provide SONET/SDH systems with high efficiency in the accommodation of a variety of protocols, GFP must be combined with two other technologies. The SONET networks that incorporate data traffic adaptation features have frequently been referred to as next-generation SONET/SDH systems.

8.5 CARRIER ETHERNET

Ethernet is a family of closely related packet-oriented protocols that operate at 10 Mbit/s, 100 Mbit/s, 1 Gbit/s, and 10 Gbit/s. The lower two rates are the dominant technologies for the implementation of local area networks (LANs). The 1 Gbit/s rate is used in more advanced LANs and to connect LANs. The 10 Gbit/s rate is emerging to offer greater bandwidth in connecting LANs. While Ethernet-based LANs are a source of much of the traffic carried by transport networks, in general, the Ethernet protocol has been terminated, and the transport network is used to carry the contained IP frames.

However, the widespread popularity of Ethernet in the LAN has led to demands for Ethernet support in the WAN, or the Internet. There are two reasons: Ethernet has a reputation for low cost, which customers would like to extend into the WAN; and customers with multiple Ethernet LAN sites would like to be able to connect these LANs without getting involved in more complex protocol stacks—hence they want Ethernet-over-the-WAN service. It is in this role that we are primarily interested in Ethernet use for Smart Grid applications.

There are three approaches to the support of Ethernet over the WAN [2]. The first is called "transparent" support of Ethernet. In this solution, complete Ethernet frames are transported in point-to-point channels, which carry only Ethernet frames. This can be accomplished by carrying Ethernet frames directly in optical fiber (on in optical colors in DWDM systems). With this approach, there is very little but Ethernet present—all the overheads and operations, administration, and maintenance (OAM) services normally associated with WAN links are absent. These solutions can often be implemented at lower costs than other solutions, but the lack of effective OAM services limits the applications of this option. An alternative is to embed Ethernet frames in a SONET/SDH payload of varying sizes. For example, a SONET STS-24c (about 1.268 Gbit/s) is appropriate to carry 1 Gbit/s Ethernet. With this option, bandwidth is lost to overheads, but standard OAM services are available at the SONET/SDH level. Of course, Ethernet itself is essentially unmanaged.

The second basic approach for carrying Ethernet over the WAN is the use of switched Ethernet. With this option, the WAN is provisioned with Ethernet switches, which can direct Ethernet frames based on their Ethernet endpoint addresses. This option supports more complex networks than the transparent mechanisms, which support only point-to-point services. A switched Ethernet WAN can be as large and as complex as desired, so long as some mechanism is provided to manage the hard-coded Ethernet endpoint addresses. The Ethernet pipes between the Ethernet switches can be based on any of the transparent Ethernet solutions, for example, embedded in SONET/SDH payload.

The final class of solution for extending Ethernet over the WAN is the use of routed Ethernet networks. In this class of solution, the routing is at the IP layer. Packets enter/depart routers on Ethernet links, but are routed in the normal manner for IP frames. Links between routers may or may not be Ethernet. In other words, this solution is indistinguishable in essence from normal IP routed layers. The use of Ethernet for access and perhaps for inter-router links may reduce costs. This solution is the only one of the three that offers universal IP access to the data frames, as only in this solution is the traffic not isolated to an Ethernet island of isolated traffic.

Carrier Ethernet has been defined as an ubiquitous, standardized, carrier-class service and network defined by five attributes that distinguish it from familiar LAN-based Ethernet:

- *Standardized Services.* Standardized point-to-point, point-to-multipoint, and multipoint-to-multipoint LAN services that require no change to the existing LAN devices including TDM clients. Provide wide range of bandwidth and quality of service (QoS) options that facilitate building converged voice, video, and data networks.
- *Scalability.* Ability to support millions of users with a diverse set of applications running at the same time. From 1 Mbit/s to 10 Gbit/s bandwidth services in granular increments across metro and wide area networks over a wide range of transport network infrastructure from multiple service providers.
- *Reliability.* Ability to recovery from network failure automatically within a short period of time that could be as low as the well-known 50 ms SONET protection time. Provide high quality and availability of services that are comparable to the existing TDM services.
- *Quality of Service (QoS).* Provide a rich set of bandwidth and quality of service options through service level agreements (SLAs) provisioning that is based on committed information rate (CIR), frame loss, delay, and delay variation characteristics.
- *Service Management.* Ability to rapidly provision end-to-end Ethernet services and provide full carrier-class OAM capabilities.

Carrier Ethernet is designed to be commercial services offered by service providers and therefore needs to be associated with a clear definition of service parameters like legacy TDM services. Service availability, delay, jitter, and packet loss may all become part of service level agreements (SLAs). Historically, Ethernet is not a service interface, but rather a networking solution for local area networks, and is short of meeting stringent SLAs. New service architecture needs to be introduced to provide a clear demarcation point between user and network. The use of carrier Ethernet in Smart Grid networks remains uncertain at the time of this writing.

8.6 CONCLUSIONS

The Smart Grid communication scheme must address multiple requirements. While core utility operational networks can be based on a number of technologies, the most prevalent is next-generation SONET, also known as multiservice provisioning platforms (MSPPs). Packet-switched networks are also gaining attention in the utility market. Next-generation SONET is attractive, as it can support IP and Ethernet applications and legacy services simultaneously. It augments the functionality of the existing SONET networks and enables its evolution to IP/MPLS by providing effective Ethernet transport. Both packet-switched networks and Ethernet over SONET are considered carrier grade services.

Optical Ethernet is a connection-oriented protocol that provides "carrier grade" performance for mission-critical applications, including high reliability, QoS, provisioning, and security. The challenge is to combine these features with the costeffectiveness and simplicity of Ethernet. A typical MSPP implementation involves Ethernet over SONET technology. MSPP is a proven and mature carrier-class infrastructure offering robust reliability, protection, and operations, administration, and maintenance (OAM). The key future technology under development is an MPLS-based Ethernet network that uses MPLS as a circuit-oriented layer spanning the entire optical Ethernet and SONET network. While it is certain that a Smart Grid communication scheme will involve some optical communication technologies discussed in this chapter, the extent and manner in which these technologies will actually be deployed in massive roll-out by utilities remains to be seen.

What is clear at this point is that communications networks within the utility company environment have to undergo a major transformation. The control and communications networks must extend to all customer points if real-time demand-side management is to be used effectively. Similarly, if customers are to be encouraged to become generators, an additional data stream must be aggregated to the network extending to the home. And if both distribution company and customer control are to be supported, data requirements increase even more.

At its edge, the network does not need to have very high capacity. But it must have high service penetration. One can envision a mix of technologies at the edge, with an aggregation into a core network that is built on optical fiber. Most utilities have implemented large, privately owned and operated telecommunication transport networks, supporting both fixed and mobile voice and data communications for its operational as well as corporate functions. Although these two major areas typically make use of the same facilities, their needs are quite different in terms of bandwidth, traffic, availability, performance, and security and communications protocols. Utility companies require fiber at aggregation points to collect information from large groups of houses, and at data centers to manage the interchange of information for real-time decision making and control. Aggregation points may be local substations, where fiber, for example, may also be a good choice as a passive temperature and monitoring sensor for electricity cables.

REFERENCES

[1] K. Iniewski, *Convergence of Mobile and Stationary Net-Generation Networks*, Wiley, 2010.

[2] K. Iniewski, C. McCrosky, and D. Minoli, *Network Infrastructure and Architecture: Designing High-Availability Networks*, Wiley, 2008.

[3] P. Sobotka, R. Taylor, and K. Iniewski, "Broadband over power line communications: Home networking, broadband access, and smart power grids," in K. Iniewski (Ed.), *Internet Networks: Wired, Wireless, and Optical Technologies*, CRC Press, 2010.

9

NETWORK LAYER ASPECTS OF SMART GRID COMMUNICATIONS

Kris Iniewski

9.1 INTRODUCTION

As outlined in previous chapters a Smart Grid network will need to integrate electronic meters in either a wireless or wireline fashion into the two-way communications network. Once the meter network is installed and tied into the grid system, real-time load monitoring and new aggregated power loads can be viewed and managed instantly across the grid: at the substation, feeder, neighborhood, and even at the household level. Tracking usage, managing peak loads, sending real-time market signals to the consumers, and verifying conservation programs are now all critical in advanced metering infrastructure (AMI) related benefits. In this chapter we will discuss a vision of the future network and its ability to carry large amounts of data at the high level of the OSI protocol stacks.

The advent of Smart Grids, other autonomic systems, cloud computing, and the presence of huge storage capacity at the edges of the network will change significantly the networking paradigms. Efforts in the past several years have focused

Smart Grid: Applications, Communications, and Security, First Edition.
Edited by Lars Torsten Berger and Krzysztof Iniewski.

on exploiting the progressive penetration of computers in the network to make the network more intelligent. A simple economic drive motivated this evolution: the network is a central resource whose cost can be split among the users. It makes more sense to invest in the network to provide better services to low cost, low intelligent edges.

However, we are seeing increasing shift of processing, storage, and intelligence from the network to the edges, to the terminals using it. Smart Grid network is a primary example of that trend. A vision for the network of the future is a collection of very high capacity pipes, several terabit/second each, having a meshed structure to ensure high reliability and to decrease the need for maintenance. This network likely terminates with local wireless drops. These wireless loops will be built using either small femto cells, traditional Wi-Fi, emerging LTE 3G services, and long range cells covering rural areas such as WiMax as discussed in more detail in Chapter 6. Combined wireless traffic that includes AMI data will be carried on by standard TCP/IP networks as they exist today.

9.2 TCP/IP NETWORKS

Smart Grid networks will likely use TCP/IP networks in some shape or form. While we have no intention to explain TCP/IP here, there are complete books devoted to that topic, see Stevens [1, 2], for example; we intend to provide some insight here to explain challenges in Smart Grid implementations.

9.2.1 TCP/IP Protocol Stack

The TCP/IP protocol suite uses a simplified layered structure of four layers, instead of the seven layer OSI stack. This reduction in layer count is accomplished mostly by paying less attention to the various higher layers of the OSI stack and lumping them into one layer called the *application layer*. The TCP/IP protocol stack is organized into layers as shown in Figure 9.1.

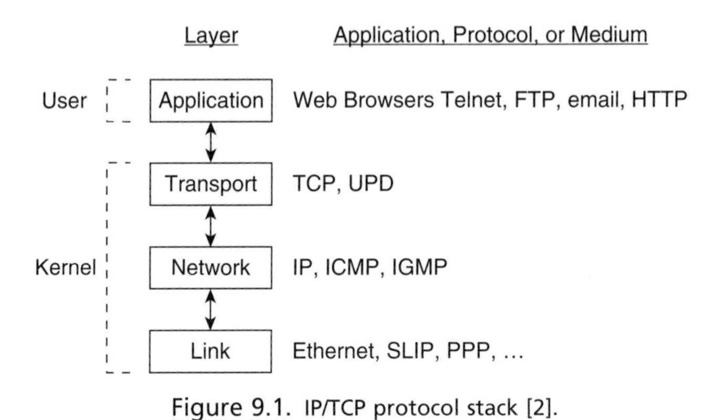

Figure 9.1. IP/TCP protocol stack [2].

From the bottom up, these layers and their primary roles or responsibilities are

- *Link Layer.* The principal responsibility of the link layer is to move packets from point to point over direct links. The link layer consists of the physical means of signaling, the electrical or optical devices that drive the physical link, the definition of the data formats used on the link, the definition of the protocol's signaling and state transitions that control the use of the link, the hardware that controls the link (usually called a network interface controller (NIC) card), and the operating system components and device drivers that interface to the NIC card. Fast Ethernet is an example of a link layer protocol.
- *Network Layer.* The principal responsibility of the network layer is to move packets across the network from source to destination. The network layer consists of the protocol elements such as header formats, signaling, and endpoint state transitions, which serve the purpose of moving packets through the network, from endpoint through multiple internal nodes to endpoint. The TCP/IP network layer provides a simple unreliable datagram service—that is, it neither recognizes endpoint-to-endpoint flows nor explicitly attempts to maintain a quality of service. Instead, it makes a best effort to correctly deliver each independent datagram, which it is given by the *transport layer*. The Internet protocol, or IP, is the dominant network layer of the modern Internet, and shall be the basis for much of the material in this book.
- *Transport Layer.* The principal responsibility of the transport layer is to establish and manage end-to-end communications flows and to provide a reliable service. The transport layer controls the movement of data between two endpoints. This is achieved by a distinct layer of protocol, which consists of header format definitions, signaling between endpoints, and endpoint state transitions. The transport layer views the channel provided by the network layer as an abstract pipe through a network "cloud." The layer has two primary responsibilities: to provide a reliable service over the inherently unreliable network layer, and to regulate the flow of information between endpoints over that pipe in such a manner as to maximize throughput and to minimize congestion and packet loss in the pipe. The transmission control protocol, or TCP, is the dominant transport layer of the modern Internet.
- *Application Layer.* The application layer consists of all and any software that makes use of the transport, network, and link layers to achieve endpoint-to-endpoint and process-to-process communications in support of some user purpose. The best known application layer entities are FTP (file transfer protocol) for moving files from endpoint to endpoint, Telnet for remote login services, and SMTP (simple mail transfer protocol) for email services. Web browsers (e.g., Firefox or Internet Explorer) are all applications that support a range of Web services which depend on a range of application layer protocols such as the familiar "http" service.

TCP is a connection-oriented service. Thus, connections must be established when needed and torn down when their purpose is completed. The process works as follows: one end proposes to the other that they establish a connection, the recipient of the proposal agrees with an acknowledgment message, and the original proposer acknowledges the acknowledgment. In the process, the two endpoints exchange the port numbers they intend to use and the maximum segment size (MSS) that each will tolerate, and the starting sequence numbers that will be used to refer to progress in transmitting the two unidirectional byte streams that constitute the duplex interchange. A variant on this scheme occurs when both endpoints actively request a connection.

IP, at the network layer, offers no flow control mechanisms. IP sources transmit whenever they receive a segment from the transport layer. Networks and sinks can be overwhelmed by overly aggressive traffic loads, resulting in high loss rates and poor overall service. Some layer of the protocol stack must assume responsibility for flow control, in order that networks can deliver reasonably efficient and reliable services. It is clear that networks cannot rely on the application layer for flow control discipline. The solution to prevent network chaos must come from somewhere. We have only one remaining layer as a source of discipline—the transport layer. TCP at the transport layer does take up this challenge, and provides several interacting mechanisms to enforce a reasonable level of flow control in the face of network limitations. TCP's flow control mechanisms can be divided into two types: receiver-based flow control and transmitter-based flow control. Receiver-based flow control's primary role is to prevent receiver buffer overflow. Transmitter-based flow control's primary role is to further limit the flow in response to network conditions.

IP's routing algorithm results in one path for all traffic between any two endpoints. While the use of the shortest path makes sense in the case of a single connection, it can be a very poor choice in a congested network. The problem is that nearly-as-short paths are ignored, and their carrying capacity is lost. Thus, too much traffic is forced onto the best link, resulting in congestion, while nearly-as-short links starve for traffic. In addition, any attempt to preserve QoS (e.g., for VoIP) is destroyed by the congestion, while ideal channels exist for such traffic on the nearly-as-short links.

Congestion is bad for networks, as it leads to packet loss and compromises QoS for sensitive services. We would like to avoid congestion entirely. Yet TCP deliberately seeks congestion. It ramps up its transmission rate (within the constraints of the receiver flow control system) until it discovers congestion by seeing packet loss. TCP uses this mechanism to seek ever higher transmission rates within available capacity, which is a good goal. But it has no way to avoid increased transmission rates before congestion happens. So TCP breeds congestion. Furthermore, after congestion, the transmission rate of the source drops dramatically. This tends to leave the pipe unfilled after congestion events which have affected numerous TCP flows. During such periods, the formerly congested will be underloaded. This mechanism would probably be acceptable if it were not for the increasing demands to support higher QoS for applications like video links and VoIP. Such services cannot be useful if their TCP/IP connections are constantly subject to higher queuing delays, congestion, and packet loss.

9.2.2 Quality of Service (QoS)

While IP was designed to support a modest system of QoS differentiation via the *type of service* (TOS) bits in the IP header, these features were ignored by implementers for many years. As a result, attempts to add QoS services at the IP level today are compromised: QoS requires end-to-end support, and the occasional new router cannot force all the other routers in the Internet to begin respecting the TOS bits. Consequently, it is essentially impossible to make use of IP's TOS facilities at this late point in time for Internet-wide applications. In addition, modern QoS standards require more support than IP's TOS facilities can provide.

TCP/IP has many well-known shortcomings, mostly related to difficulties in achieving high level quality of service (QoS). But it has been one of the most important foundations of the Internet and its continuing growth. It has provided decent functionality and an enormous communality of technology, which has enabled a huge range of new applications. These protocols have won a beachhead in our lives that is essentially impossible to dislodge. There are far too many TCP/IP conversant bits of technology for us to willingly throw away because we want to replace TCP/IP with something better. Any evolution to solutions better than TCP/IP can only take place in an incremental manner, which remains compatible in important functionalities with the existing TCP/IP solutions.

9.2.3 IPv6

The IPv4, main Internet protocol today uses 32-bit addresses and can support 4.3 billion publicly addressable devices on the net. With 7 billion people worldwide, clearly that number is not sufficient, and in fact we are very close to running out of available address space. Therefore, the Smart Grid, while supporting the current IPv4 architecture, needs to adopt the next-generation Internet standard known as IPv6 in corporate and home networks. Available for more than a decade, IPv6 has been slow to catch on until recently because there has been no concrete need that compelled companies to spend money upgrading their routers, servers, and applications to support it.

IPv6 uses a 128-bit addressing scheme and can support an inconceivably huge amount of devices: 2 to the 128th power. IPv6 also features built-in security and enhanced network management features when compared with IPv4. Future Smart Grid implementations have to use IPv6 equipment in the future. In the short term, energy utilities could use private IPv4 addresses hidden behind network address translation (NAT) boxes for Smart Grid projects, but this approach is more complicated and has a greater risk of error in network management.

9.2.4 TCP/IP for Wireless Networks

With the ever-increasing convergence of the Internet and wireless modes of access, pressure has steadily been increasing on the networking community to improve the performance of TCP, the Internet's most widely used transport protocol, when being carried by wireless networks. For this reason, TCP performance enhancement in wireless access networks with a wireline core is an important ongoing area of research. The hostile nature of the wireless channel and the mobile nature of wireless users interact adversely with standard TCP congestion control mechanisms, often causing a drastic reduction in throughput. How TCP/IP deals with these issues is presently unclear, but introduction of performance-enhancing proxy mechanisms or new switching technologies like multiprotocol label switching (MPLS) might help.

9.3 MULTIPROTOCOL LABEL SWITCHING (MPLS)

MPLS uses label switching to forward packets through the network. As a result, MPLS introduces the opportunity of controlling traffic within a mesh of interconnected IP routers basically by enhancing IP with a traffic engineering capability. By replacing ATM switches with MPLS capable IP routers, the ATM layer can be removed from the networking stack. MPLS as a packet switching solution is not expected to replace the WDM core because of the scalability and much lower cost of optical switching where large volumes of traffic are to be switched along the same route. However, MPLS can play a major role also in the WDM core and its integration with the outer part of the network as the protocols of the MPLS control plane can be adopted as the control plane for the optical network.

The basic idea underlying MPLS is to add a label to packets to be used for routing instead of the destination address carried in the IP header. The label was originally added as a means to enable a lookup operation faster than the longest prefix matching required on addresses when routing IP packets. As integration technology advanced and router manufacturers refined their ASIC (application specific integrated circuit) designs, IP addresses longest prefix matching became possible at wire speed, thus eliminating the need for a different type of lookup. Consequently, the original label purpose was meaningless, but the importance of another capability enabled by the label became predominant: facilitating traffic engineering. The key MPLS feature in facilitating traffic engineering is the separation of the routing functionality in the control plane from the one in the data plane, which is made possible by the presence of a label in each packet. This enables network operators to deploy MPLS-based IP networks with traffic engineering and fast fault recovery. More information about MPLS can be found in Minei and Lucek [3].

9.4 CONCLUSIONS

The Smart Grid communication network will use TCP/IP transport protocol and IPv6 addressing space. What is clear at this point is that communications networks within the utility company environment have to undergo a major transformation. The control and communications networks must extend to all customer points if real-time demand-side management is to be used effectively. Similarly, if customers are to be encouraged to become generators, an additional data stream must be aggregated to the network extending to the home. And if both distribution company and customer control are to be supported, data requirements increase even more.

REFERENCES

[1] W. R. Stevens, *TCP/IP Illustrated, Volume 1: The Protocols*. Addison-Wesley, 1994.

[2] W. R. Stevens, *TCP/IP Illustrated, Volume 2: The Implementation*. Addison-Wesley, 1995.

[3] I. Minei and J. Lucek, *MPLS-Enabled Applications: Emerging Developments and New Technologies* (Wiley Series on Communications Networking & Distributed Systems). Wiley, 2011.

10

SMART GRID SENSING, AUTOMATION, AND CONTROL PROTOCOLS

Wolfgang Mahnke

10.1 INTRODUCTION

When controlling a power grid it is necessary to know the state of the power grid, for example, how much current is on a power line or whether a breaker is open or closed. In order to automatically or manually control the power grid, it is essential to have sensors sensing the state of the grid as well as actors (e.g., breakers) changing the state of the grid.

In a Smart Grid there will be many decentralized energy sources with some of them uncontrollably varying their capabilities (e.g., wind and solar power providing power based on weather conditions). In addition, the behavior of the consumer needs to be considered and can, to a certain degree, be controlled through the use of demand response and dynamic pricing (as outlined in Chapters 1 and 3). This increases the need to have an accurate and prompt picture of the current state of the grid and to control the actors fast in order to control the state of the grid.

Smart Grid: Applications, Communications, and Security, First Edition.
Edited by Lars Torsten Berger and Krzysztof Iniewski.

Handling Smart Grids requires sensing various applications in classical areas like power generation, substation automation, and transmission and distribution systems, but also sensing end consumer behavior (via smart meters) or sensing data for weather forecasts. Depending on the application the sensors might be local on a site like in a traditional power plant, distributed by a virtual power plant aggregating several solar power collectors or wind mills, or heavily distributed like in an HVDC transmission application (high-voltage direct-current transmission), distributing power over some thousand kilometers.

The time frequency to exchange information differs very much depending on the application. When trading energy on the energy market, the frequency is very low. For example, at the European Energy Exchange the lowest time span is currently 60–75 min before delivery at the intraday market [1]; at the OTC (over-the-counter) market it goes down to 15 min. In North America, some ISOs/RTOs (Independent System Operators/Regional Transmission Organizations) maintain location-based marginal pricing at 5 min intervals. In contrast, a FACTS (flexible alternating current transmission system) application, enhancing the controllability and increasing the power transfer capabilities of an AC transmission system [2] (see also Chapter 1), needs to react in milliseconds.

The amount of information exchanged also varies depending on the use case. For example, a simple binary value is needed to exchange the information whether a breaker is opened or closed. On the other hand, creating reports on the performance of a power plant might require access to a large amount of data representing the history of measured values.

In the area of Smart Grids there are several players like utilities providing and/or distributing power in connected power grids, with large international players like E.ON or EDF as well as small municipal energy suppliers, industrial consumers, and end consumers and the suppliers of hardware, systems, and solutions. In order to have a functional Smart Grid it is necessary to exchange information inside and between different systems of the Smart Grid. To achieve this, it is required that the systems that need to exchange data agree on common mechanisms for information exchange. To avoid point-to-point integration scenarios the most suitable solution is to standardize the information exchange mechanisms for similar application scenarios in order to support multivendor installations like Smart Grids. To exchange information more than just the communication protocols need to be defined. Information interoperability requires that systems understand the meaning and structure of the information exchanged and that the business objectives, procedures, and workflows are aligned with and support the information exchange. The Interoperability Context-Setting Framework (aka GWAC Stack) was developed by the GridWise Architecture Council to frame the requirements for system interoperations [3] as shown in Figure 10.1.

(1) *Basic connectivity* describes the mechanism to establish a physical and logical connectivity of systems, like Ethernet over twisted pair or Ethernet over fiber optic as addressed in Chapters 6–8.

(2) *Network interoperability* defines the exchange messages between systems across a variety of networks, like IP, TCP, or UDP as introduced in Chapter 9.

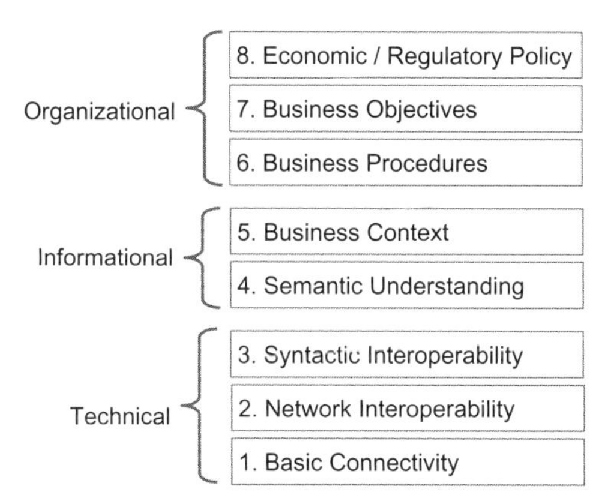

Figure 10.1. Interoperability layered categories. From Gridwise Architecture Council [3].

(3) *Syntactic interoperability* defines the understanding of data structures in the exchanged messages, like XML and HTML. Those three categories describe the *technical aspects* of interoperability.

(4) *Semantic understanding* defines the understanding of concepts contained in the message data structures. Examples include models based on XML schema definitions but also standards addressed in this chapter like IEC 61850 or the Common Information Model (CIM) in IEC 61970 / IEC 61968.

(5) *Business context* defines the relevant business knowledge that applies semantics with process workflow, meaning that, for example, the content of a contract between two parties is applied and restricts the exchanged information. Those two categories define the *informational aspects* of interoperability.

Organizational aspects of interoperability include (6) *business procedures*, (7) *business objectives*, and (8) *economic and regulatory policy*. Those aspects are not addressed in this chapter. In addition, the GridWise Architecture Council [3] speaks about cross-cutting issues of interoperation like system evolution, quality of service, and system preservation that are also not addressed in this chapter.

In the following subsections some general considerations on communication (addressing the technical aspects of Figure 10.1) and information modeling (addressing the informational aspects of Figure 10.1) will be given before protocol standards in the domain of Smart Grids are introduced. As just seen, the requirements on information exchange might be different in different use cases of the Smart Grid, for example with respect to the frequency, bandwidth, and distribution.

10.1.1 Communication

The base to communicate is providing a physical connection between two systems. In the very old days of automation, this was not even done by electricity, but sensing and control was done pneumatically [4]. Later, it changed from pressure to electricity with 4–20 mA, and this is still a common communication mechanism. However, modern communication protocols rely on Ethernet-based communication, independent of the physical layer (e.g., wired or wireless bridged—see Chapters 6–8). In this chapter the focus is on Ethernet-based protocols, as they are the base for Smart Grid communication in sensing, automation, and control. Several important communication characteristics will be described. These include:

1. Communication models
2. Type of data exchanged
3. Real-time support
4. Frequency
5. Latency
6. Throughput
7. Granularity of data
8. Triggering mechanism
9. Reliability
10. Security

Having the base connection established, there are different *communication models*. In general, there are unicast, multicast, and broadcast (see Figure 10.2). Unicast is used in client–server architectures or peer-to-peer networks, where one communication partner is connected to one other communication partner. Multicast typically uses a publish–subscribe model, where the publisher provides information and sends it to all subscribers that have subscribed to the data. In a broadcast model the information provider sends a message to everyone in the network. For example, this can be used for time synchronization. Different protocols use different communication models.

Communication protocols differ in the *type of data* exchanged. In sensing, automation, and control typically *current data* are exchanged, like measurements coming from the sensors or operator data like changing a setpoint or opening/closing a

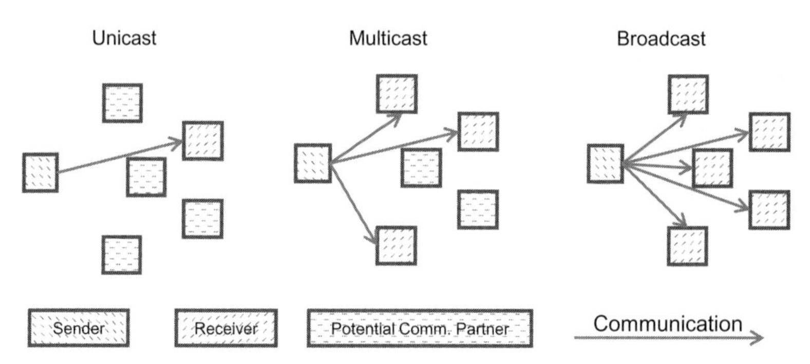

Figure 10.2. Communication models.

breaker. In addition, configuration data can be transferred, to populate a graphic display as well as configuring the measurement unit of a sensor. More advanced applications allow the exchange of *meta data*, meaning data describing the measurements, for example, by identifying the type of device used to generate the measurement, which will be presented in more detail in Section 10.1.2. Besides current data, which are typically provided per time interval, *alarms and events* provide infrequent information, for example, when a device fails. Whereas events just occur, alarms have a state and can be active or not. Operators need to acknowledge an alarm and take appropriate actions (e.g., exchange/repair the device that failed). Operators can also have the ability to add *comments*, for example, on an alarm but also as an operators note on measurements or meta data, or by providing an electronic version of an operators logbook or diary—describing events like "plant was shut down due to maintenance interval." Finally, there are *historical data*, allowing one to display or to analyze the trend of measurement values (numerical logs) as well as exposing the sequence of events that had happened in the past.

In sensing, actuating, and control, real-time characteristics play an important role. Depending on the application, it is important to have some measurements in a specific amount of time and it is important to know when each measurement was done. The *real-time support* of a protocol can deliver hard real-time, meaning the data is guaranteed to be available in a certain amount of time. That does not necessarily mean the frequency of data exchange must be very high. In theory, it could be hours or more, but typically the frequency needs to be high as well. To allow Ethernet-based protocols to provide hard real-time, specific precautions need to be made in order to avoid collisions at Ethernet level, for example, by specific hardware (e.g., switches) or making sure that no other protocols run on the same Ethernet connection. However, hard real-time constraints typically only apply on the lowest level of control, where closed-loop control needs to react very fast. When different subsystems exchange data, having a fast and reliable communication is typically sufficient. Nevertheless, also in those areas some real-time-related requirements exist. First, when real-time related data is handled, the time when a voltage was measured or some breaker was closed needs to be known. Therefore, the data needs to be connected with a timestamp. In addition, failures need to be considered. This can be communication failures, hardware failures like a broken device, or software failures when generating a calculated measurement value.

Therefore, real-time-related data needs to be associated with a quality identifying when the measured value is valid (good quality), or cannot be received anymore (bad quality), or might not be reliable anymore (uncertain quality). These are also true for historical data. On the other hand, some data exchanges do not require hard real-time or real-time-related data handling, for example, when reading the configuration of a device.

As already mentioned, *the exchange rate* or *frequency* of data exchange is important. A very high frequency is given when data is exchanged in the area of milliseconds like needed in closed-loop control when, for example, controlling a generator turning with 3000 rpm (revolutions per minute). A high frequency is in the area of seconds, for example, to provide data for an operator workstation, where some data is updated every second or every 5 or 10 seconds. Other data is exchanged infrequently, for example, alarms and events or switching a breaker. When dealing with very high frequency some protocols support oversampling or fault recording. In the first case, not every sampled data is transferred directly but several samples are transferred at once. In the second case very high frequency data is stored locally and only transferred once if a failure occurred so the application has a very high resolution log of what happened before the failure.

The *latency* is the time it takes starting from the sender having the data to send to the receiver actually receiving the data. The *throughput* is the amount of data that can be exchanged per time interval. Often the throughput can be increased by transferring larger chunks of data but this increases the average latency of the communication. Latency and throughput are not only influenced by the protocol but also by hardware (e.g., bandwidth of an Ethernet connection). Thus, the sections about specific protocols will not provide numbers or do other investigations unless the protocol does not support a reasonable latency and throughput.

The *granularity of data* can differ when exchanged by a protocol. It can have a high granularity when a single measurement value is considered (e.g., Boolean value identifying the state of a breaker, typically considered together with a timestamp and quality). Medium granularity can be considered for data containing a descriptive text (e.g., some protocols transfer an event with a descriptive string) or a number of measurements are put together (e.g., the numeric log of a measurement considering the values of the last couple of days). A low granularity is reached when some configuration files are exchanged or data representing the image or raw data of a complex measurement made by an analyzer.

The *triggering mechanisms* of the communication might differ based on the purpose. The configuration of an operator display needs to be read only once when the display is started. The measurements of the display need to be updated every x seconds, so the display needs to get them very frequently. In the first case *polling* the data, that is, having the receiver of the data request the data, is an appropriate approach. In the second case it would be preferable if the sender *publishes* the data frequently without the need of the receiver to trigger this every time, in order to reduce overhead. In addition, the client typically does not need to get data every time interval, but only if the value has changed. If the data is published every time interval, it is called *time-triggered*; if only changes are published it is called *exception-based* or event-triggered. In the case of polling you can also distinguish between time-triggered, which is typically meant by polling, and exception-based, for example, based on an action of the user on the user interface. In case of analog values (e.g., measured temperatures) the client might not be interested in every change, but actually only in significant changes as analog measurements tend to change a little bit with every measurement. For example, when measuring the outside temperature it does not matter if the temperature is 23.124 °C or 23.123 °C. But when it changes to 24.012 °C the receiver should know. In this case a deadband can be used defining what changes should be published.

Similar observations can be made for writing values. The operator will change the state of a breaker very infrequently; however, some applications might write a setpoint very frequently, for example, to optimize the startup of a boiler in a power plant. In the case of reading or writing a value only once, the best approach is to use a simple write operation. In the case of frequent writes it might be worth preallocating the resources and only reference some handles in the communication.

The *reliability* of communication can be increased by different mechanisms. On the protocol level there can be acknowledgments of received messages and rerequesting lost messages, sequence numbers to identify lost messages, heartbeats, and so on. In addition, redundancy can be used to increase reliability. Here, protocols can support failover without losing data and without the need to have two active connections.

Finally, a feature becoming more important in general and in the Smart Grid world in particular is *security* as discussed in Chapters 11–13. Whereas in the old days communication in automation was separated and not connected to the Internet at all, nowadays this is not true anymore. Systems do not only get connected via phone cable and modem for remote service operations: systems become connected to the Internet (behind a firewall). Communication protocols need to pass through firewalls and provide secure communication over the Internet. That means they need to deal with authorization and authentication. This part of security is not that hard for protocols; it is a nontrivial task from the organizational perspective. If systems from different players have to interact, user credentials or username/password need to be configured and maintained across the borders of a company. Another aspect of security is making sure that no one can manipulate messages sent from authorized users (so-called man in the middle attacks). For example, the operator might set a breaker to close and the attacker changes it to open. To avoid this, messages need to be signed so a manipulation can be identified on the receiver site. If the transferred data needs to be secured—meaning no one else can read them—the messages also need to be encrypted.

10.1.2 Information Model

The purpose of an information model is to bring semantic understanding to the data and thus exchange information rather than data by addressing the informational aspects of Figure 10.1. An information model represents concepts, relationships, constraints, rules, and operations for a chosen domain of discourse [5]. Standardizing information models brings interoperability to the next level, providing not only a standardized data exchange but an information exchange with standardized semantic.

To get a better understanding, a short theoretical view on modeling in general is given. There is a four layer meta model hierarchy defined by the OMG (Object Management Group) [6]. As shown in Figure 10.3, Level 0 contains the actual data (e.g., a specific event). Level 1 contains the model—in our context it would be called information model. The model describes the information about the data, for example, the type of an event. Level 2 contains the meta model, giving the foundation when defining a model. For example, an audit event type of the information model can be

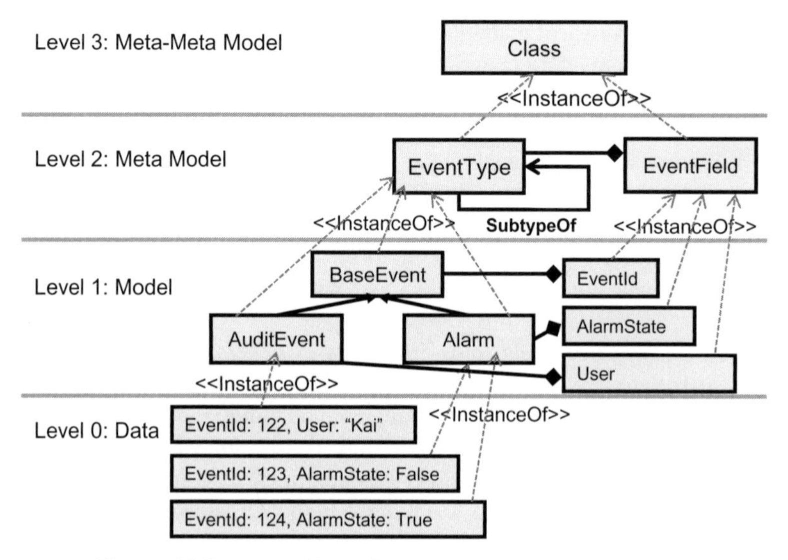

Figure 10.3. Exemplifying four-level meta model hierarchy.

an instance of a construct called event type in the meta model. Thus, the meta model has one construct called event type and the model can have several different event types, like audit event or alarm. Level 3 finally is the meta-meta model used to define meta models. In theory, one meta-meta model is enough to define all different kinds of meta models, like the Meta Object Facility (MOF) [7].

To stay on the practical level, a meta-meta model is not considered anymore and the focus is on the data with their information model and the meta model to describe information models. The second level is still important as some standards allow extending existing information models with domain- or vendor-specific information. In other words, when creating a subclass in your information model you are already extending the information model and need to understand the concept of a class and subtyping, which are defined in the meta model. Some standards, on the other hand, only provide a fixed information model and thus the meta model does not matter anymore. When taking the example of Figure 10.3 and defining the Level 1 information model as fixed, the communication does not have to consider event types having event fields and thus information from the meta model, but just needs to transfer base events, audit events, and alarms, all with a fixed set of fields. The information that a base event has an event id does not need to be queriable but is part of the protocol. When the Level 1 model of Figure 10.3 is not fixed, that is, new event types can be added having additional fields, the communication needs to transfer events in a more generic way as something having fields. In addition, the receiver of eventing information needs to know what event fields and types exist in the information model. Of course, there are some grey zones in between like having a fixed extensibility as in the example of an event field with an array of vendor-specific key-value pairs.

Looking at information models from a semantic point of view, there can be models mainly defining the structure of the data with little or no semantics. For example, they could just provide measured data with a quality and a timestamp, but no semantics on the measured data or maybe only some minor information like the engineering unit. In this case, typically the information model is fixed and cannot be extended. Thus, the protocol has a simple model. When the information model provides more information, it can either be complete, trying a "whole world approach" where every potential part of information is already put into the model, or it can be extensible, allowing one to extend the model with vendor- or domain-specific information. The advantage of the first approach is that sender and receiver know exactly what information they exchange; the disadvantage is that there is no flexibility if there is more information available than exposed in the model. The second approach provides more flexibility but requires querying the information model; thus sender and receiver can exchange the information model with which they deal. In those more advanced standards it is possible to add standardized information models based on the base standard, as done in IEC 61850 or OPC UA (see Sections 10.2.1 and 10.2.3).

10.2 PROTOCOLS AND STANDARDS

The communication protocols described below have been identified by NIST (U.S. National Institute of Standards and Technology) as standards having relevance for Smart Grids [8]. In Figure 10.4 an overview over different applications in Smart Grids is given along with potential standards to use. Power generation is done in

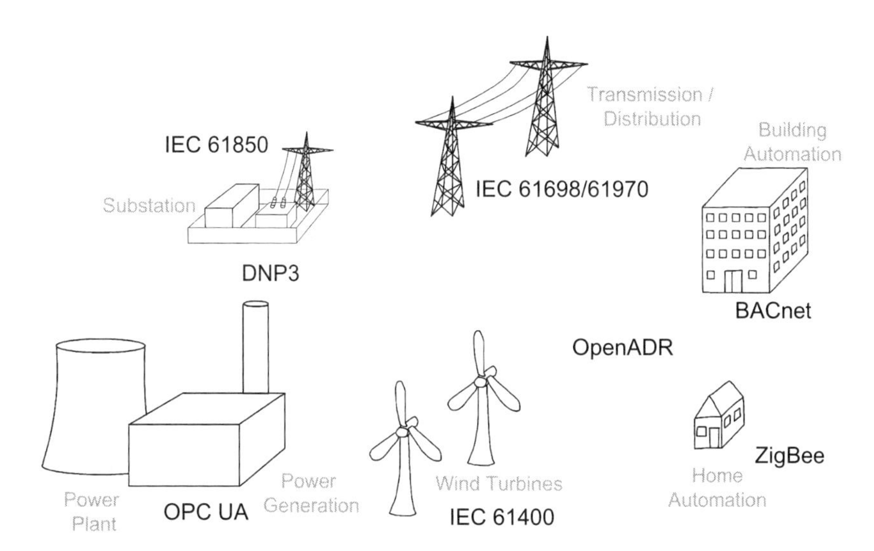

Figure 10.4. Overview over different applications and used standards.

traditional power plants running with coal, gas, or nuclear power. Here, a distributed control system (DCS) is used having controllers (PLC—programmable logical controller) controlling the process of power generation. Ethernet-based protocols that can be used in this area include OPC UA. For regenerative generation of energy like wind power the IEC 61400-25 (communications for monitoring and control of wind power plants) can be considered. The substation transforms the generated voltage and uses IEDs (intelligent electronic devices). In RTUs (remote terminal units) it allows remote access to the substation. DNP3 is often used in RTUs and IEC 61850 is an appropriate standard in substation automation. SCADA (supervisory control and data acquisition) systems can supervise the substations accessing the RTUs. For distribution and transmission the IEC 61968 and IEC 61970 play an important role. Smart Grids not only address the producing site but also the consumption on client site. ZigBee is used in home automation and BACnet in building automation. OpenADR is used as the communication protocol between utilities and consumers, providing demand response signals, so customers can, for example, reduce energy consumption if needed.

10.2.1 IEC 61850

10.2.1.1 Motivation. IEC 61850 (communication networks and systems in substations) was developed to provide interoperability between intelligent electronic devices (IEDs) for protection, monitoring, control, and automation in substations [9]. It is based on concepts and fundamental work done for the UCA (Utility Communication Architecture) [10].

IEC 61850 not only addresses communication but also information modeling, tailored to the needs of the electrical power industry. In addition, it defines an XML-based configuration language standardizing the engineering/configuration of substation automation devices.

Although IEC 61850 originally addresses substation automation, there are already additional information models defined based on the IEC 61850 model for other domains like wind turbines in IEC 61400-25 or hydroelectric power plants in IEC 61850-7-410.

10.2.1.2 Overview. The IEC 61850 consists of several parts addressing different issues. The different parts are summarized in Figure 10.5. Parts 1 to 5 give an overview and define requirements, including requirements on hardware (part 3) and engineering (part 4).

The meta model of IEC 61850 with concepts like logical nodes and data classes is defined in parts 7-1 and 7-2. The SCL (substation configuration language) defined in part 6 is used to configure data sources as well as receive information on the data sources in order to access them. The technology mapping from the abstract services to a concrete technology is given in 8-1 as well as 9-1 and 9-2. The first one defines a mapping to MMS (Manufacturing Messaging Specification, ISO 9506, protocol on top of TCP/IP) and Ethernet for Generic Object Oriented Substation Events

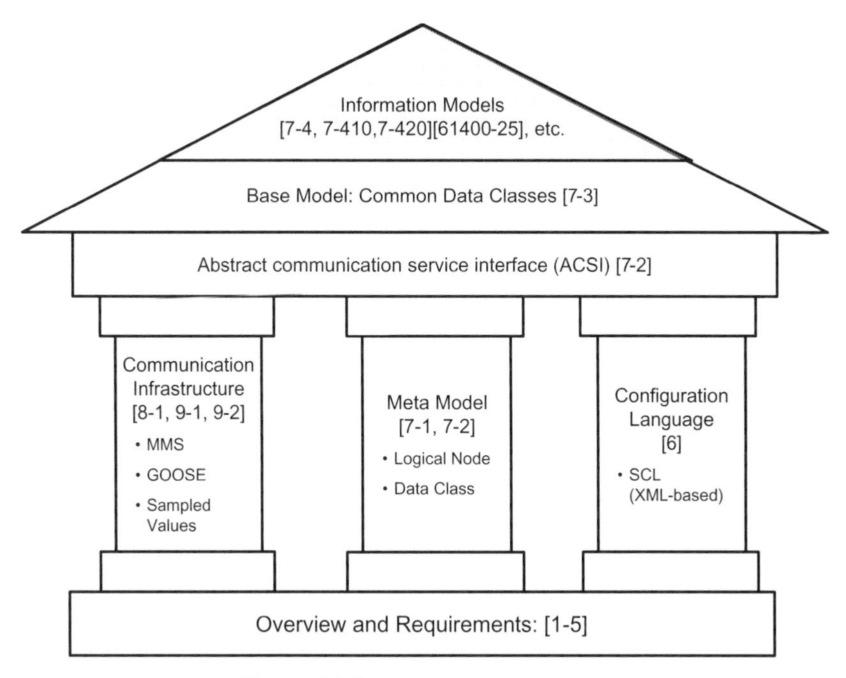

Figure 10.5. Overview of IEC 61850.

(GOOSE). Part 9-1 defines a mapping of sampled values to a serial connection and 9-2 to Ethernet. On these three pillars—communication infrastructure, meta model, and configuration language—the abstract communication service interface (ACSI) is built. The abstract interfaces provide the possibility to add or exchange existing technology mappings of the communication infrastructure. It is already planned to add a mapping to web services.

The base model defined in part 7-3 defines common data classes for status information, measurand information, analog settings, and so on. On top of these common data classes domain-specific information models are defined in terms of logical nodes and data classes. In 7-4 the domain of substation automation is addressed with logical nodes for switchgear, power transformers, and so on; 7-410 addresses hydroelectric power plants having logical nodes for turbines or hydropower dams. The domain of distributed energy resources is addressed in 7-420 with logical nodes like unit generator and fuel cell controller.

The information model of a device can be accessed online by using the appropriate services to browse and read the information or offline using the SCL of the device. Using the same mechanisms the configuration can be changed—however, it is up to the device what mechanisms it supports.

10.2.1.3 Information Model. IEC 61850 provides powerful information modeling capabilities. In IEC 61850, information can be exchanged using the ACSI (together with a technology mapping) or using a SCL-based document. With respect to the information model, the capabilities of the SCL are more powerful than using the ACSI. The ACSI focuses on concrete data exchange. In this section first the model for the ACSI is given and then the additional capabilities provided by the SCL are introduced.

The meta model of IEC 61850 that builds the foundation for IEC 61850 information models is summarized in Figure 10.6. Please note that the figure only contains those parts of the meta model that are accessible using the ACSI. The SCL targets more concepts, as will be described later in this section. The meta model contains constructs representing the information model (measured value, setpoint, sequence of events, etc.) as well as constructs representing the configuration of the communication (called information exchange model in IEC 61850-7-1 [11]), like the configuration of a buffered report control block that is used by one client or the associated applications. In this section the focus is on the information model concepts; the information exchange model is touched on in the next section (marked in Figure 10.6 in the Information Exchange Model box).

As shown in Figure 10.6, each data source in IEC 61850 starts with a server having files and logical devices. Each logical device represents a physical device. A server must have at least one device but can have many, for example, when representing a whole substation. Each logical device consists of many logical nodes, having at least a logical node 0 containing some general purpose information like the name plate of the logical device. The logical node is based on a logical node class. However, the logical node class is only referenced by name and not otherwise accessible using the ACSI. The logical node classes are defined in information model standards for the IEC 61850 like 7-4, 7-410, and 7-420 or IEC 61400-25. Logical node classes define the semantic and can be very generic like GGIO just representing generic IOs or more specific like XCBR representing circuit breaking capabilities of a switch. The logical node contains several constructs of the information exchange model and in addition at least one data object. The allowed data objects of a logical node and whether they are mandatory, optional, or constrained are defined by the logical node class. Data objects are defined by common data classes, either defined in 7-3 for the general applicable common data classes, or in the parts defining the logical node classes if a new or very specific data class needs to be defined. The second edition of 7-3 adds more common data classes based on additional information models like 7-410 and 7-420. Like the logical node classes the common data classes are only referenced by name. Common data classes are defined for various things like integer status, measured values, or analog settings. They do not define a very strong semantic; this is done for the data objects in the context of their logical nodes defined by their logical node classes. For example, the OpCnt data object of the logical node class XCBR (circuit breaker) represents the operation count of the circuit breaker. OpCnt uses the integer status data object class. For simplicity, the name of the data object actually represents the semantic and thus the OpCnt is used by several logical node classes needing an operation count. The data objects finally contain data attributes and may be nested (containing other data objects). Data attributes have a type that can be simple like an integer or Boolean but can also be complex and built from simple data types.

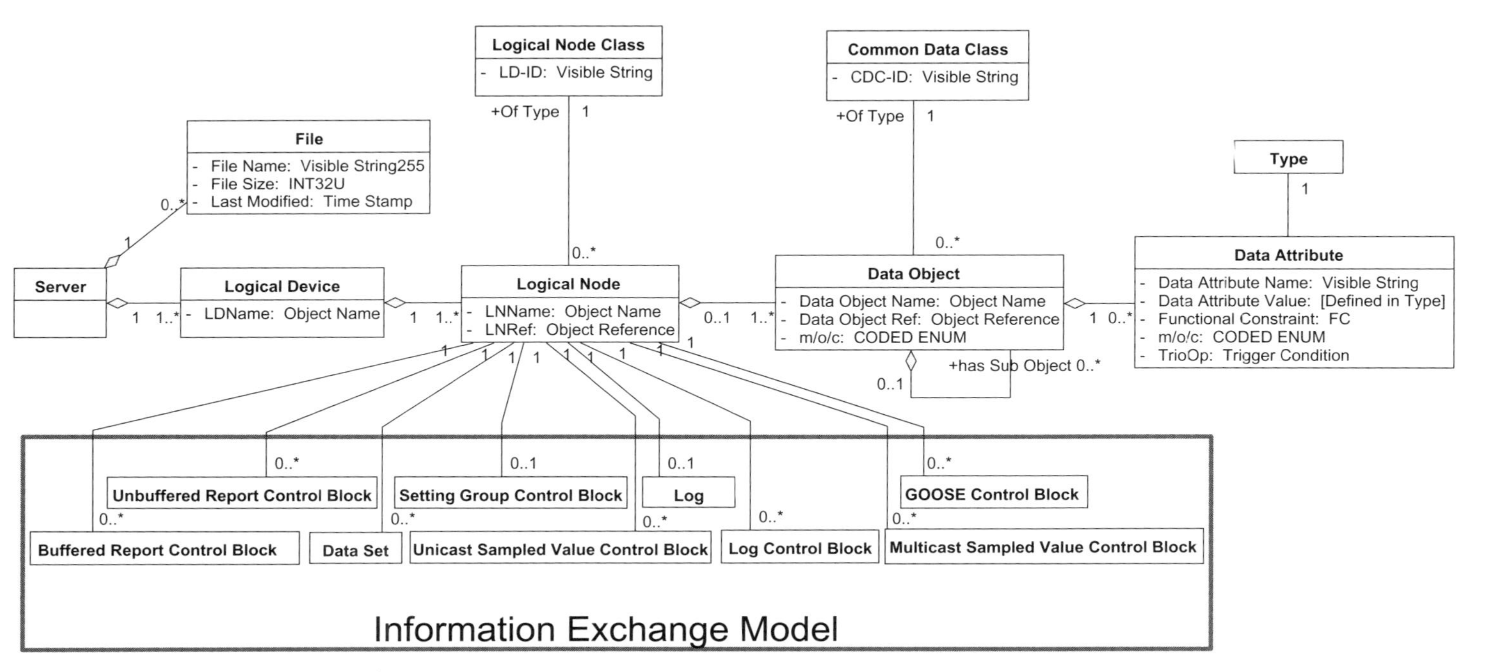

Figure 10.6. IEC 61850 model in UML notation as accessible by 7-2 ACSI.

Data objects are grouped by functional constraints. IEC 61850 defines several functional constraints like status information (ST), measurands (MX), settings (SP), configuration (CF), or description (DC). Generic user interfaces often make use of functional constraints and group the data objects based on them.

The IEC 61850 does not define mechanisms for a class hierarchy of logical nodes. However, it does define a common logical node class with mandatory, optional, and constraint data objects. Each logical node, for example, must support the mandatory data object "Beh" representing the current behavior of the logical node, that is, whether the logical node is on, off, or in test mode, as defined by the common logical node class in 7-4.

As mentioned before, the ACSI does not define mechanisms to access logical node classes. As logical node classes are defined by the specification, it first might not seem to be necessary to gain that information. However, as new information model standards are based on IEC 61850, a client built without the knowledge of those new or upcoming information model standards cannot make use of that information. In addition, the logical node classes defined in the specification contain not only mandatory but also optional data objects and also data objects that can be used several times using an index as postfix like IND1, IND2, IND3, . . . representing several indications in GGIO. As mentioned before, the IEC 61850 has no class hierarchy. However, in the SCL a logical node is defined by a template and the template can be applied several times. The template is not only defined by name but contains a definition including the data objects and reference to the logical node class it is based on by name. Thus, a client can use the SCL file to receive the necessary type of information to build applications, like defining a faceplate for a type of circuit breaker which is applied several times.

10.2.1.4 Communication. IEC 61850 provides different technology mappings for communication. In 8-1 the mapping to MMS as well as Ethernet-based GOOSE messages are defined. In 9-1 the mapping of sampled values (SVs) to a serial connection and in 9-2 the mapping of SVs to Ethernet are defined. The mapping to serial connections does not play a role in practice and will probably not be maintained in the future and thus is not considered here.

The different technology mappings fulfill different functional requirements as well as nonfunctional requirements.

- The SVs provide very fast cyclic communication of sampled values, like 4000 samples per second. As IEC 61850 is based on Ethernet without any specific real-time behavior (like reserved time slots) it does not support deterministic real-time behavior but can be considered to be very fast and provides real-time-related data on a high accuracy time-synchronized basis between different data sources. SVs can be provided by multicast, allowing several receivers to access the same data only put once on the wire or by unicast only sending the data to one receiver.
- GOOSE messages also provide a very fast communication mechanism. GOOSE messages are sent when an event occurs. GOOSE messages are sent by multi-

cast. There is no acknowledge-mechanism by the receivers of the messages. In order to guarantee that a client receives an event when something changed, the GOOSE message is sent several times in a high frequency after the change. Like SVs, GOOSE does not provide deterministic real-time-but is very fast and provides real-time-related information.

- MMS uses a client–server model and is used to browse, read, and write the information of the data source. In addition, it supports reporting capabilities (buffered and unbuffered) and access to logs containing historical data. MMS is based on TCP/IP and while not as fast as GOOSE and SVs it is still fast and provides real-time-related data.

Using Ethernet SVs and GOOSE are limited to an Ethernet segment, whereas MMS uses TCP/IP and can pass through routers and so on. However, there is already Ethernet-hardware available for passing GOOSE messages to other segments (e.g., for substation to substation communication).

In Figure 10.7 the model of accessing data in IEC 61850 is described. The data source contains data organized in logical nodes, data objects and data attributes. Those data can be directly read or written using MMS.

On top of the data, different data sets can be defined, referencing parts of the data. In Figure 10.7 *DataSet1* is referencing *DataObject2* of *LogicalNode1* and *DataAttr1* of *DataObject1* in *LogicalNode2*. When a data object is referenced, all containing attributes are included. Thus, data sets group the data into new groups; and data can be contained in several data sets. The data sets can be preconfigured

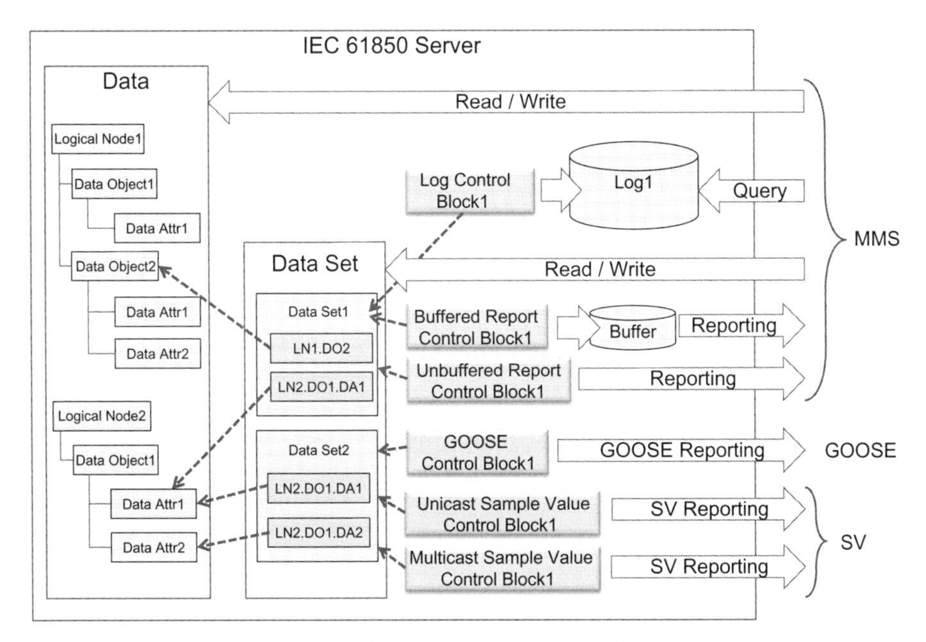

Figure 10.7. Communication model in IEC 61850.

using SCL or dynamically defined using an MMS service. Not all servers support both mechanisms.

By using MMS, the data sets can be used to read and write data. However, the more important role of data sets is to use them as the data source for different reporting mechanisms. A log control block uses the data of a data set to store the history of the data in a log. Clients can query the log using MMS. Using buffered or unbuffered report control blocks, servers send out data to a client. Here also MMS is used as the protocol. The difference between buffered and unbuffered is that in the first case the server buffers the data for the preconfigured client so that the client can receive the data later, even if it is not connected when the data should be sent. In the unbuffered case any client can subscribe to reports dynamically, but the changes that occur while the client is not connected will get lost. Buffered and unbuffered control blocks reference a data set to define what data they report. A data set can be used by several control blocks. The report control block configuration specifies when a report shall be triggered and can be set for quality changes, data changes, data updates, integrity, or general interrogation. The latter two are triggered outside the data to receive the current state from time to time (integrity) or request the state by the client (general interrogation), for example, when reconnecting. When changes trigger the report, only changed values of a data set are returned. Thus, it makes a difference whether a data set contains a data object or all attributes of a data set individually. In the first case all attributes are returned if one value changes, whereas in the second case only the changed one is returned. As the attributes can represent, for example, value, quality, and timestamp, it is important to select the data object if quality and timestamp should be received as well.

A GOOSE control block specifies what data set a GOOSE message should be referencing. It can be enabled or disabled. When enabled, it sends all data of the referenced data set using the Ethernet-based GOOSE communication.

The unicast and multicast sample control blocks define what data should be sent by the server in a periodic way (cyclic communication). When a control block references a data set all data will be sent even if they have not changed. The sampling rate defines how often the data should be sent. In the case of unicast, the control block also contains information about the client to which the data should be sent.

In addition to the above-mentioned access of the data, the MMS mapping also supports browsing capabilities to access the structure of the data (logical devices, logical nodes, data objects, etc.) and the configuration (data sets, control blocks, etc.) and to read and write files in the IEC 61850 server.

Security is not directly addressed in IEC 61850, but by referencing the IEC 62351 (Power System Control and Associated Communications—Data and Communication Security) as discussed in Chapter 12. At the moment only the MMS communication is addressed using TLS (transport layer security) as the security layer. Securing GOOSE and SV is under development. Both technology mappings can be considered to run in an Ethernet segment and thus to only provide very fast communication in local networks where security is not that important at the moment. However, these local networks will need to be secured in the future.

10.2.1.5 Status. The first version of IEC 61850 was released between 2003 and 2005. In 2010, new versions of many parts were released or are in the final state before release. These contain clarifications as well as additional content. For example, part 7-3 about common data classes contains additional data classes used in modeling other domains.

IEC 61850 is already supported in various products and applied in many applications [12]. There is a high request for IEC 61850 from customers and IEC 61850 can be seen as *the* prospective protocol in substation automation. As other specifications based on IEC 61850 are already defined addressing wind turbines, distributed energy resources, or hydroelectric power plants, IEC 61850 might become a much more widely used specification in the future.

10.2.2 IEC 61968/IEC 61970

10.2.2.1 Motivation. The IEC 61970 [energy management system application program interface (EMS-API)] defines an information model with common objects in the area of transmission to provide a common semantic. In addition, different exchange profiles are defined or under development as, for example, the profile for network application models in IEC 61970-452. The profiles are based on RDF (Resource Description Framework). The IEC 61970 also defines an abstract API for data exchange that is platform and technology independent and can be used on different operating systems and with different programming languages as well as in different database systems [13]. A web service mapping is under development in IEC 61970-502-8.

The IEC 61968 (application integration at electric utilities—system interfaces for distribution management) extends the model with focus on distribution networks and enterprise systems. Standard mechanisms for intrautility systems integration and B2B (business-to-business) integration based on standard IT integration technologies—JMS (Java message services) and Web services—are under development.

10.2.2.2 Overview. The Common Information Model (CIM) aims to allow the exchange of information of the status and the configuration in the electrical grid. It is defined in IEC 61970-301 with focus on transmission and extended in IEC 61968-11 addressing the distribution. The CIM is defined and maintained in the Unified Modeling Language (UML) and the specification and implementation artifacts are automatically derived from the UML model.

The IEC 61970-501 defines a standardized exchange format for CIMs based on RDF, which is based on XML. This allows exchanging human- and machine-readable documents. The IEC 61970-45x series defines profiles as subsets of the CIM and the RDFs for information exchange. This permits automatically checking if exchanged documents are compliant to the profile. Most profiles are under development like IEC 61970-452 (network applications) and IEC 61970-456 (solved power system state), whereas IEC 61970-453 (graphics) has already been released.

The Component Interface Specification (CIS) defined in the IEC-61970-40x series specifies abstract interfaces to exchange data—including mechanisms to exchange current data, events, and historical data. However, it does not define a technology mapping to a concrete communication protocol. In 61970-502-8, a Web service mapping is under development.

The IEC 61968 defines an Interface Reference Model (IRM) in IEC 61968-1. Based on the IRM it defines abstract interfaces in terms of message types for different purposes, like network operations (IEC 61968-3), asset management (IEC 61968-4), maintenance and construction (IEC 61968-6), customer support (IEC 61968-8), and meter reading and control (IEC 61968-9). Technology mappings are under development in IEC 61968-1-1 to JMS and in IEC 61968-1-2 to Web services.

The IEC 61968-13 defines a CIM profile for power distribution and the mapping to RDF. Like the IEC 61970-452 it is used for bulk data exchange as required for configuration (similarly as SCL is used in the case of substations), using RDF syntax. All other profiles in IEC 61968 are expressed in XML schema syntax and are used for operational exchanges among enterprise-wide systems.

10.2.2.3 Information Model. The IEC 61970 defines its information model in the CIM and derives profiles from it for different purposes of information exchange. The IEC 61968 extends the CIM.

The CIM is based on UML; that is, it uses the UML as the meta model. Thus, the CIM uses object-oriented technologies and is defined in terms of classes using inheritance, associations between classes, and so on. It focuses on the static capabilities of UML; that is, no behavior in the form of activity diagrams, sequence diagrams, and so on is defined in the model. The main concepts of the UML used in the CIM are summarized in a simplified view in Figure 10.8. Packages can be nested

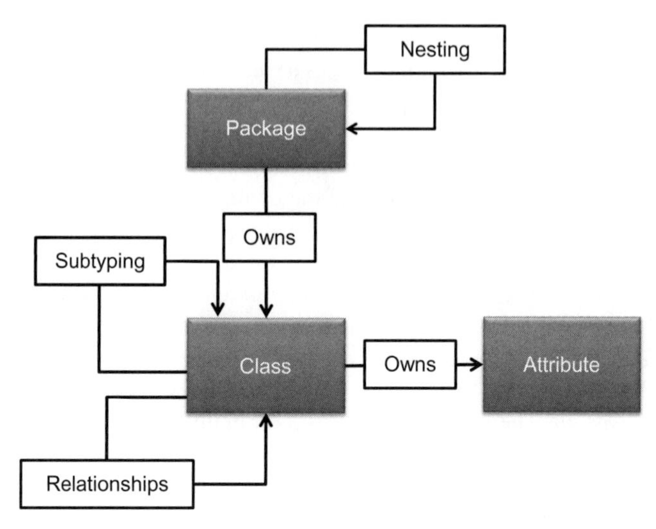

Figure 10.8. Simplified view on UML meta model used by CIM.

and contain classes. Classes have attributes and relations to other classes. Inheritance is supported by subtyping.

The CIM is organized in different packages as defined by IEC 61970-301:

- *Core*—addressing core issues like naming and equipment
- *Topology*—modeling how equipment is connected
- *Wires*—addressing electrical characteristics of networks
- *Outage*—modeling current and planned network configurations
- *Protection*—addressing protection equipment
- *Meas*—containing dynamic measurement data exchanged between applications
- *LoadModel*—addressing the system load as curves and seasons and day types
- *Generation* with its subpackages
  - *Production*—modeling production costs
  - *Generation Dynamics*—addressing prime movers like turbines and boilers
- *SCADA*—addressing data points located in remote units like RTUs
- *Domain*—serving as data directory defining data types for other packages

Additional packages have not yet been released by the IEC specification but are already included in the UML model provided by the CIM User Group [14]:

- *OperationalLimits*—defining a model for operational limits of equipment
- *Equivalents*—modeling equivalent networks
- *ControlArea*—addressing area specifications for different purposes
- *Contingency*—modeling contingencies

The IEC 61968-11 adds the following packages addressing distribution: Common, Assests, AssetsDetail, AsssestsModels, WiresExt, Work, Customers, Metering, and PaymentMetering.

Each package contains classes with attributes and relationships between the classes, also to classes of other packages. In addition, the classes span a class hierarchy.

In Figure 10.9 an excerpt of the CIM is shown modeling a steam turbine defined in the Generation Dynamics package. The *SteamTurbine* class inherits from *PrimeMover* and thus from *PowerSystemResource* and *IdentifiedObject* defined in the Core package. The *IdentifierObject* defines attributes for identification purposes like name. The *PowerSystemResource* adds semantic and several associations to other classes like *Measurement* (not shown in the figure). The *PrimeMover* defines an association to *SynchronousMachine* of the Wires package (attributes not shown in the figure) and defines an attribute *PrimeMoverRating*. The *SteamTurbine* finally defines several attributes describing the time constraints and pressure outputs of the turbine. In addition, an association to the *SteamSupply* is given.

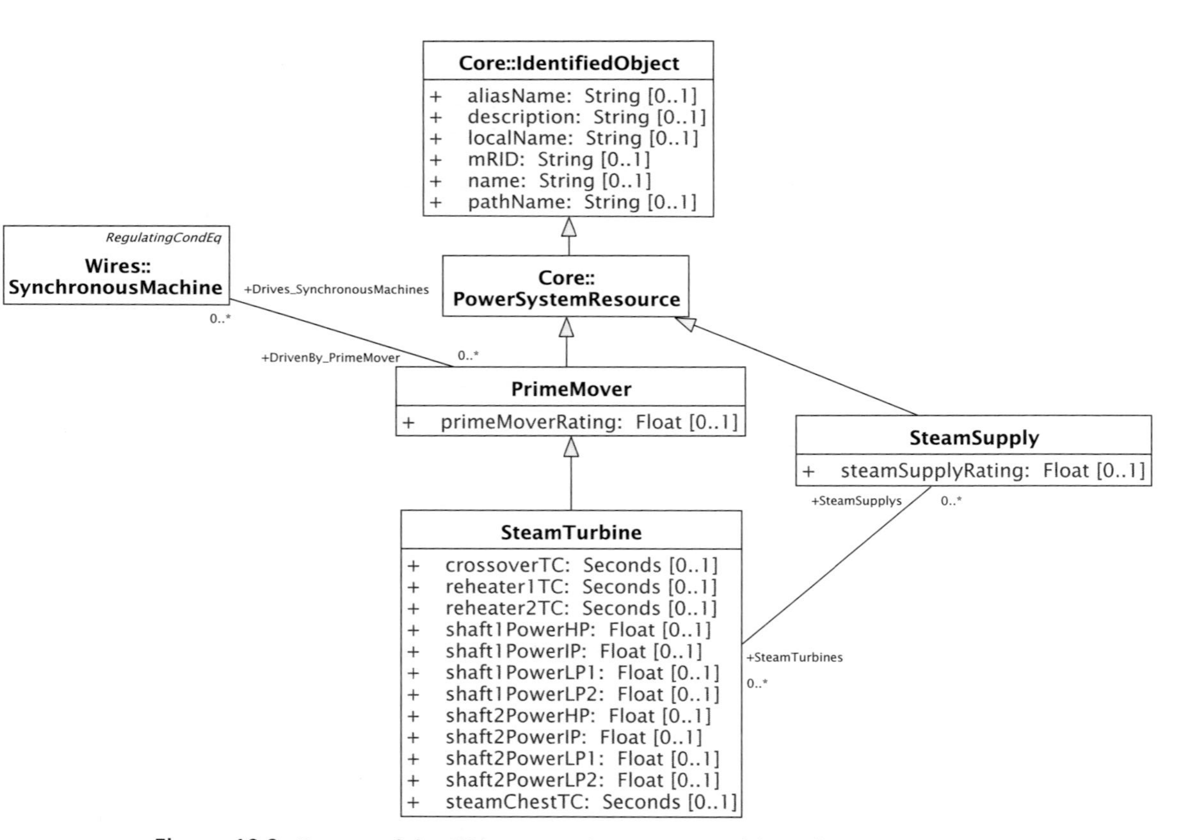

Figure 10.9. Excerpt of the CIM representing a steam turbine using UML notation.

In the current model [14] the IEC 61970 part of the model defines more than 300 classes and the IEC 61968 more than 200 classes. Thus, the different profiles are very important to address only the required subset when exchanging information.

In a way the CIM is a closed world model—new packages and classes as well as additional attributes and references to existing classes need to be added to the standard using the IEC processes. However, it is allowed to extend the local model with non-CIM packages and exchange this information using the standardized exchange mechanisms. Thus, vendors can add additional information to the model when needed.

10.2.2.4 Communication. CIM-based data exchange can be done to exchange model information—the format of the model is standardized and uses XML [15]. The mapping to XML is defined in IEC 61970-501 using RDF syntax. Thus, the exchange can be done file-based and any file-based exchange protocol like FTP (file transfer protocol) can be used. Profiles are defined like IEC 61970-453 for CIM-based graphics exchange or under development like IEC 61970-452 defining the equipment model and IEC 61970-456 defining a model for state variables, measurements, and topology. The same approach is used for IEC 61968 as defined in IEC 61968-13.

In addition, data described in the CIM can be exchanged. Therefore, the CIS defines abstract interfaces for the following purposes:

- Generic data access (GDA) in IC 61970-403
- High speed data access (HSDA) in IEC 61970-404
- Generic eventing and subscription (GES) in 61970-405
- Time series data access (TSDA) in 61970-407

The functionality of those interfaces allows browsing, reading, and writing as well as exception-based subscriptions of current data, historical data, and alarms and events. This functionality is very similar to classic OPC (Microsoft COM/DCOM based interfaces). The specification references classic OPC as potential implementation technology as well as DAIS DA, DAIS AE, and HDAIS (CORBA-based specifications similar to classic OPC). A Web service mapping is currently under development in IEC 61970-502-8, but in general the focus of IEC 61970 is more on the bulk-like data exchange using the different profiles of the 45x series.

The IEC 61968 defines its own set of abstract interfaces tailored to different aspects of the distribution domain. Based on the purpose (network operations), a subset of the functionality provided by the IEC 61970 is provided by the IEC 61968, tailored to more specific interfaces like sending outage records, whereas IEC 61970 CIS is more generic and provides interfaces to read values or sending events.

In IEC 61968-1-1 a technology mapping of those interfaces to JMS and in IEC 61968-1-2 to Web services is under development. Implementations of those draft specifications have already been interoperability tested in 2010 [16].

10.2.2.5 Status. The release of IEC 61970 started in 2004 with the glossary (IEC 61970-2) and continued until 2008 having all 4xx parts mentioned in the sections before. Other parts are still under development, like IEC 61970-502-8 defining the Web service mapping. The release of IEC 61968 started in 2003 with parts 1 and 2 and the latest release was the CIM (part 11) in 2010. Other parts are under development, like 61968-1-1 defining data exchange using JMS and 61968-1-2 using Web services. The standards are starting to become more widely implemented [8].

10.2.3 OPC UA

10.2.3.1 Motivation. Classic OPC*—the predecessor of OPC UA—is well accepted and applied in industrial automation. There are over 22,000 OPC products on the market from more than 3500 companies, including all major automation vendors [17]. As a consequence, almost every system targeting industrial automation implements classic OPC. With its flavors Data Access (OPC DA), Alarms and Events (OPC A&E), and Historical Data Access (OPC HDA), classic OPC provides interoperability between automation components like controllers and HMI (human–machine interface) running in Microsoft Windows environments.

OPC Unified Architecture (OPC UA, [18, 19]) unifies these existing standards and brings them to state-of-the-art technology using service-oriented architecture (SOA). By switching the technology foundation from Microsoft's retiring COM/DCOM to Web service technology, OPC UA becomes platform-independent and can be applied in scenarios where classic OPC cannot be used today. OPC UA can run directly on controllers and intelligent devices having specific real-time-capable operation systems, where classic OPC would need a Windows-based PC on top to expose the data. It can also be seamlessly integrated into Manufacturing Execution Systems (MES) and Enterprise Resource Planning (ERP) systems running on Unix/Linux using Java applications and still fits very well in the Windows-based environment where classic OPC lives today.

Security is built into OPC UA by authentication mechanisms and signing and encrypting messages as security requirements become more and more important in environments where automation is not running separated in an isolated environment but is connected to the office network or even the Internet and attackers start to focus on automation systems [20]. OPC UA provides a robust and reliable communication infrastructure having mechanisms for handling lost messages, failover, heartbeat, and so on. With its binary encoded data it offers a high-performing data exchange solution.

OPC UA scales very well in different directions. It can be applied on embedded devices with limited hardware resources as well as on very powerful machines like mainframes. Of course, an application running on limited hardware can only provide

*OPC stands for OLE for process control. Today it is no longer used as an acronym as modern OPC specifications do not rely on Microsoft's OLE (and thus COM) technology anymore.

a limited set of data to a limited set of partners, whereas an application running on high-end hardware can provide a large amount of data with several decades of history for thousands of clients. In addition, the information modeling capabilities scale. An OPC UA server might provide a very simple model or a very complex model depending on the application needs. An OPC UA client can make use of the model or only access the data it needs and ignore the meta data accessible on the server.

With its information modeling capabilities, OPC UA offers a high potential for becoming the standardized communication infrastructure for various information models from different domains. Several information models are already defined based on OPC UA making use of the generic and powerful meta model of OPC UA. OPC UA has built-in support to allow several different standardized information models to be hosted inside one OPC UA server.

10.2.3.2 Overview. OPC UA defines the communication infrastructure and the meta model for defining OPC UA based information models. Based on the meta model, the OPC UA services define operations like read, browse, and write, which are transferred via the communication infrastructure (see Figure 10.10). The base OPC UA Information Model defines the base types and is refined by Information Models for Data Access, Historical Access, Programs, and Alarms and Conditions. In addition, vendor-specific information models or standardized information models can be put on top of the base information model or derived information models.

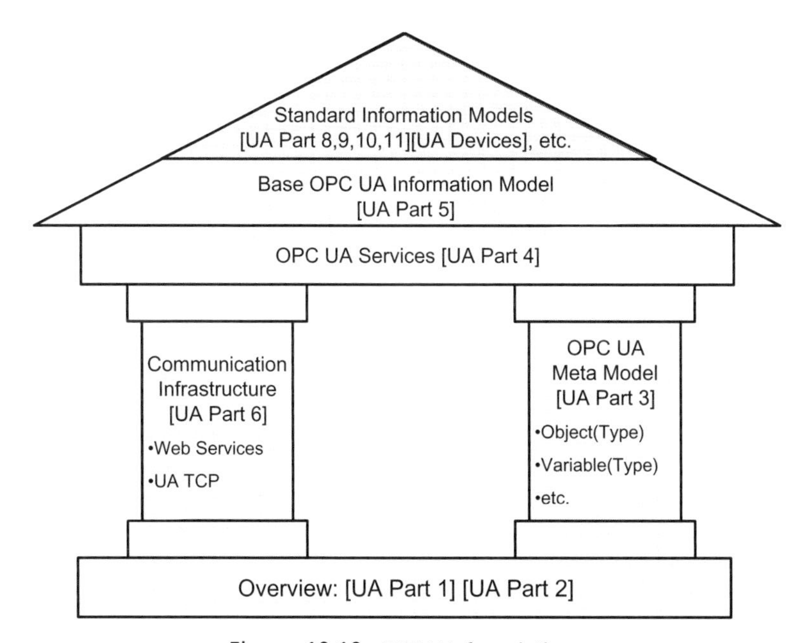

Figure 10.10. OPC UA foundations.

In the communication architecture OPC UA offers technology mappings to HTTP/SOAP based Web services as well as a mapping to its own protocol on top of TCP/IP called UA TCP. Data can be encoded using XML for better interoperability to pure Web service application, but it can also be binary encoded for high-performing data exchange.

OPC UA applications can use communication stacks available for different programming languages provided by the OPC Foundation or commercial software development vendors. Alternatively, when using the Web service mapping, it is possible to generate the communication stack based on the WSDL (Web service description language) of OPC UA. With the concept of abstract service definitions it is easily possible to add another technology mapping when a new outstanding base communication technology appears in the future. This would only require extending the stacks supporting another technology mapping, not reengineering the applications using the stacks.

10.2.3.3 Information Model. OPC UA provides very powerful information modeling capabilities. The base principles of OPC UA information modeling are [18]:

- *Using object-oriented techniques including type hierarchies and inheritance.* Typed instances allow clients to handle all instances of the same type in the same way. Type hierarchies allow clients to work with base types and to ignore more specialized information.
- *Type information is exposed and can be accessed the same way as instances.* The type information is provided by the OPC UA server and can be accessed with the same mechanisms used to access instances.
- *Full meshed network of nodes allowing information to be connected in various ways.* OPC UA allows supporting various hierarchies exposing different semantics and references between nodes of those hierarchies. The same information can be exposed in different ways, providing different ways to organize the same information depending on the use case.
- *Extensibility regarding the type hierarchies as well as the types of references between nodes.* OPC UA is extensible in several ways regarding the modeling of information. Besides the definition of subtypes it allows one—for example—to specify additional types of references between nodes and methods, extending the functionality of OPC UA.
- *No limitation on how to model information in order to allow an appropriate model for the provided data.* OPC UA servers targeting a system that already contains a rich information model can expose that model "natively" in OPC UA instead of mapping the model to a different model.
- *OPC UA information modeling is always done on the server side.* OPC UA information models always exist on OPC UA servers, not on the client side. They can be accessed and modified from OPC UA clients. An OPC UA client is not required to have an integrated OPC UA information model and it does not have to provide such information to an OPC UA server.

This allows providing very simple as well as very complex and powerful information models. The base concepts—the meta model of OPC UA—is summarized in Figure 10.11. A node can be connected by references. Each node has attributes like a name and id. There are different node classes for different purposes, for example, representing methods, objects for structuring the address space, or variables containing current data. Each node class has special attributes based on their purpose. The variable, for example, contains a value attribute.

OPC UA supports simple data types for current data like integer and Boolean, but also enumerations, and structured data types composed of other data types. Each value can be a (multidimensional) array with specific mechanisms for accessing parts of an array only. The set of supported data types is extensible. Clients can get the encoding of the data type from the server to interpret the data on the fly.

Events in OPC UA have several fields like *id*, *timestamp*, and *message*. Unlike other discussed protocols, OPC UA as well as classic OPC allow putting localized text messages in the event. This simplifies the implementation of an alarm or event list in the HMI but, on the other hand, increases the amount of exchanged data. OPC UA provides an extensible event type hierarchy using subtypes for categorizing the events and to add additional event fields. Clients can filter on the categories and specify the event fields they want to receive.

OPC UA allows standardized information models based on OPC UA. In Figure 10.12 an example is given. The base is the base OPC UA Information Model, providing base object and variable types and in addition a folder type to organize the address space. The data access information model refines the base variable type providing places to put information about engineering units or precision of a variable. The devices information model defines the structure of how to represent devices, for example, by defining how parameters of a device are structured. The analyzer devices model defines classes of analyzer device types and their parameters and methods. Finally, a server-specific information model provides concrete analyzer device types representing a concrete analyzer type of a specific vendor.

The drawing in Figure 10.12 represents the standardized notation of OPC UA to document OPC UA information models, having rectangles for objects, shadowed rectangles for object types, and so on. The references between nodes also have a specific semantic representing different reference types, for example, for subtyping.

The devices model was initially developed to describe devices in the context of field device integration (FDI) [21]. It was adapted by the working group defining analyzer devices and later also by a combined working group with PLCopen to define an OPC UA based information model of IEC 61131-3 (Programmable Controllers—Part 3: Programming Languages) languages.

10.2.3.4 Communication. OPC UA communication is based on a client–server model. That means that a client is initiating the communication and after a communication channel and session is created, the client requests and the server responds to the requests. However, OPC UA provides a publishing model where the server is able to immediately respond to changed data or abnormal conditions (e.g., alarms) and to send those data to the client [18].

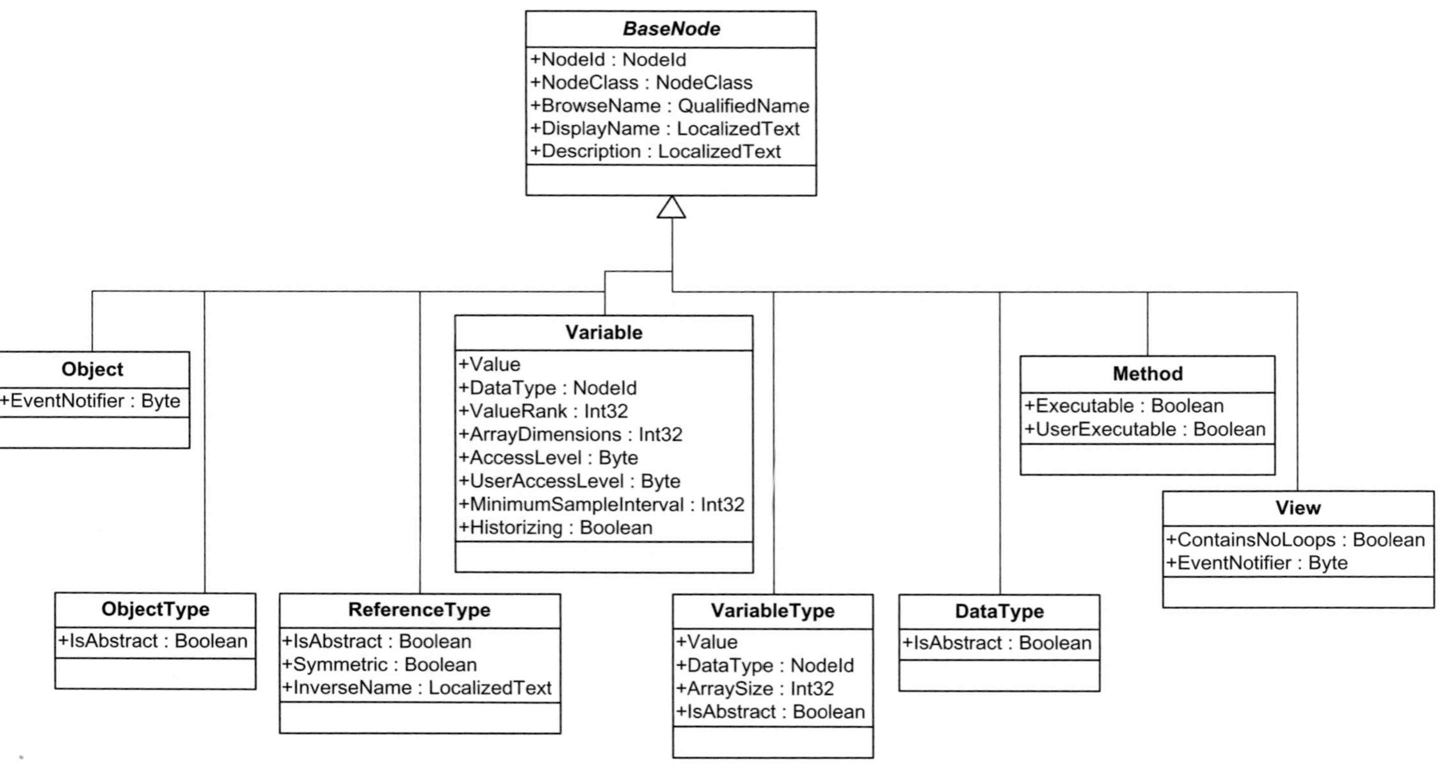

Figure 10.11. OPC UA meta model.

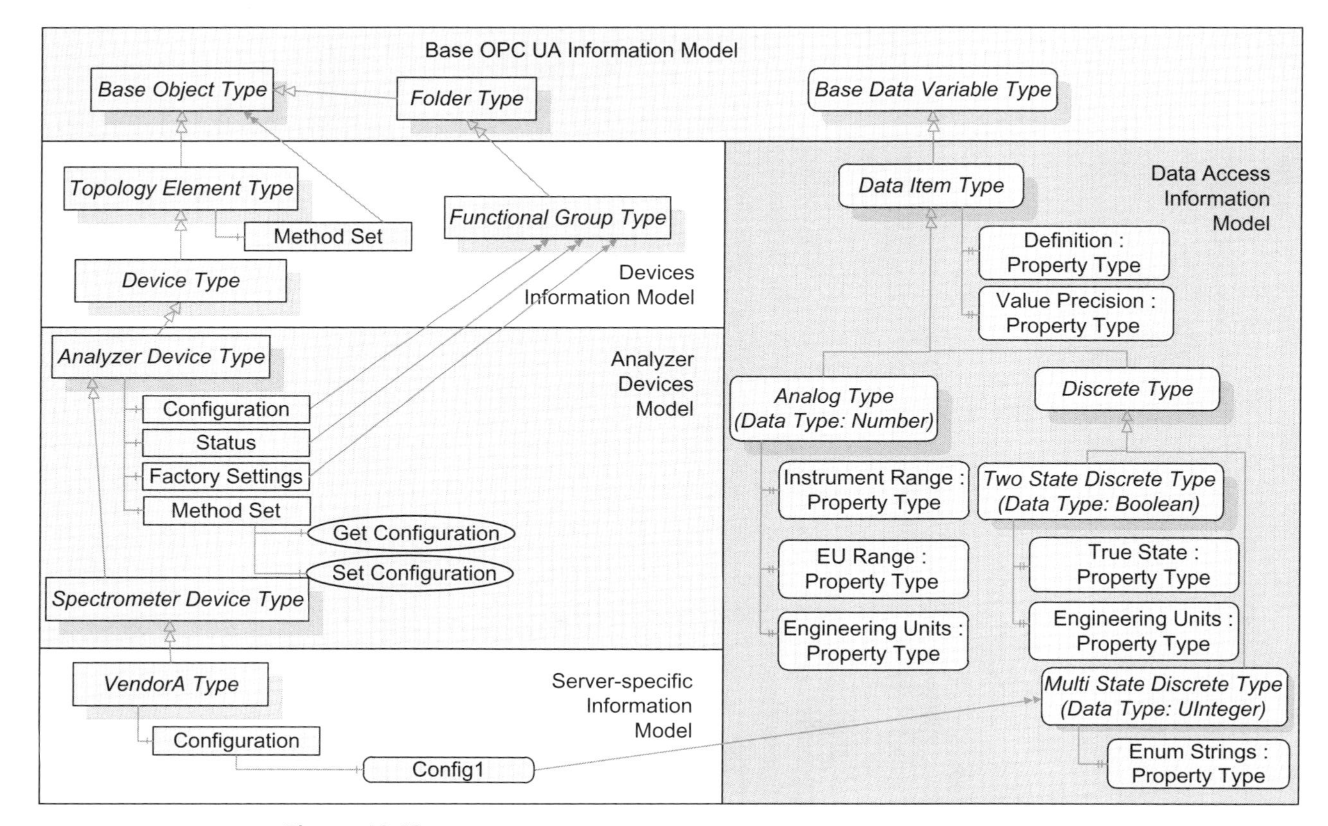

Figure 10.12. Derived standardized information models based on OPC UA.

OPC UA allows exchanging current data, alarms and events, and historical data for both, adding comments and in general calling methods, as well as meta data. OPC UA does not support deterministic, hard real-time, but it has built-in support for real-time-related data by providing a status (quality) and timestamp for current data and alarms and events as well as historical data.

To access the meta data it allows browsing and querying the address space and reading the attributes. Optionally the meta data of a server can be manipulated by a client by writing attributes and changing the structure of the address space (adding/deleting nodes and references). This is done by OPC UA operations. Currently, there is no standardized configuration file format like SCL in IEC 61850 to reconfigure an OPC UA server.

Current data can be read and written (e.g., setpoints). For frequent updates of the current data polling can be used by frequently reading the data. However, the more appropriate approach is the publish mechanism, where the server reports changes of the data. Clients can specify a deadband and a sampling interval indicating the highest frequency they want to receive on changes. OPC UA always works exception-based, that is, only changed data is transferred to reduce the amount of data transferred. By providing a reliable communication infrastructure, clients know that the last value received is still valid as long as the connection was not broken, which would be indicated to the client. OPC UA supports oversampling. That means that the client does not get every change immediately but several changes in a lower frequency at once. For example, the sampling rate can be a few milliseconds and the client only receives changes every 500 milliseconds.

Alarms and events are also received using the publish mechanism, allowing filtering mechanisms on event categories and values and defining what event fields should be returned. In addition, the alarm state can be read from the address space. Alarms can be acknowledged and commented using the generic method call mechanism of OPC UA. OPC UA provides advanced operations on historical data. It allows accessing the raw data as well as aggregated data for current data and allows updating the history. This is helpful, for example, when some historical data have a bad quality but complex optimization algorithms need realistic values. In that case, simulated data can be put into the history replacing the bad quality data. Historical events can be accessed as well, supporting the same filter mechanisms as for accessing current events.

The duration of the stored history can vary. It can be only some buffered values for a very short time period up to several decades. Historical data are stored and can be accessed by several clients. Only access rights may restrict the accessibility. The server stores the history independent of clients connected to the server. In addition, OPC UA provides buffering data requested by a client. This allows a client to subscribe to some values, disconnecting from the server and later (e.g., a day later) reconnecting to receive the buffered data.

Taking the above described functionality into account, the granularity of exchanged data can go from high, reading a single value, over medium, receiving several changed values at once, to low, subscribing to events or accessing the history.

However, the intention of OPC UA is to avoid a high granularity and thus reduce the number of round trips.

10.2.3.5 Status. OPC UA was released in 2009 in a version containing all information needed to implement interoperable OPC UA products after having the first parts of OPC UA released in 2006. OPC UA is new on the market and only a few hundred products have been released. However, there are not only stacks and SDKs provided by the OPC Foundation but also commercial SDKs on the market so that the OPC UA product base is expected to grow fast. OPC UA is in the state of being published as an IEC specification (IEC 62541). Three parts have already been released; others are in their final state.

Looking at the history of OPC, it was developed coming from the field of process automation. Nevertheless, classic OPC is applied in the power domain as well. In the NIST release [8], it is considered as the only standard from process automation suitable for Smart Grids. With its technology based on SOA and security built-in and additionally being capable to communicate over the Internet and passing firewalls, it has the potential to be highly applied in Smart Grids. By creating standardized Information Models based on OPC UA suitable for Smart Grid applications, this potential can be even increased. Using wrapper and proxy solutions provided by the OPC Foundation it can already be integrated into any classic OPC application like HMIs.

10.2.4 DNP3

10.2.4.1 Motivation. DNP3 (or DNP 3.0) was developed to achieve open, standards-based interconnectivity between substation computers, remote terminal units (RTUs), intelligent electronic devices (IEDs), and master stations for the electric utility industry [22].

10.2.4.2 Overview. Communication partners in DNP3 are called master and outstation. The outstation provides data and the master receives the data using a peer-to-peer connection. A master can communicate with several outstations and be the outstation for another master, aggregating the data of its outstations.

DNP3 not only supports data acquisition but also executes commands like close or trip a circuit breaker, or set analog output values to set a regulated pressure or define a desired voltage level. In addition, it provides the possibility for file transfer and time synchronization.

10.2.4.3 Information Model. The protocol itself does not provide a readable or browseable information model. However, DNP3 defines documentation describing the functionality of a DNP3 device, the DNP device table and the DNP implementation table. The first one defines the principal protocol functionality a device supports, like writing analog values. The second specifies the objects a device supports [23]. However, those documents are not machine readable [8] and thus masters must be configured by hand based on the documentation.

10.2.4.4 Communication. DNP3 can run over a serial connection or TCP/IP. It supports splitting messages into several fragments, error detection, addressing, and so on to exchange messages between receivers and data sources.

The preferred way to exchange frequently changing data is polling; that is, the receiver reads the data from the data source. However, to notify the receiver about unsolicited responses, the data source can spontaneously transmit a response without having received a specific request [24].

By having different classes the receiver can request all data of a data source (class 0) and later only request changes of class 1–3 data and thus reduce the amount of exchanged data. The classes 1–3 also provide a way to prioritize more important data. Thus, the granularity of transferred data can vary, going from a high granularity where one or a few values are read in one request to a low granularity when files are exchanged or all data is read at once (class 0 requests).

DNP3 further distinguishes between static data and event data. The main difference is that event data is always associated with a flag containing status information like "data source restarted," "source is started," or "communication is lost with an underlying data source." Static data can be associated with a flag. Events can be associated with a timestamp, static data cannot. Events are buffered and not deleted before confirmed from the master.

10.2.4.5 Status. DNP3 has a high market penetration, especially in North America, but also in South America, Australia, and China [25]. Developed in the early 1990s by GE Harris Energy Control Systems, ownership of the specification as well as further development was handed over to the DNP User Group. Without security features, DNP3 does not fit into Smart Grid scenarios. Therefore, an IEEE working group was created in cooperation with the DNP3 Users Group. This resulted in creating the IEEE 1815 specification in July 2010. It provides a secure DNP3 based on IEC 62351-5 (see Chapter 12) and, in addition, provides device data profiles in a format that can be mapped to IEC 61850 object models [26].

10.2.5 BACnet

10.2.5.1 Motivation. BACnet (building automation and control networks) [27] was developed to provide a common, vendor-neutral communication standard in the area of building automation. BACnet not only addresses communication over different mechanisms like twisted-pair cables, Ethernet, IP, or LON (see Section 10.2.8 for a summary on LON), but also defines an information model with object types in the area of building automation like accumulator, pulse convertor, and calendar.

10.2.5.2 Overview. The communication partners in BACnet are called BACnet-user. When the BACnet-user acts as a client it is called a requesting BACnet-user and—when acting like a server—it is called a responding BACnet-user. A device can act as server as well as client, requesting services from other BACnet servers. BACnet defines profiles for different types of devices: operator workstations, various controllers (e.g., building controller), smart sensors, and smart actuators.

The BACnet interoperability building block (BIBB) defines a set of BACnet services and devices can comply with them. There are categories for data sharing, alarm and event management, scheduling, trending, and device and network management.

A device supporting BACnet provides a PICS (protocol implementation conformance statement) describing the functionality of a device, including a list of supported BIBBs and a list of standardized and proprietary object types it supports. This includes optional properties for the object types and whether object types can be dynamically created and whether properties are restricted in their range of values and writable using BACnet services. The PICS also defines the supported protocol options.

10.2.5.3 Information Model. BACnet defines standard object types for different functionalities in the area of building automation like accumulator, pulse convertor, and calendar. Each standardized object type can have optional and mandatory properties. For the mandatory properties, it also defines whether they are readable or writable. As an example, the present value of the analog output object type is readable and writable, the out of service property is readable, and the description property is optional. Each device can support a fixed set of object types, and—depending on the device—objects of those types can dynamically be added using BACnet services.

The information model is extendable and allows adding properties to standardized object types as well as defining vendor-specific object types. However, there is no support for inheritance of object types.

The standardized object types have different purposes. Whereas the accumulator object represents "the externally visible characteristics of a device that indicates measurements made by counting pulses" [27], a trend log object is used as a kind of protocol-specific object storing historical data. Thus, there are no special services for accessing the logs but rather common services reading ranges of an array of log records. Each record contains a timestamp, status (quality), and the value (called LogDatum).

10.2.5.4 Communication. BACnet supports unicast (client–server), multicast, and broadcast communication models. It supports alarms and events, exception-based reporting of changed values, and reading and writing the information of a data source (device). In addition, it allows adding or removing objects of a device. The granularity of a read and write can be high, accessing only a single value, or low, accessing multiple values with one request. Time synchronization is built into BACnet. Discovering the objects of a device can be done by so-called Who-Has services that get answered by I-Have services. In addition, the PICS file indicates the supported object types.

The alarm and event model allows acknowledging of alarms and provides timestamps and optional text messages for events. It is possible to ask for all active alarms. The communication can be done using confirmation of messages or not. BACnet is also extensible. It allows adding properties to a standardized object type as well as adding vendor-specific object types. In addition it allows triggering vendor-specific services in a standardized way. BACnet allows the transfer of files.

Security is supported in BACnet for client–server communication. Authentication is done by handshaking and exchanging a password and messages are enciphered using a session key. However, the encryption of messages does not work for multicast and broadcast.

10.2.5.5 Status. BACnet was developed by the American Society of Heating, Refrigerating and Air-Conditioning Engineers (ASHRAE), starting development in 1987. In 1995 BACnet became ANSI/ASHRAE specification 135 and since 2003 it is also ISO specification 16484-5. Already, 468 vendor ids have been issued for BACnet [28].

10.2.6 OpenADR

10.2.6.1 Motivation. OpenADR (open automated demand response) was developed to provide a common information exchange between utilities or independent system operators and electric customers using demand response price and reliability signals. It should improve optimization between electric supply and demand by providing interoperable signals to building and industrial control systems that can automatically take action based on a demand response signals (e.g., reduce energy consumption if needed) [29].

10.2.6.2 Overview. OpenADR is a technology-neutral communication specification providing a standardized information model in the area of demand–response. The specification relies on Web services, WSDL, SOAP, and XML. This is the base for technology-neutral communication, allowing different implementations in different programming languages on different platforms.

An important role in the OpenADR architecture is played by the demand response automation server (DRAS), providing an operator interface for utilities and independent system operators on the one hand and a client interface as well as an operator interface for consumers on the other hand. Depending on the deployment scenario, not all interfaces are needed and can be exchanged by proprietary mechanisms. For example, when the DRAS is integrated into the utilities infrastructure, it does not need to use the standardized operator interface.

10.2.6.3 Information Model. OpenADR defines an information model for demand response between electric customers and utilities. It contains utility programs including bidding information and event information. In addition, it addresses transaction logs and logs of client statuses and alarms. Whereas most structures are well defined and fixed, the event information is typed and can be specified by the utilities. Clients can access the event information types by accessing the information about the programs and thus get the structure of the event information they will receive. However, the event information type just defines the structure of events. There are no object-oriented mechanisms like inheritance provided in the information model.

Alarming is addressed in OpenADR to a certain degree, providing an alarm log and specifying that operators must be informed about alarms, but there is no acknowledgment mechanism for alarms specified.

10.2.6.4 Communication. The communication is based on Web services and thus on a client–server model. The eventing mechanism allows a push and a poll model. In the case of push, the receiver of the event needs to act as a Web service and the provider of the event is acting as a client sending the events as requests. In the case of poll, the receiver acts as client to the Web service of the provider of the event. Thus, OpenADR allows polling as well as exception-based communication.

The characteristics of the communication are different for the different interfaces of OpenADR. The pure eventing client interface is just to exchange events exception-based or by polling. The client operators interface allows one to read and write bids and receive information about programs and event information types. It also supports defining settings for the eventing (e.g., opt-out mechanisms), setting test mode, and retrieving the log information of client statuses, transactions, and alarms. OpenADR does not address measurements and thus only the events contain a timestamp. However, bids, for example, also contain information about time in terms of their duration (start and end time).

Since the information model and thus the concepts of information the communication is dealing with are well defined, the access is done in specialized operations (e.g., GetBid) rather than in generic operations like browse. The granularity is middle to low; that is, there is typically a set of data accessed at once.

The exchange rate depends on the communication mode. In the polling case the rate can be configured and in the pushing case the rate is determined by the occurrence of new events. Thus, the frequency can be high or low—but not very high based on the used technology (Web services).

Security is addressed by using TLS (transport layer security—see Chapter 12) as the security layer and reliability by sending confirmation messages in a specified time-out period.

10.2.6.5 Status. Development of OpenADR started in 2002 following the California electricity crisis. The work has been carried out by the Demand Response Research Center (DRRC), which is managed by Lawrence Berkeley National Laboratory (LBNL) [29]. OpenADR has now been contributed to OASIS and the UCA International User Group. Together with the LBNL those organizations further develop the OpenADR specification [30].

Although OpenADR started out being used primarily in California [8], in general, OpenADR is starting to come into use by utilities and service providers to issue pricing signals and load control events [31].

10.2.7 ZigBee

10.2.7.1 Motivation. ZigBee is a wireless communication protocol built on top of the IEEE 802.15.4 [IEEE Standard for Information Technology—Telecommunications and Information Exchange Between Systems—Local and Metropolitan Area Networks—Specific Requirements Part 15.4: Wireless Medium Access Control (MAC) and Physical Layer (PHY) Specifications for Low-Rate Wireless Personal Area Networks (WPANs)] as detailed in Chapter 6. It targets low-power, low-cost, sensor networks in the area of home automation. It defines a network layer and in addition application profiles. Those application profiles address different domains, including home automation, health care, telecom services, and in the context of Smart Grids the most important domain of smart energy. The smart energy application profile addresses advanced metering, including load profiles and real-time consumption, demand/response and load control, pricing, and text messages.

10.2.7.2 Overview. ZigBee is based on IEEE 802.15.4 and is a wireless communication protocol that is self-organizing and allows connecting low-power devices with small battery-like wireless light switches. There are also plans to integrate ZigBee with IP [32]. The bandwidth of ZigBee is limited due to the addressed environment. Details on physical and medium access communication layers are discussed in Chapter 6.

ZigBee distinguishes three types of devices: coordinators, routers, and end devices. One coordinator per ZigBee network is the root and bridges the network to other networks, and coordinates the network including security handling in the form of being the trust center for security certificates. The router acts as a router passing messages and might take the role of a coordinator if the coordinator fails. The end device finally provides functionality, for example, like a meter.

10.2.7.3 Information Model. The smart energy profile defines a set of devices, including energy service portal, metering device, programmable communicating thermostat, load control device, and smart appliance.

The information model in ZigBee is defined in the application profile. As ZigBee addresses communication in low-power devices with limited resources, handled data is reduced to a minimum and much information is put into bitmasks. Thus, the information model of an application profile is fixed and cannot be extended. However, ZigBee can be extended in terms of information models by defining new application profiles.

In the smart energy application profile, different types of data are addressed: current data, for example, for metering information, events for different purposes like load control, alarms for the general health status of devices, and historical data on the consumption of energy. There are no concepts for acknowledging alarms or comments. Meta data can only be exchanged to a certain degree, for example, when requesting information about the profile.

PICS (protocol implementation conformance statement) defines the compliance of devices to an application profile and the supported features. PICS is typically provided as files, but those files do not contain machine-readable information.

10.2.7.4 Communication. ZigBee supports the unicast (peer-to-peer and client–server), multicast, and broadcast models. Depending on the task, different communication models should be used in the smart energy profile. Broadcast, for example, should only be used to expose pricing information.

Data can be either polled or reported. The reporting is done exception-based, providing deadband capabilities. By defining a maximum reporting interval, the reporting is also triggered after a time period. However, as this depends on exception-based reporting, it is only partially time-triggered (it does not report every 10 seconds, but at least every 10 seconds, e.g., at 0, 10, 13, 23, 33, . . . not at 0, 10, 13, 20, 30, . . .).

The granularity of data can be high, for example, only reporting a single attribute, or medium, by clustering several attributes to report. Due to the limited bandwidth, ZigBee does not address exchanging data in a low granularity. The granularity of exchanged data can only be configured to a certain degree as devices might cluster attribute reports based on their internal operations and resource consumption.

The ZigBee specification expects devices to fall asleep in order to save power. Thus, the exchange rate can be seen to be only low, although the smart energy application profile allows polling data every 2 seconds for a short period of time—but in general the frequency should be slower.

Using guaranteed time slots ZigBee supports hard real-time [33]. In addition, it transports timestamps and thus also supports real-time-related data using other communication mechanisms of ZigBee. As an acknowledgment-oriented protocol and by using timeouts for reporting, ZigBee provides a reliable communication infrastructure. Security mechanisms support certificate handling for the devices and message encryption.

10.2.7.5 Status. The first ZigBee specification was released in 2004. There are already millions of ZigBee devices installed worldwide. The first smart energy application profile (version 1.0) was released in 2008, the second, advanced version is under development.

10.2.8 Other Specifications

In addition to the above-mentioned standards, there are other standards with some relevance to sensing, automation, and control in Smart Grids. They are summarized in the following.

- *ANSI C12 Suite.* Mostly used in the North American market, the ANSI C12 suite (ANSI C12.18, 12.19, 12.20, 12.21, 12.22) defines a protocol for metering. In its different parts it defines the structure of exchanged data and different transport mechanisms like optical, modem, or a network.
- *IEEE C37.118.* The IEEE C37.118—Standard for Synchrophasors for Power Systems—defines requirements on the phasor measurement unit and communication mechanisms for phasor data exchange. The protocol can be based on Ethernet, IP, or fieldbuses, passing the defined message frames.
- *IEEE 1547.* IEEE 1547—Standard for Interconnecting Distributed Resources with Electric Power Systems—addresses the physical and electrical interconnection of distributed resources with electric power systems by providing requirements for performance, operation, testing, safety, and maintenance of the interconnection. It addresses information modeling, use case approaches, and a proforma information exchange template and introduces the concept of an information exchange interface.
- *LON—ANSI/CEA-709.2 and ANSI/CEA-852.1.* LON (local operating network) is a general purpose fieldbus used mainly in building automation. In ANSI/CEA-852.1 it defines the tunneling of Ethernet networks using UDP. There are over a million devices installed supporting LON.
- *KNX—ISO/IEC 14543.* KNX is administrated by the KNX Association and is a standard protocol for home and building automation. It supports different transport mechanisms like twisted pair, power line, radio, or Ethernet.
- *Multispeak.* Multispeak was developed by the National Rural Electric Cooperative Association (NRECA) and addresses distribution with a focus on the U.S. market. It defines an information model documented in an XML schema as well as a concrete communication protocol based on Web services and SOAP.
- *Modbus.* Modbus is a communication protocol published by Modicon in 1979 for use with its PLCs (programmable logical controllers). Different variations exist for communicating via serial port or Ethernet. Modbus is a very simple protocol without sophisticated information modeling capabilities.
- *TASE.2—IEC 60870-6 (Also Known as ICCP).* TASE.2 (Telecontrol Application Service Element 2) is a protocol to exchange time-critical control center data between different control centers. It is also known as Intercontrol Center Communications Protocol (ICCP). It is based on MMS and thus a TCP/IP-based client–server model and defines an information model for protection equipment, scheduling and accounting information, and so on.

10.3 CONCLUSIONS

Characteristics of different protocols examined in this chapter are summarized in Table 10.1.

TABLE 10.1. Comparison of Different Protocols and Standards

	IEC 61850	IEC 61968/ IEC 61970[a]	OPC UA	DNP3	BACnet	Open ADR	ZigBee
			Communication Model				
Unicast	+	NA	+	+	+	+	+
Peer-to-peer	+	NA	–	+	–	–	+
Client–server	+	NA	+	–	+	+	+
Multicast	+	NA	–	–	+	–	+
Broadcast	–	NA	–	–	+	–	+
			Type of Data Exchanged				
Current data	+	+	+	+	+	o[b]	+
Alarms and events	o	+	+	o	+	+	o
Comments	–	+	+	–	–	–	–
Historical data	+	+	+	–	+	+	+
Meta data	+/o[c]	+	+	–	o	+	o
			Real-Time Support				
Hard real-time	–	NA	–	–	–	–	+
Real-time-related data	+	+	+	+/–[d]	+	+	+
None	+	+	+	+	+	+	+
			Exchange Rate				
Configurable	+	NA	+	+	+	o	+
Very high	+	NA	–	–	–	–	–
High	+	NA	+	+	+	+	–
Low	+	NA	+	+	+	+	+
Infrequently	+	NA	+	+	+	+	+
Latency	No reliable numbers available to compare all protocols						
Throughput	No reliable numbers available to compare all protocols						
			Granularity of Data				
Configurable	+	+	+	+	+	o	o
High	+	+	+	+	+	–	+
Medium	+	+	+	+	+	+	+
Low	+	+	+	+	+	+	–
			Triggering Mechanism				
Polling	+	+	+	+	+	+	+
Exception-base	+	+	+	+	+	+	+
Time-triggered	+	–	–	–	+	–	o[e]
Reliability	+	NA	+	+	+	+	+
Security	+/–[f]	NA	+	+[g]	+/–[h]	+	+

(Continued)

TABLE 10.1. (*Continued*)

	IEC 61850	IEC 61968/ IEC 61970[a]	OPC UA	DNP3	BACnet	Open ADR	ZigBee
			Information Modeling				
Type model	+	+	+	o	+	+	o
Object-oriented	o	+	+	o	o	o	o
Extensible	o	+	+	o	o	o	–[i]
Standardized file format	+	+	–	o	o	o[j]	o
Online accessible	o	+	+	–	+	+	–
Information Model Standards defined	+	+	+	–	+	+	+

[a]Currently no released specification of IEC 61968 / 61970 contains a standardized technology mapping for communication (other than file formats without defining a transfer protocol). Thus, many table fields are marked with not applicable (NA). This will change when, for example, IEC 61968-1-1 and IEC 61968-1-2 are released, providing a concrete technology mapping.
[b]No current data in terms of measurements, but bids and schedules can be read/written.
[c]Type information is available as templates in SCL file. The structure of the instances can be accessed via MMS but not the structure of the type.
[d]Timestamps in events but not in current data.
[e]Depends on exception-based reporting by defining maximum reporting interval.
[f]Security defined for MMS in IEC 62351 for MMS; security for GOOSE and SV in preparation.
[g]Security mechanisms defined in IEEE 1815.
[h]Security supported for client–server, not for multicast or broadcast.
[i]Only extensible by defining new application profiles.
[j]XML messages are exchanged describing the type model. They could be used for offline client configuration.

The communication models of the protocols differ but all support a unicast model. The type of exchanged data differs as well based on different requirements addressed by the protocols. Only ZigBee supports hard real-time but all protocols address real-time-related data. The exchange rate of IEC 61850 is very high and ZigBee only addresses a low frequency, whereas all others support a high frequency. The granularity of data is for all considered protocols configurable to a certain degree, but ZigBee does not support a low granularity, exchanging large amounts of data at once due to limited bandwidth and OpenADR does not support a high granularity. All considered protocols support polling and exception-based triggering, some also time-triggering. Reliability is addressed by all considered protocols as well as security, although broadcast/multicast operations often do not support security today.

The information modeling capabilities of the protocols differ. Whereas DNP3 only provides very basic information modeling capabilities, the CIM of IEC 61970/61968 is a very powerful information model using object-oriented capabilities including inheritance and type hierarchies.

Issues not addressed in Table 10.1 are, for example, tool support. Protocols based on standard IT integration technologies like Web services have strong tool support compared to other protocols using their own technology (like ZigBee) or are based on specialized protocols like MMS. On the other hand, generic tool support might be less efficient than using specialized SDKs (software development kits) as provided for IEC 61850 and OPC UA.

Another topic is the dynamic of the system. Traditional protocols in the power domain are built with a static configuration in mind—that is, the source knows all its receivers in advance and perhaps only supports one receiver. Other protocols where built with a more dynamic system in mind where receivers can come and go. A Smart Grid environment is expected to be more dynamic than the power domain is today and thus must deal with the dynamic aspects.

The protocols address different domains. Whereas BACnet addresses building automation, IEC 61850 originally addressed substation automation and nowadays also has information models for other domains like wind turbines. CIM-based standards define domain models for applications and systems in the electrical utility enterprise and enterprise integration based on widespread standard IT transport mechanisms (as opposed to utility-specific protocols). Other protocols provide a generic model like OPC UA that can be adapted to different needs. Some protocols like DNP3 can be considered as legacy systems installed on various installations, but not compatible with modern protocols, providing information modeling capabilities in addition to the pure data transfer. Mappings to modern protocols [26] provide an easy to use and standardized way to deal with those legacy protocols.

In general, mappings of the different protocols can be seen as an important development. Thus, the protocols can be used in their domains and information can be exchanged over the borders of the protocols. Examples of protocol mappings include a mapping of IEC 61970 to OPC UA as proposed by Rohjans et al. [34], as well as a mapping to Multispeak as proposed by McNaughton et al. [35] and a mapping to Web services as proposed by Mackiewicz [36]. The current draft of 502-8 contains a mapping not only to Web services but OPC UA based Web services. In addition, harmonizing approaches are started like harmonizing CIM, IEC 61850 and Multispeak [8].

Mapping of different protocols in a standardized way allow interoperability of protocols and a simplified integration of installations running with different protocols as there will be in the Smart Grid environment. It is important that not only data are considered but also the information model and thus there is interoperability on the information rather than on the data only.

Recently, NIST has issued a press release summarizing the foundational Smart Grid standards [37]. Those foundational standards are a set of IEC specifications: IEC 61850, IEC 60870-6 (also known as TASE.2 or ICCP), IEC 61968, and IEC 61970 as well as IEC 62351 addressing security of the other specifications.

REFERENCES

[1] European Energy Exchange AG, "Connecting Markets—Company and Products." Available at: http://www.eex.com/en/document/72732/E_Company_2010.pdf (accessed August 2010).

[2] A.-A. Edris, R. Adapa, M. H. Baker, L. Bohmann, K. Clark, K. Habashi, L. Gyugyi, J. Lemay, A. S. Mehraban, A. K. Myers, J. Reeve, F. Sener, D. R. Torgerson, and R. R. Wood, Proposed terms and definitions for flexible AC transmission system (FACTS), *IEEE Trans. Power Delivery*, 12(4), pp. 1848–1853, October 1997.

[3] GridWise Architecture Council, GridWise® Interoperability Context-Setting Framework, March 2008.

[4] T. Samad, P. McLaughlin, and J. Lu, System architecture for process automation: Review and trends, *J Process Control*, 17, pp. 191–201, 2007.

[5] Y. T. Lee, "*Information Modeling: From Design to Implementation*," *Proceedings of the Second World Manufacturing Congress*, 1999.

[6] OMG, Unified Modeling Language (OMG UML), Infrastructure, Version 2.3, May 2010.

[7] OMG, Meta Object Facility (MOF) Core Specification, Version 2.0, January 2006.

[8] NIST, NIST Framework and Roadmap for Smart Grid Interoperability Standards, Release 1.0, 2010.

[9] K. P. Brand, C. Brunner, and W. Wimmer, "Design of IEC 61850 based Substation Automation Systems According to Customer Requirements," CIGRE PlenaryMeeting, Paris, 2004.

[10] D. Baigent, M. Adamiak, and R. Mackiewicz, "IEC 61850 Communication Networks and Systems in Substations: An Overview for Users," SIPSEP, 2004.

[11] IEC 61850-7-1, "Communication Networks and Systems in Substations—Part 7-1: Basic Communication Structure for Substation and Feeder Equipment—Principles and Models," IEC, First Edition, July 2003.

[12] ABB Review, Special Report IEC 61850, ABB, 2010.

[13] M. Uslar, S. Rohjans, T. Schmedes, J. M. Gonzalez, P. Beenken, T. Weidelt, M. Specht, C. Mayer, A. Niesse, J. Kamenik, C. Busemann, K. Schwarz, and F. Heins, Untersuchung des Normungsumfeldes zum BWMi-Förderschwerpunkt "e-Energy—IKT-basiertes Energiesystem der Zukunft" (in German), 2009.

[14] CIM User Group, "CIM UML Model—Version 13 Release," 2009.

[15] M. Power, A. Bose, T. Kostic, F. Matos, I. Naicker, J. Müller, and G. Sztrada, "System Control in Light of Recent Developments in Substation Control," Cigre Technical Brochure, 2009.

[16] EPRI, "Smart Meter Information Interoperability Test (61968 Part 9)—The Power of the Common Information Model to Exchange Messages Between Back-End Systems," Final Report, May 2010.

[17] T. Burke, "OPC and Intro OPC UA," presentation at OPC UA DevCon, Munich, 2008.

[18] W. Mahnke, S.-H. Leitner, and M. Damm, *OPC Unified Architecture*. Springer, 2009.

[19] J. Lange, F. Iwanitz, and T. J. Burke, *OPC from Data Access to Unified Architecture*. VDE Verlag, 2010.

[20] A. Ginter, *The Stuxnet Worm and Options for Remediation*. Industrial Defender, Inc., 2010.

[21] D. Grossmann, D. John, and A. Laubenstein, *EDDL Harmonisierung (in German)*. Automatisierungstechnische Praxis, 2009.

[22] D. Johnson, "Is DNP 3.0 the right standard for You?" *Utility Automation*, June 2000.

[23] G. Clarke and D. Reynders, *Modern SCADA Protocols—DNP3, IEC60870.5 and Related Systems*. Elsevier, 2004.

[24] K. Curtis, *A DNP3 Protocol Primer, Revision A*. DNP Users Group, 2005.

[25] K. Schwarz, "Comparison of IEC 60870-5-101/-103/-104, DNP3, and IEC 60870-6-TASE.2 with IEC 618501," SCC, Edition 2008-10-03, 2008.

[26] IEEE, "IEEE Delivers Critical 1815 DNP3 Standard in Record Time." Available at: http://standards.ieee.org/announcements/2010/ratification1815.html (accessed August 2010).

[27] ISO 16484-5:2007, "Building Automation and Control Systems—Part 5: Data Communication Protocol," ISO, 2007.

[28] ASHRAE SSPC 135, "*BACnet Vendor Ids*." Available at: http://www.bacnet.org/VendorID/index.html (accessed October 2010).

[29] M. A. Piette, G. Ghatikar, S. Kiliccote, E. Koch, D. Hennage, P. Palensky, and C. McParland, "Open Automated Demand Response Communications Specification (Version 1.0)." California Energy Commission, PIER Program. CEC-500-2009-063, 2009.

[30] PIER Demand Response Research Center, "Open Automated Demand Response Communication Standards (OpenADR or Open Auto-DR) Development." Available at: http://www.openadr.org/ (accessed October 2010).

[31] Microsoft, "Smart Energy Reference Architecture," October 2009.

[32] ZigBee Alliance, "ZigBee Alliance Plans Further Integration of Internet Protocol Standards," press release, 2009.

[33] J. Leal, A. Cunha, M. Alves, and A. Koubaa, "On a IEEE 802.15.4/ZigBee to IEEE 802.11 Gateway for the ART-WiSe Architecture," IEEE Conference on Emerging Technologies and Factory Automation, 2007.

[34] S. Rohjans, M. Uslar, and J. H. Appelrath, "OPC UA and CIM: Semantics for the Smart Grid," Transmission and Distribution Conference and Exposition, 2010 IEEE PES, April 2010.

[35] G. A. McNaughton, G. Robinson, and G. R. Gray, "MultiSpeak and IEC 61968 CIM: Moving Towards Interoperability," Grid Interop Forum, 2008.

[36] R. E. Mackiewicz, "The Benefits of Standardized Web Services Based on the IEC 61970 Generic Interface Definition for Electric Utility Control Center Application Integration," Power Systems Conference and Exposition, 2006. PSCE '06. 2006 IEEE PES, 2006.

[37] NIST, "NIST Identifies Five 'Foundational' Smart Grid Standards," NIST press release. Available at: http://nist.gov/public_affairs/releases/smartgrid_100710.cfm (accessed October 2010).

PART III

SECURITY

11

INTRODUCTION TO SMART GRID CYBER SECURITY

Pedro Marín Fernandes*

11.1 INTRODUCTION

Cyber security is generally defined as "the set of mechanisms that prevent cyber incidents, that actually or potentially jeopardize the confidentiality, integrity, or availability of an information system, or the information the system processes, stores, or transmits or that constitute a violation or imminent threat of violations of security policies, security procedures, or acceptable use policies" [1].

A conceptual cyber security model is, in the scope of this chapter, a risk assessment of the threats that target the assets to be protected in a Smart Grid, and recommendations to protect them. These assets normally include data in transit and at rest, and the goal is to ensure their confidentiality, integrity, and availability (CIA). As defined in the NIST document [2], mechanisms that avoid theft of secret information fall under the confidentiality paradigm, mechanisms that avoid distortion of important control commands fall under integrity, while mechanisms that prevent

Smart Grid: Applications, Communications, and Security, First Edition.
Edited by Lars Torsten Berger and Krzysztof Iniewski.
Published 2012 by John Wiley & Sons, Inc.

*Pedro Marin Fernandes is an employee of Cisco Systems, Inc. and this chapter was written with the permission of Cisco.

the communications infrastructure from being unresponsive aim at availability. The risk assessment studies the impact of a compromise of these paradigms on the protected assets.

A conceptual security model can be derived by simply mapping generic security principles to a model that represents the functional areas and transactions of the Smart Grid. The result shall differ from both the traditional information technology (IT) and the industrial control systems (ICSs) security models, and will somehow be a blend of the two: a blend not as in a sum of the parts, but as a model where both securities coexist and interoperate being mutually aware. There can be a temptation to blindly apply IT security to ICSs but the results can be dangerous, as ICS priorities are different from IT ones. For example, block encryption widely used in IT may introduce delays unacceptable in ICSs, and port scanning can cause some industrial components to malfunction. On the other hand, legacy ICS equipment were not manufactured with cyber security in mind. Security came from obscurity of field locations, proprietary and little documented protocols, impossibility of remote access, point-to-point communications, and from the fact that it was physically separated from IT networks.

Partly being a blend of ICSs and IT, Smart Grid security uniqueness is also based on:

- *Scale*, as it has the potential to be larger than the Internet, connecting a very diverse range of networks and subnetworks. Segmentation, identity management, management of encryption keys (see Chapter 13), and the integration of multiple layer 1 and layer 2 protocols are serious challenges when considered on a large scale.
- *Mix of legacy and modern devices*, with diverse security and communications capabilities, functions, life span, and levels of criticality in their roles. No matter how desirable it would be to unify the grid, the life span of power equipment obliges the security policy to cope with this diversity for, at least, a decade.
- *Hybrid model of management*: as a consequence of the previous topic, we cannot assume that there is always a stable and broadband connection to all field locations of the grid, and therefore the management of security has to comply with a hybrid distributed-centralized model: local authentication and credentials management with centralized synchronization of credentials databases, and event collection and correlation.
- *Evolving standards and regulations*: despite the efforts being undertaken by several entities, like the National Institute of Standards and Technology (NIST) [3], and the North American Electric Reliability Corporation (NERC) [4] in the United States, or the European network of SCADA security Test Centers for Critical Energy Infrastructures (ESTEC) [5] in Europe (see Chapter 12 for more details of these standards), regulations are still maturing and subsequent revisions are in process. Given the competitive nature of the industry, utilities cannot afford to wait for finalized standards to develop their Smart Grid implementation strategy.

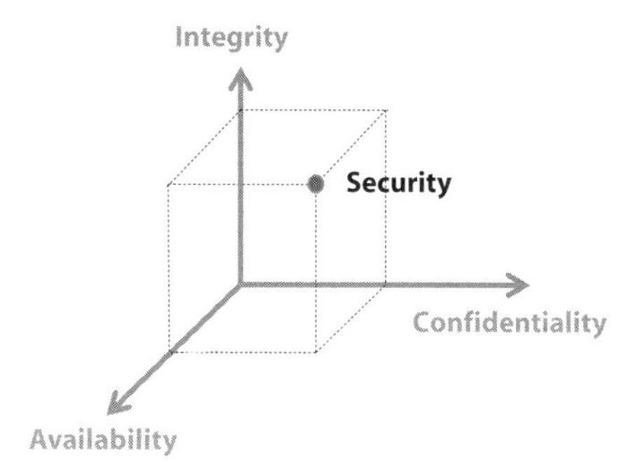

Figure 11.1. CIA security triad.

Security models in general include known concepts like CIA, Figure 11.1. The means to achieve them range from encryption, admission control, threat detection and prevention, and segmentation, to vulnerability-free software and proper information management. Confidentiality is a very dear topic to IT because of the secrecy associated with companies' information, but hardly a priority in ICSs, where the transported data are typically a series of inputs and outputs that keep systems running. In Smart Grids, however, confidentiality arises strongly in the form of privacy [6, 7] as the systems to control include the customers' smart meters, smart appliances, and home energy controllers, and therefore there is a real danger that end-user energy consumption profiles are used for nontransparent goals.

On the other hand, integrity and availability are vital in ICSs. The mere possibility of corrupted commands distorting or tearing down the flow of power justifies every algorithm in place to assure normal functioning of the grid, even if that means sacrificing confidentiality.

Admission control, in its turn, being popular in the IT world, has been poorly present in ICSs, where the restricted or local-only access to equipment often served as sufficient proof of identity. In the new era, the large number of intelligent devices that will be placed in the local proximity of users--either in the neighbor area network (NAN) or home area network (HAN)--makes remote admission control a must.

A conceptual model must include three important security concepts: *security-by-design*, *security-in-depth*, and *end-to-end security*. Security-by-design relates to the manufacturing of individual products, and assembly of systems, solutions, and architectures. It consists of designing them from scratch with security in mind, as opposed to adding security features to an already built product. Security-in-depth implies the realization that any security feature by itself is breakable with enough effort, and only multiple security barriers layered in a concentric way around protected assets can provide a security level superior to the sum of the individual parts. Each layer deters, detects, and delays intruders with different means than the neighboring ones. Protecting hosts with antivirus (AV) and host intrusion prevention systems (HIPSs) before they connect to the network, followed by admission control

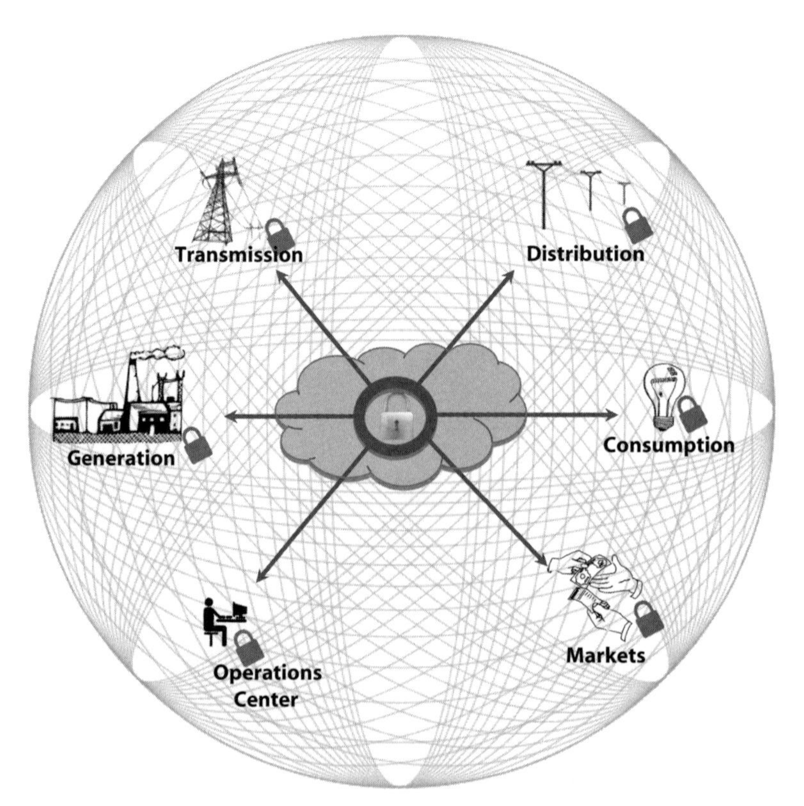

Figure 11.2. Security end-to-end, by design, and in-depth.

mechanisms at the moment of connection, firewalling the bidirectional data to the corporate servers, and analyzing the traffic patterns that flow in the core network with IPSs, is an example of security-in-depth often used in IT. Finally, end-to-end security relates to the fact that the security of a network is as strong as its weakest link. Therefore, network administrators have to maintain the same level of security in all segments of the network. Figure 11.2 represents security-by-design with a lock next to each segment in the network, end-to-end security with a secure data network, and security-in-depth with concentric circles around the whole grid.

All these concepts should blend proportionally into the utilities' security policy. As Joseph Weiss puts it, "ICS security requires a balanced approach to technology design, product development and testing, development and application of appropriate ICS policies and procedures, analysis of intentional and unintentional security threats, and proactive management of communications across view, command and control, monitoring, and safety. It is a life cycle process beginning with conceptual design through retirement of the systems" [8].

11.2 EXAMPLES

11.2.1 The North American Example

To date, the National Institute for Standards and Technology Interagency Report (NISTIR) 7628 [3] is the single most comprehensive official document that guides utilities in their quest to secure the Smart Grid.

Released in August 2010, version 1 comprises 500+ pages, where a 450-member working group from government, industry, and academia have included the previous concepts to provide a series of security requirements and implementation guidelines mainly for protection of information in transit, but also at rest. With this document NIST exemplifies a conceptual security model in the form of security requirements that are applied to every segment of the grid. It focuses on the power grid operations and not on the enterprise operations.

In a nutshell, a vast set of general security requirements was adapted to the Smart Grid specifics, and subject to a risk assessment, Figure 11.3. Companies can use the outcome of the risk assessment as a guideline for their own security policies, and the conformity assessment to evaluate its level of implementation. Even though the study by NIST is comprehensive, companies are encouraged to adapt it to their own realities, whether adding/eliminating security requirements, or replacing them by equivalent ones that better meet their internal expertise, headcount, or budget constraints.

A more detailed explanation of this model follows. In this context, *risk* is the likelihood of an unwanted outcome resulting from an incident. A risk assessment identifies assets, vulnerabilities, threats, and their impact on different segments of the Smart Grid: IT, telecommunications, and electric sectors. The risk assessment taken here as an example was derived based on four inputs:

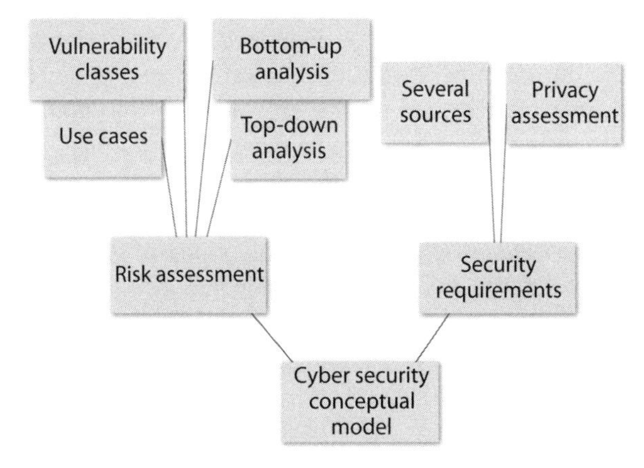

Figure 11.3. Derivation of the cyber security model proposed by NIST [3].

- Use cases, taken from electrical utilities or industry related organizations.
- Vulnerability classes, taken from published NIST documents, as well as from NERC Critical Infrastructure Protection (CIP), serve as a starting point for a more extensive list of vulnerabilities specific to the Smart Grid.
- A bottom–up analysis starts with well-identified security problems and suggests solutions for each one, like control of network admission, end-to-end encryption, or inspection of traffic.
- A top–down analysis identifies the intervenient, confines the domains, draws the main flows of information, and describes the security objectives and requirements of each flow. Overall, the top–down approach describes a logical diagram of the Smart Grid with such detail that it is valuable for more than just the study of its security.

The use cases were selected from several sources, whether nonprofit associations like Electric Power Research Institute (EPRI) [9], or utilities like Southern California Edison (SCE) [10] (see full list in the NIST website [11]).

Vulnerability in this context is a weakness in an information system, internal procedure, or implementation that could be exploited with malicious purposes. Four classes of vulnerabilities were identified:

- *Security Policy Vulnerabilities.* A security policy, defined as the set of documented procedures by which an organization operates to avoid undesirable or unforeseen scenarios, can include deficiencies that lead to security risks for the organization.
- *Platform Software/Firmware Vulnerabilities.* Errors or little security consciousness in software/firmware coding allows attackers to affect the confidentiality, integrity, and/or availability of information. Examples are buffer overflow and SQL injection.
- *Platform Vulnerabilities.* Related to the previous, these vulnerabilities can be present when the platform combines two or more of the following: application software, operating system, hardware abstraction/virtualization layer, and hardware itself. Examples are lack of or poor malware protection, late application of software patches, vulnerable application programming interfaces (APIs) between the software layers, inadequate sizing of machine resources, and insufficient alerts and log management.
- *Network Vulnerabilities.* These include all kinds of dangers related to the transmission of data, like delayed or nonconverging routing tables, erroneous allocation of quality of service (QoS) parameters, corrupted IP headers, and router malfunctions.

The bottom–up analysis identifies some specific problems and issues that are evident either by common sense or real world experience, as opposed to the top–down approach that reaches similar challenges but from a very comprehensive and exhaustive identification of the Smart Grid data transactions and their security needs.

The goal is to have complementary approaches with different implementation timelines that utilities can execute in parallel, allowing them to face major security concerns in the short term, and elaborate a detailed security policy in the long term.

11.2.1.1 The Bottom–Up Analysis. In short, the bottom–up approach identified a set of 59 very specific security challenges, mostly coming from very tangible examples in the industry. A set of specific best practices were recommended to face them. For example, admission control of utilities' employees to substation's intelligent electronic devices (IEDs), pole top equipment, and meters, or even the maintenance of an ever available and accurate time reference for synchrophasors, certificate revocation, and timestamps of logs are immediate worries of any utility. Recommendations include the use of device and network authentication, complexity rules for passwords, and careful use of shared resources like the ones located on the public Internet.

Some of these security challenges are described next.

AUTHENTICATION, AUTHORIZATION, AND ACCOUNTING (AAA). Knowing who is attempting access, where from, at what time, and using what access method are frequent criteria being probed when users try to access a certain device on the grid. Based on them, different privileges or authorizations are awarded, ranging from no access at all, read-only access, or full access. Accounting encompasses access and postaccess tracking of where the user logged in, what functions he/she executed, and when. AAA functions like the access and authorizations policies should be centrally managed and stored, but fallback mechanisms should be in place to allow personnel to access devices locally when control center communications are down. Furthermore, credentials should be specific to each user, as opposed to rolebased. AAA tasks can be applied when:

- *Maintenance Personnel Access Substation IEDs, Outdoor Field Equipment, and Smart Meters.* This equipment is expected to support local and remote access, based on application authentication protocols like Lightweight Directory Access Protocol (LDAP) [12], and/or network authentication ones such as Remote Authentication Dial-In User Service (RADIUS) [13, 14] 802.1x.
- *Consumers Access Smart Meters.* In the scenarios where the smart meter acts as the home area network (HAN) gateway, and/or as the receiver and transmitter of demand–response (DR) commands, there might be benefits in consumers accessing them. In this case, the authorization should be very limited and very different from the one granted to maintenance personnel. To avoid having to cope with credentials of millions of consumers, the authentication can be role-based. However, the smart meter is often thought of as being under the utilities' responsibility, and therefore accessible to employees only. Consumers are more likely to control their consumption/generation profiles through a home energy controller device.
- *Meters Access AMI Head-Ends.* Information exchanged from the meter and the AMI head-end is of utmost importance for grid stabilization (mainly downlink

DR commands) and billing purposes (mainly uplink reading information). Thus, there is a need for mutual authentication between the two. Without it, coordinated falsification of control commands and/or rapid rates variation could lead to grid stability problems.

- *Meters Access AMI Networks.* The networks facing the consumer, like NANs and HANs, are more susceptible to widespread compromise and denial of service (DoS) attacks than substation networks. Since the unavailability of the network is as impactful as the unavailability of the applications, AAA functionalities should be present at network level in addition to application level. NANs are expected to have a mesh topology, which presents a challenge due to the different trust models when in comparison with the more typical star topology networks. Known types of mesh trust models are based on hop-by-hop and end-to-end trust chains. The former relies on a node trusting its upstream and downstream adjacent on a path between a mesh leaf and the AMI gateway. Access control is performed at each node. The latter considers intermediate mesh nodes untrusted and a secured tunnel is created between each leaf node and the gateway. Whether one or a combination of the two is used, the algorithm must be scalable, resistant to rogue nodes, and computationally light.

Chapter 13 revisits authentication for the above scenarios.

SECURING RADIO-CONTROLLED DISTRIBUTION DEVICES. Remotely controlled switching devices that are deployed on pole-tops throughout distribution areas have the potential to allow faster isolation of faults and restoration of service to unaffected areas. Some of these products that are now available on the market transmit open and closed commands to switches over radio with limited protection of the integrity of these control commands. In some cases, no cryptographic protection is used, while in others the protection is weak in that the same symmetric key is shared among all devices.

SECURING SERIAL SCADA COMMUNICATIONS. Even if utilities are moving from slow serial links to modern wideband WANs, the transition can last for as much as a decade to cover all transmission and distribution substations. In the meanwhile, security cannot be compromised. Solutions that simply wrap a serial link message into protocols like secure socket layer (SSL) [15] or Internet protocol security (IPSec) [16, 17] may add an unacceptable message payload size that consumes bandwidth, and a computational burden that increases latency. The latter is normally bound by the following values:

- Four milliseconds for protective relaying
- Subseconds for transmission wide-area situational awareness monitoring
- Seconds for substation and feeder SCADA data
- Minutes for monitoring noncritical equipment and some market pricing information

- Hours for meter reading and longer-term market pricing information
- Days/weeks/months for collecting long-term data such as power quality information

SECURE DIAL-UP ACCESS. Today dial-up is often used for engineering access to substations, and authentication is limited to modem callback and passwords in the answering modem and/or a device connected to the modem. Passwords are not user-specific and are seldom changed.

SECURE END-TO-END COMMUNICATIONS BETWEEN SMART METER AND AMI HEAD-END. Secure end-to-end communications protocols such as transport layer security (TLS) [15] and IPSec ensure that confidentiality and integrity of communications is preserved regardless of intermediate hops.

PROTECTION OF ROUTING PROTOCOLS IN AMI LAYER 2/3 NETWORKS. As already inferred from previously identified security problems, the AMI data network presents peculiarities relative to the substations' WAN. There is an increasing likelihood that mesh routing protocols will be used on wireless links. While the IEEE 802.11i [18] security standard is supposed to grant wireless an equivalent security to one present in wired links, wireless mesh technology opens the door to some new attacks in the form of route injection, node impersonation, L2/L3/L4 traffic injection, and traffic modification. End-to-end security (like IPsec) prevents attacks such as eavesdropping, impersonation, and man-in-the-middle (MitM), but routing security is still required to prevent DoS attacks.

MANAGE LOGS OF IEDS. Several constraints like shortage of bandwidth, massive scale, or the fact that IEDs generate few types of logs, limit the availability for centralized security incident and event management (SIEM) systems to correlate logs both temporally and spatially. This is important in order to detect and respond to malicious actions by insiders or external attackers.

PROTECTION OF DIAL-UP METERS. Dial-up technology using plain old telephone service (POTS) has been a preferred method for connecting to network gear, given the low cost and predictability of the technology. However, common attacks like DoS can be triggered dialing into a bank of modems and tying up lines in a way that no other user can connect. This attack is especially easy to launch if all the line numbers start with the same prefix.

REMOTE ATTESTATION OF METERS. Given the physical proximity of meters to users, it is necessary to ensure that they are running an approved configuration and software/firmware version at all times.

OUTSOURCED WAN LINKS. Many utilities are tempted to outsource the telecommunications services to service providers (SPs) and have them connect generation plants and substations to control centers, pole-top AMI collectors to AMI head-ends, and pole-top distribution automation equipment to distribution management systems (DMSs). While this would bring immediate financial benefits, the control of a critical infrastructure such as an electrical grid would be in the hands of one or more SPs. The service level agreements (SLAs) that utilities are normally mandated to offer can be incompatible to those of SPs. Complications resulting from SCADA traffic competing with priority traffic of other SP customers, DoS attacks launched from SP customers, or generalized congestion of communications in catastrophic events (e.g., Manhattan attacks in 9/11) can potentially tear down the electrical grid. A compromise can be achieved by leveraging commercial networks from SPs in the segments of the grid that are highly distributed, and that do not compromise the bulk of the power flow, like AMI, while ensuring that SCADA traffic to substations use utility owned networks.

Side Channel Attacks. A side channel attack is based on the theft of information via a side effect of security implementation. For example, the time and power needed to encrypt a block of data are directly proportional to the length of the cryptographic key. Complementing this attack with another can lead to the guessing of some digits of the key, which, in turn, can facilitate brute-force attacks (see more key management issues in Chapter 13). Other examples are reconstructing the image of a cathode ray tube (CRT) or liquid crystal display (LCD) screen or even stealing data from a copper cable by capturing the emitted radiation from a distance. Smart Grid devices that are deployed in the field and in-home devices are at risk of side channel attacks due to their accessibility.

Tampering Field Device Settings. Numerous field devices contain settings. A prominent example is relay settings that control the conditions such as those under which the relay will trip a breaker. In microprocessor devices, these settings can be changed remotely. One potential form of attack is to tamper with relay settings and then attack in some other way. The tampered relay settings would then exacerbate the consequences of the second attack.

Absolute and Accurate Time Information. Increasingly more and more advanced applications will critically depend on an accurate absolute time reference. According to NERC [19], "these applications include, but are not limited to, Power Plant Automation Systems, Substation Automation Systems, Programmable Logic Controllers (PLC), IEDs, sequence of event recorders, digital fault recorders, intelligent protective relay devices, Energy Management Systems (EMS), Supervisory Control and Data Acquisition (SCADA) Systems, Plant Control Systems, routers, firewalls, Intrusion Detection Systems (IDS), remote access systems, physical security access control systems, telephone and voice recording systems, video surveillance systems, and log collection and analysis systems." Protocols that rely on anti-replay mechanisms, synchrophasors, digital certificates, and logging for forensic purposes are all examples of the importance of an accurate time reference.

WEAK AUTHENTICATION OF DEVICES IN SUBSTATIONS. Network authentication of devices in substations has been traditionally disregarded due to the inexistent or poor remote access. With the dawn of the Smart Grid, not only new power equipment associated with renewable energies are supposed to populate the substation, but wideband remote access is a must. Gaining unauthorized access to the substation LAN can compromise the entire grid network.

WEAK PROTOCOL STACK IMPLEMENTATIONS. Many IP stack implementations in ICSs are rudimentary compared with modern operating systems, whether in personal computers or routers. Improperly formed packets can cause some of these control systems to be locked out of the network.

INSECURE PROTOCOLS. The protocols used nowadays, like DNP3 [20] and 61850 [21], were not built with security in mind, and seldom is IEC 62351, 2003 [22] which is the security standard for these protocols, implemented. See more protolos in Chapter 12.

LICENSE ENFORCEMENT FUNCTIONS. Manufacturers often include licensing their devices as a way to enforce a certain revenue model. This can expose a series of vulnerabilities that often have to be avoided with the help of both the vendor and the utility. Examples are:

- *Misuse of Authorized Maintenance Access.* Often licenses are obtained and renewed via a direct access between the vendor and the device. Its misuse either intentionally or inadvertently can cause great damage. Related to this is the permission that vendors may have to intrude/shut down systems if nonconformities are suspected/detected.
- *Shutting Down/Limiting Features.* Some applications shut down or limit their capabilities when their license keys expire, or the contractual terms and conditions are not met.
- *Requiring the Application or Device to Access the Internet.* Many commercial applications and devices attempt to connect to public IP addresses for a variety of functions, such as to update software or firmware, synchronize time, consult a public DNS, provide help/support/diagnostic information, enforce licenses, or utilize Internet resources such as to mapping tools and search systems, etc.

The problems related to this are twofold: the unsecure channel to the Internet that can be exploited for other goals, and the behavior of said applications and devices when their access to the Internet is blocked. The former can lead to obvious problems such as unauthorized access to corporate assets, data leakage, or overutilization of the Internet access, while the latter can cause applications and devices to malfunction if they depend on external services, or IDS false positives if there are many unsuccessful retries of Internet connection.

Furthermore, such call home functions are often poorly documented. Configuration options to modify or disable them, or even information to know if they are only used at start-up or regularly in production, are often hard to find.

Patch servers are recommended for software and firmware updates, rather than having those endpoints reach out to the Internet. Unfortunately, not all applications and devices allow changing their configuration with this goal.

PATCH MANAGEMENT. The patch, test, and deploy life cycle is very different in the electrical sector than in IT. There needs to be a process whereby the risk and impact of the address vulnerability can be determined in order to prioritize upgrades. Only very stable and proven upgrades are implemented, and it can take more than a year to qualify and approve a patch before it goes into production.

SYSTEM TRUST MODEL. There has to be a clear trust chain. A trust chain is simply a set of relationships relative to a system that is always trusted. A certificate authority (CA) is normally the third party that issues digital certificates to authenticate systems, and is always trusted. An example of a chain trust would be the fact that a private network is trusted more than an open public network, since we have more control and are able to protect the former. However, there are many private networks in a Smart Grid, like the substation or the IT LANs. The latter has a higher level of trust for IT users than the Internet, but for maintenance personnel who work in substations often both have the same level of trust, and it is the substation LAN that is more trusted. On the devices side, physically accessible assets, such as pole-top equipment or smart meters, should never be completely trusted. Other factors, like incidents and logs, correlated at the security incidents and events manager (SIEM) platform, should be taken into account to increase or decrease the confidence on the data coming from said devices.

USER TRUST MODEL. The same concept applies to employees and contractors who work on the grid. A scale of trust should differentiate them, and be based on criteria such as knowledge, experience, historical behavior, and satisfaction to be working in the company or likeliness to leave. The new challenge is the mixed teams of IT and power professionals who have typically worked independently. When it comes to critical systems, it may be worthwhile considering the "two-person integrity" that prevents single-person access to key management mechanisms.

SECURITY LEVELS OF CRITICALITY. Parallel to the trust model, levels of criticality have to be assigned to the assets to be protected. From data to devices, each asset has a different impact in case of compromise and therefore a different level of criticality. Highly critical assets should be protected by highly trusted systems.

DISTRIBUTED VERSUS CENTRALIZED MODEL OF MANAGEMENT. As mentioned before, the highly distributed nature of the grid is one of the difficulties utilities will have to overcome. WAN failure cannot result in power flow complications. Therefore, a well-balanced hybrid approach where management is centralized, but break-glass authentications for critical situations or for network downtimes are important.

One of the most critical situations where we want to access field equipment is when there is a power outage, and chances are that the outage has affected the communications themselves.

Intrusion Detection for Power Protocols. IDS traffic patterns for ICS protocols (Modbus, DNP3, 61850, etc.) is something relatively new and that cannot be derived from IT protocols. It assumes a deep understanding of the protocol and its possible anomalous behaviors.

Monitoring and Management for Power Equipment. Management protocols used by power equipment are normally proprietary of legacy, and were designed with little concern about security. Standardization on open and secure management protocols such as SNMPv3 is desirable.

Security Incidents and Events Management (SIEM). Once the events and incidents have been collected securely, a SIEM system must correlate them both geographically and temporally to identify attacks and to have an overview of the grid security activity. Again, events coming from both IT and control systems, as well as their massive scale, are challenges in this domain.

Cross-corporate Security. This term refers to the security in the interactions between teams that have different responsibilities but belong to the same utility. In the AMI space, for example, the billing team is different from the one who decides what load to shed for DR, and therefore the decision of who is held accountable for the AMI head-end has to be thoroughly thought through.

Credentials Management. Due to the highly distributed nature of the Smart Grid, maintenance personnel access field equipment from many devices, be they simple terminals, laptops locally connected to the devices, or terminal emulators in PCs that connect remotely. In any case, there is not a one-to-one relationship between employee and access device, and access cannot be based on the identity of the latter only, but on a combination of user and machine credentials.

Tamper Evidence. Having a mechanism that detects when a device is being tampered with would be desirable in addition to resistance to tampering, of course. The detection would have to be in the form of an exception communicated remotely.

Legacy Equipment with Limited Resources. There is a significant discrepancy in renewal-cycle time spans of ICS and IT equipment. The former might not have the possibility for firmware upgrades or upgrades can compromise its primary function due to limited computational power, so the latter has to adapt to this reality through a layered approach with low footprint on secured assets.

Forensics and Related Investigations. It has already been said that generating and sending logs to a centralized system is not a strength of ICSs. However, whatever new equipment is introduced to smarten the grid should not follow this rule. Physical tampering along with incidents related to data traffic should be reported for correlation and used for detecting faults and forensics purposes.

Roles and Role-Based Access Control (RBAC). A *role* is a set of permissions associated with a set of credentials. An individual user may impersonate several roles. Roles clearly need to relate to the organizational structure of the utility. While roles' credentials and privileges are more manageable than users', the following has to be taken into account:

- Roles do not offer the granularity that users do in differentiation.
- Users can normally leverage the existence of repositories, such as an active directory (AD).
- Accounting is less informational if data is sorted in roles instead of users.

Limited Sharing of Vulnerability and/or Incident Information. Unlike the IT industry, there is an implicit discretion in critical infrastructure industries with respect to sharing information about vulnerabilities of hardware and software. The fear of generalized blackouts due to malicious activity, and the amount of time needed to qualify an upgrade or patch leads vendors to secrecy regarding vulnerabilities.

Data Flow Control Vulnerability Issue. Given that the power grid will not only encompass many devices but many networks and subnetworks too, the authorization will need to be at both the device and network level to avoid that simple moves of devices facilitate unauthorized access to it.

Public Versus Private Network Use. The Internet and commercially available networks, such as the mobile cellular, are going to be used by the Smart Grid at one point in time and at one place or another. When, where, and how is a matter of a risk assessment that takes into account the criticality of the data, delay requisites, and available countermeasures (e.g., firewalls, IDS/IPS, SIEM, encryption).

Traffic Analysis. Traffic analysis tools can turn out useful by identification of repetition of chunks of bytes, patterns, or packet headers only, in the case of encrypted traffic. The identity of the intervenient (e.g., IP addresses), times and frequency of communications, and information about the network diagram are examples of what can be obtained.

Broad Definition of Availability. Availability of power when demanded is admittedly the ultimate goal of cyber and noncyber activities within the Smart Grid. The problem is to translate this mantra into measureable parameters of each of its many systems. The first problem is that some systems measure availability in response times, others in throughput, others in uptime, others in software bugs, and yet others in accuracy. Second, each one has different impacts on the whole system if their "availability" decreases. Additional difficulties could be mentioned that make it extremely difficult to map a certain level of power availability to end users into individual systems' performance.

UTILITY TENDENCY FOR NONSTANDARDIZATION. Given the size and economical power of utilities, historically there has been the possibility of customization of standards. In the space of SCADA implementation only, and in North America alone, more than a hundred versions were known to be made between 1955 and 1990. It has been proved that a few standard security practices are a better way to defend a system than a variety of proprietary, little known protocols that defend by obscurity. The upcoming Smart Grid industry cannot afford going down the same path or interoperability, delay of implementation, and, of course, security would be affected.

11.2.1.2 The Top–Down Analysis. The top–down approach, being the more theoretical and comprehensive analysis of the document, provides a reusable model of the communications flows, on top of which the security requirements are applied. A more detailed description follows.

As mentioned before, the model identifies the intervenients, confines them in domains, draws the predicted flows of information among them, and describes the security requirements of each flow. For simplicity, flows are grouped in categories with similar security needs, Figure 11.4.

The seven domains are bulk generation, transmission, distribution, and customer premises (to which the power flows) and operations, markets, and service provider (where control and businesses are held). Their high level intercommunications are depicted in Figure 11.5. The domains are further broken down into 49 actors, which are entities that have a specific role in the grid, whether it is technology intervening in the physical flow of the energy, or organizations making business out of it. Actors exchange information and make decisions to execute a function in the grid. Their

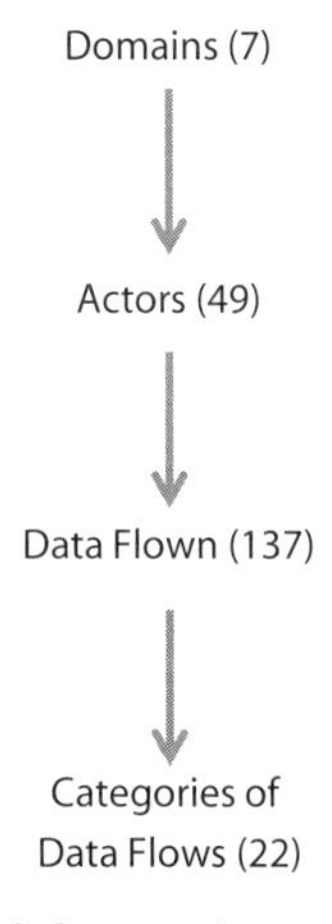

Figure 11.4. Top–down approach.

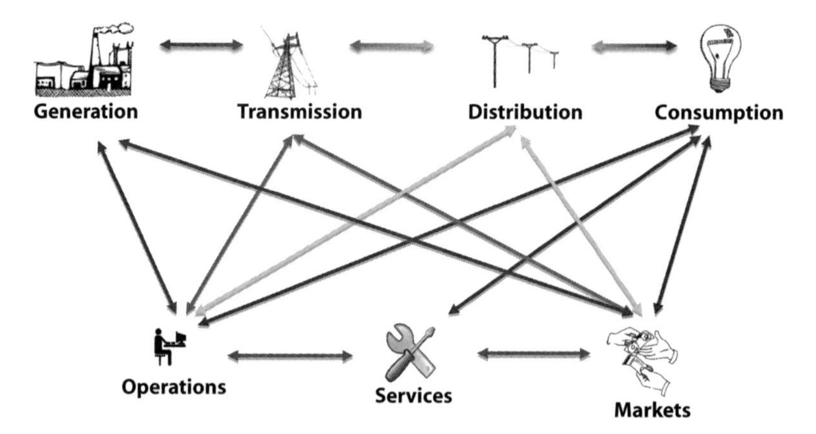

Figure 11.5. Smart Grid domains and data flows.

flows of information are where cyber security of data in transit is to be applied, and therefore they are thoroughly analyzed. In distribution, for example, NIST identified distribution remote terminal units (RTUs)/IEDs, geographic information system (GIS), field crew tools, distribution sensors, distribution data collectors, and distributed intelligence capabilities as the six actors whose communications are to be protected.

Communications needs are predicted for each actor and an intricate mesh of 137 flows of information comprises one of the assets (data in transit) to be protected. They are further grouped into 22 categories with similar security requirements. The criteria for this division is based on:

- Level of redundancy
- Bandwidth and computational restrictions
- The fact that they connect actors of the same or different organizations
- The fact that they connect actors with common/different management authority
- Type of traffic
- Type of endpoints they connect

Each category is further described with a set of 18 attributes related to confidentiality, privacy, integrity, and availability requirements as well as with the presence of legacy communication protocols and/or end-devices, remote locations, and so on. This description is used to map the security requirements to each category.

The final step in the top–down approach, and to help build the risk assessment model, is to measure the impact level of a CIA compromise. Every category is given a *low*, *moderate*, or *high* impact level for each CIA concept. For example, availability has high impact if compromised in the interface between transmission SCADA and substation equipment (category 1), medium impact on interfaces between pole-top

TABLE 11.1. Excerpt of the Allocation of Impact Levels to Categories of Flows of Information

Category	Confidentiality	Integrity	Availability
1	Low	High	High
2	Low	High	Medium
⋮	⋮	⋮	⋮
22	High	High	High

IEDs and other pole-top IEDs (category 2), and high impact on interfaces that connect security console and network routers, firewalls, computer systems, and network nodes (category 22). Table 11.1 is an excerpt of the complete impact allocation of the NIST document [3].

Once it described the risk assessment, NIST made use of other recommendations [23–25] and a privacy impact assessment (PIA) to identify almost 200 relevant security requirements.

The PIA was performed to face the detailed knowledge that utilities are expected to have on the end-users' consumption profiles. To date, energy consumption patterns were not given the same level of privacy as financial or health data because:

- Electrical meters had to be physically accessed to obtain usage data directly from buildings.
- The data showed energy usage over a longer time span such as a month and did not show usage by specific appliance.
- The utilities were not sharing this data in the ways that will now be possible with the Smart Grid.

To give the reader an idea of the extent of the lack of privacy that consumers can be exposed to in the context of a Smart Grid, Figure 11.6 pictures the outcome of nonintrusive appliance load monitoring (NALM) techniques in a residential environment [25].

With NALM techniques it can be inferred how many consumers there are in a household, when they are at home, and other behaviors. Furthermore, with the adoption of plug-in electrical vehicles (PEVs), travel locations and times can also be tracked.

The PIA conducted by NIST addresses residential users and their data, excluding commercial, industrial, or institutional energy consumers. The challenge was to create privacy principles that individuals accept, and that allow the electric sector to thrive and innovation to occur.

Based on AICPA [26], OECD [27], and ISO/IEC [28], NIST gathered a set of ten privacy principles, each one accompanied with a recommendation for utilities to cope with, as they migrate toward a Smart Grid. For example, one of the principles relates to the choices that should be available to consumers regarding the use of their energy consumption data that could be used to reveal personal behavioral

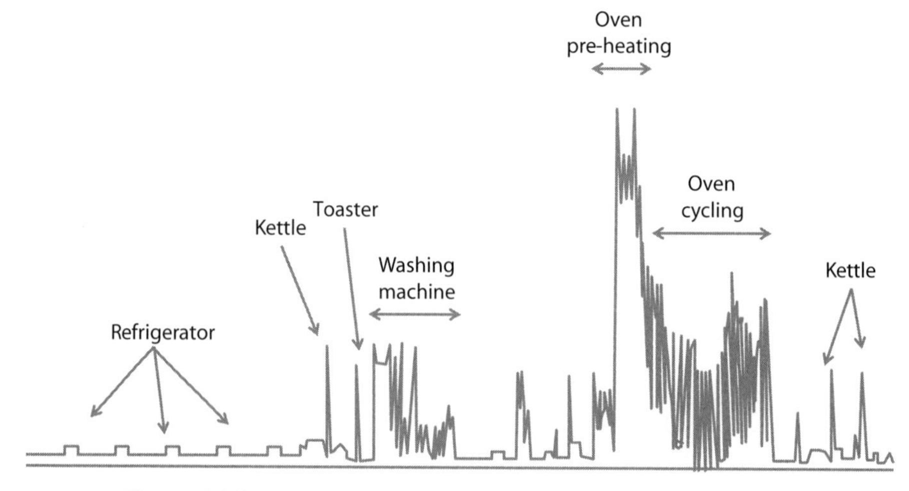

Figure 11.6. Residential power usage to personal activity mapping.

information, and the consent with respect to the collection, use, and disclosure of this information. After finding that utilities currently obtain consent only on billing related data, it is proposed that "the consumer notification should include a clearly worded description to the recipients of services notifying them of (1) any choices available to them about information being collected and obtaining explicit consent when possible; and (2) explaining when and why data items are or may be collected and used without obtaining consent, such as when certain pieces of information are needed to restore service in a timely fashion."

The 200 security requirements were divided into the following types depending on where they are applicable:

- Governance, risk, and compliance (GRC) requirements, applicable to all information systems within an organization, and typically implemented at the organization level
- Common technical requirements, applicable to all information systems within an organization
- Unique technical requirements, allocated to one or more of the logical interface categories

and the following families depending on their nature:

- Access control
- Awareness and training
- Audit and accountability
- Assessment and authorization
- Configuration management

- Continuity of operations
- Identification and authentication
- Information and document management
- Incident response
- Information system development and maintenance
- Media protection
- Physical and environmental security
- Planning
- Security program management
- Personnel security
- Risk management assessment
- Information systems and services acquisition
- Information systems and communication protection
- Information systems and information integrity

The most important piece of information of the conceptual security model is the correlation between the categories of flows of information with the above security requirements, so that utilities know what are the main concerns in each flow of data. In practical terms, this consists of a table that spans up to 10 pages in the original document, and that allocates levels of impact to the intersections. A representative excerpt is given in Table 11.2.

TABLE 11.2. Excerpt of Impacts of Security Requirements in the Categories of Flows of Information

Security Requirements	Type	Family	Categories of Flows of Information					
			1	. . .	7	8	. . .	22
1	Access control policy and procedures	GRC	Applies at all impact levels					
⋮	⋮	⋮	⋮	⋮	⋮	⋮	⋮	⋮
71	Device identification and authentication	Unique technical	High	. . .	Moderate	Moderate	. . .	High
72	Authenticator feedback	Unique technical	Low	. . .	High	High	. . .	High
⋮	⋮	⋮	⋮	⋮	⋮	⋮	⋮	⋮
165	Mobile code	Common technical	Applies at moderate and high levels					
⋮	⋮	⋮	⋮	⋮	⋮	⋮	⋮	⋮
200	Error handling	Common technical	Applies at all impact levels					

The common technical and GRC requirements apply to all categories, and sometimes they are necessary whether the company attributes a high, moderate, or low importance to that requirement in the data flow, and other times they are applicable only when two of the impact levels are relevant. The unique technical requirements apply only to one or more categories, and the impact they have on each is specified.

For example, if transmission company *TRANS* and distribution company *DIST* want to implement access control in the interfaces between their SCADA systems, the companies would start to lookup under which category this flow of information would fall, acknowledge the categories' security pitfalls, lookup what security requirements NIST recommends, and with what level of impact. After a risk assessment particular to the companies, they should evaluate if the security requirements would have to be modified or can be applied directly. In this case, these interfaces fall under category 6, which requires high data accuracy, high availability, and establishment of a chain of trust. The organization will need to review all common technical and GRC requirements to determine if any modification is needed. Also, unique technical requirements relevant to category 6 are to be reviewed. One of them, *Permitted Actions without Identification or Authentication* (Access Control security requirement number 14) is recommended with high impact. It consists of the identification and documentation of any user actions that can be performed on systems without any kind of authentication, whether in normal conditions or under certain circumstances, like emergencies.

Figure 11.7 [29] depicts yet another possible and simpler division of the main security requirements emphasizing the security management as a common part to the rest of the requirements.

11.2.2 The European Example

In parallel, the Conseil International des Grands Réseaux Électriques (CIGRE) Working Group D2.22 has published a proposal for electrical utilities security domains [30] as well as security requirements to be applicable to each domain [31].

For CIGRE, a security domain is a set of systems that are equally critical and need the same level of protection. The four security domains are listed in Table 11.3.

If we take the seven functional domains of NISTIR 7628 (Figure 11.5) and correlate them with the above CIGRE security domains, we derive a summarized version of the 22 categories of flows of information (summarized in the sense that only the seven domains were considered and not the 49 actors).

Figure 11.8 shows a perimetral view of this correlation, where it can be seen that, every time an information exchange crosses a security domain perimeter, a set of security requirements have to be applied. Information flows that have the same departing and ending domains have similar security needs, and can make use of the same category of security requirements.

The power system domains span more than one security domain. Some even connect to the public domain to take end-users' generation, consumption, and energy trade in residential areas into account.

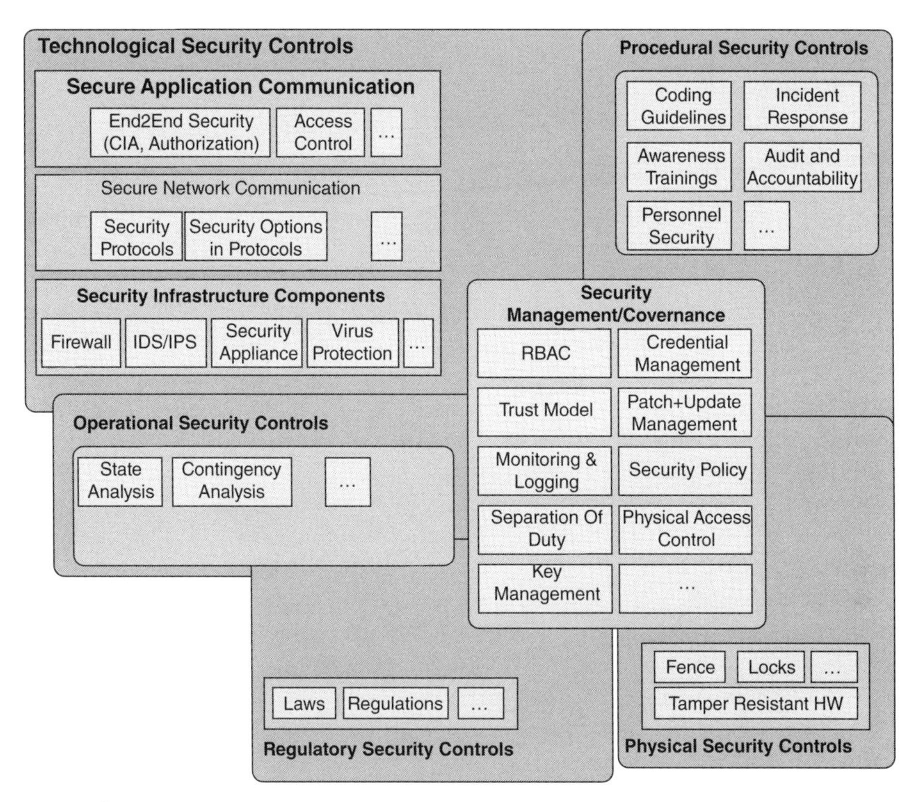

Figure 11.7. Overview of security requirements [29]. Copyright © 2011 IEC.

TABLE 11.3. Security Domains [19]

Security Domain	Required Protection Level	Applies to	Example Systems
Public	Low	Assets, supporting the communication over public networks	Third party networks, Internet
Corporate	Medium	Assets, supporting the business operation with baseline security not essential to the power system reliability and availability	Office level business network
Business critical	High	Assets, supporting the critical operation, which are not critical to power system reliability and availability	Finance network, human resource systems, ERP systems
System operation critical	Very high	Applies to assets directly connected to the availability and reliability of power generation and distribution infrastructure	Control systems, SCADA networks

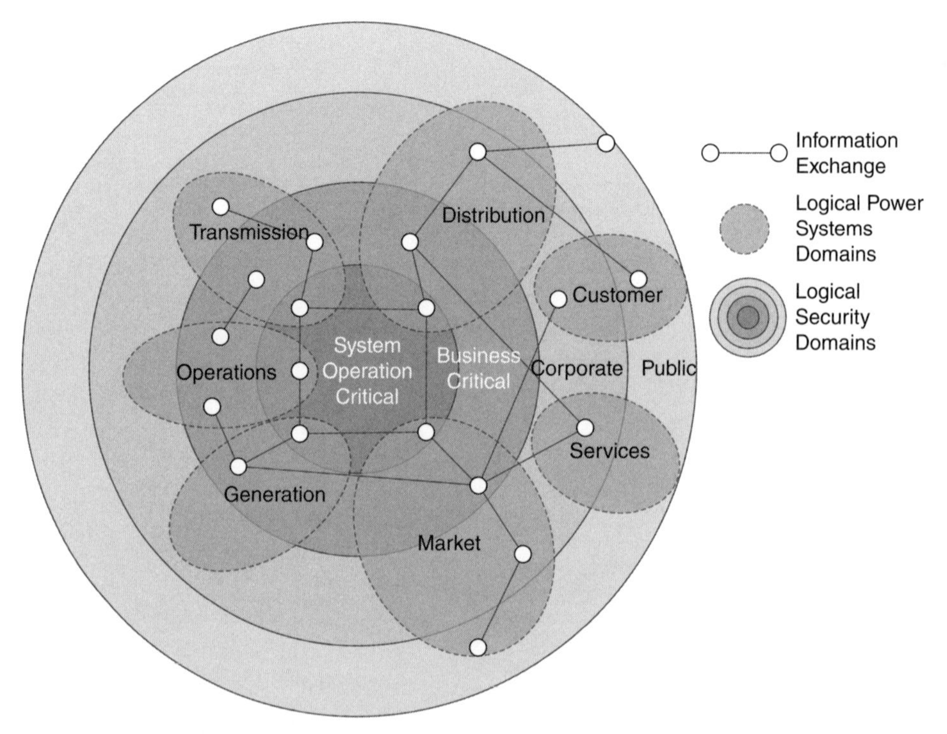

Figure 11.8. Mapping of information security domains to power system domains [29]. Copyright © 2011 IEC.

Finally, it is also of interest to map the security requirements to the security domains by CIGRE. Similar work by Steffen [29] is shown in Table 11.4.

11.3 CONCLUSION

Utilities' security policies should be revised as the grid becomes smart. Even if the ultimate goal continues to be power availability, much new hardware, software, data, and interactions with business partners appears with the new paradigm. This chapter focuses on the cyber security conceptual model as the base of a corporate-wide security policy. The old and new assets are assessed to determine their impact on normal functioning if they ever become compromised. This risk assessment, used as an input of the conceptual model, delivers a security policy for the company. Examples of security models are described, and utilities should tailor them to their reality, such as headcount, budget, internal expertise, and outsourcing policy.

All entities involved in the business, whether governments, financial institutions, utilities themselves, or end-users, have a word to say in enforcing security, as it is one of the variables with the power to transform the hype of Smart Grids into reality.

TABLE 11.4. Security Requirements Applicable to the Different Security Domains

Security Control	Realization Example, Notes	Security Domain C	B	O
Authentication	Strong user-to-device and user-to-application authentication (local and remote)	x	x	x
	Application of directory services for user authentication and single sign-on technologies (e.g., Kerberos, NTLM)	x	x	
	Support of two-factor user authentication technologies (e.g., smart token, OTP)		x	x
	Device-to-device authentication (e.g., client—server–authentication via X.509 based certificates)		x	x
	Enforcement of mutual authentication according to IEC 62351 Parts 3, 4, 5, 6 in IEC-based communication protocols			x
	(Physical) ID check through plant security	x	x	x
Access control	Definition and enforcement of a physical security perimeter supported through security personnel, locks, fences, video surveillance, and so on.		x	x
	Definition of a logical security perimeter supported through application of firewalls, remote access gateways, and so on. (for local and remote access)	x	x	x
	Enforcement of least-privilege escalation	x	x	x
	Definition and enforcement of role-based access control (RBAC)		x	x
	Application of IEC 62351 Part 8 RBAC for access controlled engineering and operation			x
	Application of secure authentication options in DNP3 communications			x
	Password policy (e.g., complexity criteria, aging, etc.)	x	x	x
	Network segmentation (applying firewalls and DMZ)	x	x	x
Integrity protection	Integrity protection of communicated data between different data network components (e.g., geographically dispersed data networks) using security appliances or security functionality in routers to protect communicated traffic		x	x
	Application of integrity protection options in SCADA and ICS protocols		x	x
	Application of IEC 62351 Parts 3, 4, 5, 6 integrity protection features in IEC-based communication protocols		x	x
	Application of secure authentication options in DNP3 communications			x
	Integrity protection of stored data (e.g., disk, tape)		x	x

(Continued)

TABLE 11.4. (*Continued*)

Security Control	Realization Example, Notes	Security Domain C	B	O
Encryption	Confidentiality protection of sensitive data during communications between different data network components using security appliances or security functionality in routers to protect communicated traffic		x	x
	Application of confidentiality protection options in SCADA and ICS protocols		x	x
	Confidentiality protection of stored live data; disk encryption for engineering and control systems		x	x
	Confidentiality protection of stored backup data (e.g., disk, tape)		x	x
	Application of IEC 62351 Parts 3, 4, 5 confidentiality protection using TLS in IEC-based communication protocols		x	x
	Access to confidential data following the need to know principle	x	x	x
Monitoring and logging	Maintenance and monitoring of computer and network security components (e.g., system event log auditing or SNMP v3 application to monitor system events)		x	x
	Anomaly/intrusion prevention/detection (IDS/IPS deployment)		x	x
	Login/account management on control and engineering systems		x	x
	Auditing/logging of all automated/scripted login sessions		x	x
	Logging of control application access, including user ID, event time		x	x
	User activity on control systems while logged in			x
Security management	Personnel risk assessment (e.g., security clearance of technical personnel)		x	x
	Separation of duty for control, operation, safety, and security	x	x	x
	Security training of personnel (incident handling, security process handling, etc.)		x	x
	System hardening according to security measurement plans	x	x	x
	Security documentation (for processes and systems)	x	x	x
	Regular security assessments of components	x	x	x
	Key and credential management (NWIP for IEC 62351 Part 9)			x
	IEC 62351 Part 7 application to describe power system objects to be managed			x
	Definition and maintenance of roles to support IEC 62351 Part 8		x	x
	Incident response (events and alarming)	x	x	x
	Backup and recovery of business and operation relevant information	x	x	x
	Patch management (organizational part)	x	x	x
	Establishment of a security organization with defined responsibilities and responsible persons	x	x	x
	Definition of identification and authentication procedures	x	x	x

TABLE 11.4. (*Continued*)

Security Control	Realization Example, Notes	Security Domain		
		C	B	O
Availability and robustness	Redundancy concepts for IT infrastructure		x	x
	Firewall concepts (DoS protection)	x	x	x
	IPD/IPS systems (DoS protection)	x	x	
	Connection limitations (bandwidth, number of connections, etc.)			
	Antivirus software/appliances, white listing	x	x	x
	Backup and restore concepts and procedures		x	x
	Emergency concept		x	x
	Quality control (e.g., NERC compliance of used products, etc.)		x	x
	Application of products supporting a secure lifecycle (product selection shall obey security requirements covering the complete product lifecycle)	x	x	x
Privacy protection	Protection of person (user, customer, etc.) related information in AMI	x	x	x
	Protection of business relevant data (e.g., generation data, system status)	x	x	

Source: Steffen [30]. Copyright © 2011 IEC. C, corporate; B, critical; O, system operation critical.

REFERENCES

[1] NIST, "Minimum Security Requirements for Federal Information and Information Systems," Federal Information Processing Standards Publication (FIPS PUB) 200, March 2006.

[2] NIST, "The Code of Laws of the United States of America," Title 44, Section 3542, February 2010.

[3] NIST, "NIST InterAgency Report 7628 v1," September 2010.

[4] NERC, "Critical Infrastructure Protection Standards," 2009–2011.

[5] ESTEC, http://www.estec-project.eu (accessed April 2011).

[6] A. Maykuth, "Utilities' smart meters save money, but erode privacy," *The Philadelphia Inquirer*, September 2009.

[7] A. Jamieson, "Smart meters could be spy in the home," *The Telegraph*, October 2009.

[8] J. Weiss, *Protecting Industrial Control Systems from Electronic Threats*. Momentum Press, 2010, p. 36.

[9] IEC, 62351-10 TR, to appear.

[10] SCE, www.sce.com (accessed April 2011).

[11] NIST, http://collaborate.nist.gov/twiki-sggrid/bin/view/SmartGrid/UseCases (accessed April 2011).

[12] IETF, "Lightweight Directory Access Protocol (LDAP): Technical Specification Road Map," RFC 4510, June 2006.

[13] IETF, "Remote Authentication Dial In User Service (RADIUS)," RFC 2865, June 2000.

[14] IETF, "RADIUS Accounting," RFC 2866, June 2000.

[15] IETF, "The Transport Layer Security (TLS) Protocol v1.2," RFC 5246, August 2008.

[16] IETF, "Security Architecture for the Internet Protocol," RFC 4301, December 2005.

[17] IETF, "Using Advanced Encryption Standard (AES) CCM Mode with IPsec Encapsulating Security Payload (ESP)," RFC 4309, December 2005.

[18] IEEE, "Part 11: Wireless LAN Medium Access Control (MAC) and Physical Layer (PHY) Specifications. Amendment 6: Medium Access Control (MAC) Security Enhancements," IEEE Standard 802.11i-2004, July 2004.

[19] NERC Control Systems Security Working Group, "Security Guideline for the Electricity Sector: Time Stamping of Operational Data Logs," November 2009.

[20] IEEE 1815 WG, "Standard for Electric Power Systems Communications--Distributed Network Protocol," July 2010.

[21] IEC Technical Committee 57, Working Group 10, "61850 Communication Networks and Systems In Substations," 2003.

[22] IEC Technical Committee 57, Working Group 15, "62351 Communication Networks and Systems in Substations," May 2007.

[23] NIST SP 800-53, "Recommended Security Controls for Federal Information Systems and Organizations," August 2009.

[24] Department of Homeland Security, "Catalog of Control Systems Security: Recommendations of Standard Developers," March 2010.

[25] Elias Leake Quinn, "Smart Metering & Privacy: Existing Law and Competing Policies," Spring 2009.

[26] American Institute of Certified Public Accounts (AICPA), "Generally Accepted Privacy Principles," January 2009.

[27] Organization for Economic Cooperation and Development (OECD), "Privacy Principles," 1998.

[28] International Organization for Standardization (ISO) and International Electrotechnical Commission (IEC), "Information Security Management Principles from the Joint Technical Committee (JTC)," *International Standard ISO/IEC 27001* October 2005.

[29] S. Fries, "Technical Report on Security Architecture Guidelines," IEC Technical Group 57, 62351-10TR, to appear.

[30] CIGRE WG D2.22, "Security Frameworks for Electric Power Utilities," December 2008.

[31] CIGRE WG D2.22, "Security Technologies Guideline," June 2009.

12

SMART GRID SECURITY STANDARDIZATION

Steffen Fries and Hans-Joachim Hof

12.1 STANDARDIZATION ACTIVITIES

Standardization in general ensures the interoperability of different vendors' products and provides a defined interface for interactions between components and systems. Within the Smart Grid standards from a variety of standardization bodies are being used. The following sections provide at first an overview about security requirements to Smart Grid deployments and their specifics. These requirements are then mirrored against the main standardization activities with respect to information technology security.

12.2 SMART GRID SECURITY REQUIREMENTS

The operational environment of the power systems information infrastructure differs from office environments or telecommunication environments in several aspects. Some specific properties of power systems include:

Smart Grid: Applications, Communications, and Security, First Edition.
Edited by Lars Torsten Berger and Krzysztof Iniewski.

- Communication between components is often asymmetric and message oriented (like multicast).
- There are strict timing requirements (down to milliseconds).
- They have a long lifetime or operation time of components (in the range of 10–30 years).
- Based on the long operation time of the components, there are strong requirements for interoperability with legacy systems and for migration concepts.
- Interoperability with legacy systems influences the maintainability of power systems and makes systems more complex, especially if security of maintenance and operation is desired.
- There are limitations of connectivity of systems or system components to a central (control) network.
- Higher availability and latency requirements leading, for example, to missing or limited patch windows within customer facilities complicate the security patch process. Often customers do not or only reluctantly accept updates in their environment.
- There is interconnection of independent entities (producers, other transport/distribution network operators).
- Computer-constrained resources preclude many information technologies.
- Field devices may be installed in physically unprotected areas. Direct access to components by maintenance personnel is costly and often impractical as devices are installed in widely distributed areas.

Security services to be supported in energy automation comprise the usual suspects:

- *Authentication.* The property that the claimed identity of an entity is correct.
- *Authorization.* The process of giving someone permission to do or have something.
- *Integrity.* The property that information has not been altered in an unauthorized manner.
- *Nonrepudiation.* The property that involvement in an action cannot be denied.
- *Confidentiality.* The property that information is not made available or disclosed to unauthorized individuals, entities, or processes.

However, in contrast to office networks, the power system environment is quite different as listed previously. Also, there is another weighting of security objectives in power systems than in office networks. Power systems are rather focused on authentication and integrity protection than on confidentiality (which is typically the main objective in office and telecommunication environments). This is especially true for energy automation control and protection communication. Nevertheless, with the increased introduction of the advanced metering infrastructure (AMI), privacy emerges as an important security requirement. Hence, confidentiality (a

	Energy Control Systems	Office IT
Anti-virus / mobile code	Uncommon / hard to deploy	Common / widely used
Component Lifetime	Up to 20 years	3-5 years
Outsourcing	Rarely used	Common
Application of patches	Use case specific	Regular / scheduled
Real time requirement	Critical due to safety	Delays accepted
Security testing / audit	Rarely (operational networks)	Scheduled and mandated
Physical Security	Very much varying	High
Security Awareness	Increasing	High
Confidentiality (Data)	Low – Medium	High
Integrity (Data)	High	Medium
Availability / Reliability	24 x 7 x 365 x ...	Medium, delays accepted
Non-Repudiation	High	Medium

Figure 12.1. Comparison of office/power system security requirements [2].

security requirement that can be used to achieve privacy) may still be required to protect the user's privacy. Figure 12.1 summarizes the differences for basic security services between office networks and power system networks. The classifications of low, medium, and high are comparable with the NISIR 7628 vol. 1 [1] impact levels.

It should be noted, that while the influences of the information infrastructure to the electrical infrastructure are obvious, there is also a feedback of the electrical infrastructure to the information infrastructure. Information about states and events or engineering data in the electrical infrastructure can be used to derive relevant input for security controls in the information infrastructure. One example is the utilization of field device configuration data for the compilation of rules for intrusion detection systems (IDSs) and intrusion prevention systems (IPSs).

12.3 SECURITY RELEVANT REGULATION AND STANDARDIZATION ACTIVITIES

The power market and the operation of power systems are strongly influenced by a large number of regulations and standards. This section gives an overview of important regulation and standardization activities relating to security in Smart Grid systems. Note that this list is not complete. For a survey on proposed standardization activities related to Smart Grid in general, the IEC and NIST activities defining standardization roadmaps are referred to (the respective documents are referenced in the following subsections).

12.3.1 ISO/IEC

The ISO/IEC (International Organization for Standardization / International Electrotechnical Commission) is involved in the development of several standards relating to Smart Grid operation and communication. The following list provides an overview about the most relevant.

ISO/IEC 27001 [3], "Information Technology—Security Techniques—Specification for an Information Security Management Systems," specifies a set of information security management requirements designed to be used for certification purposes.

ISO/IEC 62351-1 to 8 (cf. [4, 5]) is being standardized by the ISO/IEC TC 57 WG15 and defines data and communications security for power systems management and associated information exchange. It applies directly to substation automation deploying IEC 61850 and IEC 60870-x protocols as well as in adjacent communication protocols supporting energy automation, like ICCP (Inter Control Center Protocol, TASE.2) used for control center communication. A clear goal of the standardization of IEC 62351 is the assurance of end-to-end security. Currently, the standard comprises eight parts that are in different states of completion. The standard is extendable, allowing addition of further parts if considered necessary. The newest part will target the management of security credentials. Table 12.1 provides an overview about the different parts and their standardization status.

As this standard directly addresses energy automation as one main component of the Smart Grid, it is discussed here more elaborately.

Part 1 of IEC 62351 [4] provides an overview about the different parts of the standard. It also provides an overview about security services necessary to protect against certain threats from a more general point of view and their mapping to the power domain by using IEC 62351 defined security technology. Part 2 provides the

TABLE 12.1. IEC 62351 Parts [5]

IEC 62351	Definition of Security Services for	Standardization Status
Part 1	Introduction and overview	Technical specification
Part 2	Glossary of terms	Technical specification
Part 3	Profiles including TCP/IP	Technical specification
Part 4	Profiles including MMS	Technical specification
Part 5	Security for IEC 60870-5 and derivatives	Technical specification
Part 6	Security for IEC 61850	Technical specification
Part 7	Network and system management (NSM) data object models	Technical specification
Part 8	Role-based access control for power systems management	Technical specification
Part 9	Credential management	New work item proposal
Part 10	Security architecture guidelines	Draft technical report

terminology used throughout the different parts. Part 3 to 8 are directly related to dedicated protocols like IEC 61850 (IEC 62351 Part 6) and IEC 60870-5-x (IEC 62351 Part 5) and their mappings to lower layer protocols like TCP/IP (IEC 62351 Part 3) and MMS (Manufacturing Message Specification, IEC 62351 Part 4) as well as the mapping of security to the network management (Part 7) and role-based access control (Part 8). These parts utilize symmetric as well as asymmetric cryptographic functions to secure the payload and the communication link.

IEC 62351 applies existing security protocols like transport layer security (TLS), which has been used successfully in other technical areas and industrial applications, in different parts of the standard. The application of TLS provides for security services like mutual authentication of communication peers and also integrity and confidentiality protection of the communicated data. Thanks to the mutual authentication required by IEC 62351, attacks like man-in-the-middle can be successfully countered.

Part 3 of IEC 62351 defines how security services can be provided for TCP/IP based communication. As TLS is based on TCP/IP Part 3 specifies cipher suites (the allowed combination of authentication, integrity protection, and encryption algorithms) and also states requirements for the certificates to be used in conjunction with TLS. These requirements comprise, for instance, dedicated certificate context, application of signatures, and the definition of certificate revocation procedures. For the latter, the focus lies mostly on certificate revocation lists (CRLs). The application of the online certificate status protocol (OCSP) is not considered due to limited communication links within the substations. In contrast to office applications, the connections in energy automation are relatively long-lasting. This requires the definition of strict key update and CRL update intervals, to restrict the application of cryptographic keys not only for a dedicated number of packets but also for a dedicated time. Another challenge to consider is the interoperability requirements between the implementations of different vendors' products.

Part 4 of IEC 62351 specifies procedures, protocol enhancements, and algorithms targeting the increase of security messages transmitted over MMS. MMS is an international standard (ISO 9506) dealing with a messaging system for transferring real-time process data and supervisory control information either between networked devices or in communication with computer applications. Part 4 defines procedures on the transport layer, based on TLS, as well as on the application layer to protect the communicated information. For general information on MMS please refer to section "IEC 61850."

Besides TCP/IP, IEC 62351 Part 5 relates to the specialties of serial communication. Here, additional security measures are defined to especially protect the integrity of the serial connections applying keyed hashes. This part also specifies a separate key management necessary for the security measures.

Part 6 of IEC 62351 describes security for IEC 61850 peer-to-peer profiles. It covers the profiles in IEC 61850 that are not based on TCP/IP for the communication of generic object oriented substation events (GOOSEs), generic substation state events (GSSEs), and sample measured values (SMVs) using, for example, plain Ethernet. Specific for this type of communication is the usage of multicast transfer, were each field device decides based on the message type and sender if it processes the message or not. Security employs digital signatures on the message level to protect the integrity of the messages sent and also to cope with multicast connections.

IEC 62351 Part 7 describes security-related data objects for end-to-end network and system management (NSM) and also security problem detection. These data objects support the secure control of dedicated parts of the energy automation network.

Part 8 of the standard is currently in definition and addresses the integration of role-based access control mechanisms into the whole domain of power systems. This is necessary as in protection systems and in control centers authorization and stringent traceability are required. One usage example is the verification of who has authorized and performed a dedicated switching action. Part 8 supports role-based access control in terms of three profiles. Each of the profiles uses an own type of credential as there are identity certificates with role enhancements, attribute certificates, and software tokens.

A first glimpse at the current IEC 62351 parts shows that many of the technical security requirements to be applied to energy automation components and systems can be derived directly from the standard. For instance, Parts 3 and 4 explicitly require the usage of TLS. They define cipher suites, which are to be supported as mandatory. These parts also define recommended cipher suites and also deprecate cipher suites, which shall not be applied from a IEC 62351 point of view. Note that the mandatory cipher suites do not coincide with the mandatory cipher suites of the different TLS versions (1.0–RFC 2246, 1.1–RFC 4346, 1.2–RFC 5246). IEC 62351 always references TLS v1.0 to better address interoperability.

Analyzing the standard more deeply shows that several requirements are provided rather implicitly. These requirements relate mostly to the overall key management, which guarantees a smooth operation of the security mechanisms. IEC 62351 heavily applies asymmetric cryptography based on certificates and associated private keys, for example, in the context of transport layer protection (using TLS) but also on the application layer as in Part 6 to secure GOOSEs. But to apply this type of credentials, the general handling like generation, provisioning, revocation, and especially the initial distribution to all participating entities needs to be considered. This is currently underspecified, but has been acknowledged by standardization as important for the general operation and also for the interoperability of different vendors' products. As the standard is extensible, a new part describing credential handling in the context of IEC 62351 services is under development. According to the numbering this will be Part 9.

Moreover, a security architecture, necessary for building, engineering, and operating power systems, is a necessary base to ensure safety and reliability of these systems. Therefore, a further work item has been started to describe a security architecture as hands-on guideline for system engineers and operators to implement power systems securely.

Besides standard enhancements, which have become necessary through findings during the implementation of IEC 62351, new scenarios may also require the further evolvement of already existing or new parts of the standard, to better cope with new use cases.

ISO/IEC 62443 [6] is being standardized by ISO/IEC TC65 WG10 and will reflect the publications of ISA 99 (see below); that is, ISA 99 documents will be submitted to the IEC voting process. Hence, this standard is likely to be similar, if not identical, to ISA 99.

The IEC Smart Grid Strategic Group (SG3) has issued the Smart Grid Standardization Roadmap report (SMB/4175/R) [7], which encompasses requirements, status, and recommendations of standards relevant for the Smart Grid. Security is covered in detail in a separate section of this document. An overall security architecture capturing the complexity of the Smart Grid is requested. Besides this, the following recommendations pertaining to open items and necessary enhancements are listed:

- A specification of a dedicated set of security controls (e.g., perimeter security, access control)
- A defined compartmentalization of Smart Grid applications (domains) based on clear network segmentation and functional zones
- A specification comprising identity establishment (based on trust levels) and identity management
- Necessity to consider security of the legacy components within standardization
- Harmonization with the IEC 62443 standard to achieve common industrial security standards
- Recommendation to review, adapt, and enhance existing standards in order to support general and ubiquitous security across wired and wireless connections

12.3.2 IEEE (Institute of Electrical and Electronics Engineers)

IEEE 1686–2007 (cf. [8]) is the standard for substation intelligent electronic devices (IEDs) cyber security capabilities. The standard defines functions and features that must be provided in substation intelligent electronic devices to accommodate critical infrastructure protection programs. It addresses security in terms of access, operation, configuration, firmware revision, and data retrieval from IEDs. Encryption for the secure transmission of data, both within and external to the substation, is not part of this standard.

12.3.3 ISA (International Society of Automation)

ISA-99 [9] defines a framework addressing "security for industrial automation and control systems." It covers the processes for establishing an industrial automation and control systems security program based on risk analysis, establishing awareness and countermeasures, and monitoring and cyber security management systems. It describes several categories of security technologies and also the types of products available in those categories along with preliminary recommendations and guidance for using those security technologies. The standard consists of several subparts, which are in different states of completion.

12.3.4 CIGRE

CIGRE, the International Council on Large Electric Systems, published the document "Security for Information Systems and Intranets in Electric Power System." The guideline presents the work of the Joint Working Group D2/B3/C2-01 [10]. The paper focuses on the importance of handling information security within an electric utility, dealing with various threats and vulnerabilities, the evolution of power utility information systems from isolated to fully integrated systems, the concept of using security domains for dealing with information security within an electric utility, and the use of the ISO/IEC 17799 standard (predecessor of ISO 27000).

WG D2.22 is entitled "Treatment of Information Security for Electric Power Utilities." Three reports, "Risk Assessment of Information and Communication Systems" [11], "Security Frameworks for Electric Power Utilities" [12], and "Security Technologies Guideline" [13], provide practical guidelines and experiences for determining security risks in power systems and the development of frameworks including control system security domains. This is done by elaborating the specific security requirements of these types of domains, and also by giving a view of interrelated domains and high level frameworks that are necessary to manage corporate risks. Domain-specific cyber security controls are being defined and guidance is provided on how these controls can be applied to electric utility networks.

12.3.5 NERC (North American Electric Reliability Corporation)

The North American Electric Reliability Corporation's (NERC) mission is to ensure the reliability of the bulk power system in North America. To achieve this, NERC develops and enforces reliability standards and monitors users, owners, and operators for preparedness. NERC is a self-regulatory organization, subject to oversight by the U.S. Federal Energy Regulatory Commission and governmental authorities in Canada. NERC has established the Critical Infrastructure Protection (CIP) Cyber Security Standards CIP–002 through CIP–009 [13–20], which are defined to provide a foundation of sound security practices across the bulk power system. These standards are not designed to protect the system from specific and imminent threats. They apply to operators of bulk electric systems (see also NERC [21]). The profiles originate in 2006. NERC-CIP provides a consistent framework for security control perimeters and access management with incident reporting and recovery for critical cyber assets and cover functional as well as nonfunctional requirements. Table 12.2 provides an overview of the different NERC-CIP parts.

TABLE 12.2. NERC-CIP Overview

CIP	Title / Content
002	*Critical Cyber Asset Identification* Identification and documentation of critical cyber assets using risk-based assessment methodologies
003	*Security Management Controls* Documentation and implementation of cyber security policy reflecting commitment and ability to secure critical cyber assets
004	*Personnel and Training* Maintenance and documentation of security awareness programs to ensure personnel knowledge on proven security practices
005	*Electronic Security Protection* Identification and protection of electronic security perimeters and their access points surrounding critical cyber assets
006	*Physical Security Program* Creation and maintenance of physical security controls, including processes, tools, and procedures to monitor perimeter access
007	*Systems Security Management* Definition and maintenance of methods, procedures, and processes to secure cyber assets within the electronic security perimeter and not adversely affect existing cyber security controls
008	*Incident Reporting & Response Planning* Development and maintenance of a cyber security incident response plan that addresses classification, response actions, and reporting
009	*Recovery Plans for Critical Cyber Assets* Creation and review of recovery plans for critical cyber assets
010	*Bulk Electrical System Cyber System Categorization* (draft) Categorization of BES systems that execute or enable functions essential to reliable operation of the BES into three different classes.
011	*Bulk Electrical System Cyber System Protection* (draft) Mapping of security requirements to BES system categories defined in CIP-010

The draft standard CIP-011 may not lead to new cyber security requirements, but it provides a new organization of the existing requirements of the existing CIP standards. New is the classification of bulk electric systems (BESs) into the three categories low, medium, and high impact BES cyber systems and their mapping to security controls.

12.3.6 National Activities

The National Institute of Standards and Technology (NIST) is a U.S. federal technology agency that develops and promotes measurement, standards, and technology. The following subset of NIST documents directly applies to security in Smart Grid environments:

- NIST Special Publication 800-53: "Recommended Security Controls for Federal Information Systems and Organizations" [22] provides guidelines for selecting and specifying technical and organizational security controls and connected processes for information systems supporting the executive agencies of the federal government to meet the requirements of the Federal Information Processing Standard 200 (FIPS 200, "Minimum Security Requirements for Federal Information and Information Systems"). It provides an extensive catalog of security controls and maps these in a dedicated appendix to industrial control systems.
- NIST Special Publication (SP) 800-82: "Guide to Industrial Control Systems (ICS) Security" [23] provides guidance on how to secure industrial control systems (ICSs), including supervisory control and data acquisition (SCADA) systems, distributed control systems (DCSs), and other control system configurations such as programmable logic controllers (PLCs). It uses the NIST SP 800-53 [22] as a base and provides specific guidance on the application of the security controls in NIST SP 800-53. This publication is an update to the second public draft, which was released in 2007.
- NIST Special Publication 1108: "NIST Framework and Roadmap for Smart Grid Interoperability Standards" [24] describes a high-level conceptual reference model for the Smart Grid. It lists 75 existing standards that are applicable or likely to be applicable to the ongoing development of the Smart Grid. The document also identifies 15 high-priority gaps and potential harmonization issues for which new or revised standards and requirements are needed.
- The NIST Internal Report (NISTIR) 7628 [1, 25, 26] originates from the Smart Grid Interoperability Panel (Cyber Security WG) and targets the development of a comprehensive set of cyber security requirements building on NIST SP 1108 [24], mentioned previously. The document consists of three subdocuments targeting strategy [1], security architecture [25], and requirements and supportive analyses and references [26].

The *"Catalog of Control Systems Security—Recommendations for Standards Developers"* of the U.S. Department of Homeland Security [27] presents a compilation of practices that various industry bodies have recommended to increase the security of control systems from both physical and cyber attacks. The recommendations in this catalog are grouped into 18 families, or categories, that have similar emphasis. This catalog is not limited for use by a specific industry sector but can be used by all sectors to develop a framework needed to produce a sound cyber security program. It is a collection of recommendations to be considered when reviewing and developing cyber security standards for control systems. The recommendations in this catalog are intended to be broad enough to provide any industry using control systems with the flexibility needed to develop sound cyber security standards specific to their individual security needs.

The German *"Bundesverband für Energie- und Wasserwirtschaft—BDEW"* was founded by the federation of four German energy-related associations. It introduced a white paper [28] defining basic security measures and requirements for IT-based

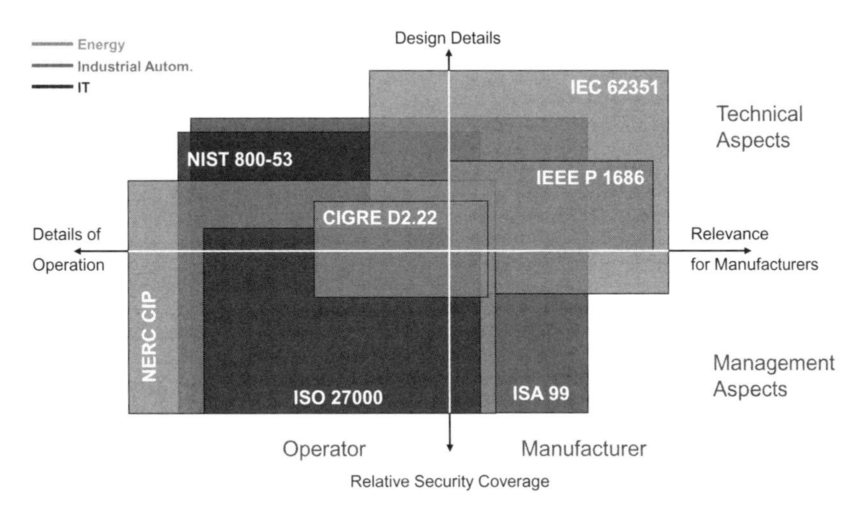

Figure 12.2. Graphical representation of scope and completeness of selected standards [2].

control, automation, and telecommunication systems, taking into account general technical and operational conditions. It can be seen as a further national approach targeting similar goals as NERC-CIP. The white paper addresses requirements for vendors and manufacturers of power system management systems and is used as an amendment to tender specification.

Within the European Union a dedicated *Expert Group of the Task Force Smart Grid* is currently working on regulatory recommendations for safety of data handling and data protection [29]. The goal of the Task Force is the identification and production of a set of regulatory recommendations to ensure EU-wide consistent and fast implementation of Smart Grids, while achieving the expected Smart Grids' services and benefits for all users involved. The goal of the Expert Group for security is the identification of appropriate regulatory scenario and recommendations for data handling, security, and consumer protection to establish a data privacy and data security framework that both protects and enables.

Selected standards and their applicability are presented in Figure 12.2 as proposed in the European funded project ESCoRTS [30]. The figure is intended to provide a better overview, for operators and manufacturers, of which standard influences their business most. The original figure has been enhanced with additional energy-related standards in the context of the technical report of the IEC regarding security architecture guidelines. The different shadings in the figure indicate the targeted audience.

While IEC 62351 addresses the energy sector, more specifically substation automation systems, NERC-CIP generally targets energy operators. While ISO 27000 and NIST 800-53 are mainly targeted to IT environments (thus targeted at protecting information), other standards such as ISA 99 or IEEE P1686 directly address (industrial) automation systems. It should be noted that NIST SP800-53 Appendix I is explicitly for industrial control systems, as is NIST SP800-82.

Standards extending to the right in the x-axis direction have relevance for manufacturers. Typically, such standards have detailed technical requirements up to the definition of special security protocols, which must be implemented by the manufacturers. In contrast, the more a standard extends to the left of the x-axis, the more it is focused on a secure operation. NERC-CIP, for example, prescribes specific actions for operators to comply with, thus providing implicit requirements to the manufacturers to support the operators.

Standards extending to the top of the y-axis list precise design details and leave little room for interpretation. IEC 62351, for instance, provides design details to such an extent that device interoperability between various manufacturers can be guaranteed.

Standards extending to the bottom of the y-axis cover a broad range of various security areas and thus can be consulted in order to get estimation on the overall security level.

12.4 TRENDS IN ENERGY AUTOMATION SECURITY

In general, one trend in energy automation is to move away from proprietary solutions. Instead, current standards are used. This goes along with reusing existing security approaches like TLS or PKI (public key infrastructure) based services instead of reinventing the wheel.

Web services, for example, are gaining more momentum in the energy domain. They have already been addressed as part of the wind power craft related standard IEC 61400-25 and it is expected that there will be a mapping for IEC 61850 in the near future. Web services enable the application of Web security mechanisms like XML security to provide encryption and integrity protection. Moreover, authorization can also be addressed utilizing the Security Assertion Markup Language (SAML). SAML allows the definition of secured tokens, to be issued by a trusted component. Currently, security is also not being addressed in the wind power standard. Nevertheless, as Web service security is already defined (by the W3C), the standard only needs to be enhanced with a mapping to the available Web security, without the necessity to define its own security mechanisms.

To ensure security interworking between installations utilizing different mappings of IEC 61850, like MMS or Web service, secure services transition functions need to be defined. Therefore, from the interworking perspective, the integration of security enhancements in MMS may provide a better base for secure interworking as it operates on the same level as Web services and already provides an end-to-end application layer connection.

Other trends include using role based access control throughout the energy automation system, from the field to the top of the energy automation pyramid, and securing devices in areas with no physical protection (e.g., phasor measurement units on poles).

12.5 CONCLUSION

Security plays an increasing role in the definition and realization of Smart Grid use cases. Especially through regulations likely NERC CIP or the BDEW White Paper, security requirements are posed to the operators of Smart Grid environments. Moreover, standardization emphasizes the work on security and Smart Grid related standards as visible in standardization roadmap documents from the IEC or NIST [24]. The importance of security within standardization is also underlined by the NIST's identification of fundamental Smart Grid standards; standard IEC 62351 is one of the five [31].

As it was seen, the general approach is to use already existing security protocols and adopt them as far as possible, without reinventing the wheel. This is important as it allows also reusing existing telecommunication products in Smart Grid scenarios and thus also supports end-to-end security from the power system domain to other domains.

REFERENCES

[1] NIST IR 7628, "Guidelines for Smart Grid Cyber Security, Vol. 1 Smart Grid Cyber Security Strategy," August 2010. Available at: http://csrc.nist.gov/publications/nistir/ir7628/nistir-7628_vol1.pdf.

[2] "IEC Technical Report on Security Architecture Guidelines for TC 57 Systems," IEC 62351-10 TR, to appear.

[3] ISO/IEC 27001:2005, "Information Technology—Security Techniques—Specification for an Information Security Management System. Available at: http://www.iso27001security.com/html/27001.html.

[4] ISO-IEC 62351, Parts 1–8. Available at: http://www.iec.ch/cgi-bin/procgi.pl/www/iecwww.p?wwwlang=E&wwwprog=sea22.p&search=iecnumber&header=IEC&pubno=62351&part=&se=.

[5] S. Fries, H. J. Hof, T. Dufaure, and M. Seewald, "Security for the Smart Grid—Enhancing IEC 62351 to improve security in energy automation control," *Int. J. Advances in Security*, ISSN 1942-2636, February 2011, available at www.iariajournals.org/security/sec_v3_n34_2010_paged.pdf.

[6] ISO-IEC 62443, Parts 1–3. Available at: http://www.iec.ch/cgi-bin/procgi.pl/www/iecwww.p?wwwlang=E&wwwprog=sea22.p&search=iecnumber&header=IEC&pubno=62443&part=&se=.

[7] IEC, "IEC Smart Grid Standardization Roadmap," Ed. 1.0—2009-12, June 2010. Available at: http://www.standards.co.nz/NR/rdonlyres/6C7E44C2-E6B6-491F-9F7A-2A206A225AC6/0/IECSmartGridStanRoadmap0610.pdf.

[8] IEEE 686-2007, IEEE Standard for Substation Intelligent Electronic Devices (IEDs) Cyber Security Capabilities, available at http://smartgrid.ieee.org/standards/approved-ieee-smartgrid-standards.

[9] ISA 99 Standards Framework. Available at: http://www.isa-99.com/.

[10] G. N. Ericsson, “Management of information security for an electric power utility—on security domains and use of ISO/IEC17799 standard,” *IEEE Trans. Power Delivery*, 20(2), pp. 683–690, April 2005.

[11] CIGRE Report, “Risk Assessment of Information and Communication Systems,” WG D2.22, August 2008, Electra.

[12] CIGRE Report, “Security Frameworks for Electric Power Utilities,” WG D2.22, December 2008, Electra.

[13] CIGRE Report, “Security Technologies Guideline,” WG D2.22, June 2009, Electra; NERC, “Cyber Security—Critical Cyber Asset Identification,” CIP-002-4, January 2011. Available at: http://www.nerc.com/files/CIP-002-4.pdf.

[14] NERC, “Cyber Security—Security Management Controls,” CIP-003-4, January 2011. Available at: http://www.nerc.com/files/CIP-003-4.pdf.

[15] NERC, “Cyber Security—Personnel & Training,” CIP-004-04, January 2011. Available at: http://www.nerc.com/files/CIP-004-4.pdf.

[16] NERC, “Cyber Security—Electronic Security Perimeter(s),” CIP-005-4 January 2011. Available at: http://www.nerc.com/files/CIP-005-4.pdf.

[17] NERC, “Cyber Security—Physical Security of Critical Cyber Assets,” CIP-006-4, January 2011. Available at: http://www.nerc.com/files/CIP-006-4.pdf.

[18] NERC, “Cyber Security—Systems Security Management,” CIP-007-04, January 2011. Available at: http://www.nerc.com/files/CIP-007-4.pdf.

[19] NERC, “Cyber Security—Incident Reporting and Response Planning,” CIP-008-4, January 2011. Available at: http://www.nerc.com/files/CIP-006-4.pdf.

[20] NERC, “Cyber Security—Recovery Plans for Critical Cyber Assets,” CIP-009-4, February 2011. Available at: http://www.nerc.com/files/CIP-009-4.pdf.

[21] NERC, North American Reliability Corporation, Standards. Available at: http://www.nerc.com/page.php?cid=2|20.

[22] NIST SP 800-53, “Recommended Security Controls for Federal Information Systems and Organizations, Revision 3, August 2009. Available at: http://csrc.nist.gov/publications/nistpubs/800-53-Rev3/sp800-53-rev3-final.pdf.

[23] NIST SP 800-82, “Guide to Industrial Control Systems (ICS) Security,” Draft, September 2008. Available at: http://csrc.nist.gov/publications/drafts/800-82/draft_sp800-82-fpd.pdf.

[24] “NIST Framework and Roadmap for Smart Grid Interoperability Standards,” Version 1.0, January 2010. Available at: http://www.nist.gov/public_affairs/releases/upload/smartgrid_interoperability_final.pdf.

[25] NIST IR 7628, “Guidelines for Smart Grid Cyber Security, Vol. 2, Security Architecture and Security Requirements,” August 2010. Available at: http://csrc.nist.gov/publications/nistir/ir7628/nistir-7628_vol2.pdf.

[26] NIST IR 7628, “Guidelines for Smart Grid Cyber Security, Vol. 3, Supportive Analyses and References,” August 2010. Available at: http://csrc.nist.gov/publications/nistir/ir7628/nistir-7628_vol3.pdf.

[27] “DHS Catalog of Control Systems Security—Recommendations for Standards Developers,” June 2010. Available at: http://www.us-cert.gov/control_systems/pdf/Catalog%20

of%20Control%20Systems%20Security%20-%20Recommendations%20for%20 Standards%20Developers%20June-2010.pdf.

[28] BDEW—Bundesverband der Energie- und Wasserwirtschaft, Datensicherheit:http:// www.bdew.de/bdew.nsf/id/DE_Datensicherheit.

[29] EU Task Force Smart Grid, Expert Group 2, "Regulatory Recommendations for Data Safety, Data Handling, and Data Protection." Available at: http://ec.europa.eu/energy/ gas_electricity/smartgrids/doc/expert_group2.pdf.

[30] ESCoRTS Project. Available at: http://www.escortsproject.eu/.

[31] http://www.nist.gov/public_affairs/releases/smartgrid_100710.cfm.

13

SMART GRID AUTHENTICATION AND KEY MANAGEMENT

Anthony Metke

13.1 INTRODUCTION AND SCOPE

Efforts to modernize the North American power grid are needed to make the grid more robust, more efficient, and less expensive to manage and operate. The new capabilities that emerge from this effort are collectively being called Smart Grid. New capabilities will depend on intelligent components, distributed control, and broadband communications capabilities. If not designed with great care, these new features would inadvertently open the grid to many serious cyber threats. Many potential threats have already been identified, which, if successfully enacted, could result in widespread blackouts and even severe damage to the grid. It has been estimated that the cost to our economy from blackouts exceeds US$100 billion [1] annually. Considering the vast size and scale of the North American power grid and the consequences associated with potential cyber attacks, it is clear that the Smart Grid will require a comprehensive security solution, and that the effort to protect the grid from cyber attack will be significant.

Smart Grid: Applications, Communications, and Security, First Edition.
Edited by Lars Torsten Berger and Krzysztof Iniewski.

The purpose of this chapter is to introduce key management issues that will likely affect the Smart Grid architecture and to propose solutions to these issues. This chapter will show that the Smart Grid community needs to develop an industry standard solution based on PKI technology.

13.1.1 Overview of Potential Vulnerabilities

As mentioned previously, the new Smart Grid will possess new features that will make power delivery more efficient. Many of these new features will make the grid more robust provided they are implemented securely. If, however, new distributed intelligence and control features are not implemented securely, then the grid could be open to a number of potentially serious attacks. This section could not list all possible attacks; however, we will list some of the main categories of new potential attacks.

First, the Smart Grid may be susceptible to all the vulnerabilities of the underlying products that make up the grid. This includes vulnerabilities in the operation systems of servers, PCs and handheld devices, as well as vulnerabilities in the underlying network equipment. Internetbankingaudits.com lists over 5000 known vulnerabilities that affect IT systems, and states that, on average, 25 new vulnerabilities are discovered each week. Additionally, the grid may be susceptible to a number of known vulnerabilities to the SCADA protocols and the IEDs, RTUs, PLCs, smart sensors, and other devices that participate in SCADA communications. Furthermore, new functions may introduce application layer vulnerabilities.

A few of the new application layer vulnerabilities are described here.

Demand/response (DR) is a new capability that offers significant efficiency benefits to the grid. With DR an energy consumer may react in real time to high power demand information, lowering his/her current power consumption. Utilities are currently testing systems where consumers allow the utility to automatically shut off the consumer's air conditioning when the demand of the system reached a threshold. Other more sophisticated systems are envisioned where the price of electricity varies with demand and the current price is periodically sent to consumers. The consumers can program into their thermostats (and various appliances) different behaviors based on varying price ranges.

One new vulnerability that this application introduces is the potential for a cyber attacker to impersonate the utility and send its own data to consumers that would force or prompt the consumers to greatly reduce power consumption. This is a significant vulnerability because coordinated and rapid drop in load must be matched by a similarly rapid drop in generation. If generation cannot react fast enough, significant damage to the grid may occur. This is not normally a problem because with large groups of consumers who operate in a relatively uncoordinated fashion, it is very unlikely that everyone will turn off their power consumption all at once. It has been estimated that an attack of this type could cause significant damage to generation, transmission, and distribution facilities, which could take weeks or months to repair.

Distribution automaton (DA) is an application that purports to improve the efficiency and reliability of the power grid. DA works by adding distributed intelligence in grid components and substations. With DA, substations can alert each other in real time to various conditions that may affect how power is routed. In one scenario, substation controllers can detect overload conditions, component failure conditions, or even possible future failure conditions (e.g., the temperature on a transformer has exceeded a threshold). The substation controller can react immediately to various situations and relay conditions to other grid components or substations. Several new vulnerabilities will likely manifest themselves here. For example, if an adversary can impersonate a substation to an adjacent second substation, he may trick the second substation into thinking that the impersonated substation is experiencing (or about to experience) an overload condition, thus causing the second substation to disconnect from the impersonated substation as a means of protecting the second substation from an expected overload condition. A single occurrence of this could cause an unnecessary blackout; a coordinated attack could cause rolling blackouts and even force actual overload conditions in a similar manner to the DR attack described before.

As the Smart Grid is still evolving, many other attacks are possible, and many other vulnerabilities have not yet been identified. Clearly, there will be very many vulnerabilities that must be addressed during the development of Smart Grid solutions.

13.1.2 High Level System Requirements

This section contains a brief summary of some of the more demanding requirements that are likely to have significant impact on any Smart Grid security architecture. The Smart Grid will, of course, have requirements for high degrees of confidentiality and integrity. However, such requirements can be met with several traditional solutions, such as using AES with IPSec. This chapter will not focus heavily on such requirements, but will instead look at requirements that are not likely to be solved by "canned" solutions.

In general, the requirements listed below are those that are likely to exceed normal high assurance enterprise systems. That is, they are the less usual requirements on the grid network that are likely to require unusual solutions on which this chapter focuses.

13.1.2.1 Scalability/Manageability. The Smart Grid will be an enormous network traversing North America. It is unclear at this time if the Smart Grid will use Internet resources or not, but regardless of this, one thing is clear: the number of devices on the Smart Grid will be enormous. The power grid consists of roughly 17,000 generation facilities, 10,000 transmission substations, and 75,000 distribution substations [2]. Each of these facilities will likely contain anywhere from dozens to hundreds of devices requiring some level of secure communications. To make matters worse, there are about 3200 utilities and 3500 nonutility service providers

[3] who will be operating and using this equipment. Lastly, the grid will connect to over 100 million homes in the United States.

For both the grid-to-grid and the AMI use case, there will be an enormous number of devices under the control of the utility. Clearly, securing the grid cannot rely on manually provisioning security credentials into all of these devices. More efficient automated methods of securing the grid will be needed.

The HAN use case may not suffer from the same problem, although there will be hundreds of millions of appliances connected to the grid, only a few appliances will be in each home. If the homeowner can be entrusted to provide minimal "management" functionality similar to Bluetooth paring, manual provisions can be used in place of the automated procedures needed in the other use cases.

13.1.2.2 Crypto Algorithm Longevity. One issue that may be somewhat extraordinary for the Smart Grid is the need for sustained operation for many years and even decades. Unfortunately, over time, many known secure protocols and algorithms have been deprecated due to increased processing power and improved attack algorithms. However, it simply is not affordable to replace grid components every few years. Therefore, as new algorithms are being deployed to protect against new vulnerabilities of older algorithms, it will be essential that critical Smart Grid components support secure means of upgrading their crypto algorithms.

13.1.2.3 High Availability/Local Autonomy. One of the purposes of the Smart Grid is to improve reliability of the power delivery system [4]. It has been estimated that power outages cost our economy between US$104 and US$164 billion annually [5]. Power losses can also pose a significant threat to our national security [6]; therefore, energy delivery systems must provide the highest possible levels of availability. Clearly, the Smart Grid architecture must employ many standard techniques for providing high availability such as redundant functionality, redundant communications paths, and automatic failover capabilities. Security components must be deployed that will not impede these techniques. Additionally, when system outages and power failures do occur, it is essential that components that are operational can communicate unhindered by lack of connectivity to other portions of the network. This requirement, known as *local autonomy*, demands that entities participating in the Smart Grid can authenticate each other and determine the authorization status of each other, without needing connectivity to remote resources.

13.1.2.4 Proactive Malware Protection. Simply put, the safety of the grid is too important to depend on reactive malware detection mechanisms, such as virus detection systems that only look for signatures of previously identified malware. It will be essential for critical components to employ sophisticated proactive malware detection capabilities.

Manufacturers will need to equip critical downloadable components with secure trust root storage and implement software signing procedures. This does not represent a major challenge for manufacturers of embedded systems. However, such functionality can be difficult to achieve on general-purpose computing platforms such as PCs or servers. Software on these systems may come from many sources. In some cases the software vendors may not have control over all the source code for the product they are selling, as they may be using third-party drivers, tools, and libraries. The industry should look into procedures used by the Nevada Gaming Commission to audit software used in gambling machines [7]. This could take the form of an industry consortium (or government agency, such as FERC) performing line-by-line inspection of software used in critical components. Multiple signatures will likely have to be verified by a device when loading software, including one from the manufacturer and one from the software validation consortium. A hardware security module would be used to store signing server public keys or certificates. Functions would likely be needed to continually ensure the validity of the running code. Clearly, this is a complex problem and other unidentified technical components may be necessary to complete a solution. Solutions like these are likely to be very complex and very expensive. This is a topic that needs much more investigation to better understand the risks and the potential solutions.

13.1.2.5 Device Attestation. Again because of the huge potential risks associated with various Smart Grid device-to-device applications, it will be essential for critical devices to know not only that they are communicating with a device that possesses the necessary credentials, but also that they are communicating with devices that have not been tampered with. Therefore, in addition to the above-mentioned malware protection, it must be possible for devices to prove to other devices that they have not been tampered with, or at least that critical subcomponents have not been tampered with. Such mechanisms are referred to as device attestation.

These requirements will have significant impact on key management within the Smart Grid, as will be shown later. It is for this reason that this chapter will focus on key management issues within the Smart Grid.

13.1.3 Review of Key Management Techniques

A brief review of key management issues is presented, and compared with the previously proposed requirements.

Key management and, in particular, *initial key provisioning* can easily be claimed to be the most sensitive part of a cryptography system. It is an area where great care must be taken to minimize exposure to threats. Also, this is an area where care must be taken to ensure that the level of effort required to support the solution does not result in excessive cost.

Let's take a minute to understand what is meant by initial key provisioning. This term refers to those keys that bootstrap the security process. These are not keys that can be derived from other keys such as session keys. These are keys from which all session keys and, in fact, all security is derived. For example, in Kerberos [8] a secret key (a.k.a. the master key) needs to be available at both the principal and the key distribution center (KDC); all other keys, such as session keys and encryption keys, depend on each initial key (i.e., the secret key) being available at the principal and the KDC. In Kerberos the KDC must be either provisioned with the principal's secret key, or given access to a server that has been provisioned with this key. The principal, which is usually a person, remembers and provides as needed a password from which the secret key is derived. Another good example of initial key usage comes from the transport layer security [9] (TLS) protocol commonly used for secure Internet transactions. With TLS, one-way authentication can occur if party A has a key pair (public and private keys), and party B knows and trusts the associated public key. Such trust is typically controlled via digital certificates. For example, party A obtains an end-entity certificate from a CA and provides it to party B (or any other party) when attempting to establish a secure session. In order to validate party A's certificate, party B needs to be securely provisioned with a CA certificate that party B inherently trusts. Party B can then determine if this CA, or any CA subordinate to it, has issued the end-entity certificate of party A. In this case, the key pair and associated certificate provisioned in party A, as well as the CA certificate provisioned in party B, can be referred to as initial keys. Of course, TLS supports mutual authentication as well, by provisioning end-entity certificates and CA certificates in both parties. As in the case of Kerberos, these initial keys are not ultimately used for data encryption of integrity, but to establish temporary session keys.

Whether using TLS, Kerberos, or some other secure protocol, or whether using symmetric key or public key cryptography, some initial keys always need to be provisioned. Initial key provisioning and management can be a difficult task. This is especially true when a large number of devices are being managed.

13.1.3.1 Issues with Symmetric Keys. All security protocols rely on the existence of a security association (SA). From RFC 2408, *Internet Security Association and Key Management Protocol (ISAKMP)*, "SAs contain all the information required for execution of various network security services." An SA can be authenticated or unauthenticated. Establishment of an authenticated SA requires that at least one party possess some sort of credential which can be used to prove its identity or attributes to others. In general, two types of credentials are common: shared secrets and public key certificates.

It is not uncommon for vendors to offer secure solutions by implementing IPsec with AES and calling it a day—leaving customers to figure out how to preconfigure all their devices with the necessary credentials. It is worthwhile to ask if it is better to preconfigure devices with shared secrets or certificates. Provisioning of shared secrets (i.e., symmetric keys) can be a very expensive process with security vulnerabilities that are not present when using public key cryptography and digital certificates. The main reason for this is that, with symmetric keys, the keys need to be configured in more than one place. This means that the keys need to be transported from the device where they were generated and then provisioned into at least one other device. Care needs to be taken to ensure that the key provisioning is coordinated so that each device receives the appropriate keys, a process that is prone to human error and subject to insider attacks. There are hardware solutions for secure key transport and loading,* but these can require a great deal of operational overhead and are typically not cost affordable for all but the smallest systems.

Another reason that symmetric key provisioning can be expensive is that usually you need to provision *N* keys into each device so that each pair of devices shares a unique key. For large systems, this can require provisioning many keys into many devices, which can be a very complex, time-consuming, and error-prone activity.

13.1.3.2 Digital Certificates. Provisioning of digital certificates can be much more cost effective, because this does not require the level of coordination that symmetric key provisioning does. With digital certificates, each device typically only needs one certificate, and one private key, which never leaves the device. Some products generate, store, and use the private key in a Federal Information Processing Standards FIPS-140† hardware security module. In systems like this, where the private key never leaves the HSM, it is not hard to see how such systems can offer higher levels of security with lower associated operational costs. Of course, this explanation is a bit simplistic, and issuing certificates is not a trivial process. For example, certificate provisioning involves several steps; including the generation of a key with suitable entropy at the subject device, the generation of a certificate signing request (CSR) forwarded to a registration authority (RA) device, appropriate vetting of the CSR by the RA, and forwarding of the CSR (signed by the RA) to the certificate authority (CA), which issues the certificate and stores it in a repository and/or sends it back to the subject. Even this is not the full story, CAs need to be secured, RA operators need to be vetted, certificate revocation methods need to be maintained, certificate policies need to be defined, and so on. Operating a PKI can also require a significant amount of overhead and is typically not appropriate for small and even some midsized systems. However, unlike the symmetric key provisioning problem, which scales very poorly, a PKI-based solution, which can have a high cost of entry, scales much more efficiently (because each node only needs one certificate) and is therefore more appropriate for large systems. In fact, the largest users of digital certificates are the civilian federal government, the DoD, and large enterprises.

13.1.3.3 Common PKI Misconceptions. A common misconception of PKI revolves around something called trust management. People often criticize PKI by saying that PKI requires one and only one root CA that everyone has to trust. This

*Companies like SafeNet and AEP offer hardware security modules (HSMs) that can be used to securely transport symmetric keys between two devices. For example, see the SafeNet Protect Server HSM product.

†FIPS are a set of information technology standards and guidelines for the operation of federal computer systems. By law, these standards are developed by the National Institute of Standards and Technology (NIST). FIPS 140-2 (version 2) is a U.S. federal government standard for cryptographic module. NIST oversees a program to validate module compliance to the FIPS-140 standard. Information about the FIPS Crypto Module Validation Program can be found at http://csrc.nist.gov/groups/STM/cmvp/index.html.

is not true. It is more common that each organization operates its own root, and cross-signs with another organization's PKI when they determine a need for interdomain interoperation. Such methods have many advantages over the single root strategy, such as allowing each organization to have much tighter control over its own interdomain trust relationships. For Smart Grid, each utility could operate its own PKI (or outsource it if desired). And those utilities that need to interoperate can cross-sign their appropriate CAs. Furthermore, it would be possible for the Smart Grid community to establish one or more bridge CAs so that utilities would each only have to cross-sign once with the bridge. All cross-signed certificates can be, and should be, constrained to a specific set of applications or use cases. Trust management is not a trivial issue and is discussed in more detail below.

Another issue that is often raised is the need for certificate revocation, and the need to determine the validity of a certificate before accepting it from a node that you are trying to authenticate. Typically, this is accomplished by having the relying party (the node that is doing the authenticating) check the certificate revocation list (CRL) or check with an online certificate status server. Both of these methods typically require connectivity to a backend server. This would appear to have the same availability issues as are typical for server-based authentication methods such as Kerberos or RADIUS [10] based methods. With these server based methods it is not possible to authenticate a node requesting a service unless you have reachability with the backend authentication server. However, this is not necessarily an issue for PKI and certificate-based authentication. Methods to mitigate the reliance on authentication servers in order to validate certificates is discussed in Section 13.1.2.3 .

One final issue that is often misunderstood is the perception that PKI can be complex and too expensive for many small and medium-sized organizations to operate. The main reason for this perception is that PKI is an extremely flexible technology. In fact, PKI is more of a framework than an actual solution. PKI allows each organization to set its own policies, to define its own CP OIDs, to determine how certificate requests are vetted, how private keys are protected, how CA hierarchies are constructed, and the allowable age of certificates, and cached certificate status information. It is exactly because of this flexibility that PKI can be expensive. Organizations that wish to deploy PKI need to address each of these and many other issues, and evaluate them against their own operational requirements, to determine their own specific "flavor" of PKI. Then when the organization decides to interoperate with other organizations, it needs to undergo a typically expensive effort to evaluate the remote organization's PKI and compare it against the local organization's requirements and determine if either side needs to make any changes and determine an appropriate policy mapping to be used in cross-domain certificates. However, it is not essential that each organization participating in the Smart Grid will have to perform each of the above tasks from scratch. An alternative would be for the Smart Grid community to define a limited set of standard policies for PKI operations within the grid. With such a set of policies it would be possible to greatly reduce the effort for each organization to operate a PKI for their internal domain, and furthermore, these standards could greatly simplify the activities associated with interdomain cross-signing or establishing local consortium bridges.

This chapter will show that these issues and misconceptions would not prevent efficient operations using PKI in the Smart Grid. And furthermore, it will be shown that a solution based on PKI can offer the type of flexible, robust, and comprehensive solution needed for Smart Grid.

13.1.3.4 High Assurance Issues and PKI. Symmetric key methods of establishing SAs can be lumped into two general categories: server-based credentials and preconfigured credentials. With server-based systems, such as Kerberos or RADIUS/AAA, connectivity to the security server is required for establishing a security association. Of course, these servers can be duplicated a few times with hopes that one of them would always be available, but considering the size of the grid, this is not likely to offer an affordable solution that can ensure that it will always be possible to establish the needed SAs in the face of various system outages. Duplication of the security server also introduces unnecessary vulnerabilities, as each server would need to be protected. Therefore, it would probably be a bad idea to push such servers into remote locations such as substations. As it is impossible to ensure that every node will always have reachability to a security server, this type of solution may not always be suitable for high availability use cases.

The preconfigured class of solution, as mentioned previously, requires that each device is provisioned with the credentials (usually a shared secret or a hash of the shared secret) of every entity that it will need to authenticate. Such provisioning could potentially provide a very highly available solution; however, this solution will likely be excessively costly, subject to human error, and encumbered with significant vulnerabilities due to the replication of all these credentials, for all but the smallest systems.

Digital certificates, on the other hand, have the distinct advantage that a first node can establish an authenticated SA with any other node that has a trust relationship with the first node's issuing the CA. This trust relationship may be direct (i.e., it is stored as a trust anchor on the second node), or it may come about due to a chain of certificates.

In the case where a chain of certificates are needed to establish trust, it is typical for devices to carry a few types of certificates. First, each node would need a chain of certificates beginning with its trust anchor and ending with its own certificate. It may also carry one or more certificate chains beginning with the TA, and ending with a remote domain's TA or CA. Lastly, it can carry its own recent certificate status. In systems where every node carries such data, it is possible for all "trustable" nodes* to perform mutual authentication, even in the complete absence of any network infrastructure.

With PKI it is important for a relying party to verify the status of the certificate being validated. Normally, the RP would check a CRL or verify the certificate status with an OCSP responder. Another method, proposed in RFC 4366, but not widely

*"Trustable" is used here to mean a relationship that exists between any two nodes for which there exists at least one chain of certificates from each node to the other nodes' TA.

deployed, involves a technique called OCSP stapling. With OCSP stapling a certificate subject periodically obtains an OCSP response (i.e., a certificate status assertion) for its own certificate and provides it to the RP. It is typical for OCSP responses to be cached for a predetermined time, as is similarly done with CRLs. Therefore, it is possible for devices to get OCSP responses for their own certificates when in reach of network infrastructure resources, and provide them to RPs at a later time.

For a complete high assurance solution, the digital certificates must carry not only authentication credentials but also authorization credentials. This can be accomplished in one of several ways. There are several certificate parameters that can be used to encode authorization information. Some options include *subject distinguished name*, *extended key usage*, the *WLAN SSID extension*, *certificate policy extension*, and other attributes defined in RFC 4334 and other RFCs. A complete analysis of which fields to use and how to use them is a large topic suitable for its own chapter. Briefly, it is worth mentioning that the DN offers many subfields which could be used to indicate a type of device or a type of application with which this certificate subject is authorized to communicate. The extended key usage field provides an indication of protocols that the certificate is authorized to use (e.g., IPSec, TLS, SSH). The WLAN SSID extension can be used to limit a device to only access listed SSIDs. The most promising extension for authorization is probably the *certificate policy (CP) extension*. The CP extension indicates, to the RP, the applicability of a certificate to a particular purpose. With such extensions in place, it is possible to use digital certificates in an autonomous mode (with no need for connectivity to a backend authorization server) for all applications requiring role-based access control.

It is possible either to encode these authorization credentials into the subject's identity certificate (which binds the subject's identity to the public key) or to encode the authorization credential into a separate attribute certificate. Typically, organizations need to weigh the benefits of needing to support only one set of certificates with the issues surrounding reissuing identity certificates every time a subject's authorization credentials change. When issuing credentials to people, this is a valid issue. For a device, it is rare that authorization credentials will need to change, and thus placing authorization credentials right in identity certificates poses few disadvantages.

With proper chains of certificates, recent OCSP responses, and authorization credentials, it is possible to provide very high assurance systems, which will allow two entities to authenticate for authorized services even when significant portions of the network infrastructure are unavailable.

13.1.3.5 Key Management Take-aways. When we compare the requirements for scalability and high availability, it is easy to see that PKI is likely to offer the best chance of meeting all of the requirements. By using digital certificates, OCSP stapling, and role-based access control (i.e., role-based authorization based on certificate attributes), it is possible to achieve authentication and authorization without any connectivity to a backend server, enabling completely autonomous SA establishment between any two nodes.

Because digital certificates can be used for more than one application and for more than one peer, they do not suffer from the *N*-squared provisioning problem from which symmetric key solutions suffer.

Because private keys never need to be transported, where systemic keys nearly always need to be transported between nodes, use of public key crypto and digital certificates enables much stronger security than is available with symmetric key crypto. It is also possible to use the public key crypto capabilities of a device to implement strong high assurance boot (HAB) and remote device attestation.

For these reasons, public key cryptography, digital certificates, and PKI are felt to offer the best possible solutions to meet the demanding requirements of the Smart Grid. Furthermore, we will show that it will be possible to couple each manufacturer's PKIs with energy service providers' PKIs to create a very strong and cost-effective security system.

13.2 AUTHENTICATION AND AUTHORIZATION ISSUES IN THE SMART GRID

When performing authentication and authorization (A&A) it is important to ask what you are authenticating. For example, when performing role-based access control it is important to ensure that the entity requesting a service has been authentically assigned to a role associated with the service being requested. In some cases you may want to ensure that the device you are communicating with is a genuine model X device from manufacturer Y, with serial number Z. That is, you may not care who owns it, but only that it is the device that it claims to be. This is useful for device management purposes, such as digital rights management. In other cases, it may be very important to know who owns the device and how it is being deployed. For example, I might want to know that a router advertising routes into my network is actually owned by me and that it is authorized to advertise such routes. For some applications such as the intelligent transportation system's collision avoidance system, you really don't care who owns the other car which is telling you that you are about to run into it. All you should really care about is that the control logic on the other car has not been tampered with. In this case, it would be possible to have certified modules that can protect themselves from malware, and only perform authentication if they have been untampered with. Let us consider each of the use case categories.

13.2.1 Grid to Grid

Organizational affiliation (for users) or ownership (for devices) will typically be required. Both characteristics shall be referred to as "affiliation." If a device is telling me to shut down a substation, I will certainly want to know if that device belongs to me or to a trusted third party or not. Furthermore, I would need to know that the device is authorized for the role that it is playing. Is it actually supposed to be sending messages indicating that a transformer is exceeding safe temperature ranges?

The actual required role specificity needs to be established. Is it enough to know that it is a SCADA device, or do I need to know that it is a SCADA device authorized to report events in substation Y? Or maybe I want to know that it is a SCADA device authorized to report temperature readings on transformer X in substation Y.

Device integrity will also be critical for some grid-to-grid operations, in which case, a remote device's integrity should also be verified prior to establishing an SA with the remote device. That is, devices may need to prove to others that they are untampered. This can be accomplished with device attestation techniques. There are many ways of performing device attestation. A recommended approach is described later in this chapter.

13.2.2 AMI

Because of the serious threats associated with the demand/response (DR) application it is essential that AMI communications are properly authenticated. Meters should therefore be required to prove that they are affiliated with the utility, that they are meters (make and model), that they are authorized to report meter readings for a given address, and that they have not been tampered with.

Meters will have to verify that received DR data is accepted only from servers affiliated with the utility with which the meter is affiliated, and that the server is authorized to participate in the DR application (you would not want a billing server to send DR data to a meter). It is also recommended that DR servers prove that they have not been tampered with to the meters. As the DR servers are likely to be general-purpose computing platforms running a variety of software from different vendors, such mechanisms of proving the integrity of the entire DR server may not be available for quite some time. An alternative would be for multiple DR servers running different OSs to both approve all DR data going to the meters. This way it is less likely that malware could have caused the DR server to send inappropriate information to the meters. Again, we are trying to protect against significant threats enabled by the DR application, which could potentially cause serious damage to the grid. It may be appropriate, in such cases, for each DR server to be administered by different groups, to prevent an insider from enabling an attack.

For billing applications it will be important to know with which building address a particular meter is associated. Meters will also have to verify that they are sending billing information to a billing server affiliated with the utility. It probably is not necessary to prove the tamper condition of the billing server to the meter.

Another possible application involves the interface between the home with a usage service provider such as Google Power Meter. This application allows for a device in the home to connect to the Google Power Meter application and obtain detailed reports and analysis. Such information is likely to help the consumer to better control energy usage.

It is unclear as to whether the meter would report this data to Google through the AMI interface or through the in-home network connection via the homeowner's ISP. Currently, both methods are being used. Utilities are currently offering smart meters that communicate to Google through the AMI network interface. Additionally, two devices are on the market [11] which can monitor a home's energy consumption and report it to Google through the consumer's ISP.

The security requirements for such applications are likely to be low, as successful attacks would only result in either the consumer receiving erroneous data about his/her usage or an adversary getting access to the consumer's power usage. Neither of these scenarios is likely to justify extraordinary steps for protection. The first scenario is likely to be detected by the consumer whose energy bill is likely to differ from the Google reports. The second scenario can be reasonably prevented by relying on TLS technology between the metering device (i.e., Smart Meter or The Energy Detective) and the Google Power Meter servers. Only server authentication would be required.

13.2.3 HAN

Within the home, appliances will be getting DR information from the meter. Consumers will want to make sure that their appliances only communicate with their meter. It would be important for the safety of the grid that a third party could not impersonate the meter to the appliances. So, who puts the credentials into the meter—the homeowner or the utility? It is recommended that the homeowner puts in the HAN credentials and the utility owner puts in the AMI credentials.

There are a few alternatives for provisioning HAN credentials. The homeowner can configure secret keys into each appliance and the meter, or the thermostat can act like a PKI and issue certificates to each appliance. Mechanisms would have to be put into place to help the homeowner vet the CSRs. This is likely to be too complex for the average homeowner. So the preferred alternative would be for the homeowner to configure a secret key or pass phrase into each of her smart appliances. There is a big problem with this technique. There would have to be methods to prevent the homeowner for using easily guessable keys or phrases. Even if consecutive sequences of numbers were prohibited, many homeowners would likely choose their address, birth date, or other easily obtainable data. Even with pass phrases it would be difficult to ensure the necessary entropy. Homeowners are likely to use popular song or film titles, or common phrases ("I'll be back", "elvis is king", "hail to the chief", etc.) all of these would have to be added to a library of disallowed phrases. The threat here is that terrorist groups may spend months or years "war driving" and building a catalog of guessed pass phrases. Then when they have gathered enough pass phrases, they could use them in a coordinated fashion to tell millions of appliances to shut down, thus successfully executing the previously mentioned DR attack, potentially causing irreparable damage to the grid. Clearly, we cannot let this happen. One solution would be to force homeowners to change their key or pass phrase every few months. Clearly, this is not going to happen.

Another alternative is to have keys automatically rotate coordinated either by the meter or the thermostat. In this case, the homeowner would not know the key after the first rotation. Now if she added a new appliance to the network, she would have to rekey all the nodes. Clearly, this would not be popular with consumers. Rather than having a single key configured by the homeowner, each appliance would have a unique key shared with the meter. Diffie-Hellman* (DH) could be used in this case. Each time a new appliance is installed, both the appliance and the thermostat (or meter) could be put into "learning" mode; they discover each other and perform DH key agreement. There is, of course, a risk that an imposter will be able to successfully act as a man-in-the-middle (MITM), but this would require the MITM device to remain dedicated to this home, keep performing the protocol, and keep alive the signaling and perform necessary key rotations periodically. This would preclude an adversary from learning the key and moving along to another location to learn more keys. Hence this is a recommended approach.

Open Issue: Should the thermostat or the meter be the central control point that maintains SA with all other devices? Recommendation is that the thermostat is the control point that supports an SA with each appliance and the meter. This puts control where it should be—in the homeowner's hands. Meters are typically located outside and may be susceptible to greater threats, such as improperly being put into learning mode. Therefore, it is also recommended that the meter can only be put into learning mode by the affiliated utility. The meter would receive a signed assertion from a server with a chain of certificates going back to the meter's IA TA. This assertion would only have to be sent to the meter, when a new meter is installed or when a new thermostat is installed. In this way when a homeowner bought a new appliance, the consumer would typically only have to push a button on the thermostat and on the new appliance to force each into learning mode, during which they would perform a DH exchange. Symmetric keys would then be rotated at a rate dependent on the cryptographic methods being used.

13.3 ARCHITECTURAL CONSIDERATIONS AND RECOMMENDATIONS

This section provides specific recommendations, as well as identifies issues for further consideration.

13.3.1 Malware Protection

There are two categories of devices for which malware protection should be considered: embedded computer systems and general-purpose computer systems.

*Diffie–Hellman is the name of a popular key agreement protocol based on public key cryptography, which allows two devices to each securely derive the same symmetric key without sending information between the devices that would allow a third party to derive the same key. Each device must initially contain a public key pair.

Embedded systems are computer systems that are designed to perform a specific task or set of tasks. They are intended to run only software that is supplied by the manufacturer of the hardware. In contrast, general-purpose systems are intended to support third-party software purchased by the specific consumer who purchased the system. A PC is an excellent example of a general-purpose system. A microwave oven and a cable television set top box are examples of embedded systems. This problem of malware protection should be considered separately for each category.

For embedded systems the problem of protecting the system against the installation of malware can readily be solved with high degrees of assurance. First and foremost, the manufacturer must implement secure software development processes; many standard models for such processes are well established in the industry [12]. Second, if the device is intend to be field upgradeable the manufacturer must provide a secure software upgrade solution. One prominent method of doing this is to manufacture the embedded system's hardware with secure storage containing keying material for a software validation. Typically, the hardware is configured with the public key of a secure software signing server (or a CA that issues a certificate to a signing server) operated by the manufacturer. With this key, the device can validate any newly downloaded software prior to running it. Such a proactive approach can provide higher levels of assurance than can be obtained with a reactive approach such as a virus checker. Additional security can be obtained by validating the software each time the device boots up. Such techniques are referred to as high assurance boot (HAB). HAB techniques typically rely on core software in the hardware security module (HSM) to validate boot-block code. The boot-block code then validates the OS, and the OS in turn validates the higher level applications. Each validation step is performed with public key or keys preinstalled in the HSM.

For devices that are intended to run for long periods of time (e.g., years) without booting, it is useful to have a method of performing secure software validation on running code. It is possible to have background tasks that can periodically perform such functions without disrupting the operations of the device.

It is further possible to couple such background software validation steps with other operational aspects of the device, such that if the device is found to be compromised, secure hardware on the device (needed to bring up and maintain security associations with remote entities) will prevent the local device from establishing and maintaining security associations with the remote entities. This is covered in more detail in Section 13.3.2.

Each HSM for a critical device that will support secure software download should be manufactured with a device management trust anchor certificate. When the manufacturer wishes to allow a piece of code to be downloaded into the device, the manufacturer will bundle the software with the model number for which the software is intended, along with a release date or software serial number. The bundle would then be signed with an authorized signing server. The signed bundle along with the signing server certificate (or certificate chain rooted at the device management trust anchor) can then be forwarded to the device for download. When the device receives a software update bundle, it should verify the signature on the bundle and the chain of certificates back to the preconfigured device management TA certificate. The device should ensure that the software being loaded is at least as new as the software that it would be replacing. Optionally, if the device has an accurate real-time clock, the device could check the date in the bundle and ensure that it is not older than policy would allow. For example, policy may require that software updates must occur within one month of signing. This is comparable to requiring the device to have access to a CRL (for the signing server) that is no more than one month old.

For general-purpose computing devices, such mechanisms, that only allow software approved by the manufacturer to run, have not been popular. Consumers of PCs or servers typically feel that they should not be restricted by the manufacturer from loading any software that they want, even if it means having to put up with malware attacks. The predominant means of protecting general-purpose computers has been to use malware detection and removal software typically referred to as antivirus software. One of the most effective tools that the antivirus software uses to detect malware is a "signature" dictionary. The term "signature" is being used here in a different manner than in the rest of the chapter. Here a "signature" refers to a pattern of known recognizable malicious code (as opposed to cryptographic signatures). With the signature dictionary, only known viruses can be discovered and removed. Such methods are not helpful in protecting against new or unknown viruses. Clearly, with the stakes so high, the Smart Grid needs a better solution than the reactive antivirus dictionary approach.

To make matters worse, the rapid adoption of cloud computing, and development of sophisticated Internet-based applications, has resulted in the widespread deployment of a number of "mobile code" technologies. Mobile code is a name for code that is downloaded and run on your PC, typically by your browser, without the user's knowledge. Examples of mobile code include ActiveX, Flash animation, Java, JavaScript, PDF, Postscript, and Shockwave. The DHS Control System Security Program recommends tight controls on mobile code in critical control systems for the nation's critical infrastructure and key resources (CIKR) [13].

One solution to this problem would include the adoption of, and adherence to, strict code signing standards by Smart Grid suppliers and operators. Mechanisms for enforcing such standards on general-purpose computers, such as PCs, have been put forth by the Trusted Computing Group [14]. Such standards should cover all critical devices including field-deployed units, such as RTUs and IEDs; network devices such as router, switches, and firewalls; and control center equipment such as servers and user consoles. The standards should cover embedded systems, as well as general-purpose computers, their operating systems, drivers, and applications, as well as all mobile code. That is, no mobile code should be allowed to run on a critical PC or server that has not been signed by an authority that is able to determine the trustworthiness of the code. Considering that it is certain that hardware and software elements for critical components of the grid will come from many different providers, it is likely that a trust management framework will have to be established for Smart Grid. Resources such as the National Software Reference Library [15] could be used, such as has been proposed by NIST for validating software in voting systems [16]. This framework will likely require the establishment of a set of criteria that are to be met by vendors who wish to sell critical components to Smart Grid operators. Additionally, it is likely that one or more accreditation organizations will need to be established to audit suppliers to determine whether they are meeting the specified criteria.

It would be possible for meters to require that the utilities (the legal owners of the meters) approve all software downloads. In this case the meter should require both a signature by the manufacturer signing server, as well as a signing server with a chain of certificates back to the utilities' TA. This would give utilities an opportunity to ensure that meters are not loaded with new software until the utility, its agent, or an accreditation lab has thoroughly tested the new software.

To some, these measures may seem extreme, but when we consider what is at stake, and the large potential for vulnerabilities related to malware in the Smart Grid, it is hard to imagine any other practical way of providing complete malware protection in the grid.

13.3.2 Device Attestation

Device attestation is a technique for proving, to remote devices, that a particular device has not been tampered with. To support device attestation, critical devices should be designed with a hardware security module (HSM) capable of performing high assurance boot, secure software update, and real-time operational system integrity checks. The same HSM must be capable of storing the necessary keys and performing all cryptographic functions needed to authenticate to other entities. The HSM should be capable of detecting if it has been tampered with and disable authentication activities when tamper events are detected. Lastly, this module should be able to tear down any previously established SAs when tampering is detected. The level of tampering detection required will need to be defined for each type of device. The FIPS-140* standard [17] should be used, and a specific FIPS level (as they are often referred to) should be specified for each end point in an application class.

13.3.3 Holistic PKI Model

This section describes a holistic PKI model that could be adopted by the Smart Grid community. This model is a first pass attempt to meet the requirements listed above. The model should be applied to grid-to-grid and AMI use cases.

13.3.3.1 Certificate Scoping. Some critical devices will require two types of device certificates. These two types of certificates are said to have different scope. Each type of certificate is explained below.

*FIPS-140 defines security requirements for cryptographic module, and include requirements for four levels of security over each of 11 different requirement area. Among other things, FIPS 140 specifies requirements for tamper detection, and prevention.

DEVICE MANAGEMENT CERTIFICATE SCOPE. For some critical devices it is not sufficient for the grid operator to install his own certificates. It would be further required that manufacturers install at a minimum the manufacturer's root-of-trust for device management [also known as the device management trust anchor (DMTA)]. This DMTA or one of its subordinates would be used to sign approved software from the manufacturer. Such operations are needed to help prevent malware from getting into critical components. Such mechanisms also serve to protect the manufacturer from liability associated with third-party or unapproved software running on devices.

For devices supporting device attestation techniques, accredited manufacturers can factory-install device attestation certificates in each Smart Grid device. These device attestation certificates are similar to device management certificates, in that they are used to assert the device's manufacturer, model, and serial number. However, device attestation certificates, when used to perform authentication, would further indicate that the device has not been tampered with. These certificates would be nearly identical to device management certificates with the exception that they would contain an attribute that indicates that the manufacturer asserts that this device will not perform the needed authentication protocols if the device determines that it has been tampered with. These certificates, coupled with the appropriate authentication protocol, can be used by the energy service provider to ensure that the device is exactly what it claims to be, prior to loading IA certificates that indicate the affiliation and role of the device.

INFORMATION ASSURANCE CERTIFICATE SCOPE. Furthermore, critical devices will require operator issued certificates, which can be used to authenticate the device and authorize the device for appropriate service. We refer to these certificates as belonging to the operator's IA domain. These certificates would include, at a minimum, a device identity certificate and the operator's TA certificate. Because DMTA certificates can be installed at the time of manufacture, these certificates can be used to lock the device to a specific operator (for a period of time), ensuring that only the designated operator (the actual customer who purchased the device from the manufacturer) can load its IA TA certificate. Mechanisms such as the IETF TA Management Protocol (TAMP) can be used for this purpose. Such a mechanism can help support the efficient, cost-effective, and secure loading of TA certificates by the operator. Loading the TA certificates on a device is clearly one of the most important security operations for a device. If unauthorized TAs can be loaded onto critical grid components, it would be difficult, if not impossible, to guarantee the secure operation of the device.

WiMax certificates, by comparison, do not prove ownership or role; they can only be used to prove that the entity with the corresponding private key is the entity listed in the certificate. An AAA server must then be queried to obtain the authorization credential of the device. Such methods require access to the AAA server for authentication and authorization to function properly and hence are not able to meet the high availability requirements of the Smart Grid.

In summary, the DM certificates protect the manufacturer's assets, and can be used to identify the make, model, and serial number of a device; while the IA certificate identifies the owner of a device, along with authorization attributes approved by the owner.

13.3.3.2 Secure TA Installation. Before a device has a TA certificate installed, it is considered unsecured. Once a device has a TA certificate installed the device is now under control of the TA. As mentioned above, it is possible for a device to have both a DM TA and an IA TA. The DM TA is used primarily for software validation, while the IA TA is used for pretty much everything else. However, the DM TA can also be used to securely install the IA TA certificate into a device. One great advantage of such a mechanism is that the device can be secured in the factory during manufacturing by loading the DM TA certificate into the device.

We would not want to require a manufacturer to load in the utility's IA certificate at the factory as this would affect the manufacturing process and would likely be a costly effort. Luckily, we can achieve the same level of protection if the device is manufactured with a DM TA and a mechanism to only allow the DM TA to authorize the loading of the IA TA certificate.

High assurance devices should require a signed assertion by the manufacturer's DM TA in order to load an IA TA certificate. Preferably the signed assertion indicates the serial number (or a range thereof) of the meter(s) or other devices intended to load a given IA TA certificate. Device manufacturers would provide utilities with such an assertion for the utilities' chosen CA.

13.3.3.3 Smart Grid Trust Model. There are many possible models to operate a PKI. Some think of PKI as a hierarchy with a single trust root with many subordinate CAs; others think of the Internet model, where there are many PKI service providers and anyone can choose any service provider to get their certificates. Neither of these common models are appropriate for Smart Grid.

When multiple organizations are endeavoring to provide a rich web of connectivity that transcends across large systems with many devices, the strict hierarchy model can quickly be eliminated. This is because it is typically very difficult to get everyone involved to agree on one entity, who they can all trust, and the policies under which this "trusted" party should operate. Just as importantly, this model relies on the absolute security of this central "root of trust" because breach of the central root destroys the security of the whole system. Also, we can quickly eliminate the Internet model. This model is intended to provide consumers with a reliable method to know with whom they are communicating. This model does not provide any indication that the remote party is authorized to participate in Smart Grid use cases, nor is there any ability for service providers acting within this model to assert specific authorization attributes, which would be necessary to meet the Smart Grid's high availability requirements.

This leaves mainly the mesh model and the federated trust management model. A full mesh model is likely to be too expensive, and completely unnecessary. It is hard to see any reason to require all parties to develop a trust relationship with each other.

The federated model brings together the best features of a hierarchy and a mesh. PKI federation is an abstract term, usually taken to mean that a domain controls (whether owned or outsourced) its own PKI (components and policies), and that the domain decides for itself its internal structure (usually, but not always, a hierarchy), and that the domain decides when and how to cross-sign with other domains (whether directly or through a regional bridge). Such a federated approach is really the only reasonable solution for large interdomain systems.

Small utilities could outsource their PKI. It is worth mentioning that this is not necessarily the same as going to a public PKI provider such as Verisign, and getting an "Internet model" certificate. With the Internet model, a certificate mainly proves that you are the rightful owner of the domain name listed in your certificate. For Smart Grid this is certainly not sufficient. Certificates should be used to prove ownership, as well authorization credentials. Smart Grid certificates should be issued under Smart Grid sanctioned policies and should carry authorization credentials when appropriate, such as for high assurance use cases.

13.3.3.4 Operator Internal Topology. The design of a PKI topology is a very important issue. Many options are available. Too few or too many components (CAs and RAs) can result in a system that does not scale well, is too costly to administer, and suffers from unnecessary vulnerabilities. Topology choices may likely be dependent on internal organization structure. Each operator (utility) or energy service provider should be allowed to operate their PKI with control over their internal PKI topology, with few restrictions.

There are a number of different ways that a utility may choose to distribute CA functionality within its PKI. It would be possible to use one CA and RA for issuing certificates to humans, and another CA and RA for issuing certificates to devices. Alternatively, one could divide CAs and RAs by the geographic location of the entities for which they issue certificates. For example, CA1 handles all the certificates issued to devices in the north portion of my servicing area, while CA2 handles certificates issued to devices in the south. Another approach would be to use one CA to issue certificates to devices in substations and another to issue to devices in control centers, and another to issue certificates to residential meters.

In general, utilities should be permitted to use a topology that fits their operational requirements. However, there are likely to be reasonable limits that should be mandated by the Smart Grid community. For example, it may be required that in order to issue certificates with a high assurance policy, that the issuing CA is also not the trust root (i.e., TA). In some cases, it may even be required that the issuing CA be subordinate to an intermediate CA, which is subordinate to the trust root. Also, it may be required that a CA that issues high assurance certificates can also not issue low assurance certificates.

To establish these topology requirements the Smart Grid community will need to establish industry standard PKI policies. These policies will include allowable PKI topologies and topology constraints.

13.3.3.5 Cross-Signing. Cross-signing is the process by which trust can be extended between independent PKI domains. Here a "domain" typically refers to a legal entity or business such as an enterprise, a utility, or a government agency.

However, a PKI domain can also be associated with a department within an enterprise or utility, or even a consortium of multiple utilities. A more technical definition of a PKI domain is: an organization that independently operates its own PKI under a unified policy management authority.

Cross-signing is a process where two CAs in separate PKI domains issue certificates to each other. In this way an RP in domain-1 can build a chain of certificates from subjects in domain-2 through domain-2's trust root, to domain-1's trust root. Since RPs in domain-1 trust domain-1's trust root, they can trust the subject at the end of this chain. Care must be taken to impose proper constraints when issuing cross-signed certificates to other agencies to appropriately constrain the trust being extended. Policy and name constraints can be added to the certificate issued to the other agency's CA, imposing these constraints on the other domain. For example, I may want to allow only certificate holders with high assurance certificates to have access to my network.

Cross-signing can typically be a complex process. In simple terms, this is because each domain needs to understand the other domain's certificate issuance policies to ensure that the other domain's certificate holders are given the appropriate access permissions in the local system. This would normally require a process known as policy mapping, which can be quite involved and expensive. This effort would typically require signing legal interoperability agreements and may even require audits of the PKI operation of each domain to be conducted.

In the Smart Grid where many organizations work together to supply the nation's power, there may be need for a significant amount of cross-signing between organizations. Two methods can be used to reduce the burden, and hence the cost, of this activity. First, certificate standards can be implemented across the Smart Grid community to ensure that each domain issues certificates with common formats and common policies. Second, bridging can be used to reduce the number of cross-signing operations necessary.

13.3.3.6 PKI Bridges. A PKI bridge is a CA that only cross-signs with other CAs. In this way a large number of organizations can extend trust to each other through a single cross-signing step between each agency and the bridge.

Bridges and cross-signing effectively create a PKI topology that extends between and across domains. Some people feel that there should be a single bridge for the entire Smart Grid community; this would have the benefit that all agencies could interoperate with each other. However, such a mechanism would likely be too difficult to manage securely; furthermore, such a mechanism would likely be plagued with as many political issues as a PKI that uses a single national CA. A more reasonable approach would be for local or regional consortiums to establish their own bridges, which would provide bridging services for their own members. Because such consortiums would have well-defined memberships of reasonable size, secure administration of the consortium's PKI bridging service would be a manageable effort.

Sometimes local bridging is not sufficient, and it may be necessary for the bridges of two or more consortiums to cross-sign. When several consortiums are cross-signing, a regional bridge may be appropriate. Similarly, it may someday be required for regional bridges to cross with a national bridge.

Such a federated topology differs significantly from the top–down topology where all trust is extended from a single central national CA, through a hierarchy of subordinates. First, the federated model is a network of peers rather than a hierarchy. The security of the central hierarchy model fails if the central CA is ever compromised. In the federated model, each entity controls its own security. A compromise in an external CA may open a possible threat against other domains; however, those domains have the capability to mitigate this threat by revoking or suspending certificates issued to the compromised domain. Another significant difference between the two models is that in the federated model each domain can control what type of certificates it issues. And each domain can constrain what type of certificates it accepts from foreign domains. In the central hierarchy model the root CA determines what types of certificates its subordinates can issue, and in turn these CAs determine what types of certificates their subordinates can issue. With such a model, utilities would be at the mercy of the higher layers to decide what types of certificates it could issue.

13.4 CONCLUSION AND NEXT STEPS

Virtually all parties agree that the consequences of a Smart Grid cyber security breach can be enormous. New distributed intelligence capabilities in the grid can introduce significant new vulnerabilities, which could potentially cause substantial damage to distribution, transmission, and generation facilities. Considering the incredible size of the threat and awesome potential consequences from cyber attack, the Smart Grid cyber security protection requirements must be extreme. The grid will require a comprehensive security plan that encompasses virtually all aspects of grid operations.

Initial bootstrapping (provisioning) symmetric keys (i.e., shared secrets) in the Smart Grid will be too expensive for many use cases and can increase the likelihood of human error, which can introduce cyber vulnerabilities. It is recommend that the Smart Grid community should develop an industry standard solution based on PKI technology for grid-to-grid and AMI use cases. For home area networks (HANs) it is recommended that communications be secured through a procedure similar to the Bluetooth v2.1 Pairing Protocol, which uses elliptical curve Diffie–Hellman to establish shared symmetric keys.

Additionally, recommendations are made for managing trust among the various Smart Grid operators (i.e., utilities). These include recommendations that utilities either manage their own PKI or outsource their PKI operations to an industry approved service provider, who provides PKI services with individual designated trust roots for each utility. This is quite distinct from the Internet model, where a few providers offer a few trust roots, all of which are trusted by any user.

Manufacturers of critical components will also have a role to play in the overall trust management architecture. In addition to implementing the required security protocols and algorithms, some critical components may require hardware security modules to support strong key protection, high assurance boot, and/or device attestation. This will require devices to be issued with a manufacturer's trust anchor certificate, which would be used in validating new software that the device is instructed to download. Furthermore, by providing devices with immutable device-identity certificates, manufacturers can enable more cost-effective provisioning of utility issued IA certificates for these devices.

Clearly, there is a great deal of work that is needed to ensure the security of the Smart Grid. Below are just of few of the steps needed to advance that security. In general, I have listed items that I believe are not getting the attention that they deserve.

- Develop a device certificate lifecycle management* (CLM) interface specification, so that device vendors can build products that will interoperate with SG PKIs.
- Definition of Smart Grid roles and responsibilities to be used for authorization attributes in digital certificates [similar to the National Incident Management System (NIMS)†]
- Establish software signing standards for manufacturers of software upgradable components on the Smart Grid. Different standards may be required for different assurance levels.
- Establish HAB and device attestation standards for Smart Grid components. Again, different standards may be required for different assurance levels.
- Smart Grid common model certificate policy and Smart Grid PKI accreditation requirements.
- Creation of a PKI accreditation board and rules for independent third-party testing and accreditation organizations.
- Certificate distribution management. Different chains of certificates are needed to authenticate entities from various external domains—the process of constructing a certificate path from the certificate subject back to each RPs trust anchor. This is especially true if an organization has many cross-signing agreements with external organizations or bridges. The problem appears to be even more daunting when you consider that, for high assurance reasons, it is desired

*CLM also known and certificate managment is the process of requesting, provisioning, revoking, and managing digital certificates. Additional information can be found at http://www.ietf.org/rfc/rfc2510.txt.

†National Incident Management System, developed by the Department of Homeland Security under the authority of HSPD-5, provides a template and a set of best practices for enabling various federal, state, and local government agencies to work together efficiently. See http://www.fema.gov/emergency/nims/AboutNIMS.shtm.

that for any certificate subject and any RP, the total number of certificates needed for mutual authentication would always be available between these two nodes without the need to go out to a repository. Work needs to be done to ensure that these certificate paths and the associated certificate status would be available for any nodes requiring high availability.

Undoubtedly many other issues will be discovered, and the task at hand of creating a set of Smart Grid industrywide PKI standards will not be a small task. However, once an initial set of such standards are in place, it would be possible to truly provide a scalable and efficient-to-operate, secure trust management system. This system could then be used for bootstrapping virtually all security associations needed for the Smart Grid.

REFERENCES

[1] "The Cost of Power Disturbances to Industrial and Digital Economy Companies," Consortium for Electronics Infrastructure to Support a Digital Society, and EPRI Initiative. See Figure 3.8. Available at: http://www.epri-intelligrid.com/intelligrid/docs/Cost_of_Power_Disturbances_to_Industrial_and_Digital_Technology_Companies.pdf.

[2] U.S. Energy Information Administration, Independant Statistics and Analysis, 2008. Available at: http://www.eia.doe.gov/cneaf/electricity/epa/epat1p2.html.

[3] US Energy Information Administration, Independant Statistics and Analysis, 2008. Available at: http://www.eia.doe.gov/cneaf/electricity/page/gen_companies/codesp1.html.

[4] Energy Independence and Security Act of 2007, Sec. 1301. "Statement of Policy on Modernization of Electricity Grid."

[5] "The Cost of Power Disturbances to Industrial and Digital Economy Companies," Consortium for Electronics Infrastructure to Support a Digital Society, and EPRI Initiative. See Figure 3.8. Available at: http://www.epri-intelligrid.com/intelligrid/docs/Cost_of_Power_Disturbances_to_Industrial_and_Digital_Technology_Companies.pdf.

[6] GAO Report to the Committee on Governmental Affairs, U.S. Senate, July 2002, "Critical Infrastructure Protection: Federal Efforts Require a More Coordinated and Comprehensive Approach for Protecting Information Systems."

[7] http://gaming.nv.gov/forms/frm141.pdf.

[8] RFC 4120, "Kerberos Network Authentication Service (V5)." Available at: http://www.ietf.org/rfc/rfc4120.txt.

[9] RFC 2246, "The TLS Protocol." Available at: http://www.ietf.org/rfc/rfc2246.txt.

[10] See RFC 2865, "Remote Authentication Dial In User Service (RADIUS)." Available at: http://tools.ietf.org/html/rfc2865.

[11] Such as http://www.currentcost.com/powermeter/ and http://www.theenergydetective.com/store/.

[12] N. Davis, *Secure Software Development Life Cycle Processes*. Software Engineering Institute, Carnegie Mellon University.

[13] Catalog of Control Systems Security: Recommendations for Standards Developers, DHS, September 2009.

[14] D. Challener, et al., *A Practical Guide to Trusted Computing*. IBM Press.

[15] http://www.nsrl.nist.gov/.

[16] http://www.nsrl.nist.gov/vote.html.

[17] The Federal Information Processing Standards (FIPS) are available at: http://csrc.nist.gov/publications/PubsFIPS.html.

PART IV

CASE STUDIES AND FIELD TRIALS

14

HYBRID WIRELESS–PLC SMART GRID IN RURAL GREECE

Angeliki M. Sarafi, Athanasios E. Drougas, Petros I. Papaioannou, and Panayotis G. Cottis

14.1 INTRODUCTION

This chapter presents a large-scale pilot wireless-broadband over power line (W-BPL) network deployed in Larissa, a rural area in central Greece, during July–August 2007. The motivation for this implementation was to provide broadband Smart Grid services to the Greek utility aiming at controlling the high energy consumption of irrigation pumps.

The area of Larissa is the biggest vast plain in central Greece, comprising mostly agricultural fields. The medium voltage (MV) grid consists of 20 kV, three-phase, tree type overhead wires; 70% of it was developed before 1965. Due to the lack of telecommunications infrastructure near the agricultural fields and the high cost of a wireless solution, the MV grid constitutes the only infrastructure exploitable for communications. The power grid in the deployment area contains an HV-to-MV substation wherefrom stem the MV lines feeding the villages and the agricultural loads.

Smart Grid: Applications, Communications, and Security, First Edition.
Edited by Lars Torsten Berger and Krzysztof Iniewski.

The agricultural loads mainly consist of high-energy-consuming irrigation pumps, the operation of which leads to frequent blackouts during periods of high energy demand. Since these irrigation pumps are not in constant operation, the Public Power Corporation (PPC) and the farmers have agreed on service level agreements (SLAs), which involve reduced tariffs for the irrigation pumps but reserve the right for the PPC to turn off power supply during peak hours. However, enforcing the SLAs was hampered by two characteristics. On the one hand, manual control of the pumps is not feasible over this vast geographical area; on the other hand, the MV circuits feeding the irrigation pumps also provide electricity to nearby villages, making the option of switching off power supply to these circuits practically impossible. Hence, the Greek utility decided to try a Smart Grid solution to implement remote monitoring and control of the irrigation pumps and, consequently, continuous network availability. To transform the MV grid into a Smart Grid, the hybrid W-BPL technology was adopted. The main motivation for the adoption of BPL technology was the absence of already deployed communications infrastructure due to the small population of the area. On the other hand, other alternatives for the implementation of a Smart Grid platform could neither guarantee network scalability nor afford broadband applications like VoIP, Internet access, and grid surveillance. Moreover, as will be shown in the following sections, W-BPL is the only technology supporting real-time monitoring of power supply and consumption, being able to provide competitive Smart Grid services.

The Smart Grid network installed in Larissa proved efficient in providing network management and monitoring in the attempt to: (1) optimize the power quality offered, (2) reduce energy consumption during peak hours by saving tens of megawatts, (3) accurately detect and prevent outages, and (4) reduce service recovery time and cost by enabling VoIP communication and enhancing physical surveillance. Finally, W-BPL technology will enable the seamless incorporation of alternative energy providers into the MV grid, allowing and promoting free competition in compliance with relevant regulations.

14.2 NETWORK DESIGN AND IMPLEMENTATION

The high data rates supported have rendered BPL technology an alternative choice for last-mile access, extending the coverage of the telecommunication infrastructure. The total area of the network deployment encompasses the 107 km end-to-end distance covered by the two MV lines where BPL units [1] were installed. The second MV line is depicted in Figure 14.1. The BPL units were organized into cells since a cell-based architecture allows network scalability.

This section provides a short description of PHY and MAC layers as they were implemented in the deployment. The cell-based architecture of the BPL network is then analyzed along with the presentation of the basic network management issues. Finally, access through the BPL network is examined.

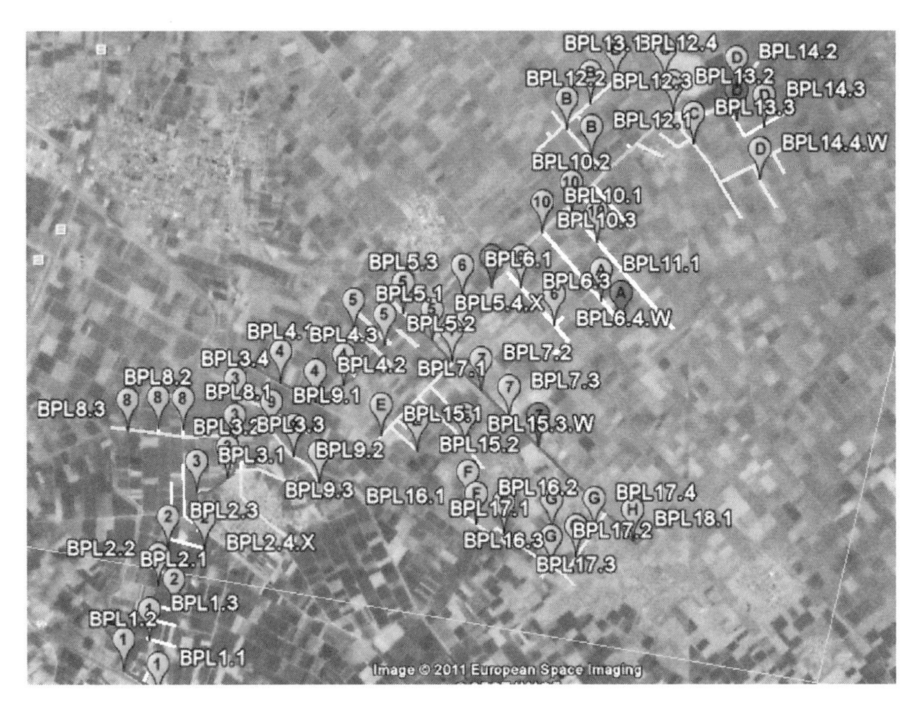

Figure 14.1. BPL units along the second MV line of the W-BPL network deployed in Larissa by courtesy of the Greek utility (PPC).

14.2.1 PHY and MAC Specifications

A broadband access network is deployed by installing W-BPL units at appropriate poles of the overhead MV grid [2]. The distance between adjacent BPL units varies since it depends on the transmission characteristics of the power line link and the QoS requirements of the supported Smart Grid or broadband services. BPL propagation is significantly impaired by severe, frequency dependent, attenuation [3, 4, 6], which exhibits a lowpass behavior, as already presented in Chapter 7. Also, to ensure network connectivity over an MV line, discontinuities due to various kinds of branching along the end-to-end transmission path require an accurate evaluation in order to extract information about the necessity of signal repeaters and the exact location where they should be installed [7]. In addition to frequency-dependent attenuation, BPL signals suffer from nonperiodic, impulse noise that severely reduces the SNR along certain BPL links. Another type of noise met in overhead MV-BPL transmission is narrowband noise due to interference from broadcasters, amateur radio users, or other radio emissions. Although this type of noise is characterized by higher power spectral density, adaptive spectral notching seems to be an effective countermeasure to eliminate interference either to or from MV-BPL networks [7]. In a realistic MV/BPL network, spectral notches are observed in the end-to-end attenuation due to reflections and multipath propagation caused by branches and junctions [3–5]. Since the respective signal deterioration depends on the number and the electrical length of the branches, urban areas are more severely affected by multipath than rural areas, as, in the latter areas, the MV grid is sparsely branched. Besides branching, impedance mismatches are also caused by the existence of insulators, transformers, MV circuit breakers, and other devices mounted on the MV network.

To implement the BPL PHY layer, the FFT-based OFDM multichannel transmission technique is used [8], separating the frequency range 3–34 MHz into 1024 subcarriers, where up to 64-QAM is used. Moreover, since licensed (primary) wireless services vary with the location and/or the general environment (urban, suburban, or rural), spectral masking/shaping adaptive to local traffic conditions may block/modify BPL operation at frequencies where and/or when required. This is a very encouraging characteristic of OFDM modulation since it controls interference without having to use transmit notch filters [8] or to reduce the total injected power.

To assure network synchronization and acceptable QoS, adaptability of the PHY layer is made feasible by employing adaptive modulation and coding. The BER-bandwidth trade-off is that the MV-BPL links subjected to adverse transmission conditions operate at lower useful data rates due to FEC coding with error-resilient code rates and low-order modulation schemes.

A time division duplexing (TDD) approach is the token-based transmission MAC scheme currently implemented in BPL networks. A BPL unit denoted as the head end (HE) acts as a master to a number of time division repeaters (TDRs) and is responsible for token generation. The last node in the sequence of BPL units constituting a BPL cell is the customer premises equipment (CPE). The HE also acts as a QoS controller assigning priorities to low-delay service classes (SCs). Both types of units, HE and TDRs, regenerate the signal extending the BPL access network coverage. Together with the physical discontinuities of the power line channel, the token-based transmission scheme imposes a limit on the maximum number of BPL units in an autonomous BPL network. Specifically, with regard to the token-based MAC scheme, since the transmission time is divided among all BPL units in a cell, the respective number of BPL units should be limited to ensure that each node has adequate timeslots for efficient transmission during one token passing round. Hence, a cell-based architecture is preferred, where each BPL cell consists of a limited number of BPL units—usually up to six as depicted in Figure 14.2—rendering the operation of each cell independent of the rest of the BPL network. The BPL units inject the signal into a phase of the MV line through inductive coupling.

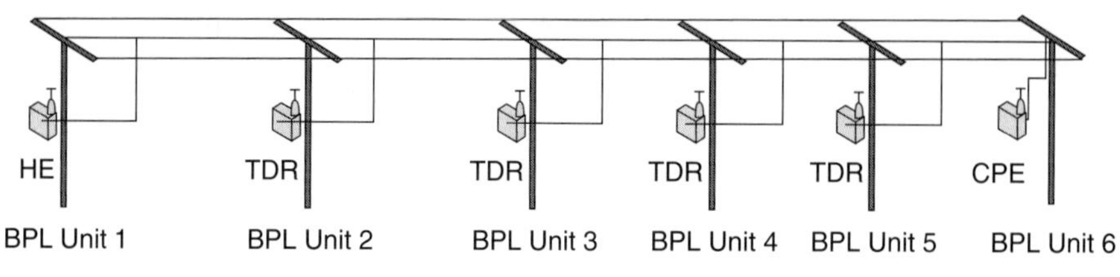

Figure 14.2. A BPL cell consisting of six BPL units.

14.2.2 Cell-Based Architecture for W-BPL Networks

As the W-BPL network deployed over the area of Larissa is cell based, every cell comprises a single HE responsible (1) for token generation and (2) for forwarding the traffic generated by the cell to the rest of the network, acting as the cell traffic aggregator. Although every cell operates independently of the rest of the W-BPL network, mutual interference between adjacent cells may exist. To minimize intracell interference, specific segments of the 3–34 MHz spectrum are assigned to each cell, under the constraint that no interference is caused between neighboring cells operating at overlapping frequencies. The frequency planning for the Larissa network led to the formation of three subchannels of 10 MHz to ensure that neighboring cells operate in nonoverlapping channels.

Cell dimensioning is critically affected by grid topology. The existence of physical discontinuities along the power grid or of segments where high signal reflections exist—caused by other already installed devices such as insulators, transformers, and MV circuit breakers—impair the transmission conditions. Hence, "bad" links of low SNR [4] appear in certain cells, where error-free transmission is not guaranteed. In such cases, BPL cells are formed aiming at avoiding links identified as bad.

Usually, wireless connectivity between the cells is provided. BPL deployments in rural areas such as that in Larissa favor the wireless interconnection of BPL cells exploiting the line-of-sight (LOS) connectivity of neighboring cells. The BPL units used in the Larissa project have been developed by Amperion Greece [1] and operate according to OPERA specification [8]. Each unit has a wireless interface used to enable point-to-point connections to neighboring cells. The HEs of the cells are responsible for routing the respective traffic to the NOC (network operating center), wherefrom access to the network gateway (G/W) is provided. Traffic generated from a BPL cell toward the NOC can be routed either directly via a wireless link, if LOS from the HE to the NOC exists, or through other BPL cells. The overall link from the HE of a BPL cell toward the NOC may involve both wireless and BPL transmission. Hence, a hybrid wireless–BPL platform is formed.

Finally, every BPL unit includes additional wireless interfaces to form wireless links to neighboring units. This offers link availability in case the BPL link becomes temporarily unavailable (i.e., in case of an outage or of any sort of physical BPL disconnection) and assures continuity of network operation. The BPL units are powered at LV making use of an MV-to-LV transformer. To provide power backup, a battery is incorporated in the BPL units employed [1]. By switching to battery mode in case of grid failure, several hours of communications backup are allowed. This constitutes a critical property of the W-BPL communications platform.

14.2.3 The Network Operating Center (NOC)

The complete network architecture of the NOC is presented in Figure 14.3. Being the main network management point for the utility, the NOC is installed in the HV/MV substation. Its position has been chosen so that the backhaul router can be

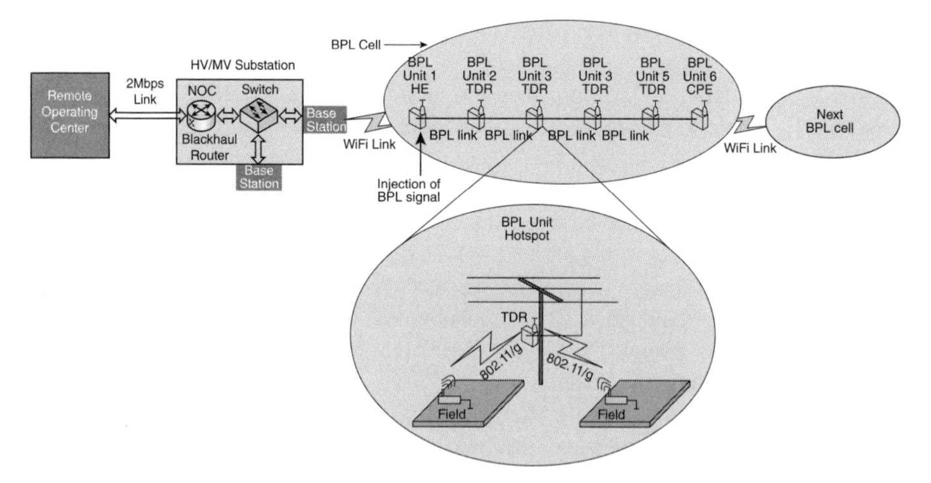

Figure 14.3. W-BPL network architecture.

connected to the local point of presence of the communications network. The traffic generated by the W-BPL network is routed to the Internet, thus allowing the remote management of the Smart Grid. For this purpose, a VPN link is established connecting the NOC to the central office of the utility. In addition to collecting the W-BPL traffic, the central switch located at the NOC is responsible for the interconnection of the W-BPL network to the application servers installed at the NOC, for both Smart Grid and broadband services.

W-BPL technology does not restrict the ways the NOC may be connected to the W-BPL network. As previously mentioned, in the rural area of Larissa, wireless connection seems a suitable choice offering continuous connectivity and QoS assurance. The BPL units installed along the two MV lines are grouped into cells. The HEs of the cells located closer to the HV/MV substation are responsible for routing the respective traffic to the NOC, acting as gateways for the two MV-BPL branches. These two gateways are wirelessly connected to two base stations installed at the NOC, which are directly connected to the central switch.

14.2.4 Last-Mile Access

Access to the BPL network can be provided in many ways. However, Wi-Fi wireless access or BPL access via the LV grid is usually preferred. As to access via the LV grid, severe problems exist due to the high noise level and to impedance mismatches since different types of LV cables are used. Another drawback is that the installation of a BPL modem at the customer premises is required. The wireless option adopted for the last hop connection of the end-users to the W-BPL network was the IEEE 802.11b/g option introduced in Chapter 6, since the BPL units employed [1] can serve as Wi-Fi hotspots extracting the signal from the MV power lines and converting it into IEEE 802.11b/g signal. Accordingly, the authorized Wi-Fi enabled end-users or Smart Grid devices can easily be connected to the W-BPL network, benefitting from the low interference existing in a rural environment. In the Larissa project, the Wi-Fi option was selected since a single BPL unit could offer low-cost and efficient coverage over an area covering 200 m around the unit.

14.3 SMART-GRID APPLICATIONS OFFERED IN LARISSA

The hybrid W-BPL network deployed in Larissa offers a wide variety of services that can be distinguished into two main categories: (1) services for the grid operator, mainly focused on Smart Grid solutions for monitoring and managing the power grid; and (2) services for third-party users, mainly focused on broadband applications. While the first category of services was the motivation for the utility to deploy the W-BPL trial under consideration, the second category may prove equally important in rural, sparsely populated areas.

14.3.1 Grid Monitoring and Operations Optimization

The efficiency and cost effectiveness of the operations related to the power transmission grid constitute the main driving force behind the Smart Grid revolution [9]. A large part of grid operations are related to network outages, aiming either at preventing them or at handling them when they occur. Grid operators strive to minimize the occurrences and the consequences of outages, in order not only to avoid revenue loss and customer dissatisfaction, but also to comply with the respective regulatory legislation. An efficient grid monitoring system must perform three main functions related to network outages: detection—immediate notification when a fault occurs, prediction—early notification about an imminent fault; and *fault recognition*—detailed identification of an imminent or an ongoing event.

BPL technology offers unique advantages with regard to the communications infrastructure required to implement Smart Grid services compared to other available solutions, such as the wireless networks discussed in Chapter 6. These advantages are justified by the following reasons:

- A BPL network should be deployed, owned, and maintained by the grid operator and not by a third-party telecom provider, since security and availability are primary concerns, given the importance of grid monitoring.
- The BPL links follow the branches of the power grid. Thus, coverage of even the most remote areas is assured.
- As BPL units are coupled to the power grid itself, they can perform many functions—such as fault recognition—that a solely wireless infrastructure cannot. The grid operator exploited many of these advantages in the trial W-BPL deployment in Larissa.

An essential capability for network monitoring offered by the W-BPL deployment in Larissa is fault detection, achieved by monitoring the beacon packets (in the form of periodically generated SNMP traps) constantly relayed by every BPL unit to the NOC. When, for any reason, a power line fails, the two BPL units on either side of the point of failure have no longer a wired communication connectivity. Due to the hybrid W-BPL operation, the beacon signals will be relayed via the backup wireless link. This change is diagnosed through the packets payload. If this situation lasts for more time than expected as a result of usual intermittent communication link failures, a power line failure alarm will be raised. This alarming procedure has two significant advantages over traditional fault detection methods: (1) the notification reaches the NOC almost instantly via the high-speed BPL network and (2) the position of the fault is very accurately determined, since it can only be located along the branch joining the two BPL units that have lost wired connectivity, usually a line length of roughly 500 m. Hence, though this method does not reduce the number of faults, it reduces significantly the fault detection, location and repair times, which are vital measures of operation efficiency.

To increase the power grid reliability and reduce the number of network outages, the BPL network can be connected to a number of control devices, consisting of data loggers connected to appropriate voltage and current sensors. The specific components used in the Larissa trial were manufactured by Eltek Corporation and are commercially named "squirrels" [11, 12]. Such devices are capable of monitoring the quality of the three-phase MV power flowing through the respective line segment by measuring and reporting on (1) voltage and current values that exceed specified levels, (2) high harmonic distortion, (3) unbalances between the three phases, (4) power values close to the overload threshold, and so on. Constant reporting of such events signifies that a network outage may be imminent, due either to a line fault or to a power overload. Thus, BPL technology provides the system with a critical network outage prediction capability, giving the grid operator an early warning either to perform preventive actions or to enforce planned outages of noncritical loads employing the demand-side management system to be described next.

The "squirrels" may also contribute to the formation of a revenue assurance mechanism destined for the identification and reduction of nontechnical power losses, that is, losses due to unauthorized and unbilled access to the power grid. Generally, this situation is common in many regions and constitutes an important problem for the Larissa grid operator. Through comparison of the average power flow values via the various line segments provided by the "squirrels" with the power meter readings from the same area, relatively high discrepancies may be discovered, indicating points where energy theft might be taking place.

Besides the use of "squirrels," an alternative network outage prediction system has also been tested in the Larisa trial based on the experimentally verified observation that, for a time period before a breakdown, a faulty power line component—for example, transformer, switch, insulator—induces significantly higher noise power levels in the BPL communications spectrum. Hence, the BPL link across this line experiences unusually low SNR values. By comparing periodic SNR measurements reported by the BPL units with the respective historical records and, hence, determining the links that exhibit abrupt SNR deterioration, imminent component faults may be predicted, which, otherwise, might have not been so.

Regarding the QoS requirements concerning fault detection and outage prediction, only a relatively small bandwidth must be reserved, since only a few bytes are sufficient for the transmission of SNMP beacons, "squirrel" reports, and SNR readings of BPL units. Furthermore, the above applications are delay tolerant since they are not based on real-time data but on average values. On the other hand, as fault detection and outage prediction are considered critical applications for grid stability, the respective availability specifications are very strict.

An important step concerning services related to network monitoring and operations would be to enhance network fault detection and outage prediction by providing the capability to differentiate between faults that are imminent and faults that have already occurred, that is, providing a fault recognition service. Experimental work is underway with promising results, in the attempt to correlate specific types of line faults to associated noise patterns in the BPL frequency spectrum [10]. Such noise patterns, emitted before or after a fault takes place, may be interpreted as a signature for this type of fault. Assuming that the BPL units are able to record such noise patterns when they happen, the recordings can be forwarded to the NOC and be compared to a noise pattern library, providing information on a specific type of fault, either imminent or having taken place.

14.3.2 Demand-Side Management

Demand-side Management is a strategy for more efficient power generation that addresses the well-established fact that the demand for electrical power is not constant over the period of a day. It has been extensively discussed in Chapters 1 and 3, where it was pointed out that shaping the power demand curve, aiming at distributing power demand uniformly over time, often leads to a significant efficiency improvement and reduces the requirements for new power plant installations.

The shaping of the power demand curve is accomplished mostly by exploiting the cooperation of consumers through publicity campaigns and/or by specifying financial incentives such as adaptive, time-of-day-based tariffs. A simpler, though not always feasible, solution is to switch off any noncritical loads during the expected peak demand hours. This is feasible only if (1) noncritical loads (i.e., devices that can be disconnected from the power network without causing significant problems) are connected to distinct electrical circuits and can be switched off independently and (2) these loads can be switched on or off by the grid operator as required or scheduled, in an efficient and reliable way.

In fact, distinct circuits for noncritical loads are encountered in the rural region of Larissa, where the trial W-BPL network was deployed. Specifically, the MV network of the region feeds a significant number of high-power irrigation pumps that consume high levels of energy when they operate and are connected to the MV power grid via dedicated MV/LV transformers. These pumps are mostly used in the summer, for a few hours per day. It has been observed that most farmers switch on their pumps during the summer peak hours (around midday), thus overloading the local transmission grid. However, these pumps can be considered as noncritical loads, since irrigation could easily be done during off-peak hours without causing any problems to the agricultural process. A reliable and efficient management system for noncritical loads has been implemented in Larissa based on the W-BPL network installed there, which provides the grid operator with communication channels from the NOC to virtually every irrigation pump in the covered area.

To implement a demand-side management scheme for the electrical power required by the irrigation pumps, the trial BPL deployment employed remote terminal units (RTUs) connected to the power supply line of every pump. The RTUs are wirelessly linked to the BPL network (thus overcoming the signal discontinuity introduced by the MV/LV transformer) and thus establish a data communication path to the NOC. The RTUs are properly enabled to collect near real-time electrical measurements at the connection points (voltage, current, power consumption, etc.), to monitor line quality and, more importantly, to remotely switch on and off the power supply line of the pumps. Hence, the grid operator has a real-time picture of the total power consumed by the pumps and can also intervene when this consumption affects critically the stability of the transmission grid. This intervention may affect either all the RTUs or just a part of them. Also, it may be proactive, with planned switch-offs during peak hours, or reactive, when real-time monitoring indicates excessive power demand. These intentional outages cause a minor inconvenience to the irrigation pump users, an inconvenience alleviated by the reduced probability of a wider area blackout, which would take place if the power demand was not monitored and controlled.

With regard to its QoS requirements, the RTU network for demand-side management is considered as a low bandwidth application for the same reasons concerning the network monitoring applications already discussed. Delay tolerance for the RTUs is rather low, taking into account that a near real-time picture of the power consumption is required. Finally, the critical character of this service sets a strict availability requirement.

14.3.3 Broadband Services

A different family of services may be offered by taking advantage of the relatively large bandwidth provided by the W-BPL technology. These services are mainly user-oriented, since Smart Grid applications have small bandwidth requirements and could easily be served by a narrowband channel. For Wi-Fi users, the W-BPL network can be made to resemble an IP pipe either to the Internet or to another user. Hence, virtually all applications available on a LAN or on the Internet can be provided.

The only broadband service implemented so far in the Larissa W-BPL trial is VoIP offered to the support and maintenance personnel of the grid operator. A Wi-Fi VoIP device is issued to every maintenance unit, connecting to the various BPL units via the wireless interface. A VoIP application server has been installed at the NOC, allowing the units in the field to communicate with the NOC personnel for reporting and instructions. This VoIP server could easily be upgraded either to allow unit-to-unit calls or to route calls to external VoIP and public switched telephony network (PSTN) connections, if necessary. This free communication option has partly replaced the costly GSM/3G mobile calls, thereby reducing operational costs. Practically, Wi-Fi connectivity to the BPL units is always available in the proximity of the MV power lines, where the maintenance personnel need to communicate. Such an availability is not always guaranteed by the GSM/3G network, especially in remote areas.

As already stated, almost any IP-based application can be offered via the W-BPL platform. However, it is reasonable to limit the discussion to services that make use of the main advantage of the system, namely, its ability to reach remote rural areas. MV power lines are present in the smallest villages as well as in a number of remote buildings (warehouses, agricultural premises, etc.). DSL and 3G broadband have not yet achieved such a penetration and are not expected to achieve it cost effectively. Last-mile access, from the termination of an MV line to the user CPE, is achievable via the wireless interface of the BPL units.

An extension of the broadband services in such remote areas would be to make VoIP services available to the public. It is not evident, however, that this would give any advantages in areas where the traditional PSTN is almost as widely deployed as the power grid. Probably, a more interesting application would be providing broadband Internet access in remote areas. The demand for broadband Internet is constantly rising and BPL access may prove an effective solution for remote rural access, complementary to the other options discussed in Chapter 6.

Finally, a BPL network may be effectively used as the communications infrastructure for implementing physical security and remote management applications in isolated buildings and installations by providing broadband connectivity to alarm systems, IP controlled surveillance cameras, building management systems, and so on. Such monitoring and security services are of critical importance to owners of remote renewable generators, thus securing and facilitating the exploitation of sustainable energy sources.

Regarding the QoS requirements, broadband applications require significantly more bandwidth than Smart Grid applications. In general, the delay and availability requirements vary: Internet access requires as much bandwidth as available but on a best-effort basis, with few QoS guarantees. VoIP service, on the other hand, requires relatively smaller bandwidth but with guaranteed resource allocation for the whole duration of the connection. VoIP is also very sensitive to delay and, especially, to delay variations (jitter).

14.4 KEY LESSONS LEARNED

14.4.1 Issues Related to the Site of the BPL Deployment

14.4.1.1 Rural or Urban Areas. The site of the deployment seems to be the most critical factor when deploying a W-BPL network. Both transmission channels used in this deployment, the BPL channel and the wireless channel, are severely affected in urban areas [3, 7]. BPL transmission is affected by multipath fading caused by reflections at junctions and terminations. This phenomenon becomes significant in urban environments, leading to severe SNR deterioration of BPL signals. With regard to network design, cells of smaller size should be created and BPL units should be placed at smaller distances, thus increasing the installation cost.

Regarding the interconnection of BPL cells, wireless transmission is not a preferable choice in urban environments. The absence of LOS between neighboring cells and multipath propagation adversely affect the interconnection of cells located far from the NOC. In such cases, alternative transmission media should be considered. On the other hand, wireless transmission remains the best option when rural areas are considered.

14.4.1.2 EMC Constraints. The operation of BPL in the 3–34 MHz band may cause or suffer from electromagnetic interference (EMI) to or by local primary wireless users. In such cases, the OFDM modulation used at the PHY layer of BPL systems allows the dynamic isolation of the subchannels that are involved.

14.4.1.3 Purpose of the Deployment. The primary motivation for the deployment of a BPL network can be either to support Smart Grid services or to extend the communications infrastructure offering broadband access to end-users. Each purpose may lead to a different network design. If providing Smart Grid services is the primary objective of the deployment—which is usually the case—a BPL cell may cover large areas, avoiding the installation of a high number of BPL units. This is because most Smart Grid services are low bandwidth, non-real-time, applications. Moreover, the number of grid points that must be monitored and/or controlled is known prior to the BPL network deployment. Therefore, when designing a BPL cell, the majority of RTUs should be served by the BPL units close to the HE of the cell, since these units are less affected by the presence of bad links.

In case broadband access is the primary objective of the deployment, the end-users density is the key characteristic affecting the BPL network planning. When the last-hop access to the BPL network is wireless, careful site survey is required to guarantee end-users coverage.

14.4.2 Issues Related to the Condition of MV Grid

14.4.2.1 Noise Conditions. The appearance of low SNR levels along BPL links is an indication of significant transmission impairments and should appropriately be taken into account. Low SNR levels observed constantly over a BPL link indicate that it is a bad link unable to support reliable transmission. On the other hand, SNR variations might indicate the existence of a faulty component along the MV line.

14.4.2.2 MV Grid Topology. The topology of the MV grid critically affects the formation of BPL cells. The grid topology should be carefully examined to determine any bad links. Generally, such links should be avoided by activating appropriate wireless links. Otherwise, a bad link indicates the need for cell segmentation. In case a bad link should be incorporated in a cell, it should be positioned close to the end of the cell since, then, its bandwidth would be shared by fewer users.

14.4.3 Application Related Issues

14.4.3.1 End-to-End Security Constraints. Incorporating a robust and scalable security architecture is a major concern when designing modern communication networks. In the case of W-BPL networks, security is essential both for broadband services, where end-user privacy must be guaranteed, as well as for those Smart Grid related services that are expected to have a critical impact on grid stability and availability. In fact, it has already been mentioned that grid operators may favor W-BPL networks over alternative wireless metropolitan area network (MAN) technologies because the latter option will result in critical information, such as network status reports or automatic meter reading (AMR) billing data, being transmitted via a third-party and possibly via not reliable routes, beyond the control of the grid operator.

Communications via the two possible PHY layers in the Larissa trial are always encrypted. As to the Wi-Fi interface, the standard WEP protocol defined in IEEE 802.11 has been used. However, as discussed in Chapter 6, WEP has proved to be inefficient in meeting wireless security goals and has been superseded by the WPA/WPA2 protocols. Future deployments of W-BPL networks should, therefore, be upgraded accordingly. For the BPL interface, traditional wired-network security mechanisms are employed, based on the Diffie–Hellman key-exchange algorithm and 3DES encryption. Furthermore, end-user authentication is possible through the installation of a RADIUS server enabling the Extensible Authentication Protocol (EAP).

It is important to note that even though the above encryption protocols have proved to be fairly robust, they are not immune to attacks by competent intruders. Moreover, in remote and isolated areas, where the W-BPL network is primarily intended to operate, it is not possible to prevent physical access. Therefore, it is possible that a malicious eavesdropper with adequate equipment may have access to virtually all the encrypted traffic via the Wi-Fi or even via the wired BPL link. The issue of BPL security is still an open issue. Further research in this area is required if critical services are to be supported on a large scale.

14.4.3.2 End-to-End QoS Assurance. QoS assurance is a key issue in BPL networks since various applications have varying QoS requirements. These applications are grouped into various service classes, each of them assigned a fixed MAC priority. This approach favors real-time applications, like VoIP, or critical Smart Grid services. Also, fairness for all end-users served by the W-BPL network should be provided. This remains an open issue since the BPL cells operate independently and, possibly, under different transmission conditions. A trade-off between fairness and cell utilization should be achieved, so that even the less favored cell nodes with regard to propagation conditions meet the QoS requirements and offer reliable access to their end-users.

14.5 CONCLUSIONS

The W-BPL trial deployed in Larissa during 2007, the first of its kind in Greece, has provided valuable insight from both a Smart Grid as well as a telecommunications perspective. The main conclusion is that W-BPL networks constitute the ideal communications platform for Smart Grid applications as well as a viable and, in many cases, competitive alternative for middle and last-mile broadband Internet access in rural and remote areas. From the grid operator point of view, the fact that the BPL network is mounted on the electrical grid guarantees ubiquitous presence and cost efficiency together with the opportunity to deploy critical applications aiming at predicting and/or preventing network outages, a feature of this technology that is unique. From the point of view of broadband access, providing Internet access and related applications via W-BPL networks is probably the only cost-effective solution available to a significant part of the population in remote and rural areas of developing countries.

A large number of future extensions for this trial deployment could be considered. Among the more interesting of them are:

- Installing RTUs on a number of other loads of the grid in addition to the irrigation pumps, for example, on the noncritical, high-power-consumption loads, such as factories and hospitals, or on distributed renewable energy generators, such as household wind turbines and roof-mounted photovoltaics, to deal with the issues described in Chapter 1:
- Using the W-BPL network as a telecommunication infrastructure for AMR applications (described in Chapter 1)
- Replacing the "squirrel" control devices with more advanced, time synchronized phasor measurement units (PMUs), which would provide a real-time description of the condition and the dynamic behavior of the grid
- Extending the offer of broadband services according to the various scenarios referred to in Section 14.3.3

BPL communications is a relatively new technology; therefore, a number of research issues related to BPL network design are open. Issues of particular interest to W-BPL network deployments, such as the trial network discussed in this chapter, have to do with accurate modeling of the adverse BPL communications channel created over the complex grid topology, resource allocation and efficient MAC schemes, efficient techniques for QoS differentiation between Smart Grid and broadband applications, and robust security architectures. The recently published IEEE 1901-2010 BPL standard [13], which provides a common reference framework for the PHY and MAC layers, is expected to drive research toward innovative solutions.

REFERENCES

[1] *Amperion Greece S.A.* Available at: http://www.amperion.gr (accessed April 14, 2011).

[2] A. Sarafi, G. Tsiropoulos, and P. Cottis, "Hybrid wireless-broadband over power lines: A promising broadband solution in rural areas," *IEEE Commun. Mag.*, 47(11), pp. 140–147, November 2009.

[3] A. G. Lazaropoulos and P. G. Cottis, "Transmission characteristics of overhead medium voltage power line communication channels," *IEEE Trans. Power Delivery*, 24(3), pp. 1164–1173, July 2009.

[4] A. G. Lazaropoulos and P. G. Cottis, "Broadband transmission via underground medium voltage power lines—Part I: Transmission characteristics," *IEEE Trans. Power Delivery*, 25(4), pp. 2414–2424, April 2010.

[5] L. T. Berger and G. Moreno-Rodríguez, "Power line communication channel modelling through concatenated IIR-filter elements," *Academy Publisher J. Commun.*, 4(1), pp. 41–51, February 2009.

[6] M. Zimmermann and K. Dostert, "A multipath model for the powerline channel," *IEEE Trans. Commun.*, 50(4), pp. 553–559, April 2002.

[7] A. G. Lazaropoulos and P. G. Cottis, "Capacity of overhead medium voltage power line communication channels," *IEEE Trans. Power Delivery*, 25(2), pp. 723–733, April 2010.

[8] EC/IST FP6 Project No 026920, "D20—Requirements on OPERA for Implementation of Multipurpose PLC Network Including EMS," OPERA, pp. 15–16, November 2007.

[9] S. Galli and O. Logvinov, "Recent developments in the standardization of power line communications within the IEEE," *IEEE Commun. Mag.*, 46, pp. 64–71, 2008.

[10] S. Massoud Amin and B. F. Wollenberg, "Toward a Smart Grid: Power delivery for the 21st century," *IEEE Power Energy Mag.*, 3(5), pp. 34–41, September–October 2005.

[11] E. S. Kapareliotis, K. E. Drakakis, H. P. K. Dimitriadis, and C. N. Capsalis, "Fault recognition on power networks via SNR analysis," *IEEE Trans. Power Delivery*, 24(4), pp. 2428–2433, October 2009.

[12] *Eltek Specialist Data Loggers—Products—Energy Sensors.* Available at: http://www.eltekdataloggers.co.uk/sensors_energy.shtml (accessed April 14, 2011).

[13] IEEE Standard 1901–2010, "IEEE Standard for Broadband over Power Line Networks: Medium Access Control and Physical Layer Specifications," December 30, 2010.

15

SMART CHARGING THE ELECTRIC VEHICLE FLEET

Peter Bach Andersen, Einar Bragi Hauksson, Anders Bro Pedersen, Dieter Gantenbein, Bernhard Jansen, Claus Amtrup Andersen, and Jacob Dall

15.1 INTRODUCTION

The Danish EDISON project [1–3] has been launched to demonstrate how the charging and possible discharging of *electric vehicles* (EVs), if handled intelligently, can yield benefits to the EV owner, the grid, and society. EDISON is partly publically funded through the Danish transmission system operator (TSO) Energinet.dk's research program FORSKEL. The total budget is approximately EUR 6.5 million, with EUR 4.5 million thereof coming from FORSKEL. The consortium consists of the Danish energy corporations DONG Energy and Østkraft, the Danish Technical University (DTU) CET and Risø, as well as IBM, Siemens, EURISCO, and the Danish Energy Association (DEA).

While the more progressive concepts such as using the EVs as energy storage or for regulating services using V2G show great promise, the EVs potential as a controllable load could be seen as the low-hanging fruit in EV integration. The smart

Smart Grid: Applications, Communications, and Security, First Edition.
Edited by Lars Torsten Berger and Krzysztof Iniewski.

charging concept, where the charging of EV batteries is delayed or advanced in time based on energy costs, grid constraints, or renewable contents, has great potential and is the initial focus of EDISON. The success of smart charging, however, relies on a suitable and standardized ICT architecture. This chapter documents the suite of contemporary communication technologies, components, and standards, which helps to facilitate smart charging in the EDISON project. The chapter is organized as follows. First, Section 15.2 describes the integration scenario used in the EDISON project and introduces the fleet operator as a new stakeholder in the power system. The objectives of the fleet operator will be described along with the requirements of the ICT architecture that supports its operation. Next, Sections 15.3 to 15.5 describe the standards and standardization work relevant to the project, a set of EDISON-developed hardware and software components and the communication technologies used to interface the main entities.

The final parts of the chapter, Sections 15.6 and 15.7, present a set of demonstration interfaces and draw conclusions on the utilization of communication technologies in EDISON.

15.2 THE FLEET OPERATOR AS A NEW CONCEPTUAL ROLE

The conceptual role of a fleet operator, which can be taken by different commercial players, is introduced to allow groups of EVs to be actively integrated in the power system. Toward the grid and market stakeholders, the fleet operator will operate as a *virtual power plant* (VPP). The virtual power plant concept describes an aggregated system in which distributed energy resources (DERs) are partly or fully controlled by a single coordinating entity. In this way, DERs can be actively integrated into the power system and market, for which individually they would be too small, in terms of power output and availability, to participate in. The concept has been demonstrated in the European FENIX project [4] and studied by Shi You et al. [5].

In the case of EDISON a fleet operator could mimic a traditional power plant by aggregating a group of electric vehicles. The fleet operator would also need to interact with each individual electric vehicle to optimize charging. The technical implementation of this concept is called the EDISON VPP (EVPP) and would be used by a fleet operator as shown in Figure 15.1.

15.2.1 Fleet Operator Interaction with Grid and Market Stakeholders

A first step in EV integration is identifying the stakeholders, old and new, that will have a role to play in interfacing the EVs with the power system and market. The composition of stakeholders depends heavily on the business models and market environments under consideration.

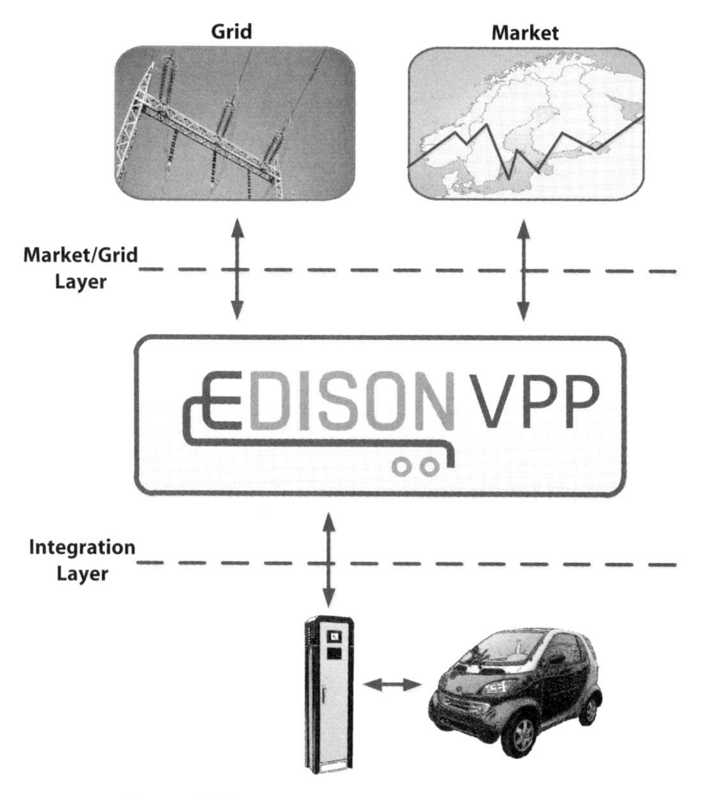

Figure 15.1. EDISON conceptual architecture.

The EDISON fleet operator integration scenario is based on the current Nordic power system and market configuration. There are obviously many other integration concepts such as near real-time markets, frequency response, or price signals in which the vehicle acts as an autonomous and intelligent agent. While such concepts are within the research scope of EDISON, the phase-1 scenario will focus on conditions as they are today and is a pragmatic first approach to EV integration. Figure 15.2 shows the market domain model in which the fleet operator interfaces the EV with the power market. Among the stakeholders in this domain is the *transmission system operator* (TSO), which controls the transmission grid and maintains the overall security of electricity supply, and the *distribution system operator* (DSO), which manages a part of the distribution grid and handles local metering. The fleet operator must maintain the appropriate balancing responsibilities when acting on the markets.

As illustrated, the fleet operator could participate either in the energy market or in ancillary services. The first project phase, however, will put its emphasis on the former and indirectly connect the EVs with the day-ahead spot market by controlling the charging in correspondence with hourly energy prices.

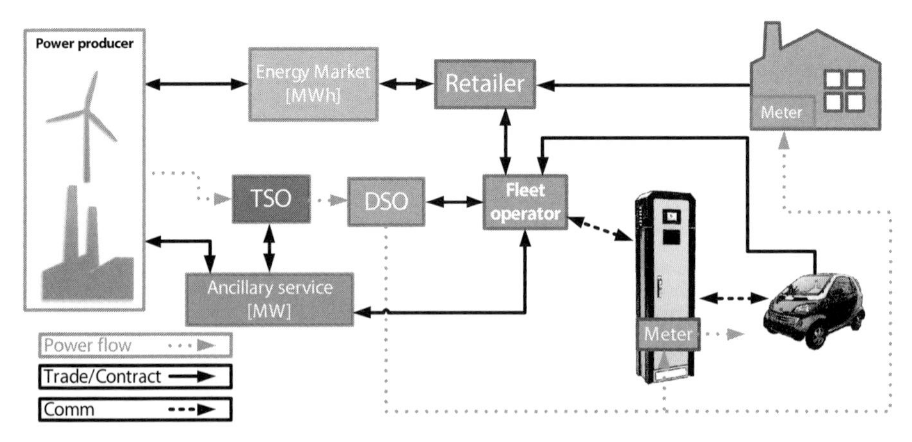

Figure 15.2. Market domain model.

15.2.2 The Objective of the Fleet Operator

Based on the market integration and stakeholder setup, as described above, the primary objective of the EVPP fleet operator is to facilitate smart charging. Under smart charging, we understand the computation of a per EV charging schedule, which is computed using some predetermined optimization targets as well as a set of constraints. The objectives of the charging schedules are primarily to ensure that sufficient energy will be delivered to the EVs such that future trips can be carried out. Sufficient energy can be further refined into energy objectives, that is, an 80% full battery or a more precise, per-trip, energy objective.

Aside from the primary objective of supplying energy for the use of driving, other objectives can be defined:

- *Minimizing Energy Costs.* Charge at the time periods with the lowest energy prices.
- *Respect Grid Constraints.* Adjust charging to capacity limitations of the distribution grid.
- *Renewable Contents.* Charge during periods when underutilized renewable energy is produced.

This results in a multiobjective optimization problem where a solution requires that a compromise between the objectives is found. For example, optimal use of renewable energy will not guarantee a minimization of charging costs. This optimization is done in the EVPP software, which is described in Section 15.4. The mathematical techniques used by the EVPP are addressed by Olle Sundström and Carl Binding in the paper "Optimization Methods to Plan the Charging of Electric Vehicle Fleets" [6], where linear and quadratic optimization methods are investigated for charging schedule generation. The paper "Planning Electric-Drive Vehicle Charging Under Constrained Grid Conditions" [7], by the same authors, add distribution grid considerations to the optimization.

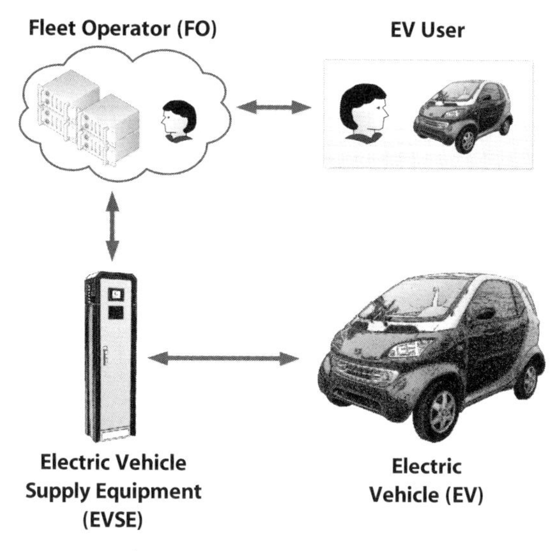

Figure 15.3. The EDISON setup.

15.2.3 ICT Architecture Setup and Requirements

A fleet operator will require a suitable ICT architecture that can connect the stakeholders and let them exchange the information necessary for smart charging. The ICT architecture in EDISON is based on the setup shown in Figure 15.3.

The figure shows the four entities directly involved in EV smart charging. In this setup the *electric vehicle supply equipment* (EVSE) facilitates the connection between EV and fleet operator. The EVSE will extract information from the EV and share it with the fleet operator. The EVSE will then receive a charging schedule from the fleet operator and follow it in the charging of the EV. Since the charging decisions are delegated to the fleet operator and communication is handled by the EVSE, this setup will support most "simple" EVs with limited computation and communication capabilities. As future EVs evolve into more autonomous and intelligent agents, the setup will most likely change. Based on the above setup the following requirements of the ICT architecture have been defined.

- *Adherence to Standards.* EDISON attempts to identify, and to some extent implement, the standards most relevant to its architecture. The chosen standards are IEC 61850 and IEC 61851 as well as the coming ISO/IEC 15118, which will be described in Section 15.3.
- *Implementation of Smart Charging Components.* EDISON must develop a set of hardware and software components that support smart charging for demonstration purposes. This includes software running on the EVSE and fleet operator platforms and the I/O components necessary to connect the EV to the EVSE. These components are described in Section 15.4.

- *Interfaces that Satisfy Basic Communication Requirements.* The protocols connecting the main entities in the EDISON setup must satisfy such requirements as interoperability, scalability, and security. Since the EVSE acts as a proxy for the EV toward the fleet operator, the main focus is on the communication between EVSE and fleet operator. The communication protocols and techniques chosen for the architecture will be described in Section 15.5 along with arguments for including them. See Chapter 10 for an overview of Smart Grid protocols.

The rest of the chapter will attempt to describe how the above requirements have been meet.

15.3 EDISON AND THE USE OF STANDARDS

The EDISON project should produce technical components that are reusable and applicable across different projects and geographies. This requires that the components, as far as possible, conform to a set of standards. By using and supporting standards, the project may also offer input and recommendations for the continued standardization process.

This section describes contemporary standards on which the EDISON ICT architecture is based. As seen in Figure 15.4, these standards can be split into two groups: the ones used for performing (1) EV-to-EVSE and (2) EVSE-to-fleet operator communication, respectively.

IEC 61850 is the communication standard used by EDISON. This standard is not specific to EVs, but supplies the necessary components to describe and send relevant data between EVSE and fleet operator. The IEC 61851 focuses on the

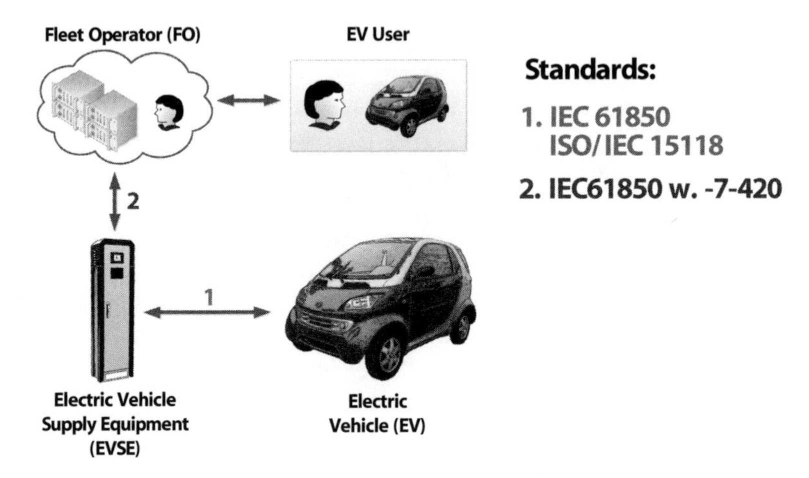

Figure 15.4. Standards used in EDISON.

physical connection and charging of an EV. The future ISO/IEC 15118 deals both with the physical interconnection and a high level communication protocol that, as opposed to IEC 61850, will be oriented toward data and services specific to EVs. Although the ISO/IEC 15118 standard will not be ready for implementation by EDISON, its relevance to EV integration justifies a brief mentioning in this section.

15.3.1 Standards Between Electric Vehicle and Electric Vehicle Supply Equipment: IEC 61851 And ISO/IEC 15118

Between the EV and the EVSE, the following two standards with special relevance to EDISON have been identified.

15.3.1.1 IEC 61851. The "IEC 61851—Electric Vehicle Conductive Charging System" standard was first published in 2001 and has been released in a 2nd edition in 2010. It describes the charging of EVs using different AC or DC voltages over a conductor using on- or off-board equipment.

The main topics of the standard are:

- General system requirements and interfaces
- Protection against electric shock
- Connection between the power supply and the EV
- Specific requirements for vehicle inlet, connector plug, and socket outlet
- Charging cable assembly requirements
- EVSE requirements

"General System Requirements and Interfaces" covers the definition of four different charging modes that an EVSE can support. These modes vary in the currents they support, safety requirements, and location of the charger (in the EV or in the EVSE). EDISON must, for instance, support mode 2 to allow for up to 32 amperes with three phases charging using an onboard charger.

IEC 61851-1 supports the plugs defined by IEC 62196-2. IEC 62196-2 specifies the requirements for plugs, socket outlets, connectors, inlets, and cable assemblies. Among the plugs to adhere to IEC 62196-2 is the Mennekes EV plug, which was developed and tested during the German e-mobility projects and is close to becoming a common European standard.

Another important component of the standard is the definition of simple EV–EVSE communication via a control pilot wire using a *pulse width modulated* (PWM) signal with a variable voltage level. This allows for the definition of different "states," which are listed in Section 15.5 within the description of the EV–EVSE interface.

The use of IEC 61851 is relevant to EDISON for several reasons. First, the safety recommendations described by the standard could be essential in having the developed components approved for live demonstrations. Second, adherence to standards

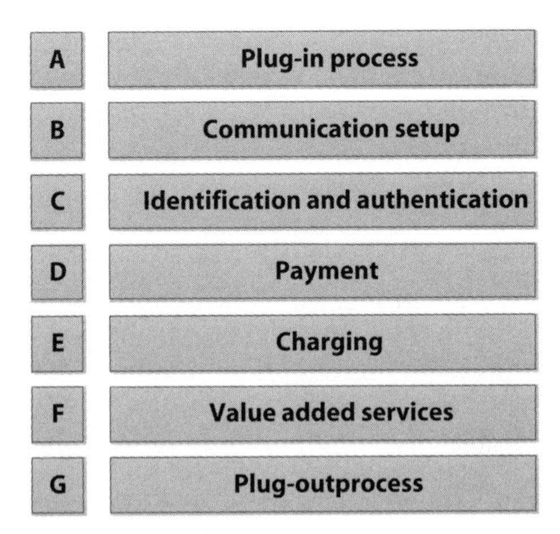

Figure 15.5. The IEC/ISO 15118-1 use case elements.

such as IEC 61851 is one of the prerequisites for roaming, allowing EVs of various brands to use EVSE from different manufacturers.

15.3.1.2 IEC/ISO 15118. The standardization process involving "IEC/ISO 15118—Vehicle to Grid Communication Interface" is still ongoing. The purpose is to make a standard for scenarios that require advanced communication between EV and EVSE.

The standard is divided into the following three parts:

1. IEC/ISO 15118-1 (General Information and Use Case Definition)
2. IEC/ISO 15118-2 (Message and Protocol)
3. IEC/ISO 15118-3 (Physical Layer)

The first part (15118-1), which describes the use cases and terms and definitions, is currently in a Committee Draft (CD) stage and the following use case elements have been identified (Figure 15.5).

In the EDISON project IP-based communication will be implemented, including the use case elements above. The "Value added services" element could include smart charging as defined by EDISON. Adherence to IEC/ISO 15118 would further benefit roaming.

15.3.2 Standard Between Electric Vehicle Supply Equipment and Fleet Operator: IEC 61850

For a couple of decades, the IEC 61850 standard has been one of the preferred ways to relay information and control within the domain of substation automation. Lately though, with the added support for distributed energy resources in the form of the

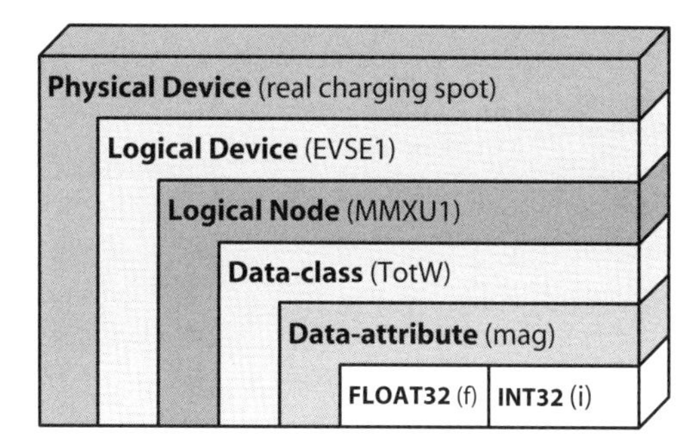

Figure 15.6. The hierarchical structure of an IEC61850 model.

IEC 61850-7-420 substandard, it has moved out of the substation domain to see much wider use.

As illustrated in Figure 15.6 the IEC 61850 standard is highly modular and consists of a large collection of hierarchical building blocks, with which almost every feasible piece of electrical equipment can be modeled. Starting from the top, the device is represented by a logical device. This in turn consists of a series of logical nodes representing various components within this device, and for every layer the granularity becomes even finer. This continues toward the bottom of the structure, where basic data types like strings, Booleans, and integers or floating point numbers make up the final link.

The structure illustrated in Figure 15.6 shows a part of a charging spot model, specifically the *total consumption* (TotW). TotW, which is a data class of type *Measured Value* (MV), is contained in an MMXU logical node. The latter is defined in the basic IEC61850 standard and is used to represent power system measurements. Because TotW is an MV class it contains the data attribute for magnitude (mag), which in turn contains the floating point value in question.

15.3.2.1 IEC 61850 with -7-420 Extension. To move the use of the IEC 61850 standard beyond that of the substation environment for which it was designed, an extension called IEC 61850-7-420 was developed to add the necessary logical nodes needed for communicating with *distributed energy resources* (DERs). Wherever possible, the extension makes use of the existing logical nodes; the standard defines nodes for generation and storage devices, including reciprocating engines, fuel cells, microturbines, photovoltaic arrays, combined heat and power units, and batteries. While IEC 61850-7-420 has been released as an international standard, development of the extension is an ongoing process and logical nodes are being redefined as well as added. During the course of the EDISON project, a proposal was made to extend the standard with logical nodes for both a charging spot (DCHS) and an electric vehicle (DBEV).

15.3.2.2 IEC 61850 Energy and Power Schedules. Sometimes trying to enforce instant control over distributed energy resources (DERs) is not desired and

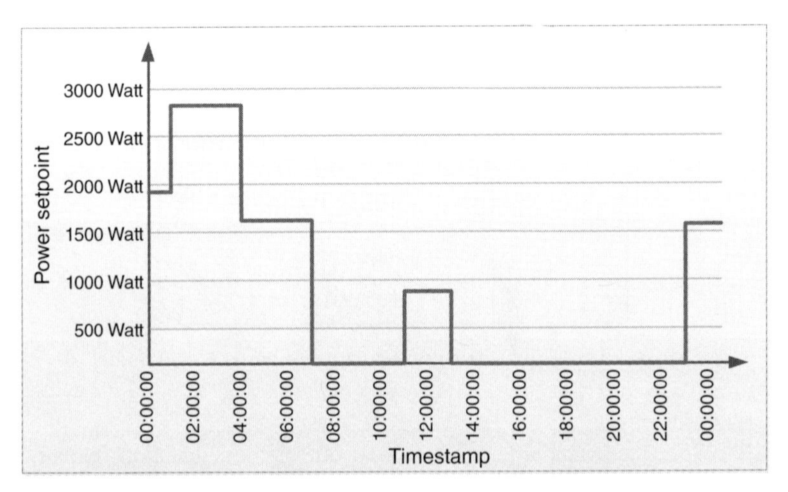

Figure 15.7. Illustration of an EDISON charging schedule.

for this reason the IEC 61850-7-420 extends the standard with the logical nodes for handling absolute and relative timed energy schedules. Consisting primarily of a group of arrays, one of the logical nodes allows for the definition of a series of power setpoints and ramp types together with individual timestamps or time offsets. The usage can vary greatly, allowing the scheduled production of generators, the load of pumps, or, in the case of the EDISON project, the charging schedules for the electric vehicles. Figure 15.7 illustrates how such a charging schedule might look, defining the load the vehicle charger should draw from the grid at the various times specified in the schedule. Though not implemented as yet, support for vehicle-to-grid is easily done with the power schedules by simply stating negative power setpoints.

15.4 SMART CHARGING COMMUNICATION COMPONENTS

This section describes the software and hardware components that have been developed in EDISON to implement the communication interfaces and facilitate smart charging.

As illustrated in Figure 15.8, the components covered in this section will be the I/O board located in the EV and the EVSE, the IEC 61850 compatible server, and the EVPP software used by the fleet operator. The following sections will cover these components in turn.

15.4.1 The IEC 61850 Server

Early on, in the course of the work done in EDISON's work package 3, the IEC 61850 standard was chosen as the main communication protocol between the EVPP and the EVSE.

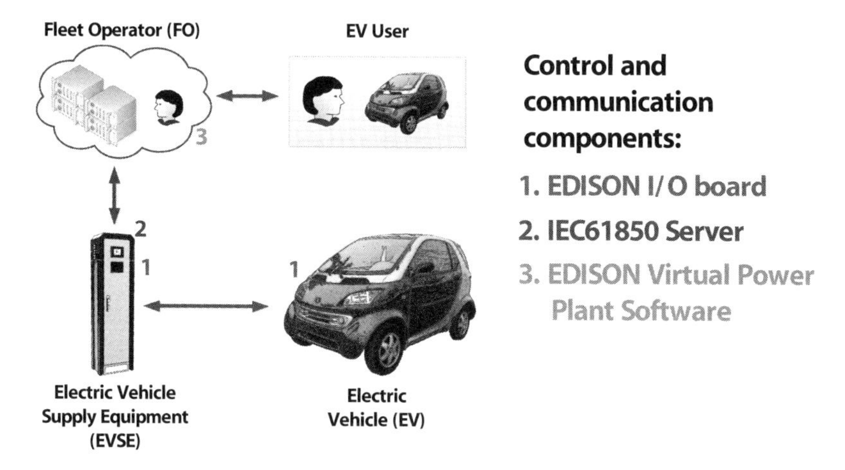

Figure 15.8. EDISON main components.

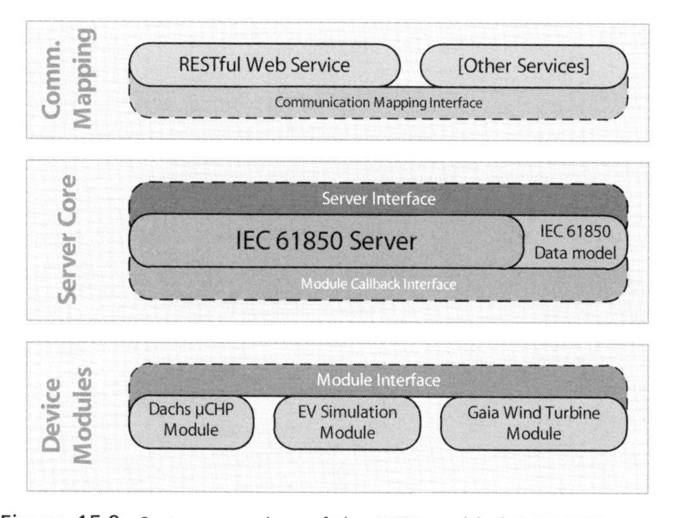

Figure 15.9. System overview of the REST enabled IEC 61850 server.

In order to aid development and testing, and to get rid of proprietary dependencies, an IEC 61850 enabled server was developed. To further facilitate the interoperability between parties within EDISON, the server was designed with a mapping of the standard to so-called *representational state transfer* (REST) web services. The use of REST services is described in greater detail in Section 15.5.2. Due to its modular construction of the IEC server, it can, however, be extended to support other protocols should the need arise. As seen in Figure 15.9, which gives an overview of the primary server components, the setup comprises three layers. In the middle is the server core, sandwiched between the communication layer and the device modules. The same modularity that enables additions of communication protocols also allows the server to support any number of devices through the use of device-specific plug-in modules. In the course of the EDISON project, modules have been written for anything from photovoltaics, wind turbines, micro combined heat and power units, and of course charging sports and electric vehicles.

Successful tests have been carried out with all of the above running off the same server instance. Because electric vehicles have not always been available when needed, and in the quantities needed, a simulation was developed as a specific plug-in device module, enabling any client to access the simulated vehicles and charging spots through the IEC 61850 protocol—as if they were real.

15.4.2 The EDISON VPP

The EDISON VPP is a piece of server-side software that coordinates the behavior of a fleet of EVs while communicating with external power system stakeholders. To illustrate the internal workings of the EVPP, it can be useful to group its functions into three groups: data, analytics, and logic. This is done in Figure 15.10, which also differentiates between the aggregated and the individual level of EV management.

Basically, the EVPP handles the EV fleet as an aggregated group when acting and optimizing toward market players (upper interfaces), but will have to take individual considerations into account when handling the behavior of a single car (lower interfaces). The three main functional groups perform the following:

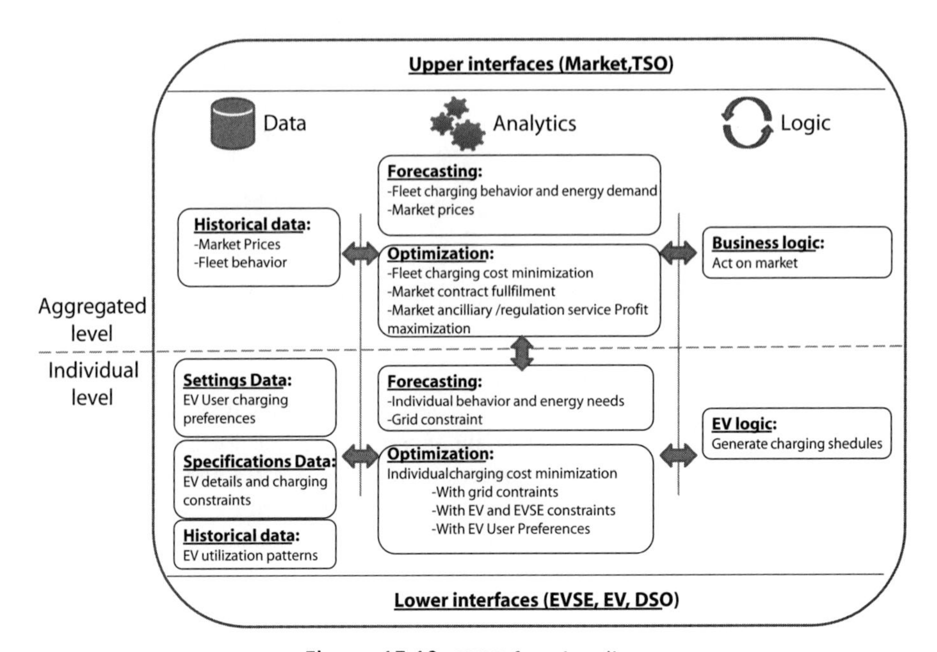

Figure 15.10. EVPP functionality.

- *Data.* This group stores previous market prices and fleet behavior on an aggregated level, enabling better forecasting and optimization for acting on the power market. On an individual level, data are stored that describe the service level agreement between the EVPP and an EV owner, for example, to which degree the EVPP should control the charging process. EV hardware specifications, like battery size and supported charging powers, are also stored. Finally, the EVPP stores the EV users plug-in habits, that is, where, when, and for how long the EV is typically connected to the grid for charging. These parameters are all vital for individually optimizing the charging of an electric vehicle.
- *Analytics.* Analytics means the mathematical computations necessary to support the logic of the EVPP. Forecasting relies on historical data to predict market prices on an aggregated level, which supports better bids and strategies. Forecasting also determines future individual EV usage patterns. The latter helps the EVPP predict when the EV user will need the EV for the next trip and can thus better estimate the time period available for smart charging. Such a prediction can be based on the statistical methods of exponential smoothing or using the Markov chain approach.

  Optimization is used to minimize charging costs of the EVs on both the aggregated and individual level. The individual optimization is limited by the constraints introduced by the distribution grid, EV specifications, and EV user energy requirements. On the aggregated level, profit maximization can be done when acting on the regulating and reserve markets. Such optimization can be achieved through stochastic or linear programming.
- *Logic.* The logic defines the main operational goals of the EVPP, namely, to act on the power market to generate savings or revenue for itself and its clients, and to intelligently manage the charging behavior of EVs through individually tailored charging schedules.

The operation of the EVPP is illustrated through the EVPP panel interface, which is described in Section 15.6.

15.4.3 The EDISON I/O Board

The EDISON I/O board was developed to allow testing of the interface described in the IEC 61851-1 standard, by handling the initial signaling between EV and EVSE. Furthermore, it facilitates the communication with the internals of both the EV and the EVSE and the exchange of information between EV and EVSE.

Figure 15.11 shows the conceptual overview of the features of the I/O board. The board itself has virtually no processing power and in common terms only provides a means of transportation (the envelope) without knowing what is being transported (the content). Hence, an external controller is needed to utilize the features of the board and handle the business logic in the application. As the features needed by an EV and an EVSE are slightly different, the board shall be configured to be used

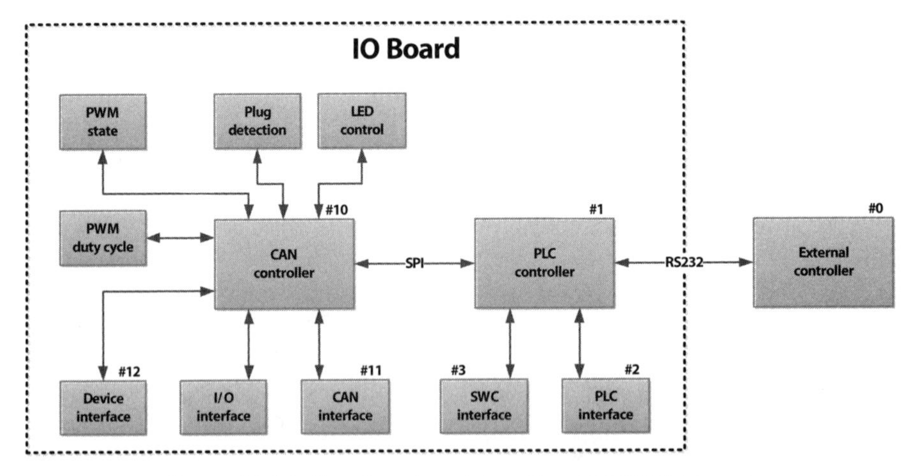

Figure 15.11. Functional modules of the EDISON I/O board.

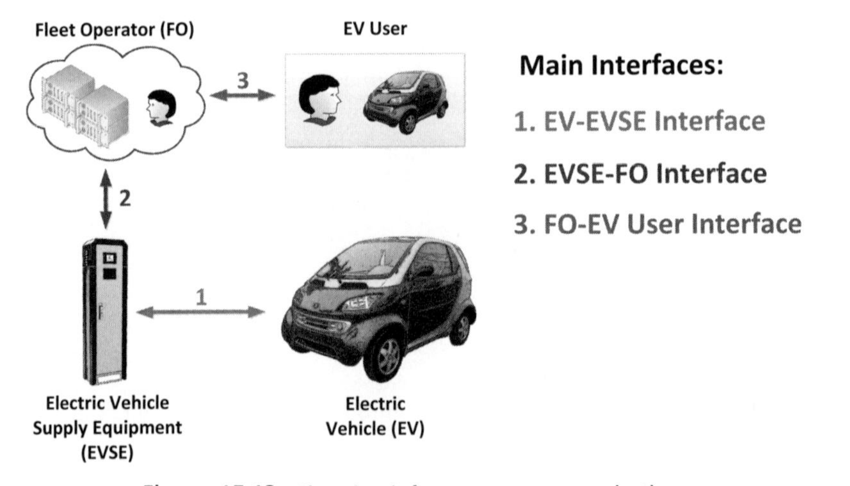

Figure 15.12. Charging infrastructure communication.

specifically for one or the other. As an example, the PWM duty cycle module provides duty cycle selection in EVSE mode and duty cycle detection in EV mode.

15.5 CHARGING INFRASTRUCTURE COMMUNICATION

The main interfaces investigated are those between the EV, the EVSE, the fleet operator, and the EV User, all of which can be seen in Figure 15.12.

The following will describe the main protocols and technologies used in the interfaces that connect the stakeholders.

15.5.1 Interface Connecting EV to EVSE

The question of how to connect the EV with the EVSE is both a question of which physical medium to choose and which standards and protocols to follow. The two main communication technologies that could be considered are wireless and wireline, as discussed in Chapters 6 and 7, respectively.

The use of wireless technologies is researched in several Smart Grid applications to allow for network communication. Wireless technologies such as GSM, GPRS, and 3G are well tested and are valid options for transferring data to and from EVs. The upcoming 4G technology will increase support for applications where high data rates are required. If, or when, all EVs require connectivity for features beyond managing the charging, a constant Internet connection supplied by, for example, 4G could become a necessity.

If, however, the EVs only need Internet connectivity for the purpose of smart charging, such a connection only needs to be maintained for the duration of the electrical connection. In other words, the fleet operator primarily needs to communicate with the vehicle when it is plugged in and it could be practical to use the physical medium already linking vehicle and EVSE, namely, the power cable. *Power line carrier* (PLC), where data is carried on a conductor, is the technology primarily explored by EDISON and the standards. It has been chosen in EDISON for a scenario where EVs would implement an I/O board, as described in Section 15.4, and would not have any permanent wireless network connection. A RENESAS chip is installed using a proprietary PLC technology.

The two standards investigated by EDISON, IEC 61851 and IEC/ISO 15118, both concentrate on using a wired medium but have different focuses on the EV-to-EVSE communication. The communication can, for AC charging using an on-board charger, be divided into initial signaling as described in IEC61851-1 and a high-level protocol-based communication, which is going to be standardized in ISO/IEC 15118.

The initial signaling in IEC 61851 has the purpose of indicating the state of operation between the EV and the EVSE.

State A—No vehicle connected

State B—Vehicle connected, not ready for energy flow

State C—Vehicle connected, ready for energy flow, ventilation not required

State D—Vehicle connected, ready for energy flow, ventilation required

State E—Vehicle connected, charge spot fault

State F—Charge spot not available for action

The EVSE will also be able to signal the maximum charging current back to the EV, in order to protect the EVSE's circuit breaker, allowing for simple load control by an external energy controller or operator.

A high-level IP-based communication protocol, as part of ISO/IEC 15118, would be required for more sophisticated services, including exchange of contract-ID, charging schedules, charging status, and value added services.

15.5.2 Interface Connecting EVSE to Fleet Operator

With the predicted increase in electric vehicle penetration in the coming years and the ever increasing number of players in this field, the use of standards, especially for communication, is one of the primary resolutions of the EDISON project.

For the connection between EVSEs and the fleet operator, the well-tested IEC 61850 standard was chosen. In a bold move, mainly to ease interoperability between parties and facilitate quick prototyping, the traditional MMS standard (ISO 9605) was abandoned in favor of RESTful web service.

15.5.2.1 IEC 61850 Using REST Services. As illustrated in the IEC 61850 section the data model is a hierarchical tree structure and in order to be able to navigate this model, every element has a path that uniquely identifies its position within the structure. All paths are absolute, meaning that they list every element from the root of the structure to the element in question—exactly as seen in a file system and as illustrated in Figure 15.13.

Traditionally, the IEC 61850 standard is paired with a communication standard called the *manufacturing message specification* (MMS), which is extensively used in some industries where it is also known as ISO 9605. Being a binary protocol, MMS requires a detailed understanding just to get started, unless of course one uses a prebuilt API. Unfortunately, only proprietary solutions seem to exist, which further hinders interoperability. In order to better facilitate the communication among components in EDISON, an implementation was developed which enabled the use of IEC 61850 through so called RESTful web services. Apart from the academic exercise, it had the added benefit of allowing IEC 61850 enabled communication across virtually every known computer platform with little effort, improving the interoperability between parties in the project.

The cornerstone in this mapping from IEC 61850 to REST lies in the resemblance between the reference path and the URL scheme used by the HTTP protocol on which REST is based. REST, which is short for representational state transfer, was first introduced in the doctoral dissertation by Roy Fielding in 2000 [8]. Fielding is a coauthor of the HTTP protocol on which the World Wide Web is built and is a cofounder of the Apache Web Server project, the most widely used web server in the world.

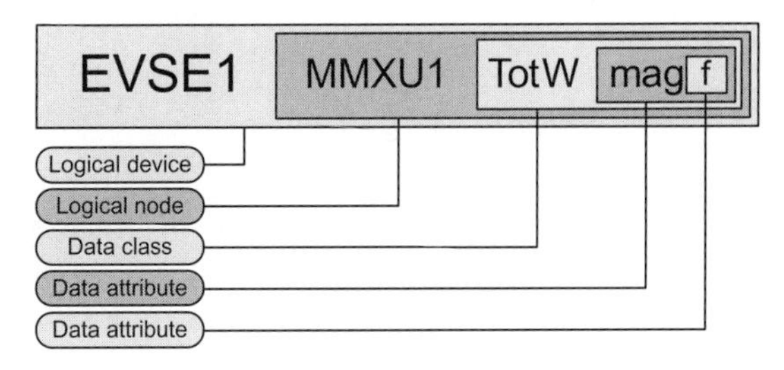

Figure 15.13. Illustration of the reference path structure for an IEC 61850 element.

Unlike the well-known SOAP services, the RESTful web services do not follow a specific standard, but rather a set of guiding principles central to which is the fact that data should be exposed as a resource. This principle closely adheres to the HTTP protocol; in fact, URL is short for *uniform resource locator*. In its simplest form, data is retrieved from a REST service by issuing an HTTP GET request for the URL representing the data one wishes to retrieve. Using this approach, access to the measurement of Figure 15.13 is a simple matter of retrieving a piece of XML from the URL http://hostname/EVSE1/MMXU1/Tot/*mag/f*, resulting in the following:

```
<DA Name="f" Type="FLOAT32" Ref="EVSE1/MMXU1.TotW.mag.f">0.5</DA>
```

Since REST services put no restrictions on the format used for transporting the data, it is completely at the developer's discretion to use whatever he/she deems suitable. If a file transfer is needed, which the IEC 61850 standard allows (e.g., for transferring configuration files or perhaps performing firmware upgrades), binary data could be the preferred option. In the case of the EDISON REST implementation, the basic format chosen is XML because it allows the relaying of hierarchical data, which means that any portion of the data model could be transferred in a single request. This has some added benefits when clients are discovering devices on the server because it allows them to retrieve the complete setup in a single request, without prior knowledge of the system. The use of IEC 61850 and REST is described by Anders Bro Pedersen et al. [9].

15.5.2.2 The Session Initialization Protocol. REST has inherited many of the benefits from the HTTP protocol, but unfortunately also suffers from one of its main drawbacks when used in a decentralized domain, which is the case with EVSEs: it is client/server based. Because many EVSEs will be attached to either private Internet connections, mobile uplinks, or the like, one cannot expect them to be reachable at a fixed network address as illustrated in Figure 15.14. One solution,

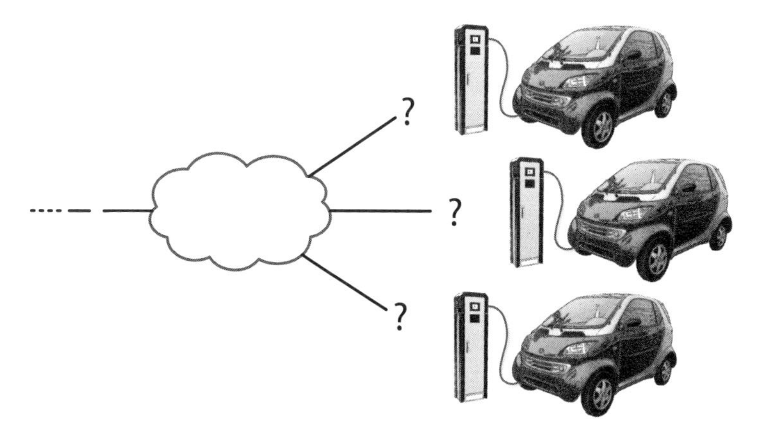

Figure 15.14. Session initiation issues in distributed systems.

the one we propose in the EDISON project, is to use the *session initiation protocol* (SIP), which was designed for use in IP-based cellular systems to solve such issues. The use of SIP has been explored by Bernhard Jansen et al. [10].

SIP, which dates as far back as 1996, is an open and incredibly flexible protocol whose primary purpose is to allow clients to locate and reach each other over the Internet. When an SIP-enabled user agent starts, it contacts an SIP registrar to register its location. When another user agent on the network needs to reach a particular user agent, it does not need to know the location of the other party in advance. Instead, it simply sends an INVITE message to the SIP proxy of its SIP domain. If the inviting user agent and the invitee are in the same SIP domain, the SIP proxy contacts the location service connected to the SIP registrar with the name in the INVITE message to look up the contact's details and then contacts the invited user agent by forwarding the INVITE message. Using the SIP proxies, the session is negotiated and set up between the user agents who, as a result, are provided with the information needed to create a point-to-point connection. Figure 15.15 illustrates the message sequencing for initializing, reinitializing, and closing an SIP session.

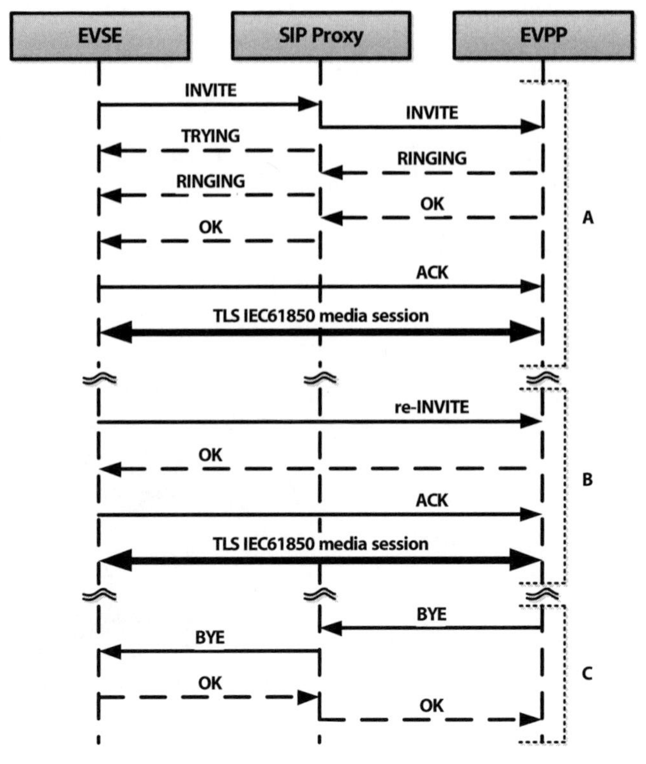

Figure 15.15. (a) SIP INVITE sequence diagram to establish an SIP dialog and a TLS/IEC 61850 session; (b) SIP reINVITE to reestablish a TLS/IEC 61850 session and (c) closing the SIP dialog.

Once the session has been initiated and the direct connection established, any type of traffic can be tunneled through. Because SIP builds on many of the same technologies as the HTTP protocol, it is also capable of using an identical security mechanism such as *transport layer security* (TLS) for encrypting the traffic.

Because SIP separates signaling and media transport, the media sessions are created and closed between requests. Most of the resources otherwise associated with keeping multiple connections open can therefore be freed. The SIP dialog, however, is kept open throughout the entire charging session and allows a quick reestablishment of a media session. As SIP dialog is connectionless, this is very resource effective. This helps to greatly improve the scalability, and effectiveness, allowing the EVPP to aggregate even more vehicles and keep communication costs low.

SIP allows the user agents in the network to be directly connected, but it does not handle barriers such as firewalls and the *network address translation* (NAT) often found in routers. To overcome these issues, the SIP protocol can be extended with additional technologies, such as the *session traversal utilities for NAT* (STUN) and the *traversal using relay NAT* (TURN).

Throughout the EDISON project great effort has been put on using standards for communication, such as SIP, TCP/IP, TLS, and IEC 61850. In this regard the SIP is considered as highly suitable to provide control and data communication between EVSE and the EVPP. As mentioned, the use of the RESTful approach helps to facilitate a much more versatile interface to the IEC 61850 server, but like all client–server communication, has some drawbacks, one of which is the ever present issue with firewalls. By combining IEC 61850 and REST with the use of the SIP protocol and NAT traversing techniques such as STUN or TURN, these issues can, to a large extent, be overcome. This allows SIP/IEC 61850 enabled EVSEs a seamless and scalable integration into an EVPP system, regardless of their location or network connection.

15.5.3 Interface Connecting EV User to Fleet Operator

For intelligent EV charging to be successful, the adherence to the EV user's driving requirements is key. User requirements can range from a general target state-of-charge for the EV to specific requirements such as having the EV at a certain state-of-charge at a certain time. The latter can be important for the user if, for example, he/she wants to leave exceptionally early the next day or go on a long trip.

To explore ways of communicating with the user, a couple of user interface prototypes, both for desktops and mobile phones, are under development within the EDISON project. The desktop interface has the form of a web site and allows the user to sign his/her vehicles up for fleet operator controlled charging, monitor the charging history, and set user specific preferences. The mobile phone interfaces can be divided into two categories: SMS and Internet based.

The SMS-based user interface enables the widest coverage, as most users have at least an SMS capable mobile device. The user then always has the ability to send a status request SMS, for example, the text "?" to a certain number, to which the fleet operator will reply via an SMS gateway, providing the latest information about the state-of-charge. Additionally, the user might receive an SMS when plugging in the EV, containing an offer for doing smart charging.

The Internet-based interaction is either in the form of a device specific application or a web site. While device specific applications probably offer the richest user experience and allow for push notifications to the device, a mobile-specific web site can reach a broader spectrum of devices. In the prototypes of these interfaces currently under development, the user can monitor the charging process, see the fleet operator's current availability prediction for the EV, and update the estimated plug-out time to meet his/her requirements, while still allowing for smart charging. It is important that the user always understands what the server-side system is allowed to do, and what state-of-charge he/she can expect when the vehicle is needed.

Apart from desktop and mobile phone based user interfaces, the EVSE and the EV could also have a user interface, which could offer similar functionality as the mobile phone interface (due to the similar screen size). A solution for these could be to simply host a browser component which displays web pages served by the fleet operator, thus allowing for communication with any fleet operator. Alternatively, device-specific applications could be used to communicate with the fleet operator via web services; this would, however, require standardization of the fleet operator APIs and would make support for different fleet operators difficult.

All of the above-mentioned user interfaces rely on the fleet operator to provide the data and do not therefore require the EV to have a wireless Internet connection.

15.6 DEMONSTRATION

This section shows some of the interfaces developed in EDISON to test and illustrate smart charging.

15.6.1 End-to-End Demonstration: From EV to Operator Panel

The EDISON VPP operator panel has been developed to demonstrate the operation of an EVPP and is implemented as a Microsoft Silverlight application hosted by a web browser. A screenshot of the panel is shown in Figure 15.16.

The interface features the following areas: "1" is a full list of EVs managed by the EVPP followed by a fleet summery in "2." When an EV is selected in "1," its status and data are displayed in the right portion of the interface. While "3" displays the selected EV's last known status, "4" displays a subset of the static information on the specific EV. The EV's location is available to the EVPP when the EV is plugged in and can be seen in "5." The graphs labeled "6" through "8" display information for a selected 24-hour period. Here "6" displays the availability periods

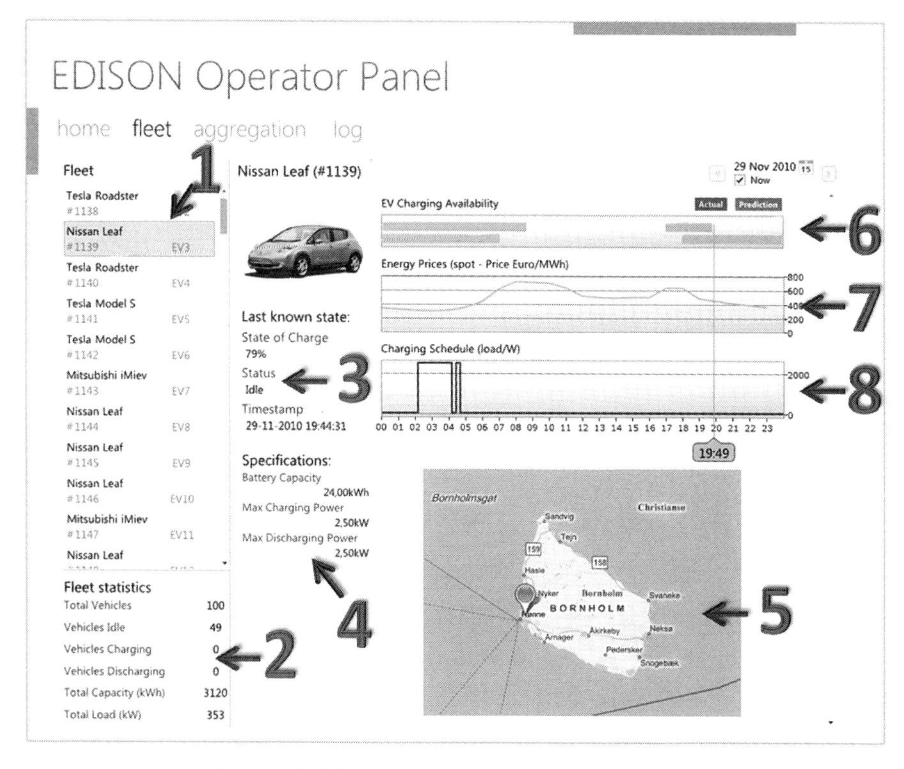

Figure 15.16. EDISON operator panel screenshot.

of the EV, that is, the periods where the car is parked and plugged in. Graph "6" shows both the recorded and predicted availability of the vehicle, illustrated by the upper and lower horizontal bars. Graph "7" displays the energy prices for the time period and "8" shows the charging schedules, which have been generated by the EVPP and are sent out to be executed by the EVSE.

The interface screenshot demonstrates that the EVPP can be connected to a set of real or virtual cars each with their own unique patterns and characteristics, and generate charging schedules that avoid charging at expensive hours. The latter can be seen by comparing prices "7" and charging periods "8" on the screenshot. The EVSE panel is also useful in retrieving and visualizing historical data.

15.6.2 Physical Demonstration Assets

The island of Bornholm has been chosen as a test site for demonstrating the technical solutions developed by EDISON. As an island, it is an isolated environment capable of running independently from the surrounding power system and it has a suitable composition of renewable energy sources, which are representative for the whole of Denmark. It is on Bornholm that the ICT architecture and its components will be

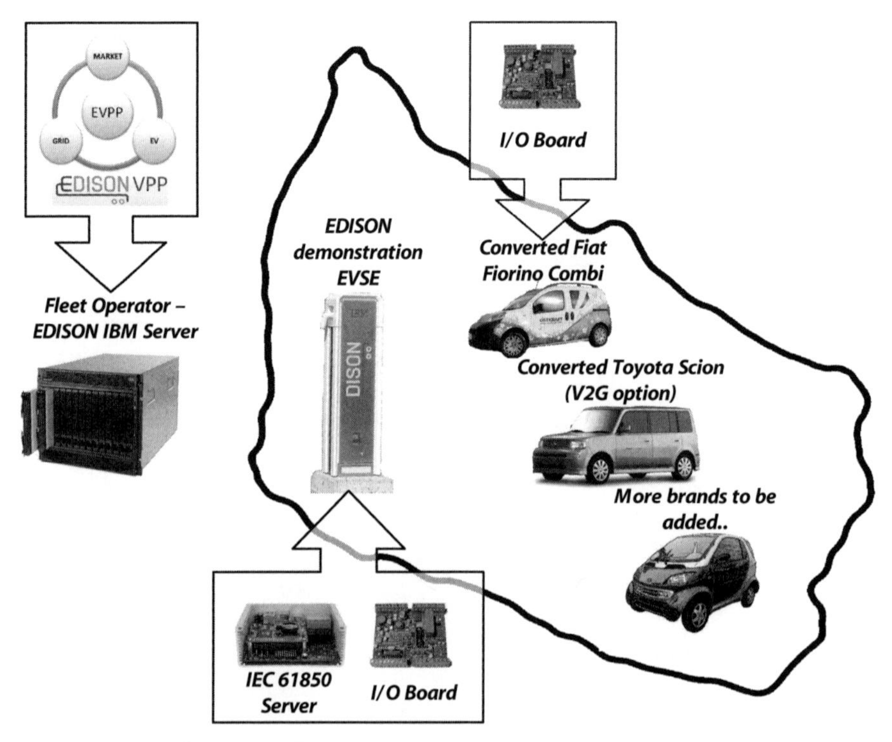

Figure 15.17. Physical demonstration assets of EDISON.

put to the test in managing smart charging for EVs of various brands and types. Figure 15.17 shows the main physical assets that will be used to validate smart charging on the island of Bornholm.

Since at least one brand of EVs (converted Toyota Scion) will support vehicle-to-grid (V2G) technology, the EDISON island demonstration can also cover advanced EV services, in which power is delivered back to the grid. Finally, roaming scenarios can be tested on the island using different EV and EVSE combinations.

Although lessons learned from the physical demonstrations will be an important part of EDISON, the project should also analyze integration scenarios where very high numbers of EVs are introduced on Bornholm. For practical reasons this requires simulations, which are the topic of the next section.

15.6.3 A Large-Scale Virtual Fleet

Massive roll-outs of electric vehicles have been predicted within the next few years and already several major automotive manufacturers are trying to get a head start: most with their individual visions for the future, what it may bring and what may be chosen as the de facto standards. Some are sticking with one type of charging socket, others another. Some are leaning toward battery swapping while others are set on fast charging and so on and so forth. On top of all the above-mentioned uncertainties, there are also questions that need to be answered regarding the impact of all these EVs on the power grid. How will the grid handle the extra load? Where and when will this dynamic load be connecting geographically? Where are the potential bottlenecks in the distribution system? How do we prevent them from occurring? The list goes on.

Common for all these questions is that they represent potential problems, for which we need a solution before the problems actually occur and the only way to test this "in practice" without causing a disaster is through the use of simulations.

For the EDISON project a very flexible EV simulation system was developed as an extension to the IEC 61850 server described earlier.

By using geographic data, demographic statistics, and recorded vehicle data from real-world experiments, large groups of virtual EVs can be created with behaviors closely resembling people's current driving habits. Because the grid impacts resulting from increased EV penetration is an important topic, it was not enough to simply simulate the consumption of a fleet of vehicles. Instead, using route data obtained from online services, the vehicles were simulated in real time as they would be driving around on the island—see Figure 15.18. In practical terms, it would have been enough to simply calculate the consumption from a given trip and then move the vehicles around, but the added effect of having moving vehicles makes for a very audience friendly demonstration.

Because the simulation runs as an extension of the IEC 61850 server, all devices are automatically made available through the IEC 61850 RESTful interface, allowing the VPP to actively aggregate the whole fleet as if they were real EVs—see Figure 15.19.

15.7 CONCLUSION AND FUTURE WORK

This chapter has explored the technologies used by the EDISON project in terms of communication standards, components, and stakeholder interfaces.

Initial testing indicates that the IEC 61850 and IEC 61851 standards are valid candidates for EV integration. Also, the upcoming ISO/IEC 15118 standard could prove very beneficial for promoting advanced EV services as well as roaming and should be followed carefully. The chapter also puts emphasis on a few selected components that were developed by EDISON partners to demonstrate smart charging. The software and hardware developed for the vehicle, charging spot, and fleet operator will serve as input to standardization work.

Another topic covered by the chapter is the specific protocols that will enable communication in EDISON. In this context we looked at how HTTP based web services and the SIP protocol fulfill certain requirements of interoperability, security, and scalability. Transport layer security (TLS), a technology used in such areas as Internet banking, has been suggested for improving security.

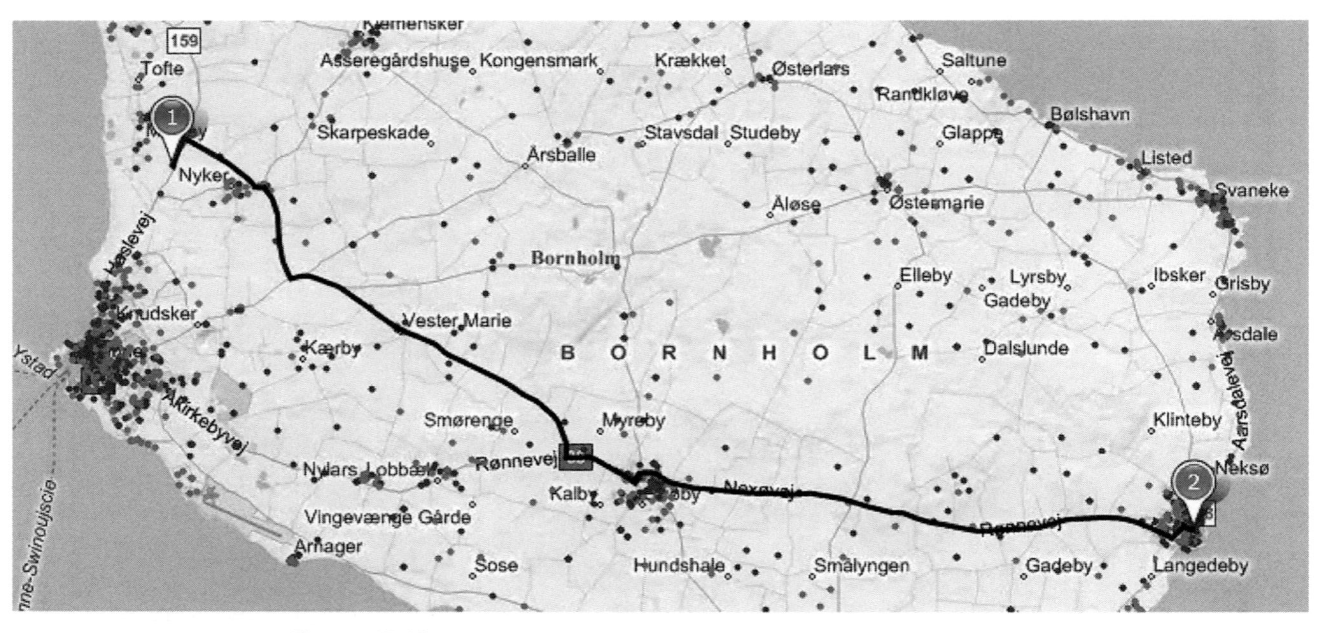

Figure 15.18. Simulated commuter route across the island of Bornholm.

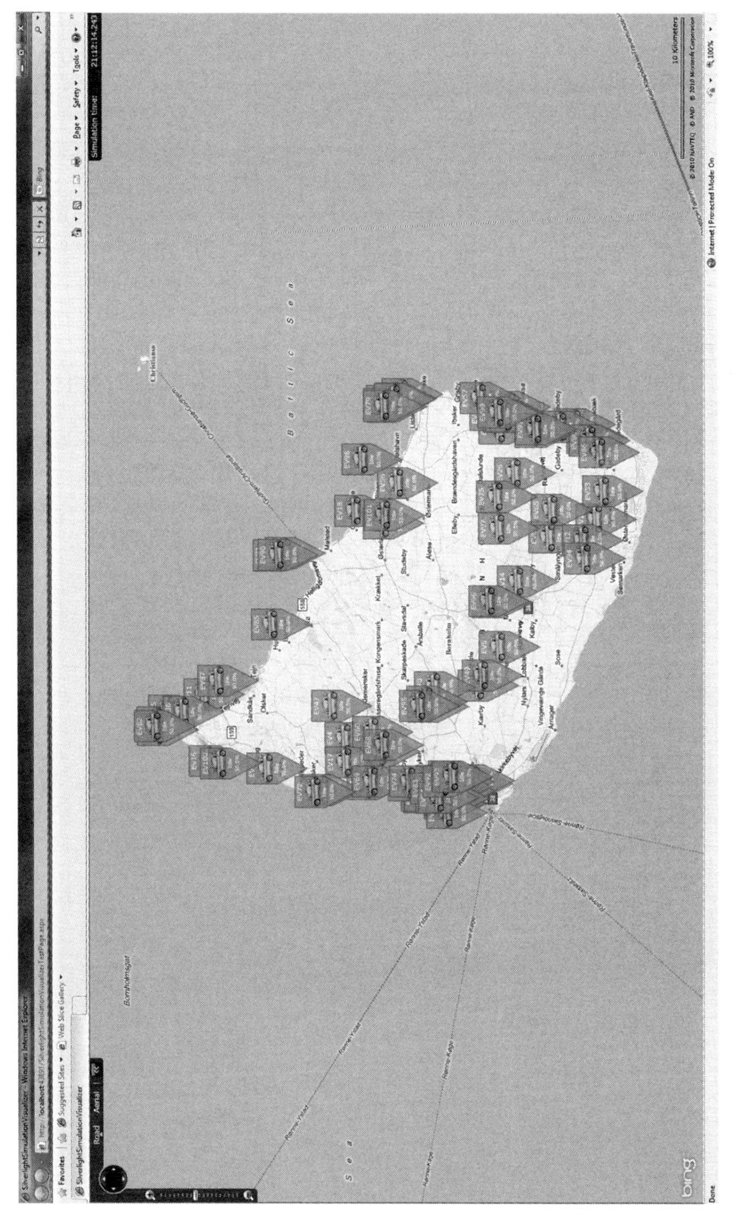

Figure 15.19. Birds-eye-view of medium-sized virtual EV fleet.

The chapter concluded with a few examples of how EDISON aims to prove and validate its work through a set of demonstrations. This includes a field test of an EV fleet on the island of Bornholm and the use of large virtual fleets simulated in software.

A suitable and standardized ICT architecture, which has been the focus of this chapter, is a vital piece in EV integration and a key part of EDISON. Other research areas such as battery technologies, fast charging, and distribution grid impacts are, however, equally important and are covered by partners in the project. Only by covering all topics relevant to society's transition to electric transportation can EDISON achieve its goal of aiding and promoting the cause of the electric vehicle.

The EDISON project will continue with its V2G research and evaluate which standards best support this type of service. While the IEC 61850 schedules presently used support V2G, the OpenV2G project [11], associated with ISO/IEC 15118, and the V2G project [12], led by Professor Willett Kempton from the University of Delaware, both represent possible alternatives and should be studied thoroughly.

The progress of EDISON is continuously updated on the project web page [3].

REFERENCES

[1] C. Binding, D. Gantenbein, B. Jansen, O. Sundstroem, P. B. Andersen, F. Marra, B. Poulsen, and C. Traeholt, "Electric Vehicle Fleet Integration in the Danish EDISON Project—A Virtual Power Plant on the Island of Bornholm," Proceedings IEEE Power & Energy Society General Meeting 2010, Minneapolis, Minnesota, USA, July 25–29, 2010. Also available as IBM RZ 3761.

[2] J. Østergaard, A. Foosnæs, Z. Xu, T. A. Mondorf, C. A. Andersen, S. Holthusen, T. Holm, M. F. Bendtsen, and K. Behnke, "Electric Vehicles in Power Systems with 50% Wind Power Penetration: The Danish Case and the EDISON Programme," presented at European Conference on Electricity & Mobility, Würzburg, Germany, 2009.

[3] EDISON project website, www.edison-net.dk.

[4] The FENIX project (Flexible electricity network to integrate the expected energy evolution). Available at: http://www.fenix-project.org/.

[5] S. You, C. Træholt, and B. Poulsen, "Generic Virtual Power Plants: Management of Distributed Energy Resources Under Liberalized Electricity Market," Proceedings of the 8th IET International Conference on Advances in Power System Control, Operation and Management, Hong Kong, China: Institution of Engineering and Technology, November 8–11, 2009.

[6] O. Sundström and C. Binding, "Optimization Methods to Plan the Charging of Electric Vehicle Fleets," Proceedings ACEEE International Conference on Control, Communication, and Power Engineering (CCPE), Chennai, India, July 28, 2010, pp. 323–328. Also available as IBM RZ 3768.

[7] O. Sundström and C. Binding, "Planning Electric-Drive Vehicle Charging Under Constrained Grid Conditions," accepted at International Conference on Power System Technology, China, October 24–28, 2010. Also available as IBM RZ 3785.

[8] Roy T. Fielding dissertation. Available at: http://www.ics.uci.edu/~fielding/pubs/dissertation/top.htm.

[9] A. B. Pedersen, E. B. Hauksson, P. B. Andersen, B. Poulsen, C. Træholt, and D. Gantenbein, "Facilitating a Generic Communication Interface to Distributed Energy Resources," Proceedings of International Conference on Smart Grid Communications (SmartGridComm), 2010.

[10] B. Jansen, C. Binding, O. Sundström, and D. Gantenbein, "Architecture and Communication of an Electric Vehicle Virtual Power Plant," IEEE International Conference on Smart Grid Communications, Gaithersburg, MD, USA, September 2010.

[11] The OpenV2G Project. Available at: http://openv2g.sourceforge.net/.

[12] The Vehicle to Grid Project—University of Delaware. Available at: http://www.udel.edu/V2G/.

16

REAL-TIME ESTIMATION OF TRANSMISSION LINE PARAMETERS

Wenyuan Li, Paul Choudhury, and Jun Sun

16.1 INTRODUCTION

Transmission line parameters include resistance, reactance, and admittance to ground, which are important input data in power system modeling, power flow computations, stability assessment, line protection design, and other relevant applications [1]. In current utility practice, a transmission line's parameters are calculated using theoretically derived formulas that are based on the information of its size, length, structure, and type [2] and are assumed to be constant in power flow modeling. Obviously, there is a difference between the calculated and actual parameters because the resistance, reactance, and admittance of lines vary in real life with environmental and weather conditions (such as temperature and wind speed) [2, 3]. Some methods have been developed to measure resistance and reactance of a line [4, 5]. However, the equivalent admittance to ground that represents reactive charging power is not directly measurable. Also, one snapshot offline measurement of

Smart Grid: Applications, Communications, and Security, First Edition.
Edited by Lars Torsten Berger and Krzysztof Iniewski.

resistance and reactance parameters is not good enough for real-time applications since environment and weather vary from time to time. The assumption of constant parameters may create an unacceptable error, particularly when the environment or weather around the line has a relatively large change. Errors in the parameters will lead to inaccuracy or even erroneous results in system modeling and analysis. Real-time estimation of transmission line parameters is essential to applications in Smart Grids.

This chapter describes a method for real-time estimation of transmission line parameters in online power flow modeling or any calculation in power system applications. The basic features are as follows:

- The method uses synchronized phasor measurement units (PMUs) with time-stamps so that time synchronization is automatically ensured.
- Invalid PMU measurements (false or erroneous information) are filtered and only reliable PMU measurements are used. This increases reliability and accuracy of estimation.
- The estimation can be performed very quickly and applied in a real-time environment.
- Updated estimates of time-varying line parameters can enhance accuracy of state estimations and power flow calculations in the energy management system (EMS) at a utility's control center, which are performed every few minutes. The updated estimates of line parameters are also important for other EMS applications that need power flow information, such as real-time contingency analysis, optimal power flow, and stability assessment.

The proposed method is based on the use of synchronized phasor measurement unit (PMU). The synchronizing PMU technology has been advanced at a very fast pace in the utility industry of both developed and developing countries since its basic concept was presented [6–9]. The applications of PMU include phasor monitoring, enhancement of system state estimator, voltage and transient stability assessments, and protection relays [10–16]. This chapter presents an application of PMU in real-time estimation of transmission line parameters [17].

16.2 BASIC CONCEPTS

The PMU devices are installed at two ends of a line whose parameters are to be estimated. The estimation includes two tasks:

1. Measurements from a synchronized PMU may include invalid data. A piece of false data that is caused by failure or malfunction of the PMU device or communication channel may not be automatically recognized from PMU measurements. Particularly, an error that is associated with only accuracy of measurement cannot be identified by the PMU itself. Fortunately, measured

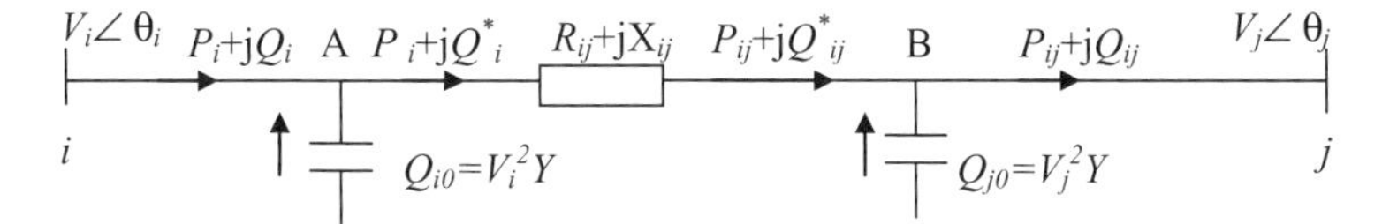

Figure 16.1. Single-line diagram of a typical transmission line.

voltage and power flow phasors of a line must satisfy the relationship defined by line flow equations. This fact enables us to identify and filter out invalid measurements.

2. The line (branch) parameters (resistance, reactance, and admittance to ground) are estimated using valid PMU measurements. The parameters vary with environment and weather (such as temperature and/or wind speed) conditions. Therefore, it is necessary to perform real-time continuous estimations of line parameters.

In the following discussion, the π equivalence of a line shown in Figure 16.1 is used to explain the presented method. This equivalence is generally sufficient. It is not difficult to extend the concept to a multiple π equivalence circuit if necessary in actual applications. In this chapter, the units of all quantities are in per unit. The concept of per unit can be found in any textbook on power system analysis. Quantities related to real or reactive power refer to the total in three phases and voltage-related quantities refer to line voltage.

In Figure 16.1, $R_{ij} + jX_{ij}$ denotes the line impedance, whereas Y represents one-half of the admittance to ground corresponding to line charging reactive power. $V_i\angle\theta_i$ and $V_j\angle\theta_j$ are the voltage phasors at the sending and receiving buses, respectively. $P_i + jQ_i$ and $P_i + jQ_i^*$ are the line power flows, respectively, before and after the charging reactive power at the sending bus i is incorporated; $P_{ij} + jQ_{ij}^*$ and $P_{ij} + jQ_{ij}$ are the line power flows, respectively, before and after the charging reactive power at the receiving bus j is incorporated; and Q_{i0} and Q_{j0} represent the charging reactive powers at the sending and receiving ends, respectively. The V_i, θ_i, V_j, θ_j, P_i, Q_i, P_{ij}, and Q_{ij} values are directly measurable through synchronized PMU devices at two ends in a real-time manner, whereas Q_i^* and Q_{ij}^* can easily be calculated using Q_i and Q_{ij} and charging reactive powers. It should be noted that initial measurements are voltage and current phasors, which can be converted to line power flows inside PMU devices. Conceptually, the charging reactive power occurs along the entire length of a line and is not measurable. However, the total charging reactive power can be calculated using the difference between Q_i and Q_{ij} minus reactive losses on the line.

Estimation of line parameters R_{ij}, X_{ij}, and Y for all lines whose parameters need to be updated is performed at a given time interval (such as every 2–5 minutes or even shorter). The PMU devices can provide synchronized phasor data at a rate of 20–40 samples per second or even faster and thus there are considerable sampling data available in a given interval. The rate of waveform sampling can be as high as 3000 samples per second or higher. The parameters R_{ij}, X_{ij}, and Y may vary with environmental and weather conditions around lines on a relatively long time scale, for example, on the order of half an hour. However, unlike measurements of voltages and line (branch) power flows, the parameters are sufficiently stable (constant or minor changes if any) over very short intervals (a couple of minutes). The benefit of frequent parameter estimation is twofold. The parameters can be updated as often as needed in a real-time manner, and their constancy within a very short time is utilized to filter invalid measurements.

16.3 FILTERING INVALID MEASUREMENTS

Sets of sampled data (measurements of voltage and power flow phasors from the PMU) are recorded at a given interval. For each set of measurements, the following data filtering process is performed:

1. The parameters R_{ij}, X_{ij}, and Y from the last estimation are used as a reference.
2. The charging reactive powers are calculated by

$$Q_{i0} = V_i^2 Y \tag{16.1}$$

$$Q_{j0} = V_j^2 Y \tag{16.2}$$

3. The reactive power flows on the line within points A and B (see Figure 16.1) are calculated by

$$Q_i^* = Q_i + Q_{i0} \tag{16.3}$$

$$Q_{ij}^* = Q_{ij} - Q_{j0} \tag{16.4}$$

4. The reactive loss on the line is estimated by

$$\Delta Q_1 = \frac{X_{ij}\left[P_i^2 + (Q_i^*)^2\right]}{V_i^2} \tag{16.5}$$

$$\Delta Q_2 = \frac{X_{ij}\left[P_{ij}^2 + (Q_{ij}^*)^2\right]}{V_j^2} \tag{16.6}$$

$$\Delta Q = \frac{\Delta Q_1 + \Delta Q_2}{2} \tag{16.7}$$

Equations (16.5) and (16.6) indicate that the reactive loss is estimated from the two buses, respectively. Equation (16.7) gives the average estimation from the two buses.

5. The parameter Y is updated using the measured reactive power flows at the two buses and the estimated line loss by

$$Y(\text{new}) = \frac{Q_{ij} - Q_i + \Delta Q}{V_i^2 + V_j^2} \tag{16.8}$$

A threshold for filtering accuracy is specified according to the precision of PMU measurements, the error transfer relationship between measurements and Y, and the range of possible small changes of Y in a given short interval. The threshold can be determined through testing and preestimation. For example, if 5% is used as the threshold, when Y(new) is larger than 1.05Y(old) or smaller than 0.95Y(old), where Y(old) refers to the value of Y in the last estimation, this whole set of measurements (V_i, θ_i, V_j, θ_j, P_i, Q_i, P_{ij}, and Q_{ij}) is believed unreliable and is abandoned. Otherwise, the following steps continue.

6. The equivalent charging reactive power at the receiving bus is updated by

$$Q_{j0}(\text{new}) = V_j^2 Y(\text{new}) \tag{16.9}$$

7. The line reactive power on the line at the receiving end is updated by

$$Q_{ij}^* = Q_{ij} - Q_{j0}(\text{new}) \tag{16.10}$$

8. The parameters R_{ij} and X_{ij} are estimated using the following method. The line power flow equation of $P_{ij} + jQ_{ij}^*$ can be expressed as

$$P_{ij} + jQ_{ij}^* = V_j \angle \theta_j \left(\frac{V_i \angle \theta_i - V_j \angle \theta_j}{R_{ij} + jX_{ij}} \right)^{\oplus} \tag{16.11}$$

where the symbol ⊕ denotes the complex conjugate operation. Separating Eq. (16.11) into real and imaginary parts yields

$$R_{ij} P_{ij} + X_{ij} Q_{ij}^* = -V_j^2 + V_i V_j \cos\theta_{ji} = a \tag{16.12}$$

$$R_{ij} Q_{ij}^* - X_{ij} P_{ij} = V_i V_j \sin\theta_{ji} = b \tag{16.13}$$

where $\theta_{ji} = \theta_j - \theta_i$. It can be derived from Eqs. (16.12) and (16.13) that

$$R_{ij}(\text{new}) = \frac{aP_{ij} + bQ_{ij}^*}{P_{ij}^2 + (Q_{ij}^*)^2} \tag{16.14}$$

$$X_{ij}(\text{new}) = \frac{aQ_{ij}^* - bP_{ij}}{P_{ij}^2 + (Q_{ij}^*)^2} \tag{16.15}$$

Similarly, a threshold for filtering accuracy is specified according to the precision of PMU measurements, the error transfer relationship between measurements and R_{ij} or X_{ij}, and the range of possible small changes of R_{ij} or X_{ij} in a given short interval. The threshold can be determined through testing and preestimation. For example, if 5% is used as the threshold, when either R_{ij}(new) is larger than $1.05R_{ij}$(old) or smaller than $0.95R_{ij}$(old), or X_{ij}(new) is larger than $1.05X_{ij}$(old) or smaller than $0.95X_{ij}$(old), this whole set of measurements (V_i, θ_i, V_j, θ_j, P_i, Q_i, P_{ij}, and Q_{ij}) is believed unreliable and is abandoned.

If the number of sets of reliable measurements within a given time interval is smaller than a specified threshold (such as 10), additional sets of sampling data should be added. If in a case, all sets of sampling data for a line in the given interval are filtered out as invalid data, a warning message should be sent to operators. Consecutive warning messages indicate that the PMU devices for that particular line may be in an abnormal situation.

16.4 ESTIMATING PARAMETERS R_{IJ}, X_{IJ}, AND Y

Each of the estimated parameters obtained in the procedure of Section 16.3 is based on individual sampling data at a time point and thus used for the purpose of filtering invalid data. The parameters should be reestimated using reliable sets of measurements at different time points in the given interval to minimize errors. It is assumed that M reliable sets of measurements are obtained after the filtering process.

The parameter Y is reestimated by the average of M estimated Y values calculated using M reliable sets of measurements in the filtering process:

$$Y(\text{estim}) = \frac{\sum_{k=1}^{M} Y_k(\text{new})}{M} \tag{16.16}$$

where Y_k(new) is the value calculated using Eq. (16.8) and the kth reliable set of measurements after filtering.

The parameter R_{ij} or X_{ij} is also reestimated first by averaging over M reliable sets of measurements obtained in the filtering process; that is,

$$R_{ij}(\text{estim}) = \frac{\sum_{k=1}^{M} R_{ijk}(\text{new})}{M} \tag{16.17}$$

$$X_{ij}(\text{estim}) = \frac{\sum_{k=1}^{M} X_{ijk}(\text{new})}{M} \tag{16.18}$$

where R_{ijk}(new) or X_{ijk}(new) is the value calculated using Eqs. (16.14) or (16.15) and the kth reliable set of measurements after filtering.

The standard deviations of R_{ij}(estim) and X_{ij}(estim) are then calculated using Eqs. (16.19) and (16.20), respectively:

$$R_{ij}(\mathrm{sd}) = \sqrt{\frac{\sum_{k=1}^{M}[R_{ijk}(\mathrm{new}) - R_{ij}(\mathrm{estim})]^2}{M-1}} \tag{16.19}$$

$$X_{ij}(\mathrm{sd}) = \sqrt{\frac{\sum_{k=1}^{M}[X_{ijk}(\mathrm{new}) - X_{ij}(\mathrm{estim})]^2}{M-1}} \tag{16.20}$$

If either $R_{ij}(\mathrm{sd})/R_{ij}(\mathrm{estim})$ or $X_{ij}(\mathrm{sd})/X_{ij}(\mathrm{estim})$ is larger than a threshold, the estimated R_{ij} and X_{ij} obtained using Eqs. (16.17) and (16.18) are abandoned and the parameters R_{ij} and X_{ij} are reestimated using the following approach. This threshold is generally selected as half of the threshold for filtering accuracy (see step 8 in Section 16.3).

Equations (16.12) and (16.13) are rewritten as

$$R_{ij} + cX_{ij} = d \tag{16.21}$$

$$R_{ij} + eX_{ij} = f \tag{16.22}$$

where

$$c = \frac{Q_{ij}^*}{P_{ij}} \tag{16.23}$$

$$d = \frac{-V_j^2 + V_i V_j \cos\theta_{ji}}{P_{ij}} \tag{16.24}$$

$$e = \frac{-P_{ij}}{Q_{ij}^*} \tag{16.25}$$

$$f = \frac{V_i V_j \sin\theta_{ji}}{Q_{ij}^*} \tag{16.26}$$

Applying the least squares method to Eq. (16.21) with the M reliable sets of measurements, we have

$$R_{ij1}(\mathrm{estim}) = \bar{d} - \bar{c}X_{ij1}(\mathrm{estim}) \tag{16.27}$$

$$X_{ij1}(\mathrm{estim}) = \frac{S_{cd}}{S_{cc}} \tag{16.28}$$

where

$$\overline{d} = \frac{\sum_{k=1}^{M} d_k}{M} \tag{16.29}$$

$$\overline{c} = \frac{\sum_{k=1}^{M} c_k}{M} \tag{16.30}$$

$$S_{cd} = \sum_{k=1}^{M} (c_k - \overline{c})(d_k - \overline{d}) \tag{16.31}$$

$$S_{cc} = \sum_{k=1}^{M} (c_k - \overline{c})^2 \tag{16.32}$$

Here, the subscript k indicates the value corresponding to the kth reliable set of measurements after filtering.

Similarly, applying the least squares method to Eq. (16.22) with the M reliable sets of measurements, we have

$$R_{ij2}(\text{estim}) = \overline{f} - \overline{e}X_{ij2}(\text{estim}) \tag{16.33}$$

$$X_{ij2}(\text{estim}) = \frac{W_{ef}}{W_{ee}} \tag{16.34}$$

where

$$\overline{f} = \frac{\sum_{k=1}^{M} f_k}{M} \tag{16.35}$$

$$\overline{e} = \frac{\sum_{k=1}^{M} e_k}{M} \tag{16.36}$$

$$W_{ef} = \sum_{k=1}^{M} (e_k - \overline{e})(f_k - \overline{f}) \tag{16.37}$$

$$W_{ee} = \sum_{k=1}^{M} (e_k - \overline{e})^2 \tag{16.38}$$

The R_{ij} and X_{ij} are finally reestimated by

$$R_{ij}(\text{estim}) = \frac{R_{ij1}(\text{estim}) + R_{ij2}(\text{estim})}{2} \tag{16.39}$$

$$X_{ij}(\text{estim}) = \frac{X_{ij1}(\text{estim}) + X_{ij2}(\text{estim})}{2} \tag{16.40}$$

In a high-voltage transmission system, R_{ij} is much smaller than X_{ij} and P_{ij} is generally much larger than Q_{ij}^*. It is possible that in numerical calculations, Eq. (16.21) is more accurate than Eq. (16.22) for estimation of R_{ij}, whereas Eq. (16.22) is more accurate than Eq. (16.21) for estimation of X_{ij}. Instead of taking the simple average in Eqs. (16.39) and (16.40), one alternate approach in an actual application is to use both Eqs. (16.21) and (16.22) as described above. If the difference (in a percentage) between R_{ij1}(estim) and R_{ij2}(estim) or between X_{ij1}(estim) and X_{ij2}(estim) exceeds a threshold, only R_{ij1}(estim) and X_{ij2}(estim) are used as the final estimates.

In the above derivation, it has been assumed that three phases in a transmission system are symmetrical and therefore a single-phase model has been used in power flow modeling. Similar to measurements in a supervisory control and data acquisition (SCADA) system, PMU devices provide separate measurements of phases A, B, and C, which may have slight differences among them. Total real and reactive power flows of three phases can be obtained by summation of the power flows that are calculated from measured voltage and current phasors of three individual phases. For voltage phasors, which are required in the calculations, one of the following two approaches can be used:

1. The voltage magnitude and angle for each of phases A, B, and C are measured and the average over the three phases is used; or alternatively, the measured voltage magnitude and angle of one selected phase (such as phase A) with the best measurement precision is used. This is the traditional method used in the existing energy management system (EMS).
2. The voltage phasor of each phase A, B, or C and the total three phase power flows are used to estimate three sets of line parameters. The final parameter estimate is the average of the three estimates obtained using voltage phasors of phases A, B, and C.

16.5 SIMULATION RESULTS

The presented method was tested using system power flow studies. Bus voltages (magnitudes and angles) and line currents (converted to real and reactive power flows) obtained from a number of power flow calculations were viewed as "measurements." The tests were conducted on several IEEE test systems and the utility system operated and planned by BC Hydro in Canada. In some cases, errors are intentionally introduced to a voltage measurement (either magnitude or angle) or a line current measurement (either magnitude or angle). An error of voltage measurement affects both voltage itself and line power flow, while an error of current measurement only impacts the line power flow. The results show that if no error is introduced, the estimated parameters are the same as those specified as input data in power flow calculations. In the cases where some errors are introduced, "measurements" with relatively large errors are filtered out and estimated parameters under the condition of a few unfiltered small measurement errors are still the same as those specified as input data in power flow calculations.

Two examples are given below to demonstrate the feasibility and effectiveness of the presented method.

16.5.1 Estimating Parameters of a Line in IEEE 118-Bus System

The IEEE 118-bus system is one of the test systems developed by the IEEE Power and Energy Society for various testing purposes. The system has 118 buses, 177 lines and 9 transformer branches, 91 load points, and 54 generating units. The data and complete single-line diagram of this test system can be found at the web site [18]. The single-line diagram of the partial system that contains the line between bus 42 and bus 49 (denoted by a dashed line) whose parameters are estimated is shown in Figure 16.2. The original three parameters of this line are as follows:

$$R_{ij} = 0.0715 \text{ p.u.} \quad X_{ij} = 0.3230 \text{ p.u.} \quad Y = 0.0430 \text{ p.u.}$$

The bus voltage phasors and line currents (converted to line power flows) obtained via a considerable number of power flows are used as measurements to estimate the line parameters. Table 16.1 presents 30 sets of sampling measurements (voltage phasors and line power flows) of the line between buses 42 and 49. Table 16.2 shows the estimates of resistance, reactance, and admittance-to-ground parameters of the line that are obtained using each set of measurements without any error introduced. Table 16.3 presents the first eight measurements with intentionally introduced errors and Table 16.4 shows the estimated parameters obtained using each of the eight disturbed measurements. Five percent is used as the threshold for filtering accuracy. It can be seen from Table 16.4 that the four sets of measurements with relatively large errors for parameter estimation (two resulting in large errors in resistance and two resulting in larger errors in admittance to ground) are filtered out. Table 16.5 presents the original and final estimated parameters of the line. Note that the percentage values in the brackets in the table refer to error percentages. It can be observed that the estimated parameters obtained using the 26 sets of measurements that include four unfiltered measurement errors but exclude four sets of filtered measurements are the same as the original parameters and those obtained using the 30 sets of measurement without error.

16.5.2 Estimating Parameters of a Line in BC Hydro System

The single-line diagram of BC Hydro 500 kV system [19] is shown in Figure 16.3. The parameters of a 500 kV overhead line named 5L96 (denoted by a bold dashed line in the figure) are estimated. The system power flow case used in the test has

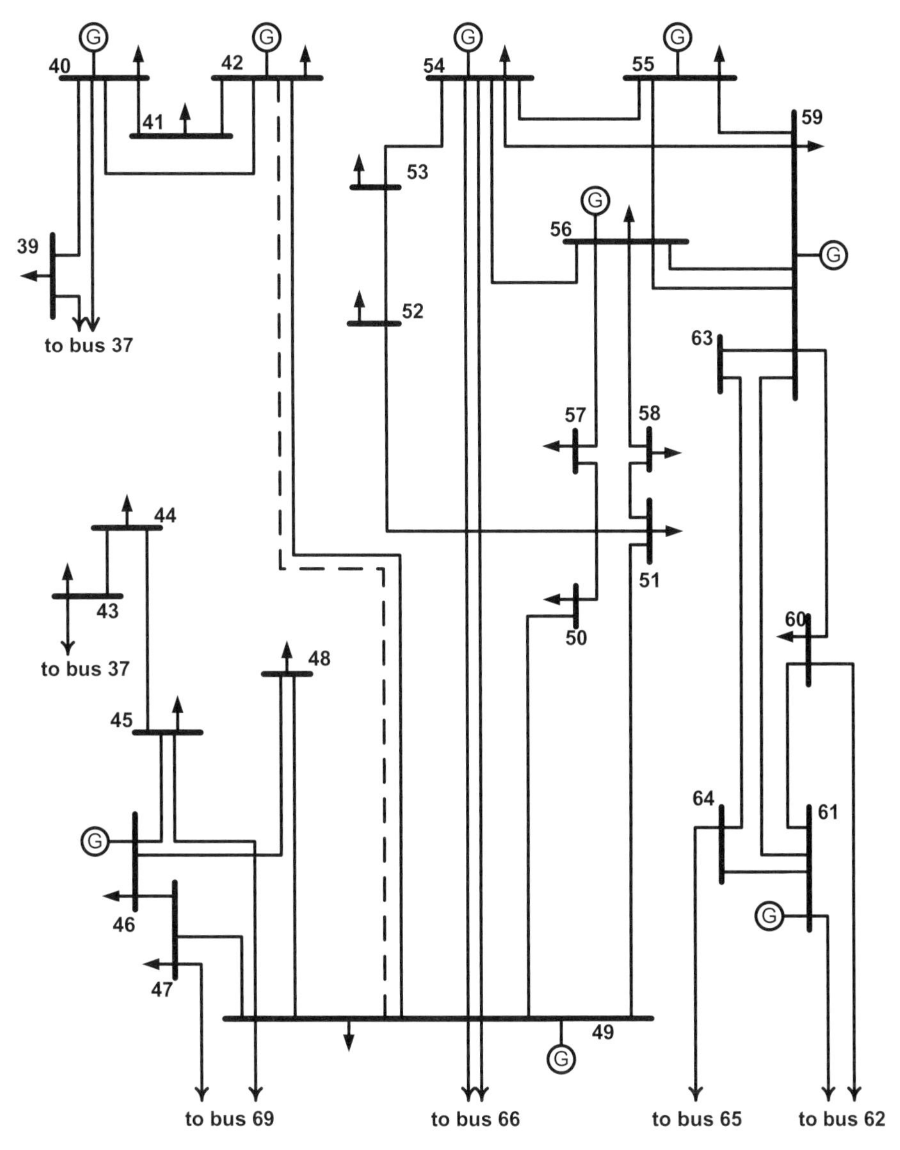

Figure 16.2. Single-line diagram of partial IEEE 118-bus system.

15,161 buses and 19,403 branches, which includes BC Hydro networks below 500 kV and partial western U.S. networks. Bus voltage phasors and line power flows obtained in power flow calculations are used as measurements to estimate the parameters of line 5L96. The system power flow cases considered have a wide range from various normal states to outage states that create heavy loading levels on line 5L96. Any change in the system operation state should not have any impact on the

TABLE 16.1. Voltage Phasors and Power Flows of Line Between Bus 42 and Bus 49 (Used as Measurements)

V_i (p.u.)	θ_i (degree)	V_j (p.u.)	θ_j (degree)	P_i (p.u.)	Q_i (p.u.)	P_{ij} (p.u.)	Q_{ij} (p.u.)
1.02500	−9.16450	0.92095	−21.60700	0.68445	0.20213	0.64841	0.12095
1.02500	−9.18850	0.92048	−21.66400	0.68607	0.20361	0.64983	0.12149
1.02500	−9. 21260	0.92000	−21.72100	0.68769	0.20510	0.65125	0.12202
1.02500	−9.23670	0.91952	−21.77700	0.68932	0.20659	0.65267	0.12256
1.02500	−9.26080	0.91903	−21.83400	0.69094	0.20809	0.65409	0.12309
1.02500	−9.28500	0.91855	−21.89100	0.69257	0.20959	0.65551	0.12363
1.02500	−9.30920	0.91807	−21.94800	0.69420	0.21109	0.65693	0.12416
1.02500	−9.33340	0.91758	−22.00500	0.69583	0.21260	0.65836	0.12470
1.02500	−9.35770	0.91709	−22.06300	0.69746	0.21411	0.65978	0.12524
1.02500	−9.38200	0.91661	−22.12000	0.69909	0.21563	0.66120	0.12577
1.02500	−9.40640	0.91612	−22.17800	0.70073	0.21716	0.66263	0.12631
1.02500	−9.43080	0.91563	−22.23500	0.70236	0.21868	0.66405	0.12685
1.02500	−9.45520	0.91513	−22.29300	0.70400	0.22022	0.66548	0.12738
1.02500	−9.48030	0.91462	−22.35300	0.70571	0.22183	0.66697	0.12795
1.02500	−9.50480	0.91412	−22.41100	0.70735	0.22337	0.66839	0.12849
1.02500	−9.52940	0.91363	−22.46900	0.70899	0.22492	0.66982	0.12902
1.02500	−9.55390	0.91313	−22.52700	0.71064	0.22647	0.67125	0.12956
1.02500	−9.57850	0.91263	−22.58500	0.71229	0.22803	0.67268	0.13010
1.02500	−9.60320	0.91213	−22.64300	0.71393	0.22959	0.67411	0.13063
1.02500	−9.62790	0.91163	−22.70200	0.71558	0.23116	0.67554	0.13117
1.02500	−9.65260	0.91113	−22.76000	0.71724	0.23273	0.67697	0.13171
1.02500	−9.67730	0.91063	−22.81900	0.71889	0.23431	0.67840	0.13225
1.02500	−9.70210	0.91013	−22.87800	0.72054	0.23589	0.67983	0.13278
1.02500	−9.72690	0.90962	−22.93700	0.72220	0.23747	0.68127	0.13332
1.02500	−9.75180	0.90911	−22.99600	0.72386	0.23907	0.68270	0.13386
1.02500	−9.77670	0.90861	−23.05500	0.72552	0.24066	0.68414	0.13439
1.02500	−9.80160	0.90810	−23.11400	0.72718	0.24226	0.68557	0.13493
1.02500	−9.82660	0.90759	−23.17300	0.72884	0.24387	0.68701	0.13547
1.02500	−9.85160	0.90707	−23.23300	0.73051	0.24548	0.68844	0.13601
1.02500	−9.87670	0.90656	−23.29200	0.73218	0.24710	0.68988	0.13655

Note: V_i, θ_i and V_j, θ_j are voltage magnitudes and angles at the two buses of the line. P_i, Q_i and P_{ij}, Q_{ij} are real and reactive line power flows at the two ends of the line.

estimation of line parameters, although heavy loading levels in outage states can create significantly different values of measurements.

Table 16.6 presents 10 sets of sampling measurements on 5L96 (voltage phasors and line power flows). It can be seen that the heavy loading cases (large power flow values) on line 5L96 correspond to large voltage-angle differences and low voltage-magnitude values. Table 16.7 shows the estimates of resistance, reactance, and admittance-to-ground parameters of line 5L96 using each individual set of measurements. Table 16.8 presents the original and final estimated parameters of line 5L96. Note that the percentage values in the brackets in the table refer to error percentages.

TABLE 16.2. Estimated Parameters Using Each Set of Measurements (No Error Introduced into Measurements)

Resistance R_{ij} (p.u.)	Reactance X_{ij} (p.u.)	Admittance to Ground Y (p.u.)
7.150819053923806E-002	0.322989968322601	4.300220526445065E-002
7.149330580993267E-002	0.323002038639029	4.300189766289475E-002
7.149046946189627E-002	0.323009052958054	4.299980527569113E-002
7.149658150866126E-002	0.322991396789738	4.300055510092178E-002
7.150472406138476E-002	0.322994653692808	4.299844120684803E-002
7.150059309021398E-002	0.322996113504715	4.299835133317951E-002
7.149665765565506E-002	0.322996425440579	4.299946261069169E-002
7.150359595814752E-002	0.322989108633325	4.300074690170377E-002
7.150254278137581E-002	0.323006488974226	4.300247211567151E-002
7.149489823420041E-002	0.323001470408724	4.299968940946629E-002
7.149109097810898E-002	0.323009477800442	4.299772505322701E-002
7.149717020207620E-002	0.322998254754971	4.299993083454870E-002
7.150410198175207E-002	0.323002420925849	4.299814438170383E-002
7.149390637544904E-002	0.323005200606151	4.300047376549218E-002
7.150129364038653E-002	0.323009639489799	4.300105320744722E-002
7.149477925376078E-002	0.323007578406282	4.299851336355507E-002
7.150033822538918E-002	0.323004930334097	4.300097996821717E-002
7.150467579167154E-002	0.322999056323605	4.300044095865730E-002
7.150930537175613E-002	0.322989863806196	4.300059970852191E-002
7.150400649437677E-002	0.323001968416419	4.299837632850823E-002
7.150718318722206E-002	0.322990614500469	4.300030297576305E-002
7.149952006321696E-002	0.323000498258261	4.299752789907323E-002
7.149265451649425E-002	0.323006951425721	4.299663995904027E-002
7.149865954872979E-002	0.323005662316445	4.300231080932703E-002
7.150161838879489E-002	0.323006100812629	4.299821434436004E-002
7.149388892505171E-002	0.323002116530079	4.300135676163106E-002
7.149668413889657E-002	0.323000221611103	4.300111705480244E-002
7.149830074128569E-002	0.322990356245569	4.299994611080062E-002
7.150330688094943E-002	0.323004133163954	4.300314836950893E-002
7.150367321102724E-002	0.322989930110879	4.300335400354474E-002

Although in a couple of cases, there is a relatively large difference between the original parameter and the estimated parameter obtained using an individual set of measurements, the final estimated parameters obtained using all the sets of measurements are still basically the same as the original ones even if no filtering process is conducted.

16.6 CONCLUSIONS

Resistance, reactance, and equivalent admittance to ground of lines are important input data in power system modeling. Traditionally, these parameters have been

TABLE 16.3. First Eight Sets of Measurements with Introduced Errors

V_i (p.u.)	θ_i (degree)	V_j (p.u.)	θ_j (degree)	P_i (p.u.)	Q_i (p.u.)	P_{ij} (p.u.)	Q_{ij} (p.u.)	Error Source
*1.03525	−9.16450	0.92095	−21.60700	*0.69129	*0.20415	0.64841	0.12095	1% error on V_i
1.02500	*−9.37230	0.92048	−21.66400	*0.68672	*0.20141	0.64983	0.12149	2% error on θ_i
1.02500	−9.21260	*0.93840	−21.72100	0.68769	0.20510	*0.66428	*0.12446	2% error on V_j
1.02500	−9.23670	0.91952	*−21.99500	0.68932	0.20659	*0.65313	*0.12008	1% error on θ_j
1.02500	−9.26080	0.91903	−21.83400	*0.69785	*0.21017	0.65409	0.12309	1% error on I_i
1.02500	−9.28500	0.91855	−21.89100	*0.69063	*0.21590	0.65551	0.12363	2% error on φ_i
1.02500	−9.30920	0.91807	−21.94800	0.69420	0.21109	*0.67007	*0.12665	2% error on I_j
1.02500	−9.33340	0.91758	−22.00500	0.69583	0.21260	*0.65764	*0.12846	1% error on φ_j

Notes: (1) I_i, φ_i and I_j, φ_j are current magnitudes and angles at the two buses of the line. V_i, θ_i and V_j, θ_j are voltage magnitudes and angles at the two buses of the line. P_i, Q_i and P_{ij}, Q_{ij} are real and reactive line power flows at the two ends of the line.

(2) The measurements with errors are prefixed by *.

(3) An error in the measurement of voltage magnitude or angle has impacts not only on itself but also on real and reactive line powers, whereas an error in the measurement of current magnitude or angle has impacts only on real and reactive line powers.

TABLE 16.4. Estimated Parameters Using Each of the First Eight Sets of Measurements with Errors (Errors Introduced on Voltage or Current Phasor)

Resistance R_{ij} (p.u.)	Reactance X_{ij} (p.u.)	Admittance to Ground (p.u.)	Error Source[b]	Note[a]
8.442778000797151E-002	0.327976565022061	4.149395128849082E-002	1% error on V_i	Filtered
7.354215645412961E-002	0.318611262005352	4.414604437753773E-002	2% error on θ_i	Unfiltered
4.650541001052875E-002	0.319639262614001	4.349025118816752E-002	2% error on V_j	Filtered
7.030523456060858E-002	0.327814339269931	4.170960516485524E-002	1% error on θ_j	Unfiltered
7.140607804964676E-002	0.323003125655167	4.276462519717335E-002	1% error on I_i	Unfiltered
7.012042547553828E-002	0.323108469786721	3.971573441817353E-002	2% error on φ_i	Filtered
7.100579512324626E-002	0.316582486741477	4.612013643859956E-002	2% error on I_j	Filtered
7.048479944309825E-002	0.323431732936217	4.495975726943799E-002	1% error on φ_j	Unfiltered

[a] "Filtered" indicates that the set of measurements is filtered by the given threshold and "unfiltered" indicates that the errors in an estimated parameter caused by the measurement error are within the threshold and acceptable.

[b] I_i, φ_i and I_j, φ_j are current magnitudes and angles at the two buses of the line. V_i, θ_i and V_j, θ_j are voltage magnitudes and angles at the two buses of the line. P_i, Q_i and P_{ij}, Q_{ij} are real and reactive line power flows at the two ends of the line.

TABLE 16.5. Original and Estimated Line Parameters

	Original	Estimated (No Measurement Error)	Estimated (With Measurement Errors)
Resistance R_{ij} (p.u)	0.07150	0.07150 (0.00%)	0.07149 (0.01%)
Reactance X_{ij} (p.u)	0.32300	0.32300 (0.00%)	0.32303 (0.01%)
Admittance to ground Y (p.u)	0.04300	0.04300 (0.00%)	0.04306 (0.14%)

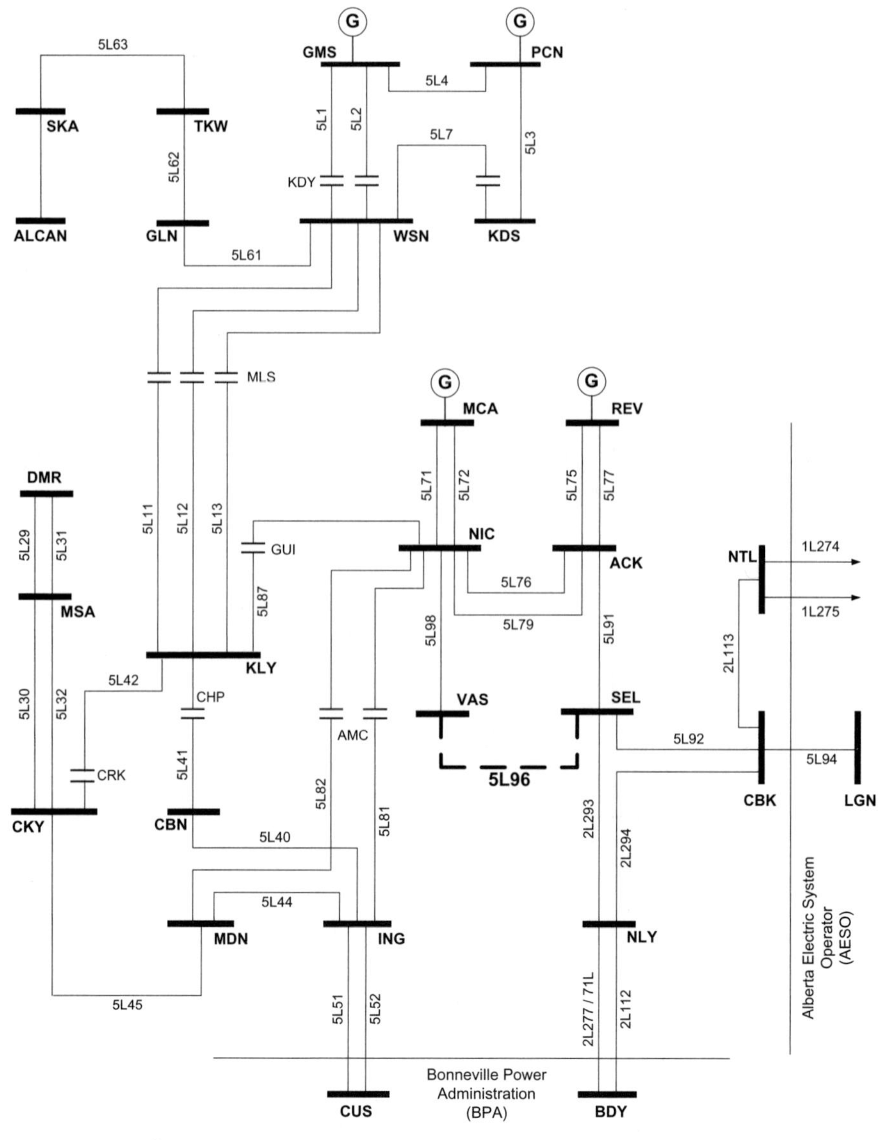

Figure 16.3. Single-line diagram of BC Hydro 500 kV system.

TABLE 16.6. Voltage Phasors and Power Flows of Line 5L96 (Used as Measurements)

V_i (p.u.)	θ_i (degree)	V_j (p.u.)	θ_j (degree)	P_i (p.u.)	Q_i (p.u.)	P_{ij} (p.u.)	Q_{ij} (p.u.)
1.07000	8.60000	1.08400	1.50000	6.57000	−1.90700	6.51000	−0.39500
1.06000	10.40000	1.07500	2.30000	7.38000	−1.86900	7.30000	−0.62900
1.04800	13.40000	1.06300	3.70000	8.65000	−1.70500	8.55000	−0.93500
1.04200	15.50000	1.05500	4.70000	9.56000	−1.47800	9.44000	−1.08000
1.03600	17.60000	1.04600	5.60000	10.47000	−1.22400	10.31000	−1.24000
0.97400	35.40000	0.95000	10.50000	18.39000	2.82700	17.82000	−3.31000
0.94300	38.50000	0.92100	11.20000	18.79000	3.25000	18.16000	−3.97000
0.93400	39.40000	0.91300	11.50000	18.83000	3.32000	18.18000	−4.14000
0.92300	40.40000	0.90300	11.80000	18.86000	3.40000	18.19000	−4.37000
0.90400	42.00000	0.88700	12.10000	18.87000	3.63000	18.17000	−4.74000

TABLE 16.7. Estimated Parameters Using Each Set of Measurements (No Error Introduced into Measurements)

Resistance R_{ij} (p.u.)	Reactance X_{ij} (p.u.)	Admittance to Ground Y (p.u.)
1.548079448990775E-003	2.164686703340432E-002	1.00634041480235
1.540257804047732E-003	2.161604192015683E-002	1.00642302851147
1.507477920127257E-003	2.158783668403971E-002	1.00748042825992
1.491795456659398E-003	2.147316672004808E-002	1.00745808660354
1.573061016193664E-003	2.149563310309445E-002	1.00788947177925
1.543499388548838E-003	2.149672277048962E-002	1.00839028571578
1.507889329436661E-003	2.153452780650219E-002	1.00544332987656
1.505684381083002E-003	2.153580979630141E-002	1.01163590422382
1.537998257741009E-003	2.149474573088756E-002	1.00884387956467
1.465168579534557E-003	2.155604011536260E-002	0.94910943742322

TABLE 16.8. Original and Estimated Parameters of 5L96

	Original	Estimated
Resistance R_{ij} (p.u)	0.00153	0.00152 (0.65%)
Reactance X_{ij} (p.u)	0.02154	0.02154 (0.00%)
Admittance to ground Y (p.u)	1.00701	1.00190 (0.51%)

assumed to be constant for online applications. In real life, however, they may vary with environmental and weather conditions, which change from time to time. Estimating varied line parameters in real time is essential in enhancing accuracy in energy management system (EMS) and Smart Grid applications. Errors in the parameters can lead to inaccuracy or even erroneous results in system modeling, system analysis, and other applications of EMSs and Smart Grids. Advancements in synchronized phasor measurement units (PMUs) and information technology provide opportunities for this objective.

This chapter presents a method for real-time estimation of transmission line parameters using synchronized PMU information with timestamps. Invalid PMU measurements (erroneous information) that may occur in PMU devices or communication channels can be filtered and reliable PMU measurements that may have small measurement errors are used in estimation. The method can be performed very quickly and applied in a real-time environment. The simulation results using the IEEE 118-bus system and the actual BC Hydro system demonstrated the feasibility and effectiveness of the presented method.

REFERENCES

[1] L. L. Grigsby (Editor–in-Chief), *Electric Power Engineering Handbook*. CRC Press and IEEE Press, 2001.

[2] *EPRI AC Transmission Line Reference Book—200 kV and Above*, 3rd edition. EPRI, Palo Alto, CA, 2005.

[3] D. A. Douglass and L. A. Kirkpatrick, "AC resistance of ACSR—magnetic and temperature effects," *IEEE Trans. Power Apparatus Syst.*, PAS-104(6), pp. 1578–1584, June 1985.

[4] L. P. Allfather, "Impedance measuring," U.S. Patent 5818245, issued on October 6, 1998.

[5] B. Bachmann, D. G. Hart, Y. Hu, D. Novosel, and M. M. Saha, "Impedance measurement system for power system transmission lines," U.S. Patent 6397156, issued on May 28, 2002.

[6] A. G. Phadke, "Synchronized phasor measurements in power systems," *IEEE Computer Appl. Power*, pp. 10–15, April 1993.

[7] M. Zima, M. Larsson, P. Korba, C. Rehtanz, and G. Anderson, "Design aspects for wide-area monitoring and control systems," *Proc. IEEE*, 93(5), pp. 980–996, May 2005.

[8] "Eastern interconnection phasor project," 2006 IEEE PES Power Systems Conference and Exposition (PSCE '06), October 29–November 1, 2006, pp. 336–342.

[9] X. Xie, Y. Xin, J. Xiao, J. Wu, and Y. Han, "WAMS applications in Chinese power systems," *IEEE Power Energy Mag.*, 4(1), pp. 54–63, January/February 2006.

[10] W. Li, J. Yu, Y. Wang, P. Choudhury, and J. Sun, "Method and system for real time identification of voltage stability via identification of weakest lines and buses contributing to power system collapse," U.S. Patent 7816927, issued on October 19, 2010.

[11] L. S. Anderson, A. Guzman-Casillas, G. C. Zweigle, and G. Bnmouyal, "Protective relay with synchronized phasor measurement capability for use in electric power systems," U.S. Patent 684533, issued on January 18, 2005.

[12] M. Venkatasubramanian, M. Sherwood, V. Ajjarapu, and B. Leonardi, "Real-Time Security Assessment of Angle Stability Using Synchrophasors," PSERC Report, Document 10-10, May 2010.

[13] J. S. Thorp, A. Bur, M. Begovic, J. Giri, and R. Avila-Rosales, "Gaining a wider perspective," *IEEE Power Energy Mag.*, pp. 43–51, September/October 2008.

[14] D. Atanackovic, J. H. Clapauch, G. Dwernychuk, J. Gurneys, and H. Lee, "First steps to wide area control," *IEEE Power Energy Mag.*, pp. 61–68, January/February 2008.

[15] N. H. Abbasy and H. M. Ismail, "A unified approach for the optimal PMU location for power system state estimation," *IEEE Trans. Power Syst.*, 24(2), pp. 806–813, May 2009.

[16] A. Carta, N. Locci, and C. Muscas, "A PMU for the measurement of synchronized harmonic phasors in three-phase distribution networks," *IEEE Trans. Instrum. Measurement*, 58(10), pp. 3723–3730, October 2009.

[17] W. Li, P. Choudhury, and J. Sun, "Method and system of real-time estimation of transmission line parameters in on-line power flow calculations," U.S. Patent 7710729, issued on May 4, 2010.

[18] IEEE power flow test cases archive. Available at: http://www.ee.washington.edu/research/pstca/.

[19] BC Hydro transmission system information, available at: http://transmission.bchydro.com/transmission_system/engineering_studies_data/studies/.

17

WAMCP STUDY: VOLTAGE STABILITY MONITORING AND CONTROL

Mats Larsson

This chapter introduces a case study concerning voltage control and protection in electric power systems.* Readers are expected to be familiar with the fundamentals of voltage control and stability, which are introduced in Chapter 1 and are treated in more detail in Larsson [1]. The described scheme is suitable for implementation on a wide area control and protection system.

17.1 WIDE-AREA VOLTAGE STABILITY PROTECTION

Voltage instability mitigation has been discussed for some time now, and some protection systems against voltage collapse are installed and in operation [2]. Most of these systems use rather simple criteria, such as low voltage, and quite rough actions, such as load shedding. Regarding the rapidly growing capability in computer and

* This chapter is based on M. Larsson and D. Karlsson, "Coordinated system protection scheme against voltage collapse using heuristic search and predictive control," *IEEE Transactions on Power System*, 18 (3), pp. 1001–1006, 2003, © 2003 IEEE.

Smart Grid: Applications, Communications, and Security, First Edition.
Edited by Lars Torsten Berger and Krzysztof Iniewski.

communication technology, the time has now come to introduce smoother schemes, where system-wide voltage levels, reactive power flows, and so on are used in a wide-area emergency control and protection system. Such a system, to be activated when the power system is in transition toward instability, must include different voltage levels and act on transformer tap changers, *automatic voltage regulators* (AVRs) on generators, as well as reactive power compensation devices available. To save a power system exposed to a disturbance, where the system survived the initial disturbance, but the system dynamics have been triggered and a transition toward instability has started, powerful and synchronized actions are required as the harm to customers has to be minimized. In such a situation (i.e., a system exposed to a long-term voltage instability), there is some time, tens of seconds to some minutes, available to counteract the transition. Basic rules are to lower the voltage on the load level as much as possible to achieve a temporary load relief. The transmission, subtransmission, and distribution system voltages should, however, be kept as high as possible to reduce the losses and maximize the line and cable reactive power generation. Furthermore, most control variables have an inherently discrete nature; for example, capacitor banks and tap changers must be switched in fixed steps and while most utilities lack direct load control schemes, load shedding must still be carried out by disconnecting whole feeders.

The scheme presented uses a network model and wide-area measurements to account for the actual power flow in the system and load models to account for load recovery dynamics, and constraints are imposed to ensure that voltage and generator current limits are not violated.

17.1.1 Power System State Prediction and Optimization

The wide-area protection scheme is based on the model predictive method described in Larsson et al. [3]. The principle is illustrated in Figure 17.1. A system model, including the load dynamics, is used to predict the output trajectories (dotted lines)

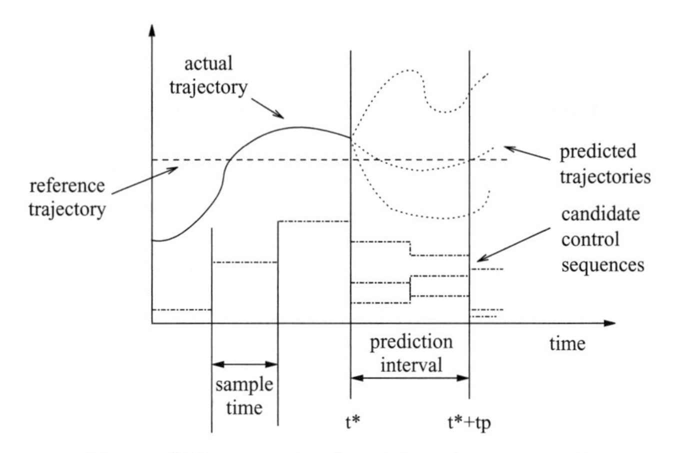

Figure 17.1. Principle of model predictive control.

based on the current state and for several different candidate input sequences. A cost function is defined based on the deviation of each predicted trajectory from a desired trajectory (dashed line). The optimal control, in the sense that it minimizes the defined cost function, is then obtained by solving an optimization problem online, and applied to the system. The interval between the current time t^* and the prediction horizon $t^* + t_p$ is referred to as the prediction interval, and is chosen based on the settling time of the slowest dynamics. Usually, the prediction interval is chosen as a multiple of the sample time.

The problem of selecting controls at the instant t^* and operating point (x^*, u^-) is formulated as the combinatorial optimization problem,

$$\begin{gathered} \min J\left(x^*, u^+\right) \\ \text{subject to } u^+ \in \mathcal{S}\left(u^-\right) \end{gathered} \tag{17.1}$$

where u^+ is the optimization variable corresponding to the new control state to be determined, whereas u^- denotes the control state presently in use, in other words before the optimization starts at time t^*. $\mathcal{S}(u^-)$ is the set of available control states. The scalar function $J(x^*, u^+)$ is referred to as a cost function and should evaluate the benefit of using the new control state u^+ such that a smaller result indicates a more desirable state. The cost function is defined as

$$J\left(u^+, x^*\right) = \int_{t^*}^{t^*+t_p} \tilde{y}^T \mathbf{Q} \tilde{y} + \tilde{u}^T \mathbf{R} \tilde{u} + P \mathrm{d}t \tag{17.2}$$

where $\tilde{y} = (\hat{y}(t, x^*, u^+) - y_r)$ is a prediction of the output deviations from the desired trajectories, provided the new control u^+ is applied during the prediction interval. The control state presently in use is denoted u^-, and thus $\tilde{u} = (u^+ - u^-)$ is the applied change of controls. The reference trajectory is denoted y_r and the weighting matrices for output errors and control variations are $\mathbf{Q}$ and $\mathbf{R}$, respectively. The scalar penalty term $P = P(u^+, x^*)$ is introduced whenever a constraint violation is predicted to occur during the prediction interval.

The predictions are obtained from a linearized system model using the linearized model predictive control (LMPC) method that is described in Larsson et al. [3]. The linearized system representation is recomputed prior to each control signal computation. Generators and their associated control systems are modeled using their quasi-steady-state representation as given in van Cutsem and Vournas [4]; that is, the short-term dynamics are neglected in the linearized model used for control signal computation. However, these dynamics are retained in the simulations.

17.1.2 Heuristic Tree Search

When solving the combinatorial optimization problem (17.1) by tree search, the possible control states are organized in a tree structure. Using terminology from the literature on search, each control state is represented by a *node* or *position* in the tree and the transition from one node to another node is called a *move*. A move in this case corresponds to the switching of a single control. The maximum number of levels in the tree is called the *depth* of a search and corresponds to the number of controls that may be switched in a single search. The *width* of the search is the number of successors that are expanded in each node, that is, the number of moves that are explored in any intermediate position. Furthermore, the *search space* of the optimization problem contains all unique control states reachable from the current state. However, in most practical cases it is not computationally feasible to explore all of the search space. The *search tree* is the part of the search space actually explored by a search algorithm.

In Larsson et al. [3], the combinatorial optimization problem (17.1) is solved using tree search with a standard *depth-first* search method as described by Russell and Norvig [5], which is an example of a *uninformed* search method that explores all of the search-space. An *informed* search method makes use of search enhancements in order to reduce the size of the search tree and consequently the computational complexity. A survey of such enhancements is given, for example, by Junghanns [6]. A short description of the enhancements used in this application are given in the following sections.

17.1.2.1 Transposition Table. It can be observed that some nodes are visited more than once during a search. For example, if there are two on–off controls and the initial state is *off–off*, there are two ways of reaching the state *on–on*, by switching either one of the two controls first. By storing previously evaluated states in a table along with some additional information about the status of the search, a transposition table may reduce the search tree considerably without sacrificing completeness of the search. A detailed study of transposition tables and their effect on search complexity can be found in Breuker [7].

17.1.2.2 Lower Bound. The cost function (17.2) can be decomposed into three components,

$$f = f_y + f_u + f_P \tag{17.3}$$

related to the output deviation, control, and constraint violation costs, respectively. All three components are implicitly determined by the control state in use. The control cost cannot decrease as progress is made deeper into the search tree, and the control cost can therefore be used as a *lower bound* on the cost function. Without sacrificing completeness of the search, the expansion of nodes can be stopped in every node where the control cost alone is greater than the total cost in any node encountered so far. However, lower bound cutoffs cannot be expected to reduce the search tree significantly when there are constraint violations, since the constraint violation term is designed to dominate the cost function when such violations are present.

17.1.2.3 Move Ordering. The purpose of move ordering is to make sure that the search space is explored in a good order, meaning an order where the number of lower bound cutoffs is maximized. The following two measures have been used to order the moves: first, the value of the cost function after the move, and second, the change in the cost function divided by the cost of move. Thus, the first move considered in each node is the one with the best evaluation according to the first criterion, the second is the best one according to the second criterion and the third is the second best according to the first criterion and so on.

17.1.2.4 Iterative Broadening. In an iterative broadening search [8], a number of successive searches are made. Ranked according to the move ordering criteria, only the best move in each intermediate node is considered in the first iteration. In the second iteration also the second best move is considered and so on. The basic idea is to quickly find one reasonably good solution by venturing deep into the tree and then use remaining time available to improve this solution. Since the size of the search tree depends exponentially on the breadth, the effort spent on the search is dominated by the last iteration. In fact, the information collected in these intermediate searches can often enable additional lower bound cutoffs, and an iterative broadening search to a certain depth and width may therefore be computationally cheaper than a single search.

Note, however, that the iterative broadening sacrifices completeness of the search. Because in most cases only a fraction of the tree search space is explored, the search is likely to miss the optimal solution unless good move ordering criteria are used. There is a potential risk that the iterative broadening finds a local minimum or a plateau of the cost function [5]. No such behavior has been detected in the work on this application, but can in such cases be circumvented by repeating the search using different initial control vectors.

17.1.3 Voltage Stability Protection Based on Local Measurements

The simulations have also been carried out with a rule-based protection scheme using only local voltage measurements. Load shedding is carried out using the scheme described in Vu et al. [9] using an assumed "lowest normal voltage" of 1 p.u.; that is, 5% of the bus load is shed at 0.9 p.u. with 3.5 s time delay, another 5% is shed at 0.92 p.u. with 3 s delay, and yet another 5% is shed at 0.92 p.u. with 8 s delay. The tap changers are controlled by constant-time relays corresponding to model D1 in Sauer and Pai [10] with a delay time T_{d0} of 40 s. Tap changers are blocked if the primary side voltage stays below 0.93 p.u for 10 s or more. All loads and tap changers are equipped with these undervoltage and tap locking relays in the schemes LPSP1 and LSPS2.

17.1.4 Test Network

The model predictive method will be demonstrated using the Nordic 32 test system (Figure 17.2) originally described in CIGRE [11], except that the generators

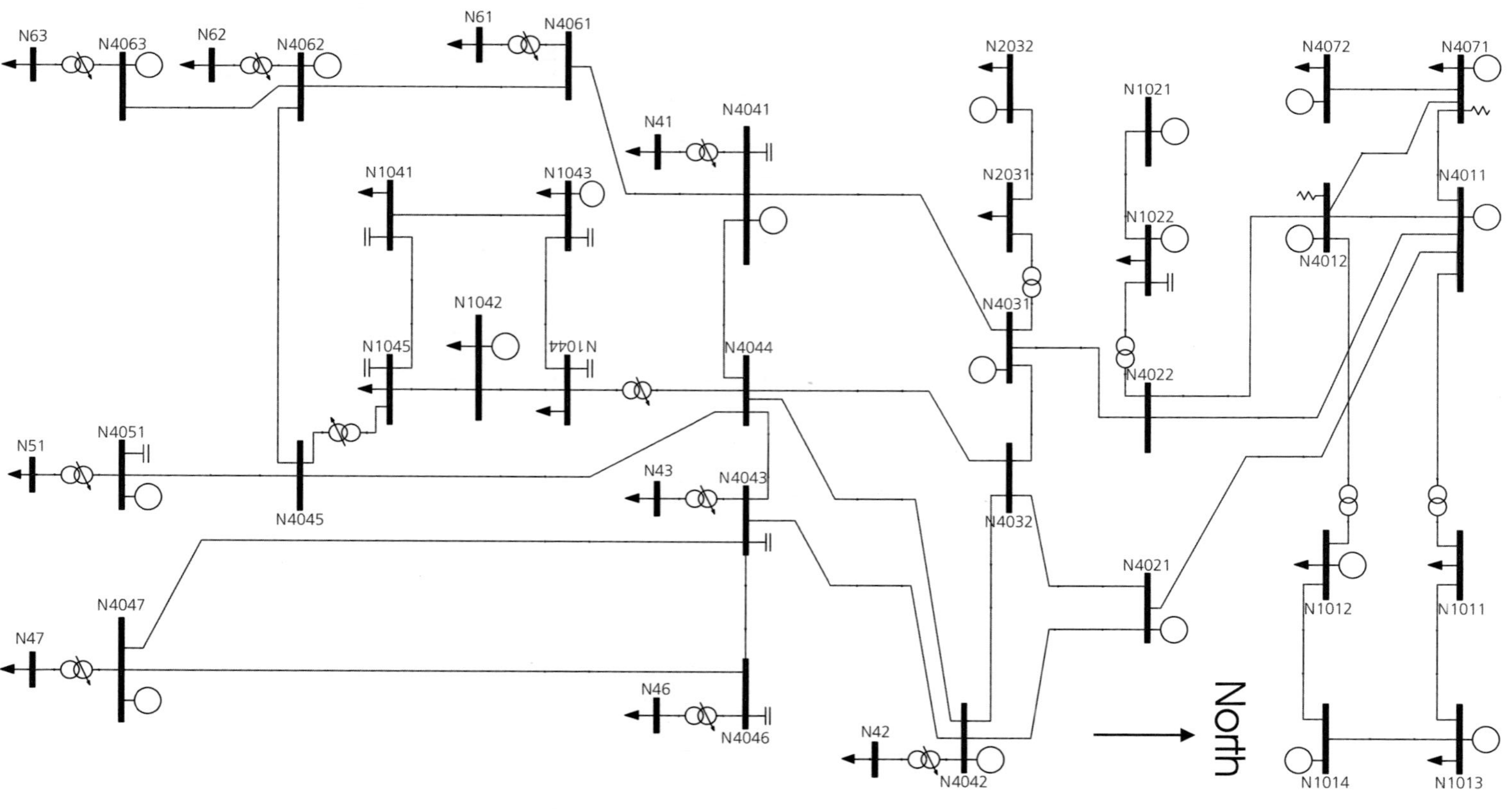

Figure 17.2. The Nordic test system from CIGRE [11].

connected to the same bus and parallel branches have been aggregated and generator saturation is neglected. Furthermore, the generator at bus N4011 is modeled as an infinite bus. The loads are modeled using the aggregate exponential recovery model [12],

$$T_p\dot{x}_p + x_p = P_s(V) - P_t(V) \tag{17.4}$$

$$P_d = k_l\left(x_p + P_t(V)\right) \tag{17.5}$$

where x_p is a continuous dynamic state that can be interpreted as a measure of the energy deficit in the load and $P_s(V) = P_0V^{\alpha_s}$ and $P_t(V) = P_0V^{\alpha_t}$ are the steady-state and transient voltage dependencies, respectively. P_d is the actual active load power and T_p is the active power recovery time constant. A similar model is used for the reactive power load with corresponding characteristics Q_d, x_q, $Q_s(V) = Q_0V^{\beta_s}$, and $Q_t(V) = Q_0V^{\beta_t}$ and time constant T_q. The scale factor k_l models load shedding and is applied to the reactive as well as to the active power load model. The parameter values $\alpha_s = \beta_s = 0.5$, $\alpha_t = \beta_t = 2$, and $T_p = T_q = 60$ s have been used for all loads.

All generators are equipped with field current limiters, and the thermal power generators (at buses N1042, N1043, N4042, N4047, N4051, N4062, and N4063) also have armature current limiters. Both types of limiters are of integral-type and allow temporary overcurrents for the first 20 s. A detailed description of the limiters can be found in CIGRE [11]. The armature current constraints are set at 1.05 p.u., whereas the field current constraints are set 15% lower than in the report in order to compensate for neglecting saturation. The initial state is given by the load flow case called *lf28* in the CIGRE report. The following components are considered as controls:

- 11 Transformers with on-load tap changers (16 steps of 1.67%)
- 22 Load shedding points (3 steps of 5% of nominal load)
- 19 Generator voltage setpoints (8 steps of 2%)

Constraints are imposed on:

- Field currents of all generators, set to the same value as the internal generator field current limiter.
- Armature current of the thermal power generators, set to the same value as the internal generator field current limiter.
- All bus voltages should be kept above 0.92 p.u.

Voltages below tap changing transformers are considered controlled and should be kept close to 1 p.u. unless transmission voltage constraints are violated. In cases where not all constraints can be satisfied, armature and field current constraints have priority over voltage constraints.

TABLE 17.1. Overview of the Scenarios

Case Number	Tripped Component(s)	Without SPS
1	Generator 4062	Collapse at 150 s
2	Line 4045–4062	Stable, no violation
3	Line 4032–4044	Collapse at 340 s
4	Line 4031–4041	Collapse at 45 s
5	Line, 4011–4021, Generator 1012	Collapse at 175 s
6	Line 1043–1044	Collapse at 201 s
7	Generator 4041	Stable, no violation
8	Line 4042–4044	Collapse at 64 s

17.1.5 Scenarios and Simulation Results

The predisturbance state *lf28* has a higher load than would be allowed by normal reliability criteria. In fact, many single contingencies will lead to voltage collapse. Simulation of the scenarios listed in Table 17.1 is used to evaluate the different schemes. The table also shows the system response without a protection scheme. In most cases, voltage collapse occurs unless emergency controls are initiated.

The simulations have been carried out with the local scheme in two versions: first, the LSPS1 where the generators control their respective terminal voltage and second, the LSPS2 where they use full line-drop compensation as described in Taylor [13], in order to make better use of their reactive capability. The coordinated scheme is used in two versions, the CSPS1 in which the generator controls have been disabled and the CSPS2 which has access to all controls. Thus, the results with CSPS1 should be compared to those with LSPS1, and the results with CSPS2 to those with LSPS2.

17.1.5.1 Case 1: Tripping of Generator at Bus N4062. In the precontingency situation, the generator at bus N4062 supplies 530 MW active and absorbs 8 Mvar reactive power. The tripping of the unit causes an active power deficiency in the south, which is compensated mainly by the hydro generators in the north and consequently increases transfer through the central region. This leads to increased losses and low voltages in the south and central regions. In response, the remaining generators in the south almost instantaneously increase their reactive power output until after 20 s, the armature current limiters of generators 1043 and 4042 are activated. Voltage support is then lost in the areas close to these buses. Load recovery and tap changer dynamics will then further depress voltages in the south until system collapse occurs, unless emergency actions are taken. Figure 17.3 shows the terminal voltage and armature current of the generator at bus N4042.

With both local schemes, the tap changing transformers in the southern region act to restore their controlled voltages after about 50 s. This depresses transmission voltages further and some tap changers are eventually blocked. Voltage decline then continues due to load dynamics and the first load shedding is ordered at about 100 s

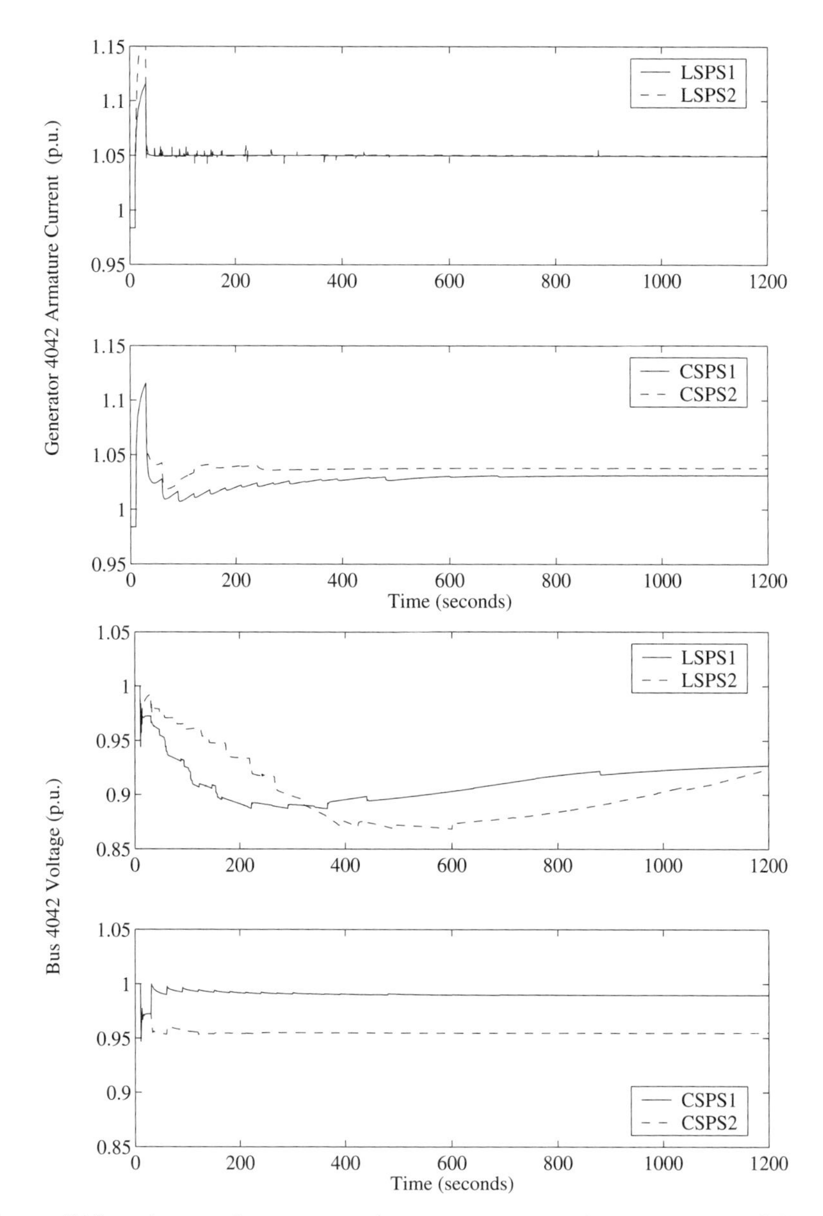

Figure 17.3. Voltage at bus N4042 and armature current of generator 4042 following tripping of the generator at bus N4062 with the different protection schemes.

with the LSPS1. Once enough undervoltage relays have been activated, the voltage instability phase ceases and voltages slowly recover close to their predisturbance values. Since the terminal voltages of generators with reactive capability to spare are kept higher with the LSPS2, reactive capability is better utilized, and therefore load shedding action is initiated. Also, less load shedding is required before the voltages become stable.

While the voltage trajectories of the CSPS1 and CSPS2 look similar, the two schemes use different means of reaching a stable state. With the CSPS1, load shedding at buses N43 and N1044 combined with backstepping of transformer N4043 (which provides a temporary load relief at bus N43) relax the armature current constraints of the limited generators. The CSPS2 decreases the voltage setpoints of the generators at buses N1043 and N4042, taking these out of their armature current limits. Generators at N2032, N4021, N4041, and N4047 have reactive capability to spare and are used to support the weak central–south region from the outside. Subsequently, the tap changers are used to restore the controlled voltages at nodes where this can be done without violating transmission voltage constraints.

17.1.5.2 Simulation Results. Tables 17.2 and 17.3 list the system responses with the two versions of the local and coordinated SPS, respectively. The local schemes stabilize the system in all cases except case 4, in which the collapse is due to instability of the short-term dynamics, not modeled in the system model used by the coordinated schemes. Nevertheless, the coordinated schemes keep a more even voltage profile than the local schemes and thereby keep the system within the (short-term) stability region even though the short-term dynamics are not explicitly modeled in the system model used by these schemes. Surprisingly, in case 3, the LSPS1 sheds less load than the CSPS1. The voltage at bus N4042 is allowed to decrease below the voltage constraint of 0.92 p.u. with the local schemes since there is no undervoltage load shedding relay at this bus. On the other hand, the CSPS1 has the option of shedding load at neighboring buses in order to avoid this violation. Therefore, the

TABLE 17.2. Simulation Results with the Local Protection Scheme

Case Number	LSPS1	LSPS2
1	Stable, 200 + j69 MVA shed	Stable, 120 + j36 MVA shed
2	Stable, no load shed	Stable, no load shed
3	Stable, 100 + j33 MVA shed	Stable, no load shed
4	Collapse at 144 s, 590 + j190 MVA shed	Collapse at 220 s, 533.0 + j168.7 MVA shed
5	Stable, 224 + j68 MVA shed	Stable, 208 + j63 MVA shed
6	Stable, 60 + j20 MVA shed	Stable, 60 + j20 MVA shed
7	Stable, no load shed	Stable, no load shed
8	Stable, 424 + j139 MVA shed	Stable, 377 + j122 MVA shed

TABLE 17.3. Simulation Results with the Wide-Area Protection Scheme

Case Number	CSPS1	CSPS2
1	Stable, 160 + j49 MVA shed	Stable, no load shed
2	Stable, no load shed	Stable, no load shed
3	Stable, 115 + j37 MVA shed	Stable, no load shed
4	Stable, 570 + j185 MVA shed	Stable, 500 + j161 MVA shed
5	Stable, 160 + j46 MVA shed	Stable, 75 + j22 MVA shed
6	Stable, no load shed	Stable, no load shed
7	Stable, no load shed	Stable, no load shed
8	Stable, 494 + j155 MVA shed	Stable, 270 + j75 MVA shed

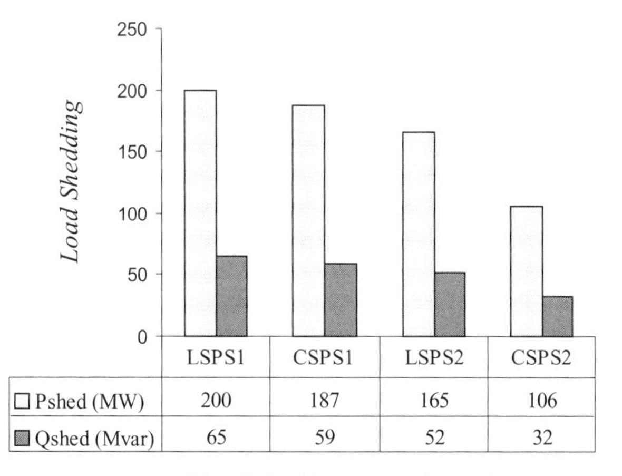

Figure 17.4. Average amount of load shedding over the eight scenarios. Nominal pre-contingency load is 10940 MW and 3358 Mvar.

amount of load shedding can be higher with the coordinated scheme. Case 6 is another interesting case, which illustrates misoperation of the local schemes—both LSPS1 and LSPS2 shed load. The simulation of the same scenario with CSPS1 and CSPS2 shows that load shedding is not necessary to avoid collapse.

Figure 17.4 shows a comparison of the amount of load shedding executed by the different schemes. The amount of load shedding by the LSPS2 compared to the LSPS1 is about 17% lower, since the generator reactive capabilities are better utilized.

Although the simulation results are slightly biased toward the LSPS1 because of less stringent handling of voltage constraints than with the CSPS1, the benefits of coordinated control when only load shedding is allowed seems to be small. The required amount of load shedding is only 6% smaller with the CSPS1 than with the LSPS1. This reduction can be ascribed to better control of the tap changers and is presumably not due to the coordination of load shedding in different geographical locations.

However, with the CSPS2, which has generator setpoint voltages included as control variables, remote generators can be used to support weak regions as an alternative to load shedding and the load shedding reduction is about 47% compared to the LSPS1 and 35% compared to the LSPS2. Also, because of better management of voltage constraints by both coordinated schemes, they are more likely to keep the trajectories of the system in a region where the short-term dynamics are stable.

17.2 CONCLUSION

A wide-area protection scheme for voltage control in the emergency state, capable of coordinating dissimilar and discrete controls such as generator, tap changer, and load shedding controls using heuristic tree search, has been presented. Eight scenarios, some of which lead to voltage collapse unless emergency control measures are taken, have been used to evaluate system protection schemes based on local as well centralized measurements. Simulation results indicate that load shedding based on local measurements is near optimal when load shedding is the only emergency control considered. Another conclusion is that the reactive capability of generators is somewhat better utilized when generators use line-drop compensation.

On the other hand, the results indicate that the protection scheme based on centralized measurements can use remote generators to support a weak region as an alternative to load shedding. The reduction in the amount of load shedding required is about 35% compared to the best local strategy. Also, the centralized scheme maintains a more even voltage profile than the local schemes and may therefore avoid short-term voltage instability that occurs with the local scheme, even though the short-term dynamics are not explicitly modeled.

REFERENCES

[1] M. Larsson, "Coordinated Voltage Control in Electric Power Systems," Ph.D. dissertation, Lund University, Sweden, 2001.

[2] "System Protection Schemes in Power Networks," CIGRE Task Force 38.02.19, Tech. Rep. 2000 (final draft).

[3] M. Larsson, D. Hill, and G. Olsson, "Emergency voltage control using search and predictive control," *Int. J. Power Energy Syst.*, 24 (2), pp. 121–130, 2002.

[4] T. van Cutsem and C. Vournas, *Voltage Stability of Electric Power Systems*, Power Electronics and Power Systems Series. Kluwer Academic Publishers, 1998.

[5] S. Russell and P. Norvig, *Artificial Intelligence—A Modern Approach.* Prentice-Hall, 1995.

[6] A. Junghanns, "Pushing the Limits: New Developments in Single-Agent Search," Ph.D. dissertation, Department of Computing Science, University of Alberta, Canada, 1999.

[7] D. Breuker, "Memory Versus Search in Games," Ph.D. dissertation, Department of Computer Science, Maastricht University, The Netherlands, 1998.

[8] M. Ginsberg and W. Harvey, "Iterative broadening," *Artificial Intelligence*, 55(2–3), pp. 367–383, June 1992.

[9] K. Vu, C.-C. Liu, C. Taylor, and K. Jimma, "Voltage instability: Mechanisms and control strategies," *Proc. IEEE*, 83(11), pp. 1442–1455, 1995.

[10] P. Sauer and M. Pai, "A Comparison of Discrete vs. Continuous Dynamic Models of Tap-Changing-Under-Load Transformers," Proceedings Bulk Power System Voltage Phenomena—III: Voltage Stability, Security and Control, Davos, Switzerland, 1994.

[11] CIGRE, "Long Term Dynamics Phase II," CIGRE Task Force 38.02.08, Tech. Rep., 1995.

[12] D. Karlsson and D. Hill, "Modelling and identification of nonlinear dynamic loads in power systems," *IEEE Trans. Power Syst.*, 9(1), pp. 157–163, February 1994.

[13] C. Taylor, "Line drop compensation, high side voltage control, secondary voltage control- why not control a generator like a static var compensator," in *Power Engineering Society Summer Meeting*, vol. 1. IEEE, 2000, pp. 307–310.

18

SECURE REMOTE ACCESS TO HOME ENERGY APPLIANCES

Steffen Fries and Hans-Joachim Hof

18.1 INTRODUCTION

Power generation, transmission, and distribution systems are characterized by the existence of two infrastructures in parallel: the electrical grid, carrying the energy, and the information infrastructure used to automate and control the electrical grid. Especially the latter is becoming more and more a critical part of power system operations as it is responsible not only for retrieving information from field equipment but most importantly for submitting control commands. A dependable management of these two infrastructures is crucial and strongly relies on the information infrastructure as automation continues to replace manual operations. Hence, the reliability of the power system strongly depends on the reliability of the information infrastructure. Therefore, the information infrastructure must be managed to the level of reliability needed to provide the required stability of the power system infrastructure to prevent any type of outage.

Smart Grid: Applications, Communications, and Security, First Edition.
Edited by Lars Torsten Berger and Krzysztof Iniewski.

As the information infrastructure is the backbone of power system control, it needs appropriate protection to ensure the operation of power systems and support the required system reliability. Information security is the base for protecting the information infrastructure against intentional and unintentional cyber incidents but needs to take the power system specifics into account [1].

18.2 CHALLENGES IN THE SMART GRID

Current challenges for the power grid include the integration of fluctuating renewable energy sources, distributed power generation, short interval feedback of users on their energy usage, user indicated demand peaks, and the foreseeable need for the integration of electric vehicles, leading to an even higher energy demand of customers at peak times. A "smarter" grid can meet many of these challenges. With the availability of pervasive IT communication services, several new use cases become possible that enhance the services to the customer. These new use cases include dynamic pricing, time of use pricing, selling local power into the grid using a marketplace, and smart metering.

Many of these use cases center around Smart Home scenarios [3]. Smart Home in combination with the Smart Grid will allow people to understand how their household uses energy, to manage energy use better, to sell energy produced by local distributed energy generation, and to reduce their carbon footprint. The energy automation standard IEC 61850 is a natural candidate to be used for communication between instances of the Smart Grid and the gateway of a Smart Home.

Figure 18.1 shows an example of an abstract system architecture of a Smart Grid including many Smart Homes. The system shown in Figure 18.1 has the following properties:

- Homes are equipped with smart meters, intelligent household devices, and energy producers (like solar cells or combined heat and power).
- Home energy gateways control the communication between the devices in a home and the Smart Grid and thus define a security perimeter. The home energy gateway hides the complexity of the in-house network from the Smart Grid. It may act as a proxy for the appliances of the home (e.g., toward other Smart Grid participants, as, for instance, the marketplace). Access to this home energy gateway needs to be restricted to a distinct group of people. This is applicable for a Smart Grid operator sending commands to the gateway or the user accessing the gateway remotely.
- A gateway operator is likely to be responsible for administration of home energy gateways and also provides connectivity for home energy gateways. As he controls the connectivity to the home energy gateway, a gateway operator may also provide authorization functions for the home energy gateway.
- The distribution network operator communicates with the home energy gateways by means of another instance (in this case, the gateway operator is this

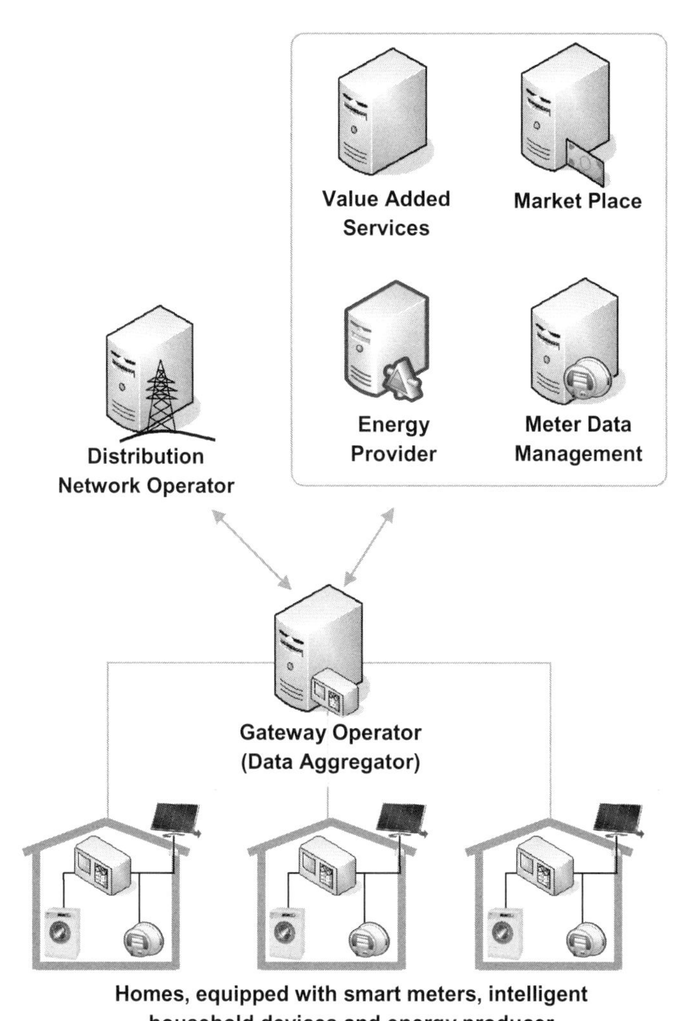

Figure 18.1. Connection of households to the Smart Grid [2].

instance) that hides the complexity of the home energy gateway management from the distribution network operator.

- The meter data management manages and processes the metering data for the various energy providers, providing them with a summary for accounting.
- At the energy market, consumers (resp. their home energy gateways) may buy energy, and energy generators sell energy; hence, the market offers a demand regulated price. An energy market alleviates the integration of distributed energy generators (e.g., solar cells).
- The Smart Grid information and communication infrastructure and the energy market are the enabler for other value added services.

When using such architecture, one important challenge is the addressability of external entities (e.g., the distribution network operator): the home energy gateways must be reachable by external entities and access to the home energy gateways must be controlled to meet security and privacy requirements. This applies to access through Smart Grid components, which belong to a utility or a smart meter operator for use cases like demand response, reactive shutoff, or similar. Also possible is the remote access of a user to his/her energy gateway at home to perform certain operations or simply check the status of energy consumption or production.

18.3 ACCESS CONTROL AND AUTHORIZATION FOR REMOTE ACCESS TO HOME ENERGY APPLIANCES

As shown in Figure 18.1, in the Smart the Grid scenario new roles and/or components may be introduced. One example is the home energy gateway operator. This gateway operator is in charge of concentrating the communication from the home energy gateways up to the control center as well as providing an easy way to the control center to reach a high number of energy gateways at once. Moreover, a gateway operator can support the discovery of home energy gateways and most likely will also offer additional services based on this discovery, like remote access for users or remote management of the home energy gateways, for example, to provide enhanced functionality or updates for installed software. The secure discovery and access to a home energy gateway is a central functionality for a secure smart energy grid.

As the home energy gateway connects both the energy appliances at the consumer's home and the communication with the Smart Grid, it is also referred to as the information communication technology (ICT) gateway.

Based on the described scenarios in the previous section, the following entities may require secure access, via both local and remote communication, to a home energy gateway:

- Energy provider (including metering / billing)
- Distribution network operator
- Gateway operator
- Energy market
- End user

Therefore, a home energy gateway has to support remote access, allowing secure exchange of measurement, supervision, and control data with the energy network.

Secure access involves the following security controls:

- *Authentication.* Mutual authentication between home energy gateway and accessing entity.

- *Authorization.* Determine access permissions. The decision whether a certain access is authorized may be made locally on the device or centrally by an authorization server. The access control enforcement takes place at the gateway itself.
- *Secure Communication.* Confidentiality and integrity of the communication is protected, for example, by using IPsec (Internet Protocol Security) or SSL/TLS (Secure Socket Layer/Transport Layer Security).

One way to handle access to the home energy gateways is at the gateway itself, decentralized. To ensure a secure communication of roles/persons allowed to access the home gateway, an access control list (containing credentials, potentially role information, associated rights, etc.) has to be managed at the gateway directly. This may be done by either the user or the gateway operator. As the user has the choice of a preferred energy provider, service aggregator, or other Smart Grid roles, there will be individual communication connections per user. Thus, user-specific configuration has to be done at each home energy gateway. Dedicated roles can initiate a secure connection to every home energy gateway based on a device certificate installed at the gateway.

The advantage of this approach is a direct accessibility for different roles or components to the home energy gateway. The disadvantage is that user-specific connection credentials have to be administered at each home energy gateway separately. Note that the gateway discovery and address resolution is neglected here. Deployment in existing home environments may not lead to the establishment of a communication channel through the existence of NAT and firewall functionality. Thus, the approach of direct access is not feasible.

An alternative approach to handle access to the home energy gateways is a central, network-based functionality for a security provider, who authenticates and authorizes potential roles and components, that wish to access a home energy gateway. Such an approach is already known from telecommunication environments like the Generic Bootstrapping Architecture (GBA) of 3GPP networks [4]. A further example can be given through server components in voice over IP (VoIP) networks, for example, in SIP (Session Initiation Protocol, RFC 3261). In the following, a central security function providing **a**uthentication—**s**ession **i**nvocation—**a**uthorization (ASIA), which leverages the existing concepts, is proposed. ASIA would be typically combined with other functions needed, for example, for the address resolution or home energy gateway discovery. It is therefore expected that the home energy gateways register with the ASIA component to publish their IP addresses. This registration connection is kept online to allow remote access to the home energy gateway and may also be used to communicate consumption data or to connect to value added services. As it is expected that the home energy gateway is operated in a home environment behind the DSL router, this permanent connection also keeps the NAT bindings as well as the opened communication ports in the DSL router. To achieve connectivity with a home energy gateway using ASIA, different modes are described in the following sections.

18.3.1 ASIA: Operation in Session Invocation Mode

In this mode, ASIA initiates a connection establishment with a requesting Smart Grid role/component.

Process

- ASIA authenticates the connecting Smart Grid role/component (energy provider, aggregator, meter data management, etc.)
- Accessing role/component provides a TAN (transaction number) for the home energy gateway to associate the home energy gateway in the direct connection later on. Alternatively, the TAN may be generated by ASIA and provided to the accessing role/component and the home energy gateway as part of a software token. This has the advantage that the software token may contain further information about a security parameter of the connection to be established between the home energy gateway and the accessing role/component. Examples are the used certificates for authentication at ASIA.
- ASIA sends the connection request from the role/component via the permanent connection to the home energy gateway including the connection parameter (address of role/component, TAN, software token, etc.) and optionally the requested command to be executed.
- The home energy gateway uses the received information to establish a direct connection to the requesting role/component. The distributed TAN can be used to associate the connection attempt with the initial request of the requesting role/component.
- After successful authentication as part of the direct connection, the information exchange can be started.

Technical Approach

- If web-based services are already used in the system the authorization process may be realized using SAML (Security Assertion Markup Language) and XML (eXtended Markup Language) security [5]. This approach supports certificate-based authorization as well as preshared key-based authorization.
- An alternative to SAML may be the application of Kerberos [6] for a purely symmetric infrastructure without the need for certificates.
- Authentication on the direct connection may be performed on a transport layer (e.g., by applying TLS—transport layer security).

Figure 18.2 shows the general protocol flow for a session invocation.

The advantage of this approach is the flexibility and scalability in changing environments, especially when the communication relations change. The approach supports a central authentication and authorization component, which keeps the administrative effort low per gateway. This may influence the used gateway hardware. The central component is not involved in the actual data exchange. The disadvantage

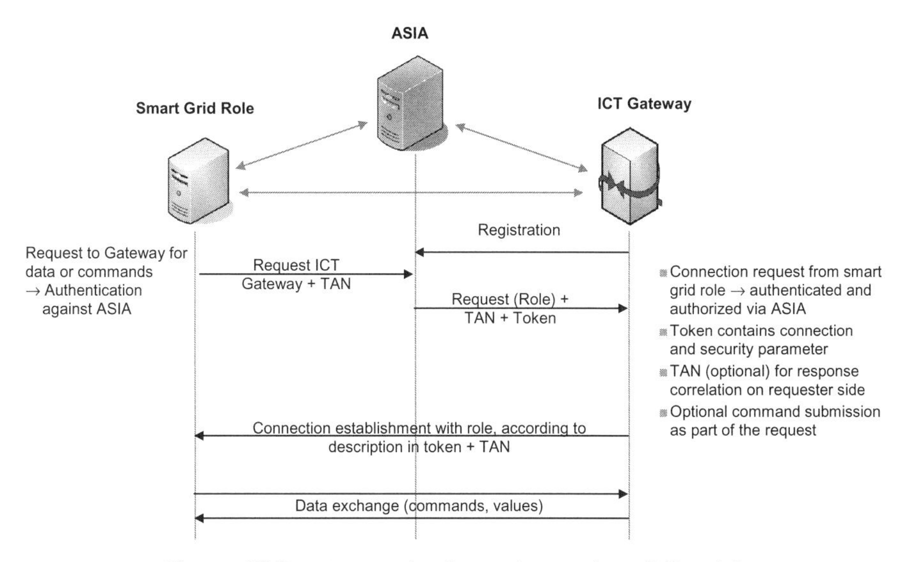

Figure 18.2. ASIA—session invocation mode, call flow [7].

is that there is no option for a direct communication establishment between the requesting role/component and the home energy gateway, which stems from the user environment, because of the missing TAN.

18.3.2 ASIA: Operation in Redirect Mode

In this mode the ASIA server provides the address location information to the requestor after successful authentication. The requestor can then directly connect to the home energy gateway.

Process

- The ASIA component authenticates the requesting Smart Grid role/component based on certificates or preshared keys alternatively.
- According to the ASIA rule set, the address information will be provided to the requesting role/component. Here, it is also possible to include a software token as proof that the role/component has been authenticated and authorized at the ASIA server. This token may include a further security parameter.

Technical Approach

- Based on the described functionality, the ASIA server would resemble a Kerberos like KDC authenticating and authorizing access to the home energy gateway. Note that this approach builds on the general accessibility of the home energy gateway in terms of addressing and connection establishment, which may be limited due to the user's environment. Figure 18.3 depicts the general call flow.

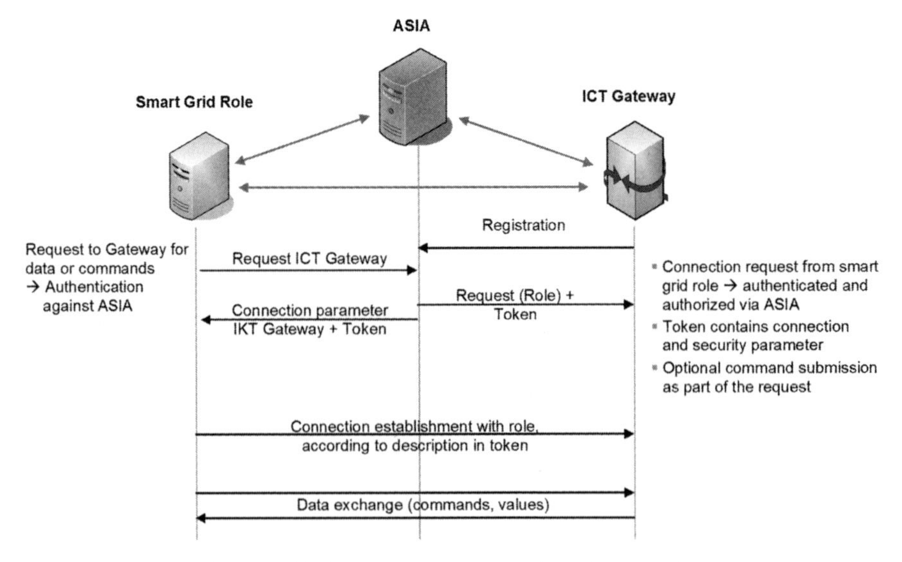

Figure 18.3. ASIA—redirect mode, call flow [7].

The advantage of this approach is the flexibility to react to changing communication relations. Like the previous approach, this solution allows for a central authentication and authorization component. The obvious disadvantage is the requirement to have direct accessibility to the home energy gateway from other components. In user environments with NAT and firewall (FW) devices, this solution will fail, as the firewall will block access to the home energy gateway. Moreover, if the user has a common DSL account the associated IP address is likely to change within 24 hours (at least in Germany).

18.3.3 ASIA: Operation in Proxy Mode

The ASIA server may act as proxy for restricting access to the home energy gateway.

Process

- The ASIA component authenticates the requesting role/component based on certificates or preshared keys alternatively.
- Based on the permanent connection between the home energy gateway and the ASIA server, the connection attempt of the requesting role/component is forwarded to the home energy gateway.
- ASIA may provide information about the authentication state and so on in a software token.
- To achieve end-to-end security also in the case of a communication proxy, application layer security mechanisms may be used. An example is the application of XML signatures to achieve integrity protection.

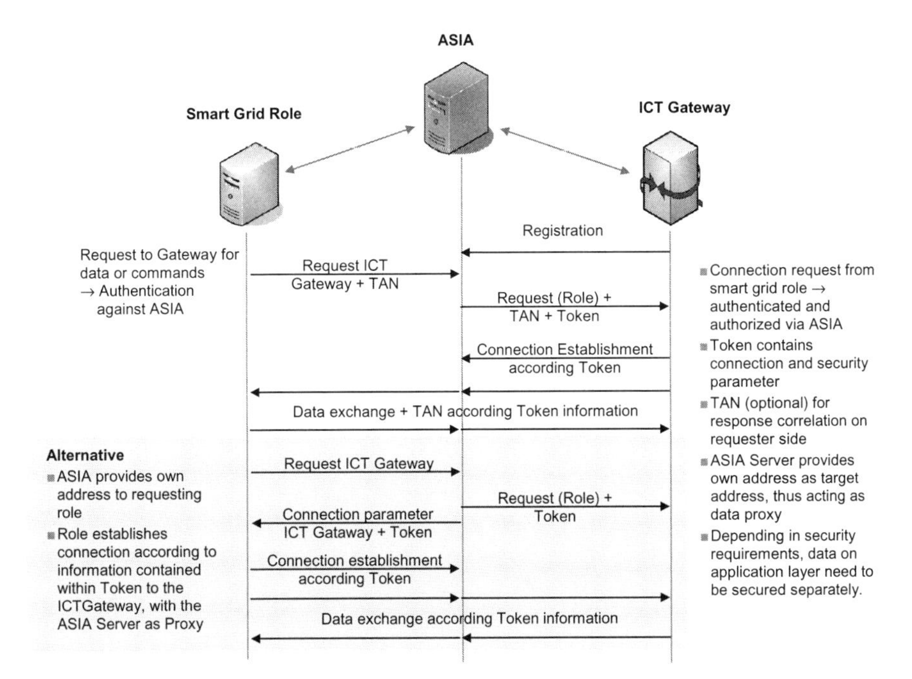

Figure 18.4. ASIA—proxy mode, call flow [7].

Technical Approach

- ASIA simply connects both links and operates as a data proxy for the duration of the connection.

This approach may be realized by using the previously described approach, with the exception that ASIA does deliver its own address instead of providing the home energy gateway address. It may be realized with the first approach letting the ASIA server connect to the requesting role/component. Thus, this approach is rather a configuration of the first two approaches with central access control. Figure 18.4 shows the abstract call flow or the connection establishment.

The advantage of this approach is the flexibility to react on changing communication relations. A disadvantage is the missing end-to-end connection establishment. Moreover, as all traffic is processed though the ASIA server, dedicated hardware requirements may arise.

18.3.4 ASIA Mode Comparison

Generally, the ASIA functionality can also be used to deliver configuration information, which communication connections are allowed or normal and which are not.

TABLE 18.1. Comparison of Solution Approaches for Home Energy Gateway Access Control

	Decentral	ASIA—Session Invocation	ASIA—Redirect	ASIA—Proxy
Discovery behind NAT		✓	✓	✓
Connectivity behind NAT and FW		✓		✓
Requires requestor server functionality		✓		
Mutual authentication on transport layer	Between ICT GW and requestor	Between ICT GW and requestor and also ASIA and requestor	Between ICT GW and requestor and optionally between ASIA and requestor	Between ICT GW and requestor and also ASIA and requestor
Local authorization	✓	✓	✓	✓
Central authorization		✓ Result may be signaled in a security token to the ICT GW	✓ Result may be optionally signaled in a security token to the ICT GW	✓ Result may be signaled in a security token to the ICT GW
Application layer authentication	✓	✓	✓	✓
Server load		Low	Low	High

This configuration information can also be used to enrich the effectiveness of the IDS/IPS (intrusion detection system/intrusion prevention system) tools, by making them aware of the network configuration.

Table 18.1 provides an overview of the different solutions discussed.

For the typical use cases, where a Smart Grid component requests access to the home energy gateway, the ASIA in session invocation mode provides the highest flexibility for the connection establishment, while considering a potential customer infrastructure. Moreover, it poses the least load requirements on the ASIA component compared with the proxy mode, which would also be a possible mode of operation. For the use case where a user remotely accesses her home energy gateway, the ASIA proxy approach is the most suitable, as the user is not expected to run a server component for the remote access to let the home energy gateway connect to the user's remote host. Hence, a combined session invocation and proxy mode is recommended for common operation.

REFERENCES

[1] F. Cleveland: "IEC TC57 Security Standards for the Power System's Information Infrastructure—Beyond Simple Encryption," WG 15 White Paper, June 2007.

[2] S. Fries, H. J. Hof, T. Dufaure, and M. Seewald, "Security for the Smart Grid—enhancing IEC 62351 to improve security in energy automation control," *Int. J. Adv. Security*, to appear.

[3] C. Müller, J. Schmutzler, C. Wietfeld, S. Fries, A. Heidenreich, and H.-J. Hof, "ICT Reference Architecture Design Based on Requirements for Future Energy Grids," First International Conference on Smart Grid Communications (IEEE SmartGridComm 2010), Gaithersburg, Maryland, USA, October 2010.

[4] 3GPP TS 33.220, 3rd Generation Partnership Project—Technical Specification Group Services and System Aspects—Generic Authentication Architecture (GAA)—Generic Bootstrapping Architecture (GBA)—Release 10, October 2010. Available at: http://www.3gpp.org/ftp/specs/archive/33_series/33.220/33220-a00.zip.

[5] XML Signature Syntax and Processing (2nd edition), W3C Rec., June 2008. Available at: http://www.w3.org/TR/xmldsig-core/.

[6] C. Neumann, S. Hartmann, and K. Raeburn, "The Kerberos Network Authentication Service (V5)," IETF RFC 4120, July 2005.

[7] R. Falk, S. Fries, and H. J. Hof, "ASIA: An Access Control, Session Invocation and Authorization Architecture for Home Energy Appliances in Smart Energy Grid Environments," The First International Conference on Smart Grids, Green Communications and IT Energy-aware Technologies, ENERGY 2011, IARIA, to appear.

INDEX

Smart Grid: Applications, Communications, and Security, First Edition.
Edited by Lars Torsten Berger and Krzysztof Iniewski.